Symposium on Aerosol Generation and Exposure Facilities (1979 : Honolulu, Hawaii)

GENERATION OF AEROSOLS AND FACILITIES FOR EXPOSURE EXPERIMENTS

Edited By
KLAUS WILLEKE

ANN ARBOR SCIENCE
PUBLISHERS INC
P.O. BOX 1425 • ANN ARBOR, MICH. 48106

230 Collingwood, P. O. Box 1425, Ann Arbor, Michigan 48106

Library of Congress Catalog Card No. 79-53420
ISBN 0-250-40293-9

Manufactured in the United States of America

PREFACE

This volume contains the technical papers presented at the Symposium on Aerosol Generation and Exposure Facilities held in Honolulu, Hawaii, April 2-3, 1979. The Symposium was held as part of the Chemical Congress 1979 which was co-sponsored by the American Chemical Society and the Chemical Society of Japan.

The purpose of the symposium was to review the state-of-the-art and recent developments in instrumentation, facilities and experimental techniques for inhalation studies with aerosols.

The proper duplication in the laboratory of air environments containing small particles is becoming increasingly important as new industrial processes and commercial products, and the utilization of new energy sources contribute to emissions into the air environment of many types of aerosols and gaseous aerosol precursors. Several types of aerosols having a broad range of chemical and physical properties are known or suspected to have adverse health effects. Damage to humans or animals may be caused by aerosol particles of a few micrometers or less in diameter after they have been depleted in the respiratory tract upon inhalation.

Section I of this volume deals with the basic concepts of aerosol generation and characterization, describes the deposition and retention of particles in the respiratory tract, and elucidates the response of the respiratory tract to inhaled pollutants.

In performing controlled experiments in the laboratory, it is important that all characteristic properties such as size, shape, density, composition, hygroscopicity and rate of dissolution are properly simulated. Section II describes various methods of aerosol generation and characterization used. This section is also of interest to the scientist, engineer and technician responsible for developing and calibrating aerosol instruments.

When using aerosols in biological experiments, one has the problem of conveying the aerosols from an aerosol generator to the point of inhalation by the exposure subject, human or laboratory animal. Section III considers practical guidelines for inhalation experiments and describes aerosol facilities which are now successfully operated.

I would like to thank the Electric Power Research Institute and the U.S. Environmental Protection Agency for financially supporting the Symposium. Each agency provided about half of the total support. Dr. J. McCarroll of EPRI and Drs. W. L. Crider, S. J. Gage, D. Gardner and S. D. Lee of EPA devoted valuable time to make the necessary funds available. The efforts by Drs. Hueter, R. Lee, F. Meadows, T. Wagner and Mr. G. J. Rausa of EPA are also appreciated. Dr. G. Glass of EPA gave an interesting keynote address in which he outlined EPA's research priorities for the next several years.

I am indebted to Professor K. Iinoya of Kyoto University, Japan, for providing valuable input, for establishing contacts with Japanese colleagues, and for co-chairing the Symposium with me. I am also grateful to Drs. M. Lippmann of New York University, B. Y. H. Liu of the University of Minnesota, R. F. Phalen of the University of California at Irvine and O. G. Raabe of the University of California at Davis for helping to organize the Symposium. I also gratefully acknowledge Dr. J. P. Lodge of Boulder, Colorado, Dr. O. Preining of the Universität Wien, Austria, Dr. D. L. Swift of Johns Hopkins University and Dr. J. Husar of Washington University for co-chairing sessions with some of the organizers mentioned above. Last, but not least, I thank the individual authors for sharing their work with me and with the readers of this volume.

Klaus Willeke

Klaus Willeke is Associate Professor of Environmental Health, Adjunct Associate Professor of Chemical Engineering, University of Cincinnati. Born in Germany, Dr. Willeke holds a BS degree in mechanical engineering from the University of New Hampshire, and earned MS and PhD degrees in aeronautics and astronautics from Stanford University. He also received a diploma from the von Karman Institute for Fluid Dynamics, Rhode-St.-Genèse, Belgium. He was a postdoctoral fellow at the Max-Planck-Institut für Plasmaphysik, Garching, Germany, and an assistant professor of mechanical engineering at the University of Minnesota.

Dr. Willeke, active in several professional societies, has published many articles for scientific journals.

I dedicate this book to my parents: To my father, Heinrich, who kindled my interest in science and technology, and to my mother, Irmgard, who helped me to understand life and people.

CONTENTS

SECTION I

BASIC CONCEPTS OF AEROSOL GENERATION AND CHARACTERIZATION AND THEIR HEALTH EFFECTS

CHAPTER 1

AEROSOL GENERATION CONCEPTS AND PARAMETERS

Virgil A. Marple and Kenneth L. Rubow

Particle Technology Laboratory
Mechanical Engineering Department
University of Minnesota
Minneapolis, Minnesota 55455

INTRODUCTION

Aerosol exposure facilities are used to determine the effect a particular aerosol may have on test subjects such as animals or humans. These facilities consist of an aerosol generation system and some means of exposing the subjects to the aerosol.

Since aerosol exposure facilities are mainly used for the inhalation of the aerosol particles, some facilities expose only the subjects' noses to the aerosol. This is generally done when it is undesirable to allow the aerosol to come in contact with the subjects, as may be the case with exposing animals to radioactive aerosols. However, in other facilities, the aerosol is introduced into a chamber completely surrounding the subjects. In either case, it is important that all subjects of any experiment be exposed to nearly identical aerosols. In this chapter, a facility which uniformly distributes the aerosol to all subjects is assumed to exist, and attention will be focused on the aerosol and its generation.

As related to exposure facilities, an aerosol can be defined as a suspension of particles in air, although other gases may be present as dictated by the objectives of the experiment. The particles may be either liquid or solid, and they may be either approximately all of the same size (monodisperse) or of a wide variety of sizes (polydisperse). In addition, there are numerous other parameters by which a particle can be characterized. These include shape, surface area, chemical composition, density, toxicity, dissolution rate, electrical charge, etc. Besides the above-listed parameters of the particle, the concentration of these particles must often be considered. Depending on the objective of the experiment, one, two or several of these parameters may be important.

After the important parameters have been identified, an aerosol generator must be selected which will hold these parameters constant, or will allow them to be varied in a manner which satisfies the objectives of the study. Some parameters, such as particle density or particle chemical composition, are easily held constant and normally will not be varied over the course of study. However, other parameters, such as the particle size distribution or particle concentration, may be varied during a study as well as be difficult to control and, thus, require monitoring.

If the values of the important parameters are not known, or fluctuate substantially about their desired values, the results of the exposure study may be questionable. Thus, well-characterized aerosol generation is an important component in most aerosol exposure facilities.

Although for a particular experiment, any of the parameters mentioned may be important, two parameters of the aerosol which are almost always important are the particle size distribution and concentration. These parameters are two of the most difficult to determine and specify, because of the varied diameters by which a particular particle can be characterized and the many ways that particle size distributions and concentrations can be expressed. Also, these parameters are of primary importance in specifying the performance of any aerosol generator.

Therefore, in this chapter, the various ways in which a particle can be sized and the different methods for expressing the size distribution and concentration will be presented. Following the discussion of these parameters will be a presentation of the general aerosol generation concepts. Because of the vast number of generators which have been reported in the literature and are commercially available, no attempt will be made to cover each generator in detail; instead, a discussion of the general concepts of generation for the different classes of generators will be given. Details of several of the aerosol generators briefly described in this chapter will be presented in following chapters.

PARTICLE SIZE DISTRIBUTION

One reason the size distribution may be difficult to define is that the size of a single particle can be specified many different ways [1]. For example, if the size distribution is being determined by microscopic examination of collected aerosol particles, the size of a particle must be specified from its two-dimensional projection.

For a liquid particle, the deposited particle will appear as a circle, and the diameter of the particle can be specified as the circle's diameter. However, it must be noted that the liquid droplet on a solid surface need no longer be a sphere, but may be flattened, depending upon the contact angle between the liquid droplet and the solid surface. This problem is normally handled by the application of a spreading factor, which is the ratio of the diameter of the droplet on the solid surface to the diameter of the original spherical droplet [2]. This factor is a function of the specific liquid and collection surface.

If the particle is solid and irregular in shape, the choices of a particle dimension are greatly increased [3]. For example, the particle shown in Figure 1 could be specified by a projected area diameter, which is the diameter of a circle with the same area as that of the particle; the Feret's diameter, which is the maximum dimension of the particle; or the Martin's diameter, which is the length of a line segment bisecting the particle into two equal areas. For the last two cases, the diameters of all particles must be measured in the same direction for the results to be statistically significant. All of these diameter definitions are valid, and all provide a different value for the particle size. There are other methods by which the projected

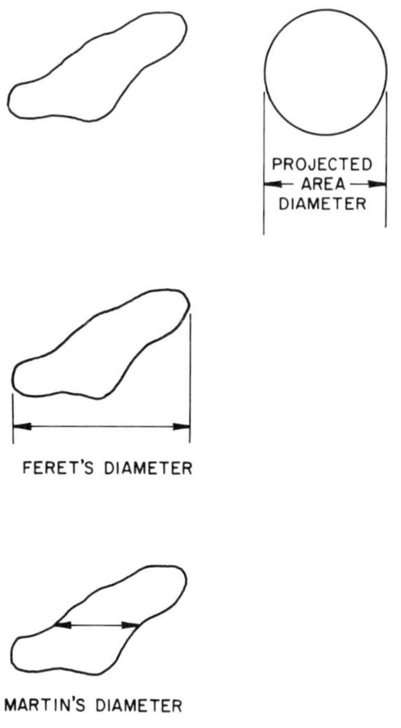

Figure 1. Particulate diameter definitions used for nonspherical particles.

area of the particle can be specified, but the three mentioned are the most common. It should be noted that specifying the size of a particle from a two-dimensional projection may be misleading, since information of the third dimension is ignored. This is especially important if the particles are flakes or rods, which have a preferential orientation on a microscope slide.

In addition to the size that can be specified from a two-dimensional projection of a particle, the size of a particle can also be defined by its aerodynamic characteristics. The most common diameters in this case are the aerodynamic diameter and the Stokes' diameter. The most widely used is the aerodynamic diameter, defined as the diameter of a sphere of unit density having the same terminal settling velocity as the particle in question, regardless of its shape and density. The Stokes' diameter is defined as the diameter of a sphere having the same density and Stokes' drag as the particle in question, or more specifically, the diameter of a sphere with the same terminal settling velocity and density as the particle in question.

Since the larger inhaled particles are deposited in the respiratory tract by impaction or gravitational settling, the particles' aerodynamic diameter is preferred to any of the geometric diameters found by two-dimensional analysis. For highly irregularly shaped particles, it may be best to determine the aerodynamic size distribution by use of some type of inertial classifier, such as an inertial impactor. However, for many cases, where the particles are fairly regular in shape or are liquid droplets, it is possible to calculate, or at least estimate, the aerodynamic diameter from the diameters determined by microscopic analysis. Even for irregular dust particles, it is sometimes possible to use a shape factor to make this conversion [1].

Once the particular diameter has been chosen to specify the size of the aerosol particles, the aerosol particle size distribution can be presented graphically as a frequency and/or a cumulative distribution. In addition, the distributions can be presented by the various weightings of number, surface area or volume, and can be plotted on linear or log scales. For greater detail than will be presented here, numerous references in aerosol science and industrial hygiene include sections on particle statistics [3-11].

To illustrate the various methods of presenting the aerosol particle size distribution, the numerical example, given in Table I, will be used. The particular distribution chosen for this example is log-normal, a distribution function often found to represent the size distribution of aerosol particles. In this table, the interval boundaries and geometric midpoints are presented in rows 1 and 2, respectively. Because of the log-normal size distribution, the interval midpoints, D_i, are the geometric means, defined as $\log D_i = (\log D_{LB} + \log D_{UB})/2$ where D_{LB} and D_{UB} are the lower and upper interval boundary, respectively. The number Δn_i, surface area Δs_i, volume Δv_i, and mass Δm_i, of particles in the respective intervals are given in rows 3, 4, 5 and 6.

For this example, it is assumed that the particles are spheres, and that the material density is $\rho_p = 1.5\ g/cm^2$. The relationships between the values in rows 3 through 6 are:

$$\Delta s_i = \alpha_s (D_i)^2 \Delta n_i \tag{1}$$

$$\Delta v_i = \alpha_v (D_i)^3 \Delta n_i \tag{2}$$

$$\Delta m_i = \rho_p \Delta v_i \tag{3}$$

where the surface area and volume shape factors, α_s and α_v, are functions of the particle shape, which for spheres are equal to π and $\pi/6$, respectively. It should be noted that only one of the quantities Δn_i, Δs_i, Δv_i or Δm_i must be input data, and the other three can then be calculated. Usually, the input data will be the number or the mass of the particles in each interval, with the number distribution being chosen as input data for this example. Due to the finite intervals being represented by midpoint particle sizes, the subsequently calculated values of Δs_i, Δv_i and Δm_i will only be close approximations.

The data for either Δn_i, Δs_i, Δv_i or Δm_i can be plotted as a histogram to give a frequency distribution of the data, as shown in Figure 2(a) for Δn_i. However, since the height of each bar represents the number of particles in each size interval, irrespective of the size of the interval, the relative heights of the bars are not an indication of the relative frequency of particles in each size range. Thus, it is common practice to calculate $\Delta n_i/\Delta D_i$ where ΔD_i is the interval size (row 7). The variable $\Delta n_i/\Delta D_i$, as tabulated in row 8 and plotted in Figure 2(b), now represents the relative frequency of the particles in the various size ranges.

It should be noted that the area under the curve in Figure 2(b) is dependent on the number of particles in the sample, making it difficult to compare two size distributions if they are different from each other in the total number of particles. Therefore, it is advantageous to normalize the distributions by dividing the value of $\Delta n_i/\Delta D_i$ for each size interval by the total number of particles in the sample $\Sigma \Delta n_i$. One thus obtains

$$f(D) = \frac{\Delta n_i}{\Delta D_i \, \Sigma \Delta n_i} \tag{4}$$

The values for the normalized frequency distribution function f(D) are tabulated in row 9 and are presented in Figure 2(c). Distributions presented in this fashion have the property that the integrated area under the curve is equal to unity, irrespective of the total number of particles in the sample.

It should be noted that calculations similar to those in rows 8 and 9 can also be made for the volume, mass or surface area distributions. The

Table I. Example Calculations for Particle Frequency and Cumulative Size Distributions

1. Interval boundary	D (μm)	0.1		0.2		
2. Interval geometric midpoint	D_i(μm)		0.1414		0.2828	
3. Number of particles	Δn_i		50		460	
4. Surface area of particles	Δn_i (μm^2)		3.1		116	
5. Volume of particles	Δv_i (μm^3)		0.07		5.4	
6. Mass of particles	$\Delta m_i (10^{-6}\ \mu g)$		0.11		8.2	
7. Interval width	ΔD_i (μm)		0.1		0.2	
8. Relative number	$\frac{\Delta n_i}{\Delta D_i}$ (μm^{-1})		500		2300	
9. Normalized relative number	$\frac{\Delta n_i}{\Delta D_i \Sigma \Delta n_i}$ (μm^{-1})		0.099		0.456	
10. Normalized relative surface	$\frac{\Delta s_i}{\Delta D_i \Sigma \Delta s_i}$ (μm^{-1})		0.0007		0.0133	
11. Normalized relative volume	$\frac{\Delta v_i}{\Delta D_i \Sigma \Delta v_i}$ (μm^{-1})		0.00004		0.0011	
12. Normalized cumulative number				0.0010		0.0930
13. Normalized cumulative surface				0.00007		0.00274
14. Normalized cumulative volume				0.000003		0.00022
15. Relative number	$\frac{\Delta n_i}{\Delta \log D_i}$		166		1530	
16. Relative surface	$\frac{\Delta s_i}{\Delta \log D_i}$ (μm^2)		10.0		385	
17. Relative volume	$\frac{\Delta v_i}{\Delta \log D_i}$ (μm^3)		0.23		18.0	

0.4		0.7		1.0		2.0		4.0		7.0		10		20
	0.5292		0.8367		1.414		2.828		5.292		8.367		14.14	
	1055		980		1705		680		102		10		2	
	928		2155		10710		17100		8970		2200		1300	
	81.9		301		2525		8050		7920		3070		3000	
	122		451		3790		12000		11900		4600		4500	
	0.3		0.3		1.0		2.0		3.0		3.0		10	
	3520		3270		1705		340		34		3.3		0.2	
	0.673		0.648		0.338		0.067		0.0067		0.0006		0.00004	
	0.0711		0.165		0.246		0.197		0.069		0.017		0.0030	
	0.0109		0.0402		0.101		0.320		0.210		0.0410		0.0120	
		0.304		0.500		0.841		0.977		0.9976		0.9996		1.0
		0.0241		0.0736		0.320		0.713		0.920		0.970		1.0
		0.0035		0.0156		0.177		0.439		0.757		0.880		1.0
	4340		6330		5660		2260		420		65		7	
	3820		13910		35580		55800		36900		14000		4300	
	337.0		1940		8388		26700		32600		20000		10000	

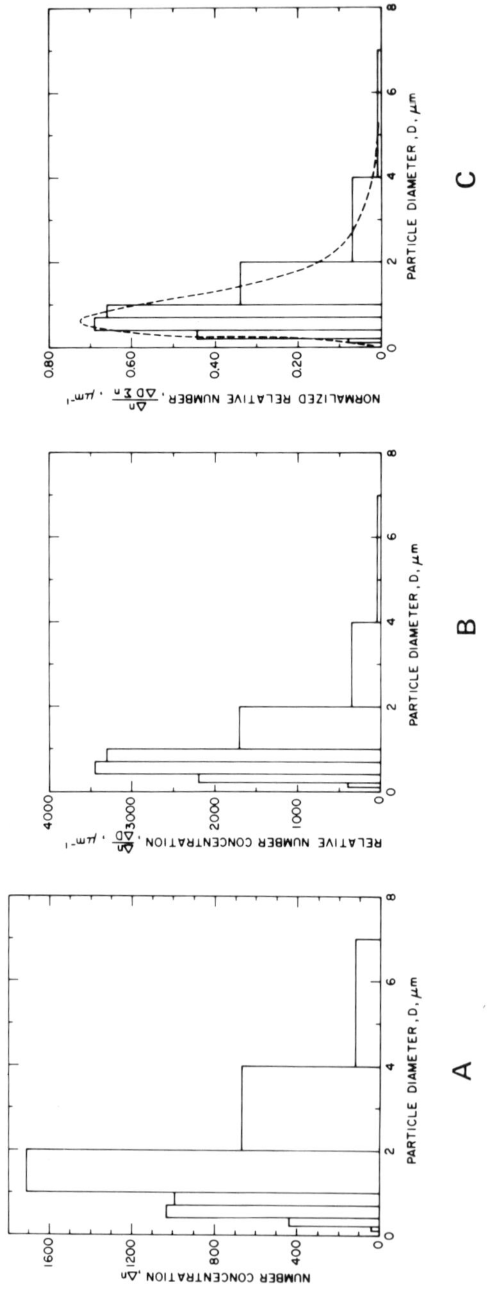

Figure 2. Particle number size distribution histograms; (a) original data; (b) relative; (c) normalized relative.

normalized distributions for surface area and volume are presented in rows 10 and 11. The normalized mass distribution is equal to the normalized volume distribution.

The size distribution can also be approximated by a smooth curve instead of a histogram, as shown inFigure 2(c). However, in doing this, some information is lost, since the interval widths are no longer presented.

Another technique for presenting the size distribution is to use a normalized cumulative distribution, as presented in Figure 3 for the number, surface area and volume distributions. Although the distributions are shown in terms of "percent of particles less than stated size," the distributions could have been represented as the "percent of particles greater than stated size." The calculated values for these curves are shown in rows 12, 13 and 14 for the number, surface area and volume distributions, respectively.

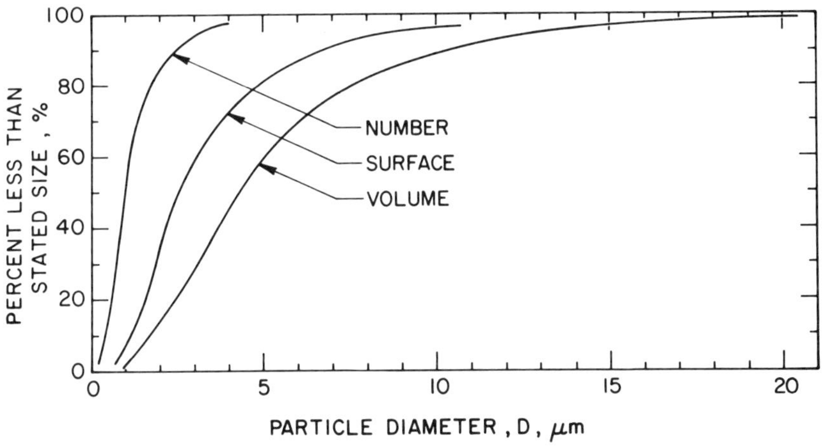

Figure 3. Cumulative number, surface area and volume particle size distributions.

It should be noted that the frequency distributions more readily show details of the distribution, while cumulative distributions more readily show differences and are often used to make comparisons.

Particle size distributions can usually be approximated by a mathematical function. Two functions which are often used to represent particle size distributions are the normal and log-normal distribution functions. For comparison purposes, typical frequency distributions for both normal and log-normal functions are shown in Figure 4(a). To determine if a size distribution is normal or log-normal, the cumulative distribution can be plotted on normal probability or log-normal probability paper, as shown in Figure 4(b) and 4(c). If the frequency distribution is normal or log-normal, it will

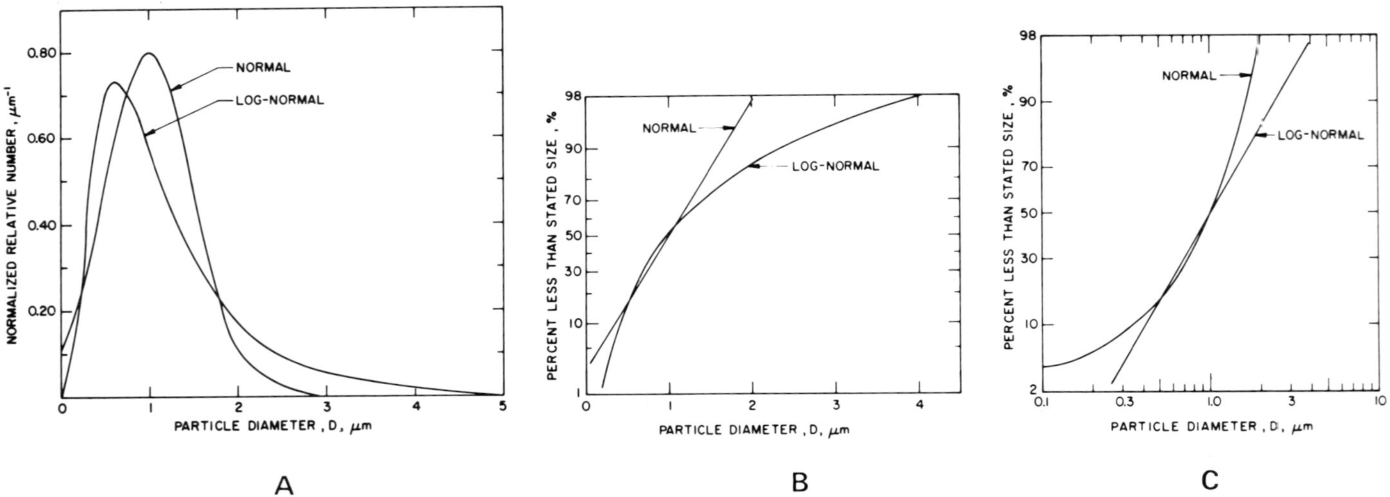

Figure 4. Comparison of the normal and log-normal size-distribution function; (a) normalized relative number; (b) cumulative number on linear-probability graph paper; (c) cumulative number on log-probability graph paper.

plot as a straight line on normal probability or log-normal probability graph paper, respectively. One should note that the log-normal distribution presented in Figure 4 represents the data used in the numerical example in Table I.

Since the particle-size distributions from aerosol generators are often log-normally distributed, it is important that the properties of this distribution be understood for the specification of aerosols generated in an exposure facility. To demonstrate log-normal distributions, the number, surface area and volume distributions (rows 12, 13 and 14 of Table I) have been plotted on log-normal probability graph paper, as shown in Figure 5.

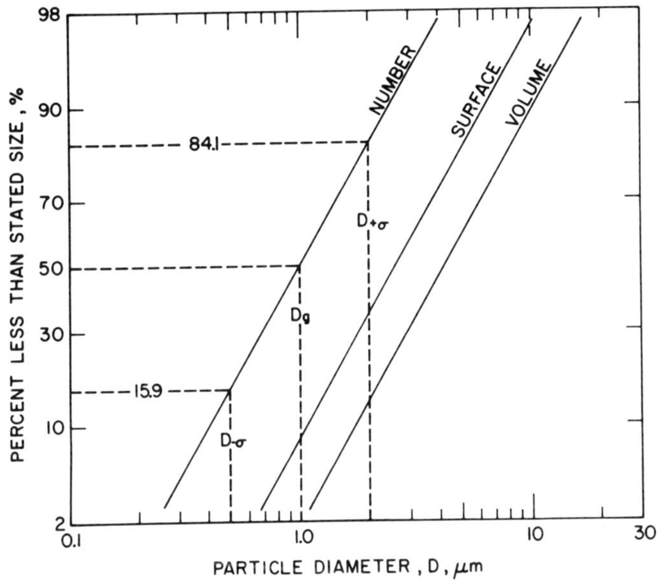

Figure 5. Cumulative number, surface-area and volume curves of a log-normal particle size distribution on log-probability graph paper.

Log-normal distributions can be characterized by two parameters: the geometric mean and the geometric standard deviation. The geometric mean or median D_g can be found graphically as the value of the particle size at 50% probability on the log-normal probability curve, or analytically from:

$$D_g = [D_1{}^{\Delta n_1} + D_2{}^{\Delta n_2} + D_3{}^{\Delta n_3} + \ldots]^{1/\Sigma \Delta n_i} \tag{5}$$

The geometric standard deviation, σ_g, is defined as

$$\sigma_g = \frac{D_g}{D_{-\sigma}} = \frac{D_{+\sigma}}{D_g} = \left(\frac{D_{+\sigma}}{D_{-\sigma}}\right)^{1/2} \tag{6}$$

where $-\sigma$ and $+\sigma$ are plus and minus one standard deviation from the geometric mean. $D_{-\sigma}$ and $D_{+\sigma}$ correspond to the 15.9% and 84.1% points, respectively, on the probability scale, as shown in Figure 5. Alternately, σ_g can be calculated from

$$\sigma_g = \left[\left(\frac{D_1}{D_g}\right)^{\Delta n_1}\left(\frac{D_2}{D_g}\right)^{\Delta n_2}\left(\frac{D_3}{D_g}\right)^{\Delta n_3} \cdots\right]^{1/\Sigma\Delta n_i} \tag{7}$$

Another important property of log-normal distributions is that if the number distribution is log-normal, then the surface area and volume distributions are also log-normal with the same geometric standard deviation, as shown in Figure 5. The surface geometric mean diameter D_{gs} and the volume geometric mean diameter D_{gv} are related to the number geometric mean diameter D_{gn} by

$$\log D_{gs} = \log D_{gn} + 2(\log \sigma_g)^2 \tag{8}$$

$$\log D_{gv} = \log D_{gn} + 3(\log \sigma_g)^2 \tag{9}$$

Thus, if the geometric mean diameter by number, surface area or volume has been measured, the value of the other two mean diameters can easily be calculated. For the given example, with $D_{gn} = 1.0\ \mu m$ and $\sigma_g = 2.0$, the resulting surface and volume geometric mean diameters are about 2.6 μm and 4.2 μm, respectively.

Finally, when graphically presenting log-normal size distribution data, plots of $\Delta n/\Delta \log D$, $\Delta s/\Delta \log D$ or $\Delta v/\Delta \log D$ vs $\log D$ on linear-log graph paper are useful, since the relative concentration in a size interval is proportional to the relative area under the distribution curve in that size interval. Thus, one can visually estimate the relative contribution of any size range to the total number, surface area or volume of particles for any aerosol distribution. The relative frequency distributions for number, surface area and volume are presented in rows 15, 16 and 17 of Table I and are shown in Figure 6.

PARTICLE CONCENTRATION

The previous discussion emphasized the particle-size distribution of a given aerosol. Of equal importance is the concentration of these aerosol particles in the inhalation chamber. One of the simplest techniques for expressing the concentration is to present the number or mass size-distribution data of Table I in numbers/cm^3 or mg/cm^3, respectively.

In many cases, the determination of concentration values is not difficult. For example, when determining the mass distribution with a cascade impactor, knowledge of the quantity of air having passed through the impactor is all

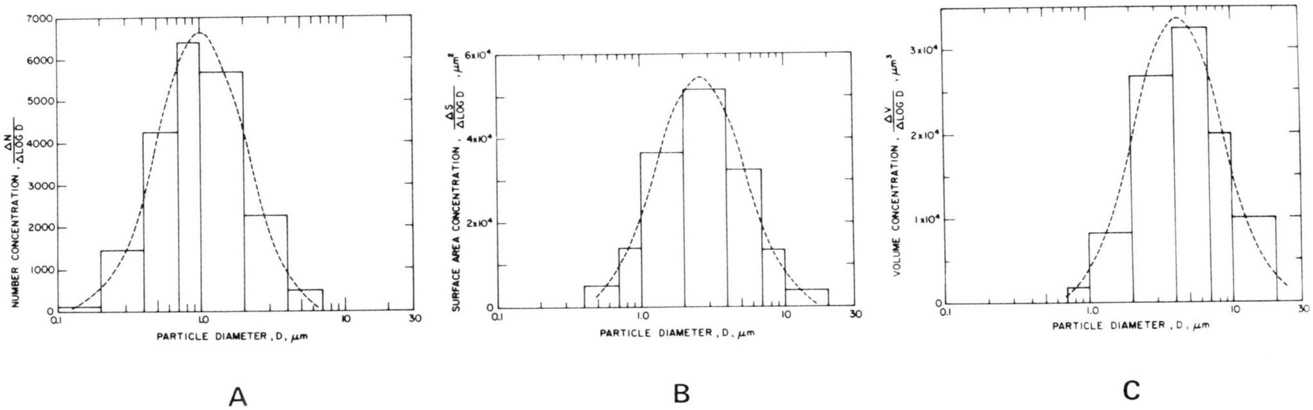

Figure 6. Relative number, surface area and volume histograms of a log-normal particle size distribution plotted on linear-log graph paper.

that is needed. When the particles are counted, e.g., by a device such as an optical particle counter, the total flow having passed through the counter determines the number concentration. However, if the number count is determined by microscopic analysis of a filter sample, determination of the number concentration must take into account the fraction of the total number of particles counted as well as the total flow of the air through the filter.

Another method by which aerosol concentrations can be expressed is by the overall output concentration of the aerosol generator. If mass- or volume-concentration output of the aerosol generator is known, this can be used as a concentration measure. However, one should be cautious to account for all losses in the system between the aerosol generator and the inhalation chamber. By knowing the size distribution and the overall concentration, the particle concentration in various size ranges can be calculated.

Although particle size distribution and concentration are important in inhalation studies, one or several of the other parameters mentioned previously may be even more important, and will dictate the specific aerosol generator to be used. For example, the particular material to be aerosolized has a strong influence on the type of generator to be chosen. Thus, the aerosol generators discussed in the following section are classified by their generation concept with the approximate ranges of mean sizes and geometric standard deviations given for each concept.

AEROSOL GENERATORS

There are numerous methods by which aerosols can be generated, and many aerosol generators have been developed and reported in the literature [5,12-18]. Two major classes of generators to be discussed here are those which generate liquid droplets and those which disperse dry powders. In each of these two categories, the more common generation concepts will be discussed and some specific examples will be examined.

Liquid Aerosol Generators

Liquid aerosol generators can be classified by the method of particle formation, which is achieved either by mechanical dispersion of the liquid into droplets, or through condensation of a vapor into droplets. In all condensation aerosol generators, the generation technique is basically the same: the liquid to be aerosolized is first heated to form a vapor, then the vapor is cooled and condensed into droplets. However, there are several different techniques which have been used to mechanically disperse liquids to form aerosols. Some of the more common methods, including air nebulizers, spinning-disc atomizers, ultrasonic nebulizers, and vibrating-orifice generators, are discussed in the following sections.

Air Nebulizers

In this type of aerosol generator, compressed air is used to atomize the liquid into droplets. In the DeVilbiss nebulizer [13], shown in Figure 7, air passes through a narrow tube to position A. The high-velocity air emerging from this tube creates a low-pressure region at the tube exit, drawing liquid from the reservoir upward through a liquid feed tube. The emitted liquid is atomized by the high-velocity air into small droplets. Large droplets impinge onto the upper wall of the curved exit duct, and the smaller droplets remain aerosolized.

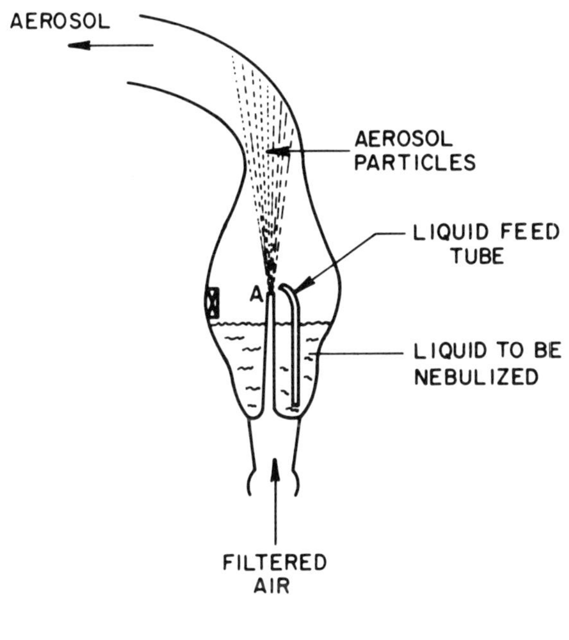

Figure 7. Schematic of DeVilbiss nebulizer.

Another common aerosol nebulizer is the Collison nebulizer [19], shown in Figure 8. Again, a high-velocity air jet is used to create a low-pressure region, causing the liquid to rise up a feed tube and be atomized by the air jet. Although only one nebulizing passage is shown in Figure 8, there may be several such passages in the same nebulizing head. A baffle surrounds the

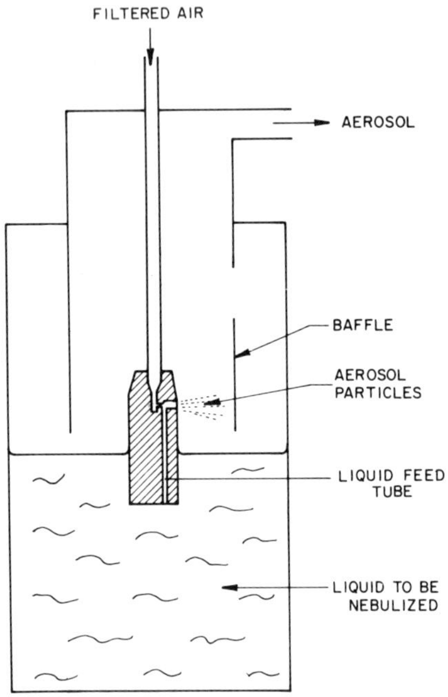

Figure 8. Schematic of Collison nebulizer.

nebulizing head, and serves to remove the large droplets by impaction. These impacted droplets drain back into the liquid reservoir, while the droplets which are not impacted follow the upward air flow and are emitted through the outlet port.

In the nebulizers discussed above, the liquid is siphoned to the point of atomization by the high-pressure air. Another technique for nebulizing the liquid is to allow the liquid to flow over the air jet, as is the case with the Babington nebulizer [20]. With this nebulizer, liquid flows over the outer surface of a hollow sphere, forming a thin film. Air supplied to the interior of the sphere flows outward through a rectangular opening in the sphere's surface, thus atomizing the liquid.

Aerosols generated by air nebulizers normally [12] have mass median diameters on the order of 1-7 μm with geometric standard deviations of about 1.7-2.5. However, the size distribution of the aerosol can be modified by any one of several techniques. For example, the liquid being atomized may be a solution of the material to be aerosolized and a volatile liquid. After atomization of the liquid, the volatile liquid evaporates from the droplet leaving a smaller particle of the material in solution. Droplet diameter reductions as high as 20 times can be obtained by this technique. The particles resulting from the aerosol solution may be either liquid or solid [20].

Another method by which the size distribution can be modified is to pass the aerosol through an inertial impactor which removes the large particles [15]. The impaction device can be designed to have any desirable particle cutoff size, but is normally used only to remove the very large particles.

Another use of air nebulizers is to disperse solid particles suspended in a volatile liquid. Upon nebulization of the mixture, the liquid is evaporated, leaving the suspended solid particles as the aerosol. The concentration of suspended particles in the liquid should be kept low, in order to avoid two or more of the solid particles in a single droplet. Thus, some of the droplets will contain no particles, which is of no consequence since the liquid evaporates, leaving only a very small residue particle.

Spinning-Disc Atomizers

In this type of aerosol generator, the liquid is atomized by being thrown from the edge of a spinning disc [16]. A schematic of such a device is shown in Figure 9. Here, the liquid is delivered to the center of the disc through a small tube. The fluid spreads onto the rotating disc and is spun off as liquid droplets from its edge.

The disc is normally about 5 cm in diameter and is rotated at speeds up to approximately 70,000 rpm by either an electric motor or an air-driven motor. The primary particles generated by such a device are fairly monodisperse [13], with a geometric standard deviation of approximately 1.1. Particle sizes generated are normally 20-100 μm but can be reduced down to 0.6 μm by use of solvents [16].

Besides the primary particles, there will also be smaller satellite particles produced. However, since the aerodynamic stopping distance is long for the large primary particles and short for the small satellite particles, the two sizes of particles can be separated by the air flow system shown in Figure 8. The smaller particles are drawn upward into the centrally located conical funnel, while the larger particles are projected further from the disc and are withdrawn through another outlet.

As is the case with the nebulizers described earlier, solid particles can also be generated with a spinning-disc atomizer by using a solution of the material to be atomized in a volatile liquid or by dispersion of solid particles in a volatile liquid.

Vibrating-Orifice Aerosol Generator

This type of generator [21] is normally used to generate monodisperse aerosols in the size range of 0.5-50 μm, with σ_g = 1.01-1.03. In the

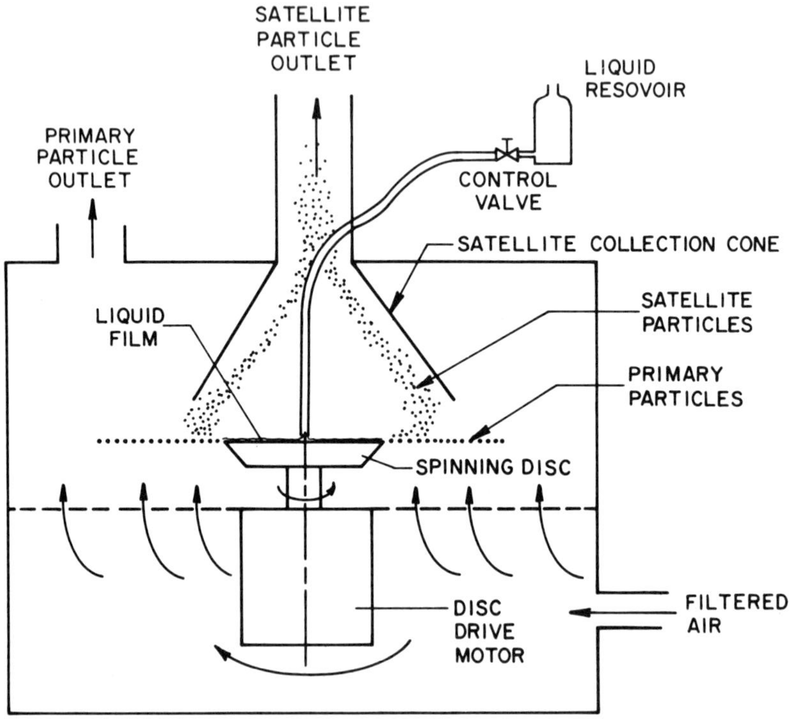

Figure 9. Schematic of spinning-disc atomizer.

vibrating-orifice aerosol generator, the liquid is forced through an orifice at a constant flowrate, as shown in Figure 10. The liquid stream passing through such an orifice becomes unstable and breaks up into droplets. By vibrating the orifice by some means such as a piezoelectric transducer, the stream is broken into uniform droplets, with one droplet formed for each cycle [22]. Thus, if the flowrate and cycle frequency are kept constant, the droplets should all be of the same size, with a very small geometric standard deviation.

Even though the generated particles are monodisperse, the coagulation of particles immediately after dispersion can lead to doublets and triplets. To reduce this problem, dispersion air is normally added to the aerosol stream immediately after generation. The dispersion-air flowrate is sufficiently high to produce turbulence near the aerosol stream, thus efficiently diluting and dispersing the particles before doublets can be formed.

Similar to the other aerosol generators described, almost any low-viscosity material which can be put into solution may be used to generate liquid or solid particles.

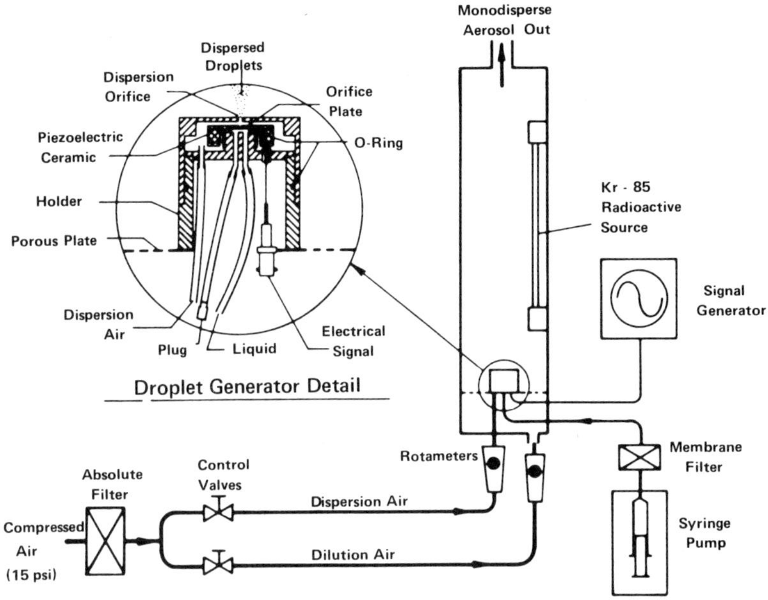

Figure 10. Schematic of vibrating-orifice aerosol generator.

Ultrasonic Nebulizers

In the ultrasonic nebulizer [23,24], a piezoelectric crystal is vibrated at high frequency and generates aerosols from a liquid, as shown in Figure 11. The piezoelectric transducer is placed at the bottom of a coupling liquid, and droplets are generated at the surface of the liquid to be aerosolized. An

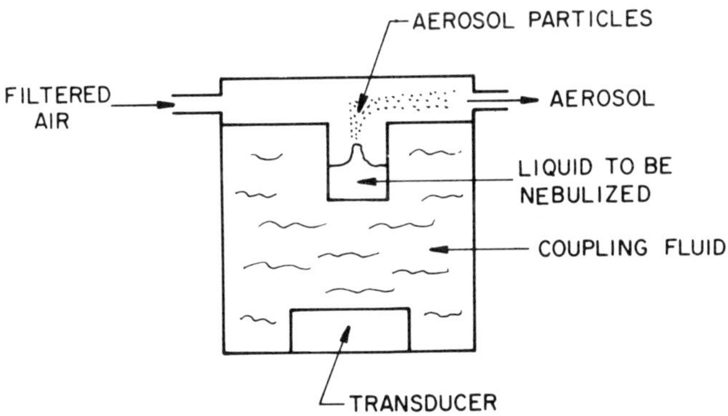

Figure 11. Schematic of ultrasonic atomizer.

air stream removes the aerosols from the generator. This generation technique provides a higher concentration of aerosols than can be obtained by most other methods. The aerosol diameter generated by such a technique [12] is normally on the order of 5-10 μm, with a geometric standard deviation between 1.5 and 1.7.

Condensation Generators

In condensation aerosol generators, the liquid to be generated is first heated to form a vapor, and then the vapor is cooled and condensed into droplets [25,26]. Upon cooling, the vapor condenses spontaneously if the supersaturation is high, or the vapor condenses onto nucleation centers introduced into the vapor. The aerosol particles generated by this method are normally fairly monodisperse, with geometric standard deviations on the order of 1.15-1.4 and particle sizes in the submicron range.

One of the earliest practical generators of this type was the Sinclair-LaMer condensation generator [27]. In this generator, the liquid was vaporized in a boiler, and the nuclei were generated by an electric arc. The nuclei and the liquid vapor were then passed into a second chamber where the mixture was reheated to ensure that any residual droplet from the boiler would be evaporated. The nuclei centers and the reheated vapor were then passed through a cooling tower where the vapor condensed upon the nucleation centers.

Another method which has been used to obtain aerosols by condensation is to atomize the liquid, evaporate the droplets and then recondense the vapor onto the nonvolatile nuclei of the evaporated droplets [25]. A schematic of one such generator is shown in Figure 12. Here, liquid atomization is achieved through a Collison atomizer. The aerosol is subsequently vaporized as it passes through a glass tube which is wrapped with electric heating tape. The vapor and the residue of the evaporated droplets then move downward into the unheated portion of the glass tube, where the vapor recondenses onto the residue particles.

Dust and Fiber Aerosol Generators

As discussed before, solid aerosols may be generated from liquid suspensions. However, if dust is to be aerosolized, it is more common to disperse the dust particles directly into the air.

Numerous different techniques have been used to disperse dust or fiber particles [12,28-42]. Many of the devices used are fairly simple: an air stream is passed through an agitated bed of dust and dust particles are carried away with the air stream to the inhalation chamber. With such devices, it is generally difficult to control parameters such as particle concentration and

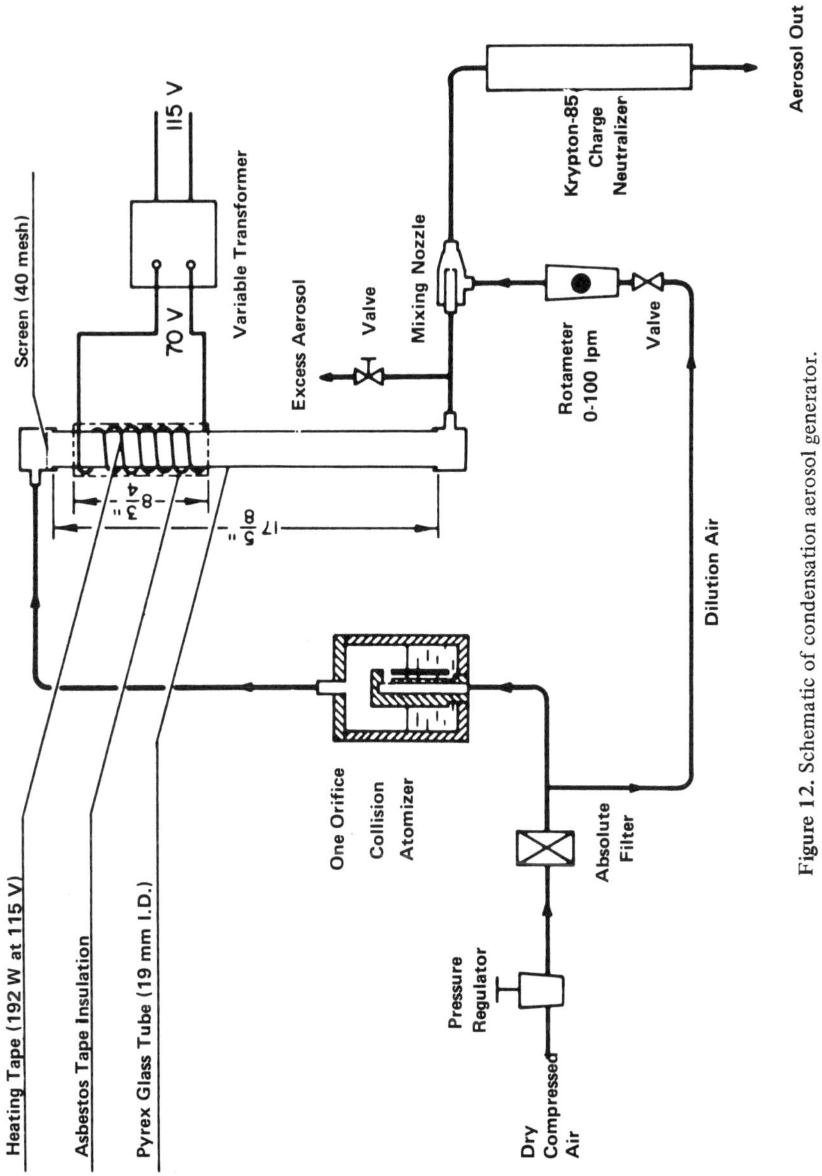

Figure 12. Schematic of condensation aerosol generator.

particle-size distribution. In several devices which are more complex, the quantity of dust being aerosolized is metered, as discussed in the following sections.

Dust Generators Utilizing Scrapers

In these dust generators, a scraper continuously slices the particles from a container of powder. In some cases, the powder is compacted into a cake, and the scraper removes the particles from this compacted cake; in other cases, the powder is held loosely in the container.

The most popular of such devices is the Wright dust feeder [28], schematically shown in Figure 13. In this device, the powder is compacted into a cup at fairly high pressures, forming a hard cake. A mechanism of gears rotates the cup across the scraper, and an air flow entrains the particles scraped from the cake.

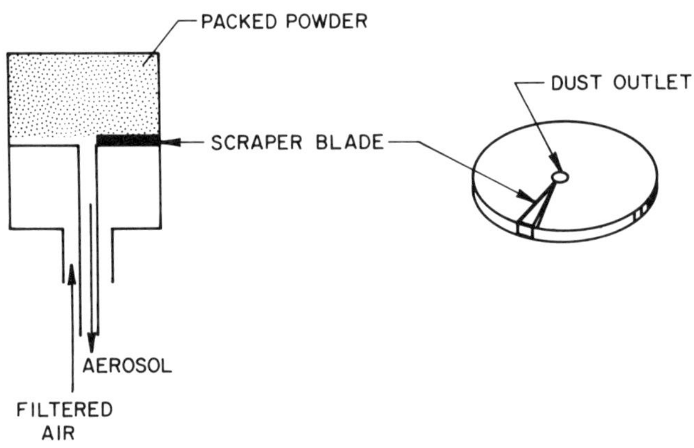

Figure 13. Schematic of Wright dust feeder.

Some generators have been constructed so that the powder is loosely packed [29]. This loose powder is advanced mechanically into a scraper which acts more as a scoop than as a cutter.

In addition, devices using the same principle as the Wright dust feeder have been designed to generate fibers [30,31]. With these generators, the fiber material, in cake form, is advanced into a chamber where knife edges on a rotating shaft slice the particles from a cake.

Fluidized-Bed Dust Generators

Several dust generators utilize a fluidized bed for dust dispersion [32-34]. Two basic methods can be used: one is to fluidize the dust material to be aerosolized, and the other is to fluidize a mixture of fairly large inert particles and smaller dust particles. The latter technique has been more successful for obtaining aerosolized dust of uniform particle size and concentration.

A fluidized-bed-type dust generator is schematically shown in Figure 14. The inert particles comprising the fluidized bed are typically brass, nickel or

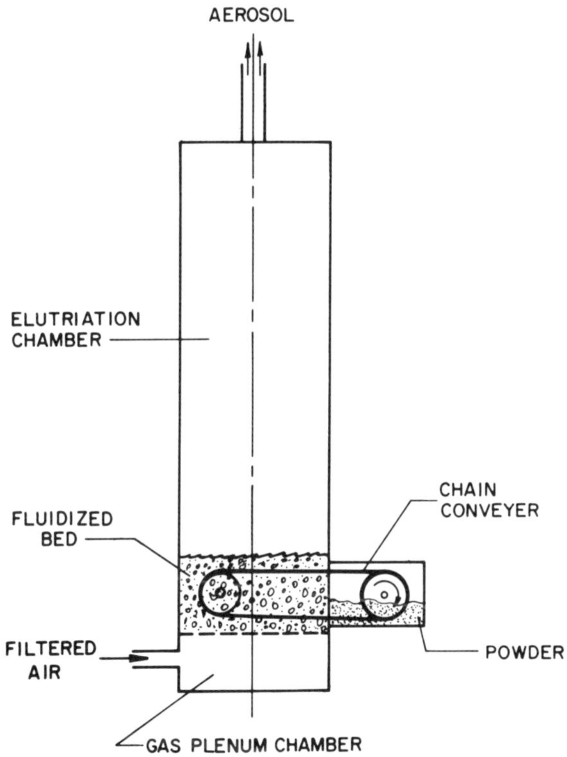

Figure 14. Schematic of fluidized-bed dust generator.

stainless steel beads of about 100 μm in size. To provide for a constant output from the generator, the dust particles are transported from a powder-holding chamber to the fluidized bed by a chain-conveyor system driven by a variable speed motor. The quantity of dust being delivered to the fluidized bed can be controlled by the speed of the chain.

In the fluidized-bed chamber, the powder is transferred from the chain to the bed by the mechanical scrubbing action of the beads. The continuous motion of the fluidized bed disperses the particles as single particles with very few agglomerates. Directly above the fluidized bed is an elutriation chamber which separates the large beads from the smaller aerosol dust particles, so that the beads remain in the generator and the redispersed dust is carried away with the air flow.

Turntable Dust Generator

This type of feeder consists of a horizontal turntable rotated by a variable speed motor [35]. A groove is cut into the upper surface of the turntable, and the dust flows into this groove from a hopper located directly above it. Agitation bars in the hopper keep the powder flowing continuously to the turntable groove. After the turntable has passed under the hopper, a scraper bar removes the excess dust from the groove to keep the level of the powder even with the upper surface of the turntable. Thus, the size of the groove and the speed of rotation of the turntable provide a method by which the powder can be metered. An aspirator, located opposite the hopper, withdraws the particles from the groove and aerosolizes them.

AEROSOL CONDITIONING

So far, only the dispersion of particles into an air stream has been discussed. Before these particles are introduced into the test chamber, it may be desirable to condition the aerosol by reducing the electrical charge on the particles and/or shaping the particle size distribution.

The mechanical action in generators may result in highly charged aerosols. The electrical charges on the aerosols are normally reduced to Boltzmann equilibrium by exposing the paticles to ionized air [43]. One technique which is commonly used is to employ a krypton-85 radioactive source, as shown in the generation systems in Figures 10 and 12. In the neutralizers shown, the aerosols are passed through a chamber which contains the radioactive source. The radiation from this source ionizes the air in the chamber, and the ions are attracted to the charged particles passing through the chamber. Although the particles will not be completely neutralized, they will be brought to Boltzmann equilibrium, which is normally satisfactory.

The size distribution of the aerosol can be modified by the removal of large particles with cyclone or inertial impaction devices. If inertial impactors are used, the size of particles removed can be controlled by the diameter of the impactor nozzle and by the flow rate through the impactor [44].

REFERENCES

1. Kotrappa, P. "Shape Factors for Aerosols of Coal, UO_2 and ThO_2 in Respirable Size Range," in *Assessment of Airborne Particles,* T. T. Mercer, P. E. Morrow and W. Stöber, Eds. (Springfield, IL: Charles C. Thomas Publisher, 1972), p. 331-358.
2. Whitby, K. T., and R. A. Vomela. "Response of Single Particle Optical Counters to Nonideal Particles," *Environ. Sci. Technol.* 1:801-814 (1967).
3. Cadle, R. D. *The Measurement of Airborne Particles* (New York: John Wiley and Sons, Inc., 1975).
4. Cadle, R. D. *Particle Size* (New York: Reinhold Publishing Co., 1965).
5. Dennis, R., Ed. *Handbook on Aerosols,* TID-26608 (National Technical Information Service, U.S. Dept. of Commerce, 1976).
6. Herdan, G. *Small Particle Statistics,* 2nd Edition (Amsterdam: Elsevier Publishing Co., 1962).
7. Hesketh, H. E. *Fine Particles in Gaseous Media* (Ann Arbor, MI: Ann Arbor Science Publishers, Inc., 1977).
8. Mercer, T. T. *Aerosol Technology in Hazard Evaluation* (New York: Academic Press, Inc., 1973).
9. Orr, C., and J. M. Dallavalle. *Fine Particle Measurement* (New York: Macmillan Publishing Co., 1959).
10. Silverman, L., C. E. Billings and M. W. First. *Particle Size Analysis in Industrial Hygiene* (New York: Academic Press, Inc., 1971).
11. Stockham, J. D., and E. G. Fechtman, Eds. *Particle Size Analysis* (Ann Arbor, MI: Ann Arbor Science Publishers, Inc., 1977).
12. Tillery, M. I., G. O. Wood and H. J. Ettinger. "Generation and Characterization of Aerosols and Vapors for Inhalation Experiments," *Environ. Health Pers.* 16:25-40 (1976).
13. Kerker, M. "Laboratory Generation of Aerosols," *Adv. Colloid Interface Sci.* 5:105-172 (1975).
14. Bailey, A. G. "Review: The Generation and Measurement of Aerosols," *J. Mat. Sci.* 9:1344-1362 (1974).
15. Whitby, K. T., D. A. Lundgren and C. M. Peterson. "Homogeneous Aerosol Generators," *Int. J. Wat. Poll.* 9:263-277 (1965).
16. Liu, B. Y. H. "Methods of Generating Monodisperse Aerosols," COO-1248-10, Department of Mechanical Engineering, University of Minnesota, Minneapolis, Minnesota (1967).
17. Raabe, O. G. In: *Fine Particles: Aerosol Generation, Measurement, Sampling, and Analysis,* B. Y. H. Liu, Ed. (New York: Academic Press, Inc., 1976), p. 57-110.
18. Fuchs, N. A., and A. G. Sutugin. In: *Aerosol Science,* C. N. Davies, Ed. (New York: Academic Press, Inc., 1966), p. 1-30.
19. May, K. R. "The Collison Atomizer: Description, Performance and Application," *J. Aerosol Sci.* 4:235-243 (1973).

20. Litt, M., and D. E. Swift. "The Babington Nebulizer: A New Principle for Generation of Therapeutic Aerosols," *Am. Rev. Resp. Dis.* 105: 308-310 (1972).
21. Berglund, R. N., and B. Y. H. Liu. "Generation of Monodisperse Aerosol Standards," *Environ. Sci. Technol.* 7(2):147-153 (1973).
22. Ström, L. "The Generation of Monodisperse Aerosols by Means of a Distintegrated Jet of Liquid," *Rev. Sci. Instrum.* 40(6):778-782 (1969).
23. Lang, R. J. "Ultrasonic Atomization of Liquids," *J. Acoustic Soc. Am.* 34(1):6-8 (1962).
24. Denton, M. B., and D. B. Swartz. "An Improved Ultrasonic Nebulizer System for the Generation of High Density Aerosol Dispersions," *Rev. Sci. Instrum.* 45(1):81-83 (1974).
25. Liu, B. Y. H., and K. W. Lee. "An Aerosol Generator of High Stability," *Am. Ind. Hyg. Assoc. J.* 36:861-865 (1975).
26. Chang, D. P. Y., and B. K. Tarkington. "Experience with a High Output Sulfuric Acid Aersosol Generator," *Am. Ind. Hyg. Assoc. J.* 38:493-497 (1977).
27. Sinclair, D., and V. K. LaMer. "Light Scattering as a Measure of Particle Size in Aerosols," *Chem. Rev.* 44:245 (1949).
28. Wright, B. M. "A New Dust-Feed Mechanism," *J. Sci. Instrum.* 27: 12-15 (1950).
29. Whitby, K. T., A. B. Algren, R. C. Jordan and J. C. Annis. "Evaluation of Air Cleaners for Air Conditioning and Ventilation–Part I. Apparatus," ASHAE J. Sect., *Heat. Piping Air Cond.* 30:171-178 (1958).
30. Timbrell, V., A. W. Hyett and J. W. Skidmore. "A Simple Dispenser for Generating Dust Clouds from Standard Reference Samples of Asbestos," *Ann. Occup. Hyg.* 11:273-281 (1968).
31. Ortiz, L. W., H. E. Black and J. R. Coulter. "A Modified Fibrous Aerosol Generator," *Ann. Occup. Hyg.* 20:25-37 (1977).
32. Guichard, J. C. In: *Fine Particles: Aerosol Generation, Measurement, Sampling, and Analysis,* B. Y. H. Liu, Ed. (New York: Academic Press, Inc., 1976), p. 173-193.
33. Willeke, K., C. S. K. Lo and K. T. Whitby. "Dispersion Characteristics of a Fluidized Bed," *J. Aerosol Sci.* 5:449-455 (1974).
34. Marple, V. A., B. Y. H. Liu and K. L. Rubow. "A Dust Generator for Laboratory Use," *Am. Ind. Hyg. Assoc. J.* 39:26-32 (1978).
35. Grassel, E. E. In: *Fine Particles: Aerosol Generation, Measurement, Sampling, and Analysis,* B. Y. H. Liu, Ed. (New York: Academic Press, Inc., 1976), p. 145-172.
36. Fuchs, N. A., and F. I. Murashkevich. "Laboratory Powder Disperser (Dust Generator)," *Staub-Reinhalt. Luft* 30(11):1-3 (1970).
37. Hounam, R. F. "The Konimiser–A Dispenser for the Continuous Generation of Dust Clouds from Milligram Quantities of Asbestos," *Ann. Occup. Hyg.* 14:329-335 (1971).
38. Deichmann, W. B. "A 'Dustshaker': An Apparatus Designed for the Production of Controlled Dust Concentration in the Air Breathed by Experimental Animals," *J. Ind. Hyg.* 26:334-335 (1944).

39. Sonkin, L. S., M. A. Lipton and D. Van Hoesen. "An Apparatus for Dispersing Finely Divided Dusts," *J. Ind. Hyg. Toxicol.* 28(6):273-275 (1946).
40. Druett, H. A., and J. McG. Sowerby. "An Apparatus for the Maintenance of a Carbon Dust Cloud of Constant Concentration," *Br. J. Ind. Med.* 3:187-193 (1946).
41. Stead, F. M., C. U. Dernehl and C. A. Nau. "A Dust Feed Apparatus Useful for Exposure of Small Animals to Small and Fixed Concentrations of Dust," *J. Ind. Hyg. Toxicol.* 26(3):90-93 (1944).
42. Borho, K. "On the Problem of Dispersion of Dusts," *Staub-Reinhalt. Luft* 33(8):306-308 (1973).
43. Liu, B. Y. H., and D. Y. H. Pui. "Electrical Neutralization of Aerosols," *J. Aerosol Sci.* 5:465-472 (1974).
44. Marple, V. A., and K. Willeke. "Impactor Design," *Atmos. Environ.* 10:891-896 (1976).

CHAPTER 2

AEROSOL SAMPLING AND CHARACTERIZATION FOR INHALATION EXPOSURE STUDIES WITH EXPERIMENTAL ANIMALS

Werner Stöber

Fraunhofer-Institut für Toxikologie
und Aerosolforschung
Munster, Federal Republic of Germany

INTRODUCTION

Atmospheric particulate matter has been of interest in natural sciences for various reasons and for a long time, but the industrial proliferation of this century has added a new dimension of concern to the old phenomena: large-scale combustion processes and other industrial activities of our modern technological society release large quantities of minute particles into the air and increase the total concentration of airborne particulate matter to an extent that, visible to the naked eye, the ill-reputed "smog" or other types of severe air pollution may occur.

The potential health hazard of these manmade atmospheric particle clouds was the primary incentive for the emergence of three new fields of applied research:

1. development of sophisticated instrumentation for the physical and chemical analysis of atmospheric aerosols;
2. assessment of the risk of air pollution to the public health by way of epidemiological studies of urban populations; and
3. experimental evaluation of toxicological long-term effects of certain air pollutants by chronic inhalation exposure of laboratory animals.

Although these fields are listed in an apparently appropriate sequence as procuring first the air pollution measuring devices, next facilitating the correlation of pollution levels and epidemiological field studies and, finally, checking stochastic evidence by detailed investigations in the laboratory, there was no such definite chronological sequence when air pollution eventually emerged as both a public nuisance and a potential health hazard. In fact, epidemiological studies with significant results were conducted in the thick of the wartime and postwar industrialization of the Western hemisphere when air pollution measurements were still rather crude [1]. Such results and the public concern over increasingly visible air pollution in industrial areas led to the determined efforts of air pollution abatement in the late 1950s and throughout the 1960s. This, in turn, required and triggered the design of better analytical equipment as needed to measure accurately the reduced levels of pollution.

Today, epidemiological studies aiming to link the trends of health parameters of the general public to differences in local air pollution levels are very frustrating to conduct because the pollution levels of today are substantially lower than 25 years ago and, thus, require large samples of the population to give such studies a fair chance of finding statistically significant probabilities for the anticipated adverse effects. Under unfavorable conditions, ambient air inhalation may still trigger allergic reactions in humans because of particular air pollutants, but, in general, acute toxicity and dramatic smog episodes with increased mortality rates of the general public appear to belong to the past in most modern industrialized countries.

However, there is still considerable concern over today's environmental pollution. The awareness is now primarily focused on potential long-term health effects due to chronic exposure, with carcinogenicity rating by far as the most important hazard to be prevented. Unfortunately enough, it was the epidemiology of occupational exposure that, occasionally in a disastrous way, revealed belated clues of the hidden dangers of certain air pollutants, but it ought to be the long-term animal experiment that should prevent such adverse epidemiological surprises.

Thus, with regard to air pollution, lifetime inhalation studies with experimental animals exposed either to model aerosols of specific components, like heavy metals or sulfates, or to a mixture of air pollutants from a single industrial source, like a stack plume, will be an important tool in elucidating the long-term inhalation effects of environmental chemicals and certain types and constituents of source emissions. Obviously, such experimental arrangements require a sophisticated aerosol technology providing constant and well-defined exposure conditions for long periods of time up to several years. This technology is a subject of its own. In addition, analytic instrumentation and monitoring equipment for the exposure aerosols will be needed to check and record the experimental conditions of the continuous inhalation studies.

This chapter describes strategies, procedures and equipment for sampling and characterizing model aerosols and particulate source emissions utilized in long-term inhalation exposure studies with experimental animals.

AEROSOL SAMPLING

In contrast to specific chemical air pollutants, like sulfur dioxide, the health aspects of aerosols are not simply of a chemical nature, but are related to a considerable extent to physical properties. While gaseous pollutants will invariably have a chance of interacting with the walls of the respiratory tract, an aerosol can only interact to the extent that it will be deposited on the walls of the respiratory system during a breathing cycle. This deposition is ruled by a number of physical and physiological parameters. In particular, the size of the aerosol particle is one of the most influential characteristics. Thus, health-related aerosol sampling implies that all particles should be sampled whose deposition in the respiratory tract is of significant proportion.

Unfortunately, significant deposition probabilities are not confined to a limited size range. On the contrary, according to their size, all inhaled particles will be deposited to some extent due to either diffusion or sedimentation and impaction processes. Figure 1 shows a typical graph of a size-dependent deposition pattern in a horizontal model duct. The pattern is quite representative. Characteristic is the minimum of the deposition probability q somewhere between 0.1 and 1.0 μm of aerodynamic particle diameter [2]. To the left of the minimum, the decreasing sizes cause an increase of the deposition probability due to the increasing diffusion velocity of the smaller particles. Thus, complete deposition is achieved when the particle sizes approach molecular dimensions. To the right of the minimum, the increase of inertia with increasing particle size enhances the deposition probability in this range. In the technical example shown, particles above some 90 μm in diameter are completely deposited by sedimentation.

Theoretical and experimental data [3-5] on the size-related probability of overall deposition in the human respiratory tract for nasal breathing are shown in Figure 2. The graph indicates that nature indeed designed the respiratory airway system as a highly effective particle filter except for the small particle sizes below some 3 μm in aerodynamic diameter. The same is true for mouth breathing, although the distribution of the deposit in the airways will be quite different. Thus, it appears that total respiratory deposition, except for some deficiency below sizes of 3 μm, is closely related to the total suspended particulate matter in the air. This situation has some bearing on health-related

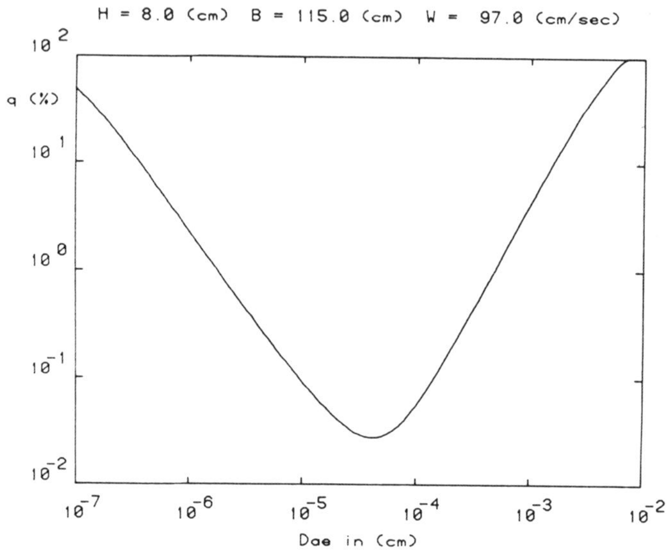

Figure 1. Size-dependent relative deposition q of aerosol particles of size D_{ae} flowing through a technical duct of height H = 8 cm, width B = 115 cm and length L = 100 cm at a mean flow velocity of 97 cm/sec.

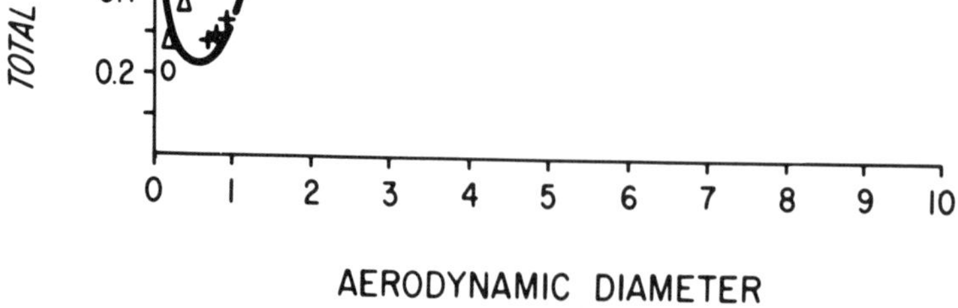

Figure 2. Relative total deposition of aerosol particles in the human respiratory tract for nose breathing (aerodynamic diameters in μm); data by Brown et al. (3)[o]; Van Wijk and Pattersen (5)[Δ]; and Dennis (4)[+].

high-volume sampling of ambient air with regard to the maximum particle size to be collected [6].

However, while ambient air sampling of aerosols has to account for the possibility of resorption of relatively large, but soluble, particles of unknown chemical nature which may be deposited somewhere in the respiratory tract, the sampling of experimental aerosols in long-term inhalation studies usually need not imply the occurrence of unknown components or big particle sizes, where the inhalation probability finally reduces the deposition. In most long-term inhalation experiments considered here, the chemical composition is either known or reproducibly constant during exposure and subject to a detailed chemical analysis. Furthermore, big particles with low inhalation probabilities are actually undesirable for long-term inhalation studies and will be eliminated by baffles confining the maximum particle sizes to below 10 μm in aerodynamic diameter.

Limiting the experimental particle size to less than 10 μm has a number of practical advantages. First of all, this is the size range where pulmonary deposition prevails, because particles of these sizes are primarily deposited in the alveolar region of the lung. Figure 3 gives recent results [7, 8] for the maximum deposition probability in this region under favorable physiological conditions.

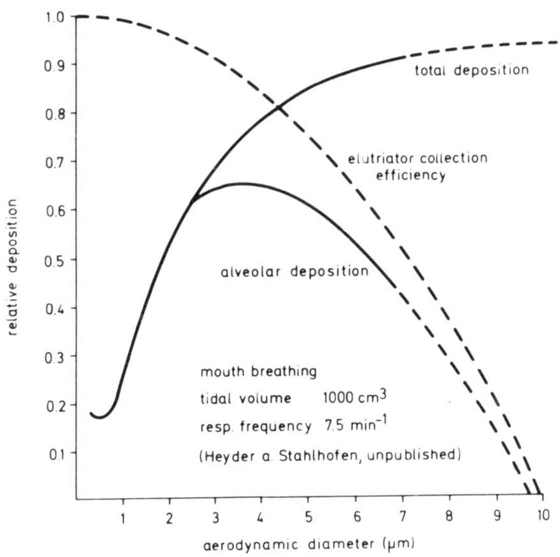

Figure 3. Total and alveolar deposition for mouth breathing under conditions of maximum pulmonary lung burdens in comparison to an enveloping elutriator collection efficiency curve for alveolar deposition.

The occurrence of pulmonary deposition is very desirable for most long-term inhalation experiments because it will usually aggravate the lung burden by accumulation. This is due to the fact that the alveoli and the alveolar ducts in the pulmonary region do not have a ciliated epithelium which provides rapid clearance of foreign material from the bronchioli and the upper respiratory tract. Instead, the clearance of the alveoli is affected by a much slower mechanism of phagocytosis by alveolar macrophages. The macrophages are mobile cells which incorporate the deposited foreign matter and, by random motion, may eventually reach the ciliary escalator. While clearance by the ciliated epithelium is normally completed within less than 24 hours, the clearance from the alveolar region is much slower and might result in a biological half-life of many months. Thus, insoluble aerosol deposits may accumulate and aggravate the irritation or other responses in the alveolar range, an effect that may be of special significance in chronic exposure studies.

Another practical advantage of the small particle sizes termed "respirable" by the British Medial Research Council [9] is the fact that experimental errors in measuring particle number and mass concentrations will be easier to avoid because, in contrast to big particles, small particles are less sensitive to anisokinetic sampling conditions which prevail for most sampling ports in aerosol exposure facilities and may cause sampling deficiencies.

Figure 4 gives a typical example of the geometry of a conventional sampling port. Since the intake flow is perpendicular to the flow of the exposure aerosol, the sampling conditions are, generally, anisokinetic because they involve a deflection of the streamlines of the flow to be sampled from. The diameter D_E of the sampling tube is usually between 0.4 and 1.0 cm and the average intake velocities v_s will range from 125 to 2000 cm/sec, depending upon the purpose and the method of measurement. Compared to these velocities, the motion of the exposure aerosol perpendicular to the sampling tube will be considerably slower. The exposure aerosols usually pass at a moderate flow rate (e.g., 80 liter/min for 48 hamsters) through the various cross sections of the conduits and the animal chambers of the exposure facility so that the linear flow velocities at the sampling ports vary between some 10 cm/sec in the ducts and less than 1 cm/sec in the animal exposure chambers.

Under these dynamic conditions, the sampling flow is not essentially influenced by the motion of the exposure aerosol and will retain the characteristics of sampling from calm air as schematically shown in Figure 5. In this case, the radial intake flow is symmetrical to the axis of the sampling tube, and, everywhere outside the tube, the intake flow velocities will remain below the value of v_S. Therefore, the stopping distance s of an aerosol particle as related to the perpendicular deflection into the sampling tube cannot exceed

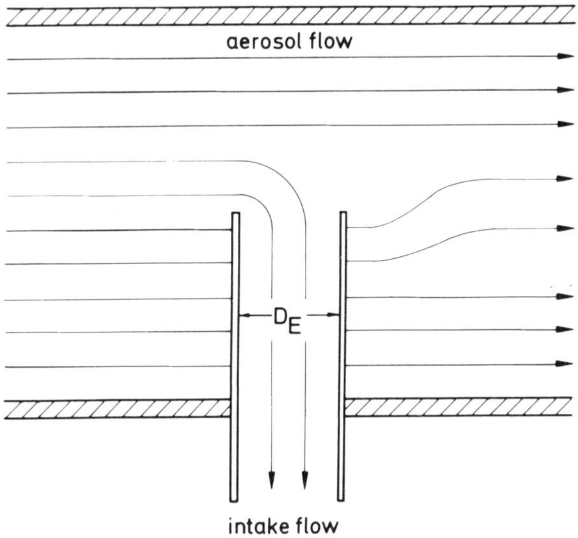

Figure 4. Geometric arrangement of sampling ports in inhalation exposure facilities.

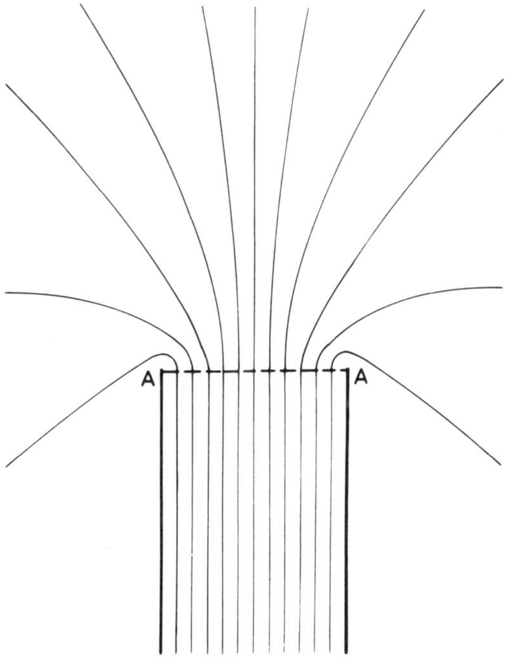

Figure 5. Intake flow streamlines in still air.

$$s = \frac{\rho_o D_{ae}^2 v_s}{18\eta\,(1 + 1/6\ \mathrm{Re}^{2/3})} \qquad (1)$$

This equation utilizes an empirical correction term of the viscosity coefficient η. According to Klyachko [10], it is valid for particle Reynolds numbers

$$\mathrm{Re} = \frac{D_{ae}\, v_s\, \rho_{air}}{\eta} < 400 \qquad (2)$$

so that it will cover all practical cases under consideration.

In these equations, ρ_o denotes unit density as defined for a particle of aerodynamic diameter D_{ae}, while ρ_{air} is the density of air.

Then, by equating the stopping distance s and the intake diameter D_E, a specific particle diameter $D_{ae\ max}$ can be estimated below which the sampling procedure is positively unimpaired by anisokinetic sampling losses due to flow deflection. The calculation shows that, even under most unfavorable conditions (v_s = 2000 cm/sec, D_E = 0.4 cm), all particles with sizes below 12 μm will, indeed, be collected without deficiencies caused by anisokinetic sampling. This indicates that aerosol dispersity is not a very critical factor for quantitative sampling from animal exposure facilities if the aerosol particles employed in the exposure studies are not essentially exceeding sizes of 10 μm diameter. Since this size range is of primary interest anyway, there is no serious limitation for the interpretation and relevance of biological data acquired with experimental exposure aerosols of this size limit.

It is instructive to check the sampling efficiency also with regard to the lower end of the particle size distribution. Very small particles are not affected by inertial forces causing sedimentation and impaction losses, but, with diminishing size, will be increasingly subject to thermal forces. Thus, diffusion to the walls of the sampling tube is a potential cause of sampling losses of small particles.

Particle deposition by diffusion in a cylindrical duct was investigated by Gormley and Kennedy [11]. For monodisperse aerosols, they derived an empirical expression relating the penetration P through a cylindrical tube of length ℓ to a unique dimensionless parameter δ

$$P = f(\delta); \quad \delta = \frac{2\pi D \ell}{\dot{V}} \qquad (3)$$

where D is the diffusion coefficient of the aerosol particles and $\dot{V}$ is the aerosol volume sampling rate. This relationship indicates that the sampling efficiency for sufficiently small particles is not dependent upon the sampling intake diameter. Instead, the length of the sampling tube is the only relevant geometric parameter.

With the empirical equation

$$P = 0.819\, e^{-1.828\delta} + 0.097\, e^{-11.15\delta} + 0.032\, e^{-28.5\delta} + 0.027\, e^{-61.5\delta} + 0.025\, e^{-375\delta} \quad (4)$$

as tested by Thomas [12] it can be shown that diffusion losses to the wall will be limited to less than 3% if

$$\delta \leqslant 2.8 \cdot 10^{-3} \quad (5)$$

Then, by inserting this numerical value into Equation 3, sampling conditions can be specified which would provide collection efficiencies of at least 97%.

According to a general expression for nonspherical aerosol particles [13], the diffusion coefficient

$$D = \frac{\rho_o D_{ae}^3}{\rho D_e^3} \frac{kT\, C(D_{ae})}{3\, \pi \eta\, D_{ae}} \quad (6)$$

is not uniquely related to the aerodynamic diameter D_{ae} as used in most aerosol studies, but it involves also the equivalent volume diameter D_e and the actual density ρ of the particle, while k is the Boltzmann constant, T the absolute temperature and $C(D_{ae})$ the slip correction of the viscosity η. However, by introducing the dynamic shape factor

$$\kappa = \frac{\rho\, D_e^2\, C(D_e)}{\rho_o D_{ae}^2\, C(D_{ae})} \quad (7)$$

which yields $\kappa = 1$ for spherical particles of unit density, the diffusion coefficient may be expressed by

$$D = \frac{k\, T\, C(D_e)}{3\, \pi\, \eta\, \kappa\, D_e} \quad (8)$$

as a function of the equivalent volume diameter alone.

Thus, for spherical particles at room temperature, Equation 5 leads to

$$\ell < 19 \times 10^6 \frac{D_e \dot{V}}{C(D_e)} \text{ [cm]} \quad (9)$$

and Figure 6 gives the corresponding minimum sampling rates $\dot{V}$ required to collect 97% of the particles of size D_e if the sampling tube has a length of ℓ = 40 cm, which may be considered a rather long sampling line in most inhalation exposure facilities. Even so, Figure 6 reveals that a low sampling rate of about 40 cm^3/min as employed in low-volume electrostatic precipitators [14] is still sufficient to keep sampling losses of 0.1-μm particles below 3%. Since in most experimental aerosol exposures, small particles of sizes below 0.05 μm are fairly uncommon and do not significantly contribute to the airborne mass concentration, it is obvious that sampling arrangements for measuring or monitoring exposure aerosols are not very critical with regard to the small-size end of the particle mass distribution.

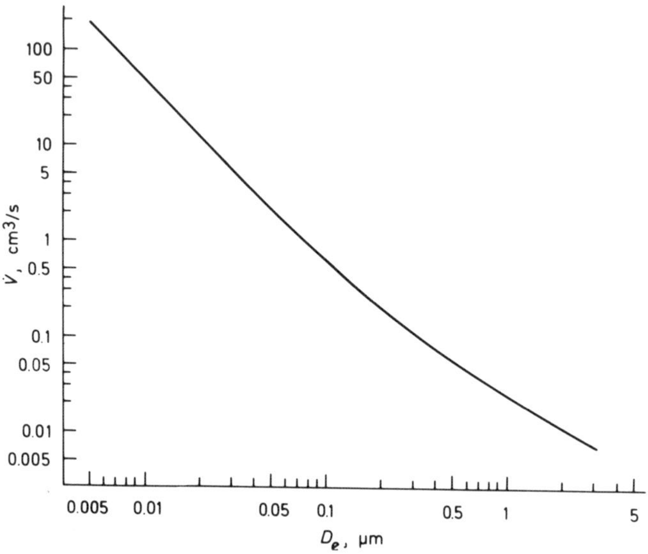

Figure 6. Size-dependence of the minimum sampling flow rates in sampling tubes of 40 cm length for achieving sampling efficiencies above 97%.

AEROSOL CHARACTERIZATION

A detailed characterization of an aerosol is, generally, a tedious and complex endeavor. This is particularly true for atmospheric aerosols in urban and industrial areas, where the airborne particulate matter may be of various origin and composition. The list of health-related parameters to be determined in air pollution control networks comprises primarily the airborne mass concentration, the size and the mass distributions of the aerosol, and the overall content of certain chemical elements and compounds. These parameters will be measured on a time-average base, usually over periods of 24 hours, and their daily fluctuations will be recorded. So far, however, these basic data were rarely acquired simultaneously to a sufficient extent. While average mass concentration measurements without size discrimination by gravimetric or indirect methods are fairly common, there are only a few systematic attempts to record simultaneously size and/or mass distribution data. The same is true for the chemical analysis of urban aerosols. One noteworthy exception is the well-conceived Los Angeles smog study [15], an intensive investigation that yielded a very detailed picture of the daily growth and decay of an urban aerosol.

Although some of the analytical equipment developed for field studies of aerosols will be very useful for aerosol measurements in experimental inhalation exposure work in the laboratory, the analytic requirements for these experiments are frequently quite different from those for air pollution surveys. In some laboratory cases, aerosol characterization is much simpler than in the field, but there are other experimental investigations, where more details of the properties of the exposure aerosol are needed than air pollution surveys would ever require.

A most convenient difference between field work and laboratory measurements lies in the fact that inhalation studies need well-defined exposure levels usually far in excess of ambient air concentrations. Therefore, the requirements for suitable monitoring equipment are mostly less stringent than for field stations, and a continuous check on the stationary state or the predesigned periodicity of the desired levels in the exposure chambers can effectively be achieved by means of conventional or specially designed aerosol photometers [16] calibrated for the aerosol under investigation.

In order to discuss other differences to ambient air measurements, it is useful to divide the inhalation exposure arrangements into three categories with regard to the type of exposure aerosol utilized. Experimental inhalation studies may deal with

1. chemically homogeneous model aerosols
2. carrier model aerosols
3. source emission aerosols

Chemically Homogeneous Model Aerosols

Inhalation experiments with aerosols of this type very often involve aerosol particles of a chemically well-defined single component, for instance a cadmium chloride aerosol generated from an aqueous solution by means of a nebulizer [17] or a benzo[a]pyrene aerosol produced by a vapor condensation process [18]. Such aerosols do not simulate actual outside exposure situations. Instead, they are model aerosols designed to be deposited to a significant amount in the lungs of experimental animals so that the specific biological effects of the airborne substance under investigation can be determined. These biological effects may either indicate an acute toxicity on inhalation or, less obviously but more frequently, a long-term toxicity after chronic inhalation exposure that finally may induce fibrotic, necrotic or functional reactions or, worst of all, cancer.

In all these model aerosols, the chemical composition is known from the outset and there is no need for additional analytic confirmation. However, if the quantitative chemical determination is simple, which for single components is more often than not the case, then this measurement may easily replace any direct gravimetric determination of the total airborne mass concentration on filter samples or the like. The same is true for the analysis of deposits of size-separating aerosol sampling devices like cascade impactors or aerosol centrifuges.

In most cases, model aerosols generated from either nebulizers [19] or vapor condensation generators [20] consist of spherical or at least isometric particles so that a determination of the size distribution of the aerosol, in addition to the overall mass concentration, will already constitute a complete aerosol characterization of these exposure models. Unless very high mass concentrations and very fine particle sizes provide for some significant coagulation during the short time between aerosol generation and exposure, the size distribution can be derived from a microscopic or electronmicroscopic statistical analysis of the distribution of the primary particle sizes of a filter deposit. If coagulation is suspected to play a significant role, it is preferable to derive the mass distribution by means of analytic devices providing aerodynamic size separation.

Frequently, condensation aerosols, particularly when produced by La Mer type generators [21], are quasimonodisperse so that the aerosol particles are of almost uniform size. In this case, the size measurements of sufficiently small particles below about 1-μm diameter can be easily determined and monitored optically by a calibrated "owl" [22]. For monodisperse aerosols of bigger particle sizes, the batch-type samples drawn onto filters or into centrifugal spectrometers and requiring a subsequent evaluation are most common.

Carrier Model Aerosols

Exposure studies with these types of aerosols are aimed at elucidating the biological effects of substances which may have an extended biological half-life in the nonciliated section of the lungs, because they are adsorbed or condensed onto insoluble aerosol particles that are preferentially deposited in the pulmonary region. A typical and important practical example is the pulmonary deposition of soot or soil particles carrying organic substances like polycyclic aromatic hydrocarbons coating the particle surface. Since the surface distribution of the adhering substance is, in general, not independent of the nature of the carrier aerosol substance, the characterization of this important type of two- or more-component aerosol has some complex aspects unless the carrier substance is available as a monodisperse aerosol. Unfortunately, due to practical difficulties, monodisperse aerosols have not yet been used in exposure studies of this type.

A great variety of inorganic and organic substances with sufficiently low solubility in the lung fluids can be dispersed in air and serve as model carrier aerosols for compounds that subsequently may be adsorbed or condensed on the particle surfaces. In practice, however, there is a preference for those carriers that, on the one hand, are advantageous from the experimental point of view and, on the other hand, will be representative for certain urban or industrial air pollution situations.

Experimental aspects in dealing with adsorbed substances favor carrier aerosol particles of high specific surface area to optimize the relative amount of adsorbed material being deposited. This preference would rule out spherical carrier particles because of their minimal surface-to-volume ratio. Spheres may be acceptable only if the surface coating can be increased by condensation, a technique of limited value because exposure conditions in this case may not significantly differ from experiments with chemically homogeneous particles. If adsorptive layers on particles with high specific surface area are most desirable, then the carrier aerosol should consist of either porous particles with internal surfaces accessible for adsorption and desorption processes or aggregates of extremely small primary particles serving the same purpose.

For practical reasons, most carrier aerosols are of the latter type leaving the aerosol analyst with the difficult problem of a proper physical characterization. Two prominent carrier aerosols of this kind are carbon black, as produced by incomplete combustion of an acetylene flame, and silica, usually generated by a pyrogenic process of burning silicon tetrachloride in a hydrogen flame. In both cases, primary particles of less than 0.02 μm will form in great numbers and immediately coagulate to aggregates of various size.

These two carrier aerosols also appear fairly representative with regard to two types of actual particulate air pollutants: carbon black will serve as a model for soot from combustion processes while silica represents a major constituent of fly ash and other inorganic airborne particles. Oxides of certain metals like iron, aluminum, lead and others continue the list of relevant carrier model aerosols. Characterizing these exposure aerosols involves the determination of both the overall mass concentration of the aerosol and the average fraction of the mass of the substance forming the surface layer. Although these measurements appear to be straightforward if an established analytic method is available, there are sometimes problems due to the fact that chemisorption or catalytic surface activity may influence the process of recovering the adsorbed substance.

For obvious reasons, the airborne size distribution of the aggregates cannot be determined by a statistical analysis of a nonselective overall precipitate under the microscope. In such deposits, the airborne identity of the aggregates is no longer preserved. Therefore, the distribution analysis has to be done by dynamic size spectrometry. Since the aggregates are often very delicate, centrifugal devices will be preferable to impaction instruments. Generally, centrifuges have a better chance of avoiding aggregate shattering, and the continuous size separation provides the possibility of identifying deposited aggregates as former airborne entities. Then, besides the size-related determination of the fraction of the adsorbed mass, the size spectrum will further permit an evaluation of the variations in shape, geometrical size and number of primary particles in the aggregates of identical diameter, which would complete the characterization of the aerosol.

Source Emission Aerosols

In recent years, laboratory inhalation exposure studies on animals increasingly made use of particulate matter from actual emission sources. In these cases, the exposure aerosol is either generated by redispersing particle precipitates collected with high-volume samplers, electrostatic precipitators and filters, or a suitable dilution of the actual emission is directly fed into the exposure chamber so that the animals breathe the complete mixture of gaseous and particulate pollutants coming from the emission source. Examples of such experiments are inhalation studies either with redispersed industrial dusts, like asbestos cement and power plant fly ash, or with actual source emissions, like automobile exhaust and chimney effluents.

For a successful correlation of the biological effects of these aerosols to their physical and chemical parameters, a very detailed aerosol characterization is necessary and constitutes a major task. In addition to the measurements common to other inhalation exposure studies, detailed analyses not only of the chemical composition of the average total airborne mass concentration but also of the dynamically separated fractions of the size distribution are required.

Depending upon the type of source, the analytic requirements with regard to the particle chemistry will vary, but the determination of all major constituents and a search for traces of suspected toxicants will be mandatory in all cases. While for inorganic materials an elemental analysis and the determination of the anions in the particles may be of special importance, the analysis for organic trace components will probably dominate in future investigations and necessitate the use of most modern analytic techniques and equipment.

Furthermore, if the airborne particles are aggregations of small primary particles or if they deviate substantially from a spherical or isometric shape, a detailed physical analysis is required characterizing the particles by packing densities and shape factors, in particular, by elongation ratios for fibrous or flaky particles. Sometimes, it is also necessary to determine the electrostatic charge of the particles.

PHYSICAL ANALYSES

Monitoring Steady State Exposure Conditions

In practically all experimental inhalation studies, the aerosol particles are of sizes that produce measurable Tyndall effects, i.e., the particles attenuate a light beam by way of light scattering. From the quantitative point of view, the Mie theory rules the phenomenon for spherical particles in the size range of interest [23]. According to this theory, which holds very well in controlled experimental tests [24], the extinction of light by scattering is a complex function of the size, the refractive index and the volume concentration of the particles. Therefore, an instrument monitoring the light scattered by the particles of an exposure aerosol is a very sensitive indicator of deviations from stationary conditions: it will respond to any fluctuation with regard to mass concentration or particle size distribution or composition. For this reason it is quite common to use aerosol light scattering photometers for monitoring the quality of the steady state of exposure levels. In general, no high sensitivity is required for the aerosol photometer. Exposure concentrations are mostly in a range where even the attenuation of the incident light beam as measured in a transmissometer arrangement will serve the monitoring purpose. A simple design measuring the relative intensity of the light scattered into an angle between 10 and 20 degrees of the forward direction of the incident light beam by means of two photo diodes was designed by Mönig [25], facilitating the recording of aerosol exposure levels between about 100 $\mu g/m^3$ and 1 g/m^3. Figure 7 shows a picture of the compact instrument and a diagram of the optical arrangement. The aerosol sampling rate is 40 cm^3/min.

More versatile than a simple aerosol photometer is a new optical instrument conceived by Höllander and Schörmann [26]. This design is called an aerosol filter photometer. It proved to be of outstanding sensitivity and permits a recording of exposure concentrations down to levels of particulate matter in ambient air. The instrument is shown schematically in Figure 8. It uses metal-coated Nuclepore® filters so that the illuminating white light will penetrate through the filter pores only. A chopper arrangement permanently compares, on a common photo diode, the intensities of the incident light beam and the light attenuated by the filter. Aerosol particles depositing on the filter will gradually obstruct the pores and further attenuate the transmitted light. Although this system needs some empirical calibration, the sensitivity is unprecedented. A surface deposit concentration of about 10 nanograms/cm^2 can easily be detected. Therefore, the sampling rate can be extremely slow and a predilution system is to be used in laboratory exposure work. Figure 9 shows a linearity check with a Latex test aerosol and Figure 10 is the actual strip chart track of a recording of ambient air particles showing intervals of sampling and standby.

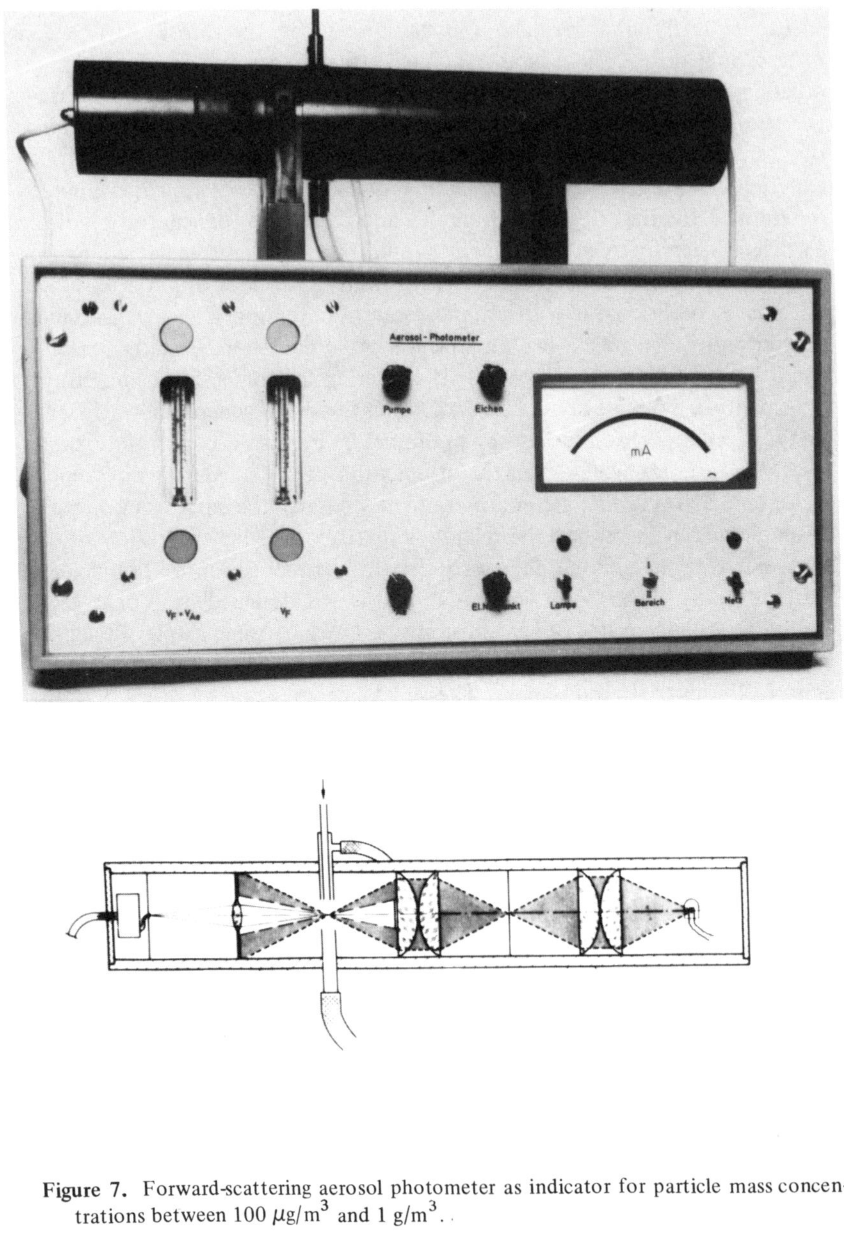

Figure 7. Forward-scattering aerosol photometer as indicator for particle mass concentrations between 100 $\mu g/m^3$ and 1 g/m^3.

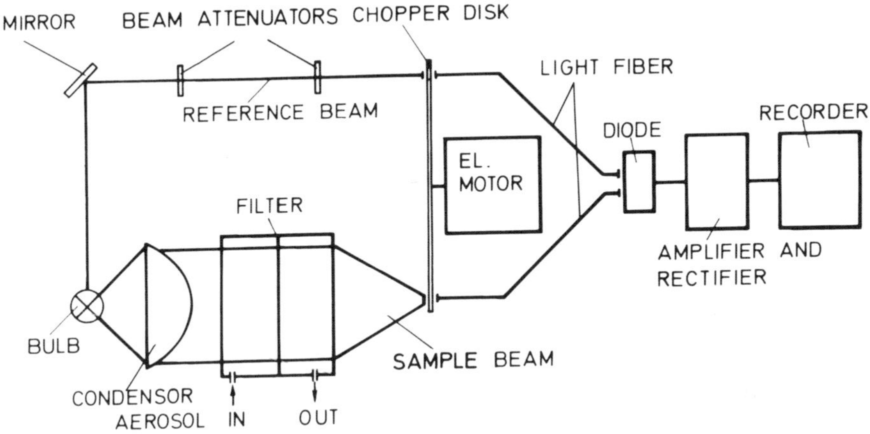

Figure 8. Schematic diagram of the aerosol filter photometer.

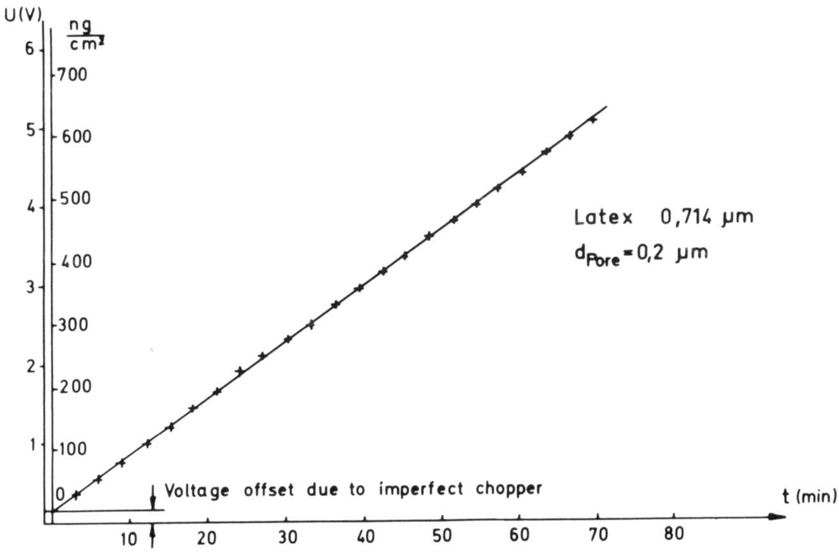

Figure 9. Linearity test of the deposition of a latex aerosol in the aerosol filter photometer.

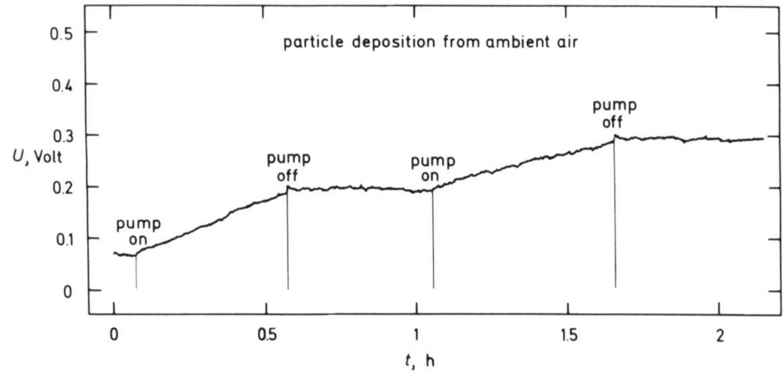

Figure 10. Ambient air particle deposition in the aerosol filter photometer.

Mass Concentration Measurements

Ordinary total mass concentration measurements in laboratory inhalation exposure work are no problem at all unless the exposure levels are extremely low. Under regular conditions, exposure aerosol concentrations will be upward of 100 $\mu g/m^3$. Thus, by using a microbalance, reliable gravimetric readings can be obtained with filter deposits of samples of about one cubic meter of exposure chamber air. This sample can easily be drawn within some 30 minutes.

In steady-state exposure runs, such gravimetric determinations can be limited to infrequent checks of an empirical calibration of a monitoring aerosol photometer. In the case of extremely low exposure levels, it is recommended to use an aerosol filter photometer as described above.

Quite frequently, inhalation exposure studies involve defined substances whose mass can easily be determined by a quantitative chemical analysis which may be more accurate than any gravimetric method. This is the case in many investigations involving heavy metal aerosols, where atomic absorption [27] or X-ray fluorescence [28] may be the analytic method of choice. Sometimes, organic compounds also offer a simple way of mass determination, e.g., the sensitive fluorometric analysis of fluorescein, a substance frequently used as a test aerosol [29].

Particle Size Measurements

The methods of measuring the particle sizes and determining the particle number frequency and/or mass frequency distribution with regard to particle size depend upon the type of exposure aerosol under investigation. For solid, compact and chemically homogeneous model aerosols as they are frequently generated by nebulization or condensation, size measurements and distribution analyses can be made by microscopic or electronmicroscopic sizing of a filter batch of suitable deposit density [30]. For isometric particles this analysis will yield the geometric diameters D_e. Then, their number frequency distribution

$$\frac{\Delta N}{\Delta D_e} = F(D_e) \tag{10}$$

can be determined with

$$N = \sum_{D_{e\,min}}^{D_{e\,max}} F(D_e) \cdot \Delta D_e \tag{11}$$

being the total number of particles evaluated. If N is sufficiently large, a density distribution function of the probability to encounter a particle of size D_e can be derived as

$$\frac{1}{N}\frac{d\,N}{d\,D_e} = \frac{d\,n}{d\,D_e} = f(D_e) \tag{12}$$

which may then be compared—and frequently fitted—to mathematical distribution functions like the Gaussian normal distribution

$$\frac{d\,n}{d\,D_e} = \frac{1}{(\sqrt{2\,\pi})\,\sigma}\exp\left(\frac{(\bar{D}_e - D_e)^2}{2\,\sigma^2}\right) \tag{13}$$

or the more adequate log-normal distribution

$$\frac{d\,n}{d\,ln\,D_e} = \frac{1}{(\sqrt{2\,\pi})\,\sigma_g}\exp\left[\frac{[ln\,(\hat{D}_e/D_e)]^2}{2\,\sigma_g^2}\right] \tag{14}$$

which does not give finite probability densities for physically meaningless negative values of D_e. Figure 11 shows typical representations.

In fitting statistical data of microscopic counts to Equation 13 or 14, the distribution parameters will not have the same physical meaning. For the Gaussian distribution, $\bar{D}_e$ represents simultaneously the arithmetic mean diameter, the geometric mean (median) diameter and the modal diameter, i.e., the diameter of the maximum probability density as defined by Equation 12. The other parameter is the standard deviation σ which serves as a measure of the dispersity of the (symmetrical) distribution. In comparison, the log-normal distribution in the form of Equation 14 denotes by $\hat{D}_e$ the median diameter only and the geometric standard deviation σ_g is a factor to the median diameter, so that

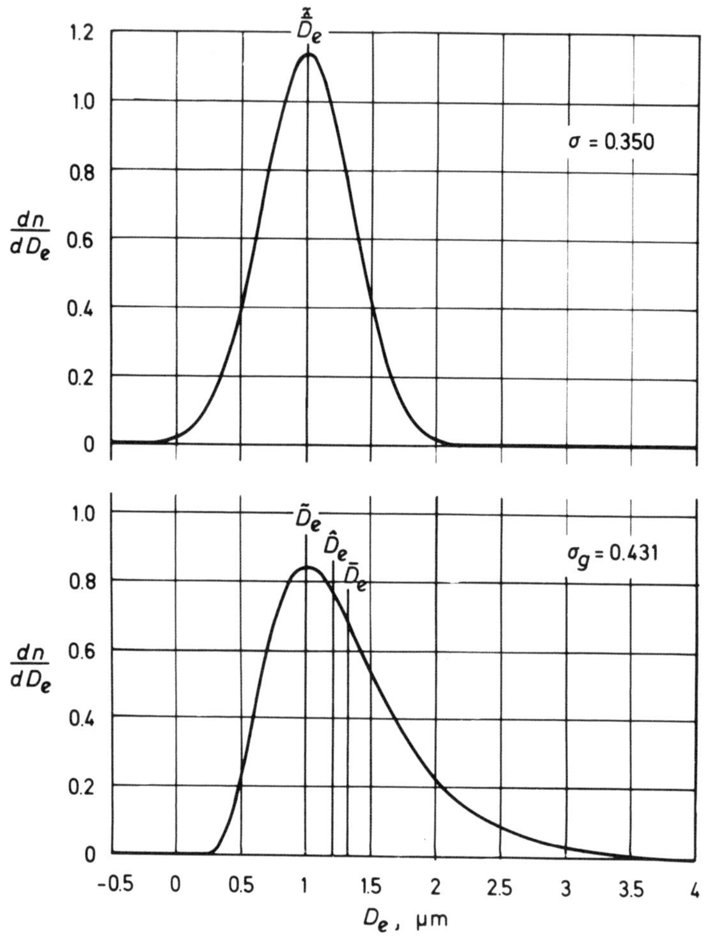

Figure 11. Mathematical distribution functions; top: Gaussian normal frequency density distribution, bottom: log-normal frequency density distribution.

$$\tilde{D}_e = \hat{D}_e \cdot e^{-\sigma_g^2} \tag{15}$$

is the modal diameter $\tilde{D}_e$, and $ln\,\tilde{D}_e + \sigma_g = ln\,D_1$, $ln\,\tilde{D}_e - \sigma_g = ln\,D_2$ defines two diameters for which the probability density

$$\frac{dn}{dD_e} = \frac{\exp\left[-\left(ln\frac{\tilde{D}_e}{D_e}\right)^2 \Big/ 2\,\sigma_g^2\right]}{(\sqrt{2\pi})\sigma_g\,\tilde{D}_e\,\exp(\sigma_g^2/2)} \tag{16}$$

as derived from Equation 14, will have decreased from the maximum value by a factor of $1/\sqrt{e} \approx 0.6$. Thus, log-normal distributions are skewed and have an extended "tail" towards larger sizes as shown in Figure 11.

The arithmetic mean diameter $\bar{D}_e$ of a log-normal distribution can also be expressed by the median diameter $\hat{D}_e$ and the geometric standard deviation σ_g. The general relationship is

$$\bar{D}_e = \hat{D}_e \, e^{\sigma_g^2/2} \tag{17}$$

As mentioned before, fitting data to the log-normal distribution is preferable to the regular Gaussian distribution. It should be pointed out, however, that log-normal distributions are not a law of nature, and a proper statistical chi-square test should be applied to check whether empirical data may actually be represented by mathematical functions of any given form.

For the correlation to lung deposition data, it is necessary to determine the mass frequency distribution of the aerosol. This may be acquired by calculations based on the number distribution provided that the particles are reasonably isometric or spherical. A formal transformation of number distribution data to mass distributions by applying a weighting factor of $D_e{}^3$ may be very misleading for elongated, flaky or irregularly shaped particles. For log-normal number distributions of isometric particles as represented by Equations 14 or 16, a regular transformation to mass distribution data is possible and yields a normalized form of the mass probability density function as

$$\frac{d\,m}{d\,D_e} = \frac{\exp\left[-\left(ln\,\frac{\hat{D}_e}{D_e} + 2\sigma_g^2\right)\Big/2\sigma_g^2\right]}{(\sqrt{2\pi})\sigma_g \; \hat{D}_e \; \exp(5/2\;\sigma_g^2)} \tag{18}$$

Here, the mass modal diameter $\tilde{\tilde{D}}_e$ is

$$\tilde{\tilde{D}}_e = \hat{D}_e \cdot e^{2\sigma_g^2} \tag{19}$$

while the mass median diameter $\hat{\hat{D}}_e$ is

$$\hat{\hat{D}}_e = \hat{D}_e \cdot e^{3\sigma_g^2} \tag{20}$$

and the mass mean diameter $\bar{\bar{D}}_e$ becomes

$$\bar{\bar{D}}_e = \hat{D}_e \cdot e^{3\sigma_g^2/2} \tag{21}$$

The size analysis described so far applies to homogeneous and isometric particles for which the time consuming batch-wise microscopic analysis can sometimes be replaced advantageously by a size-discriminating optical online measurement, if the size distribution does not spread over more than one order of magnitude.

Laser aerosol spectrometers [31] of high size resolution may be used in a size range of $0.1\ \lambda \leqslant D_e < \lambda$ of the laser wave length λ. Figure 12 shows schematically a design used for single-component submicron model aerosols. The illuminated critical volume is confined to an aerosol cylinder of 50 μm diameter and 50 μm height which permits reliable size discrimination of single particles in concentrations up to 10^6 particles per cm^3. The scattered light is collected under an angle of 40° in the forward direction. The light source is a 12 mW-He-Cd laser facilitating size measurements from 0.15 to 0.8 μm. Quasimonodisperse model aerosols can easily be monitored with devices of this kind after calibration of the optical signals counted in a multichannel analyzer.

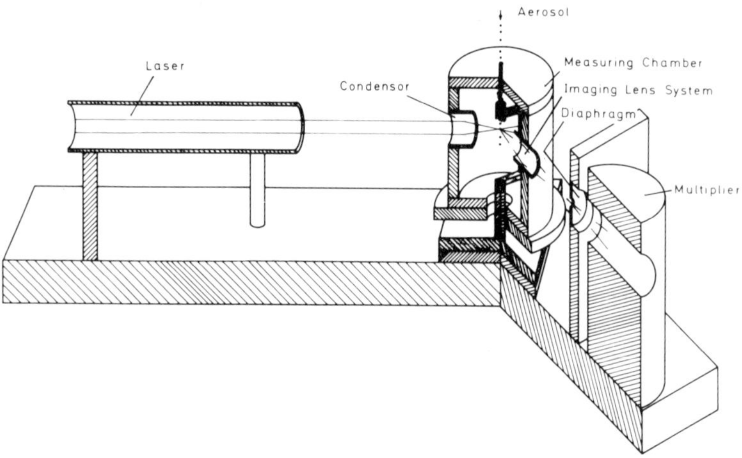

Figure 12. Schematic diagram of a laser aerosol spectrometer.

For the size range above the wave lengths of the illuminating light, a low-angle scattering instrument using the white light of a mercury lamp as an illuminating source [32] permits an online size distribution analysis between 0.7 and 6 μm of diameter. The instrument measures the light scattered into the forward lobe of the light-scatter distribution around the particles under an annular solid angle between 2.5 and 5.5°. The signal is related to the optical cross section, i.e., the projected diameter of the particles, and the influence of the refractive index of the particles is minimal.

If the particles of the exposure aerosol vary in mass density due to different composition or if they consist of irregularly shaped particles, it is necessary to measure the size distribution in terms of aerodynamic diameters which are uniquely related to the particle settling velocities. Therefore, the size separation has to be achieved while the particles are still airborne.

A variety of instruments are suitable for this purpose. They all require sampling of a proper test volume and a subsequent gravimetric analysis of the size-separated deposits. Standard procedures of that kind for cascade impactors, virtual impactors and aerosol centrifuges are quite common. Due to the relatively high mass concentrations used in inhalation exposure studies, no high volume sampling is required and low sampling rates will be acceptable that are normally unsuitable for ambient air measurements.

A particularly useful cascade impactor is the often-copied design by Mercer [33] if sensitive methods for the deposit mass determination are available. Figure 13 is a schematic diagram of this 7-stage design. The effective cut-off aerodynamic diameters between the first and the last stage cover about one order of magnitude and sampling rates between 50 and 150 cm^3/min permit the separation of particles sized from 0.12 to 5 μm in aerodynamic diameters.

Centrifugal aerosol spectrometers have the advantage of a continuous particle size spectrum produced along a deposition wall. The limited usefulness of the conifuge design [34] has left the field practically to two principal rotor designs having concentric and spiral deposition ducts, respectively, and depositing particles on removable foil strips.

Concentric duct centrifuges are advantageous because of their simple design and operation, although aerosol inlet losses can be more prominent than in spiral duct centrifuges [35]. A particularly simple instrument is a design by Hochrainer [36] which creates its own winnowing air in the first semicircle and uses it for the aerosol size analysis in the second one. Figure 14 shows the cross sections through this instrument. Aerosol entering from the center of rotation, 1, through duct 2 will be cleaned of particles in duct 4 and then be used to winnow the aerosol particles entering through duct 3. The instrument is very useful for checking model aerosols for their degree of monodispersity in the size range between 0.4 and 1.3 μm.

Spiral duct centrifuges come in different rotor dimensions [37]. Of these, the long spiral versions cover a size range of about two orders of magnitude [38] from below 0.1 μm to about 6 to 8 μm. At a common sampling rate of 1 liter/min, aerosol deposits can be collected on thin metal ribbons, plastic foils and/or electronmicroscopical grids for all kinds of subsequent investigations so that very detailed size-related physical and chemical analyses are facilitated.

In recent years, the spiral duct centrifuge was modified into a monitoring instrument by utilizing piezoelectric quartz crystals as surface deposit sensors along the deposition wall of the spiral duct [39]. Figure 15 is a photograph of the rotor of the most recent design. The rotor is housed in a thermostat-controlled casing with the drive unit. Along the outer wall of the duct, there

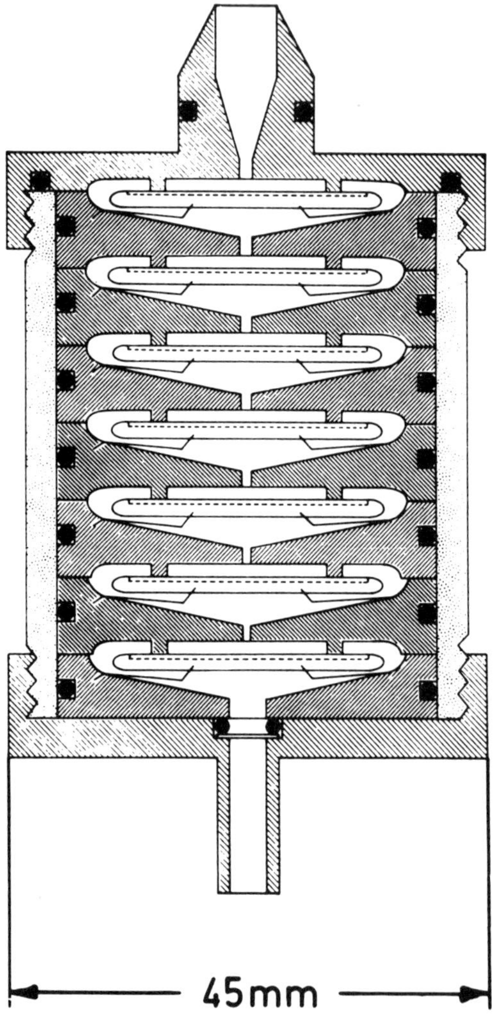

Figure 13. Cascade impactor by Mercer et al. [33].

are 14 dovetailed inserts which can be removed from the instrument. A single unit of these inserts is shown as a photograph in Figure 16. The ceramic unit holds a piezoelectric quartz sensor flush to the face of the contoured insert that forms part of the outer wall of the spiral duct when properly inserted. One of these units serves as a reference sensor at the very beginning of the spiral duct, where it does not receive any aerosol deposit. The other 13 units

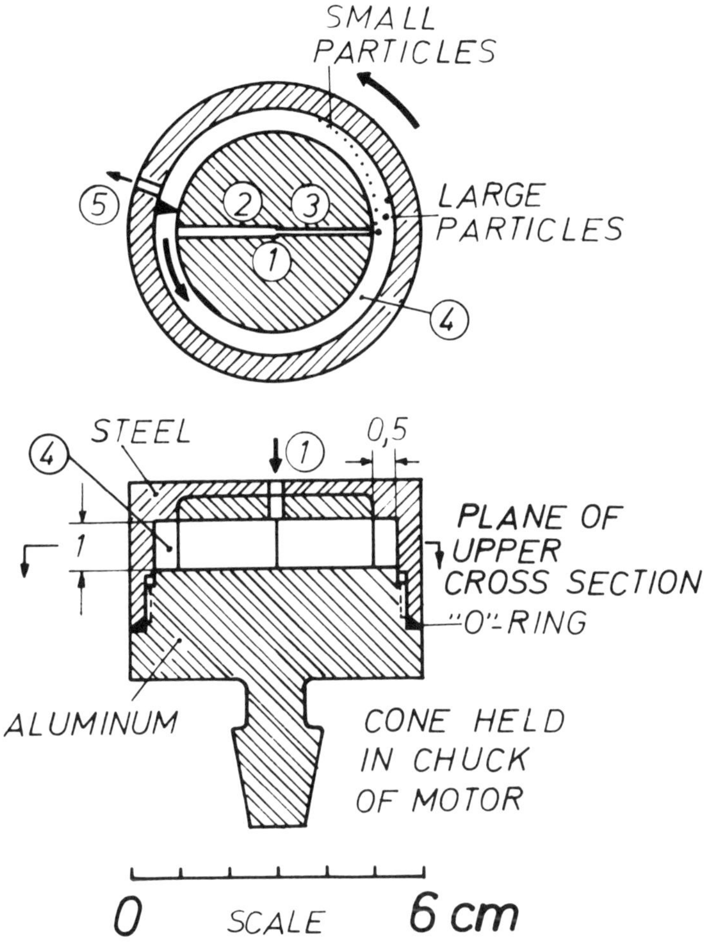

Figure 14. Midget centrifugal aerosol spectrometer by Hochrainer [36].

are spaced along the spiral duct. Their crystals receive deposits of size fractions of the aerosol which are practically monodisperse. Figure 17 shows the calibration curve of the instrument. The horizontal bars indicate the location of the quartz sensors.

When this centrifuge is sampling, the buildup of the deposits on the quartz crystals can be followed by a telemetric circuitry arrangement. This facilitates online gravimetric mass distribution analyses while the centrifuge is spinning. The sensitivity of the quartz detectors is such that it takes about one hour of ambient air sampling in a moderately polluted urban atmosphere with an

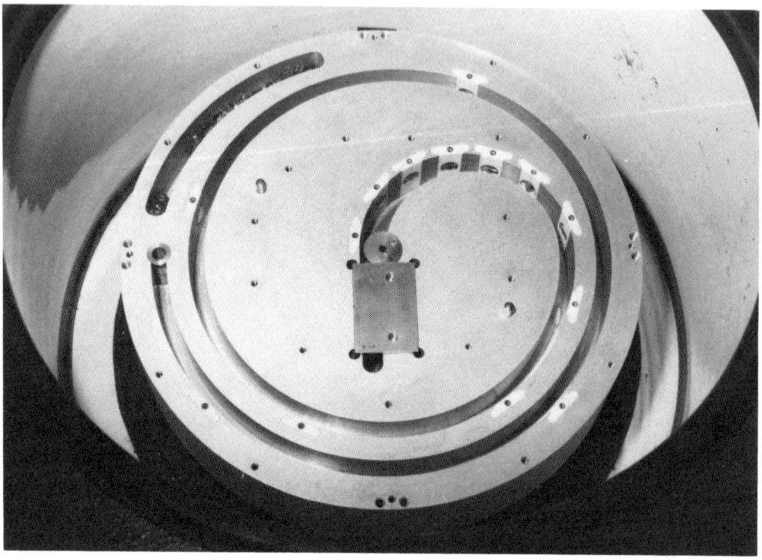

Figure 15. Rotor of a spiral duct aerosol centrifuge monitoring aerosol deposition with quartz crystal sensors.

Figure 16. Quartz sensor in plug-in mounting.

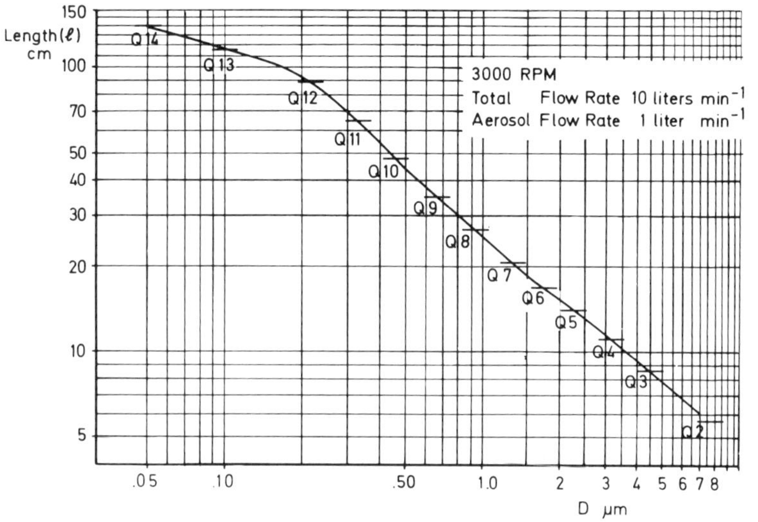

Figure 17. Calibration of the monitoring spiral duct aerosol centrifuge.

aerosol concentration of some 50 $\mu g/m^3$ to obtain readings well above the detection limit. Figure 18 shows the size distribution of a fluorescein model aerosol used at 80 $\mu g/m^3$ in a performance test to compare electronic readings with fluorometric analyses. The results are not normalized and show the actual statistical frequency density per volume unit measured over 1 to 8 hours. The slight increase of the readings with increasing sampling time reflect the gradual increase of the nebulizer output over 8 hours as verified with a filter device.

Between the quartz sensors there is considerable space left where the continuous winnowing process produces further size-separated aerosol deposits. Therefore, the monitor centrifuge can simultaneously collect these deposits on electron microscopic grids and/or chemically well-defined foils for batch analyses at the end of the monitoring period, just like in the case of the regular curved duct centrifuge without quartz sensors. When sampling for this purpose, the quartz sensors can easily be used to indicate the surface concentration collected so far. Then the sampling period can be terminated when the instrument reads the proper amount of mass deposit as required by the analytical method to be used.

Due to the excellent size resolution of the spiral duct centrifuge, it will yield the most detailed information with regard to shape factors of single

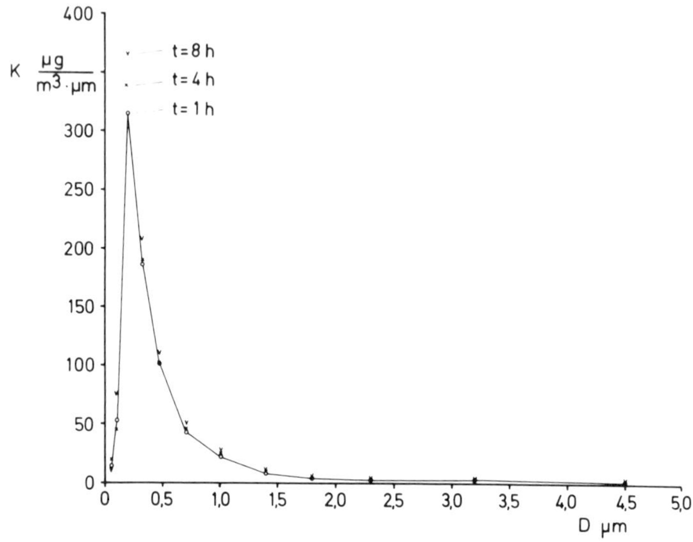

Figure 18. Monitored size spectrum of a dilute fluorescein model aerosol (80 μg/m ^{3}).

particles and packing density and contours of aggregates. Of course, investigations of this type always require the use of an electron microscope. In particular, a scanning electron microscope with a microprobe attachment is very useful for a detailed physical and elemental analysis of the aerosol samples.

A typical example of a physical configuration analysis with the spiral duct centrifuge is the study by Kops et al. [40] on aerosols generated by the exploding-wire technique. Figure 19 shows a graph where the number of primary particles in an aggregate is plotted against the aerodynamic diameter of the aggregate. It could be shown that the change in the slope of the log-log plot is associated with a change of aggregate configuration. While chain-like aggregates prevailed at low numbers of primary particles, the aggregates are clustered in case of high numbers of primary particles.

Results of this kind can only be obtained when aerodynamic size separation is employed during sampling on electron microscopic grids. Another example is the analysis of aerosols of elongated particles like asbestos fibers. The electron micrograph in Figure 20 shows a section of the deposit of an amosite fiber model aerosol in the spiral centrifuge [41]. All particles have the same aerodynamic diameter of about 1.7 μm, but this value is apparently closely related to the fiber diameter, while the fiber length varies widely. Thus, asbestos is a typical example of an aerosol, which needs more than one parameter for its proper " size" description.

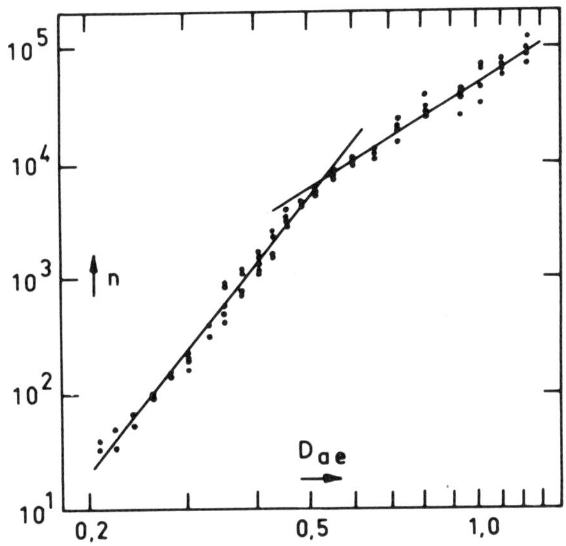

Figure 19. Number n of primary particles in aggregates of aerodynamic diameter D_{ae} (μm) produced by an exploding-wire technqiue [40].

The shape factor analysis for irregularly shaped aerosol particles always involves a statistical average over a subgroup of particles as defined by a micrograph. Since the aerodynamic size separation yields deposits of particles with the same aerodynamic diameter, it is appropriate to try to determine the dynamic shape factor κ as defined by Equation 7 as a key parameter. This, however, requires the knowledge of the particle density ρ and the equivalent volume diameter D_e. Fortunately, the particle density ρ, a major problem in the evaluation of particulate matter from ambient air, is very often known or easily determined for the exposure model aerosols unless size-dependent density changes occur for bubble-type aerosol particles [42]. While the particle density ρ may actually apply to all particles of the deposit, the equivalent volume diameter is an individual particle parameter and can only be determined as an average value for the aerosol particles of equal aerodynamic diameter. The monitoring spiral duct centrifuge provides a convenient method for this. Since the quartz sensors give a direct reading of the mass concentration M_s on the quartz surface, it is sufficient to take a microphotograph of an adjacent surface area and count the number concentration N_s.

In this way, the relation

$$\frac{M_s}{N_s} = \frac{\pi}{6} \rho D_e^3 \tag{22}$$

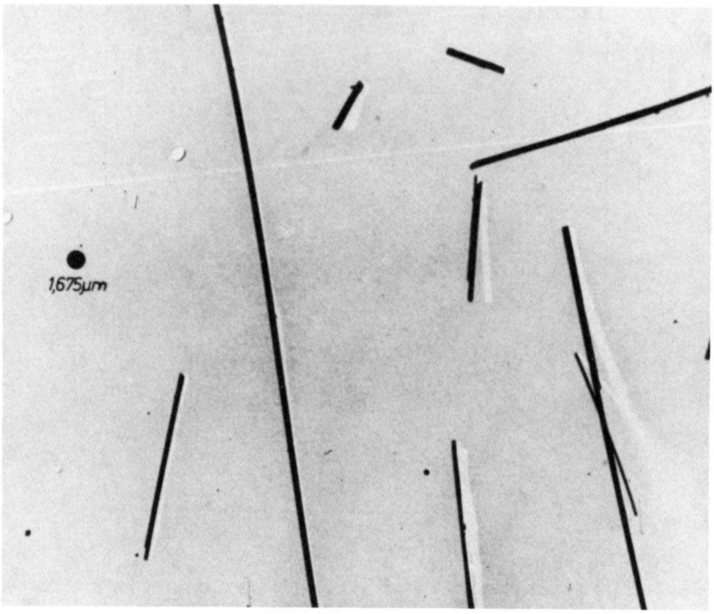

Figure 20. Amosite filter deposit in the spiral duct aerosol centrifuge at D_{ae} = 1.675 μm (size of dot).

permits the computation of an equivalent volume diameter D_e, which will be an average value for a given aerodynamic diameter. Then, when computing the dynamic shape factor by Equation 7, this parameter will also be a function of the aerodynamic diameter. Thus, the dynamic shape factor for the deposit shown in Figure 20 would be related to some average fiber length. In fact, it is suggested [43] that all dynamic shape factor data should be interpreted as prolate or oblate spheroids, i.e., elongated or flaky particles, of characteristic aspect ratios. However, this does not always seem to be adequate and it appears that the dynamic shape factors of aggregates may be better related to the packing density of the primary particles [44]. A shape factor evaluation for aerosols of uranium dioxide, thorium dioxide, low grade coal and quartz was made by Kotrappa [45,46] and discussed by Davies [47].

CHEMICAL ANALYSES

The analytical procedures for exposure aerosols depend, of course, on the chemical nature of the aerosols. Except for source emission aerosols, there is normally no phase of exploratory qualitative analysis involved, because the components of the model aerosols are usually known. As mentioned before, single component aerosols do not require any analytic chemistry unless a quantative chemical analysis of the aerosol and its size fractions is simpler and more accurate than the usual gravimetric method of mass determination of the deposits.

The situation changes immediately when the model aerosol is composed of two components which do not form aerosol particles of homogeneous composition. The carrier model aerosols are the most prominent example. Overall filter samples as well as aerodynamically separated size fractions have to be analyzed quantitatively and may require, on one or more ways, preparative steps like elution; ashing; thermal volatilization or desorption; extraction, dissolution and precipitation. Wet chemistry may not always be avoidable, but modern methods with a minimum of sample separation will be preferable. Thus, for the elemental analysis, atomic absorption spectrometry, X-ray fluorescence analysis and the related microprobe analysis on single particles are most frequently applied. The inductively coupled plasma emission spectrometry is of growing importance. A very sensitive method in this field is the anodic stripping voltametry, which is suitable for both inorganic and organic substances. Neutron activation analysis is particularly useful, in the case of defined industrial dusts with typical trace impurities, like in coal or certain minerals, being used as exposure aerosols. In most cases, composed model aerosols will not consist of more than two major compounds. However, if organic substances are involved,the possibility of surface catalysis or other types of chemical reactions has to be envisaged.

A detailed chemical analysis will be of major importance for source emission aerosols. With precollected particulate matter like mine dust or fly ash, an overall analysis including an inventory of toxic elements and compounds can be made prior to the aerosolization and animal exposure. But for a complete aerosol characterization, it is necessary to repeat these analyses on the aerodynamic size fractions sampled with impactors or centrifuges in the exposure chamber. Source emissions from combustion processes require additional analytic efforts with regard to the organic products of incomplete combustion of fossil fuels or wastes. A particular field of concern is here the occurrence of polycyclic aromatic hydrocarbons, some of which are known to be potent carcinogens. The great variety of these compounds found in case studies with automobile exhaust [48], cigarette smoke [49] and other aerosols [50] make techniques of high-resolution gas liquid chromatography possibly coupled with a mass spectrometer mandatory.

For the analysis of sulfuric acid or ammonium sulfate, two compounds originating from sulfur dioxide and quite common in certain source emission aerosols, it will be necessary to discriminate them from other sulfur compounds. For this purpose, photometric and isotopic dilution methods are available [51].

Chances are that the experimental inhalation exposure of animals to source emission aerosols will be of growing practical importance for a better evaluation of the biological effects of realistic combinations of air pollutants in our real world. Therefore, the chemical analysis of these aerosols, particularly with regard to their contents of organic compounds, will be an area of significant growth and increased sophistication in future exposure studies.

REFERENCES

1. Douglas, J. W. B., and R. E. Waller. *Brit. J. Prev. Soc. Med.* 20:1 (1966).
2. Findeisen, W. *Pflüger's Arch. ges. Physiol. Menschen Tiere*, 236:367 (1935).
3. Brown, J. H., K. M. Cook, F. G. Ney and T. Hatch. *Am. J. Publ. Health* 40:450 (1950).
4. Dennis, W. L. *Inhaled Particles and Vapours* (Oxford: Pergamon Press, 1961), p. 88.
5. Van Wijk, A. M., and H. S. Patterson. *J. Ind. Hyg. Toxicol.* 22:31 (1940).
6. Stöber, W., W. Höllander and G. Morawietz. *Health-Related Sampling of Total Particulate Matter in Air with Elutriating Filter Devices* (in press).
7. Stahlhofen, W., B. Eckardt, J. Gebhart, J. Heyder and B. Stuck. *J. Aerosol Sci.* (in press).
8. Heyder, J., J. Gebhart and W. Stahlhofen. "Inhalation of Aerosols: Particulate Deposition and Retention," Chapter 3, this volume.
9. Orenstein, A. J., Ed. *Proceedings of the Pneumoconiosis Conference, Johannesburg, 1959*, (London: J. & A. Churchill Ltd. 1960).
10. Klyachko. Otopl. i ventil. No. 4 (1934), quoted by Fuchs, N. In: *The Mechanics of Aerosols* (Oxford: Pergamon Press, 1964), p. 33.
11. Gormley, P. G., and M. Kennedy. *Proc. Roy. Irish Acad.* 52A:163 (1949).
12. Thomas, J. W. "Particle Loss in Sampling Conduits," in *Assessment of Airborne Radioactivity*, (Vienna: IAEA, 1967).
13. Stöber, W. *J. Aerosol Sci.* 2:453 (1971).
14. Morrow, P.E., and T. T. Mercer, AIHA J. 25:8 (1964).
15. Hidy, G., Ed. *Aerosols and Atmospheric Chemistry* (New York: Academic Press, Inc., 1972). p. 189.
16. Sinclair, D., and V. K. LaMer. *Chem. Rev.* 44:245 (1949).
17. Horstman, S., W. Barkley, E. Larson and E. Bingham. *Arch. Environ. Health* 26:75 (1973).
18. Hochrainer, D., and W. Stöber. *Z. Bakt. Hyg. I. Abt. Orig.* B159:163 (1974).
19. Dautrebande, L. *Microaerosols* (London: Academic Press, Inc., 1962), p. 4.
20. Rapaport, E., and S. E. Weinstock. *Experientia* 11:363 (1955).

21. Okada, T., H. Ishibashi and S. Kitani. *J. Colloid. Interface Sci.* 29:613 (1969).
22. Sinclair, D. *Handbook on Aerosols*, US Atomic Energy Comm. Washington D.C. (1950).
23. Mie, G. *Ann. Phys. Lpz.* 25:377 (1908).
24. Gucker, F. T., and R. L. Rowell. *Disc. Faraday Soc.* 30:185 (1960).
25. Hochrainer, D., F. J. Mönig, H. P. Baumert, T. Richter and N. Schwartzer. *FhG-Berichte*, Fraunhofer-Gesellschaft, Munich (April, 1978), p. 8.
26. Hollander, W., and J. Schörmann. In: *Atmospheric Pollution*, M. M. Benarie, Ed. (Amsterdam: Elsevier, 1977), p. 477.
27. Fuller, C. W. *Annual Report on Analytic Atomic Spectorscopy, Vol. 5, Review 1975* (London: The Chem Society, Burlington House, 1976).
28. Alter, J. *Chem. Technol.* 28:354 (1976).
29. Stöber, W., and H. Flachsbart. *Atmos. Environ.* 7:737 (1973).
30. Stöber, W. *Labor. Invest.* 14:892 (1965).
31. Gebhart, J., J. Heyder, C. Roth and W. Stahlhofen. In: *Fine Particles*, B. Y. H. Liu, Ed. (New York: Academic Press, Inc., 1976). p. 793.
32. Gebhart, J., and C. Roth. In: *Aerosole in Physik, Medizin und Technik*, Jahreskongreβ 1973 der Gesellschaft für Aerosolforschung, p. 41.
33. Mercer, T. T., M. I. Tillery and G. J. Newton. *J. Aerosol Sci.* 1:9 (1970).
34. Stöber, W., and H. Flachsbart. *Environ. Sci. Technol.* 3:641 (1969).
35. Tillery, M. I. *AIHA J.* 35:62 (1974).
36. Hochrainer, D. *J. Colloid Interface Sci.* 36:191 (1971).
37. Stöber, W. In: *Fine Particles*, B. Y. H. Liu, Ed. (New York: Academic Press, Inc., 1976), p. 351.
38. Stöber, W., and H. Flachsbart. *Environ. Sci. Technol.* 3:1280 (1969).
39. Stöber, W., and F. J. Mönig, *Proceedings of the 4th International Clean Air Congress, Tokyo, May 1977*, (Tokyo: Japanese Union Air Pollution Prev. Association, IUAPPA, 1977), p. 406.
40. Kops, J. A. M. M., G. Dibbets, L. Hermans and J. F. van de Vate. *J. Aerosol Sci.* 6:329 (1975).
41. Stöber, W., H. Flachsbart and D. Hochrainer. *Staub-Reinhalt. Luft.* 30:277 (1970).
42. Fisher, G. L., B. A. Prentice, D. Silberman, J. M. Ondov, A. H. Biermann, R. C. Ragaini and A. R. McFarland. *Environ. Sci. Technol.* 12:447 (1978).
43. Davies, C. N. *Nature* 201:905 (1964).
44. Stöber, W. In: *Assessment of Airborne Particles*, T. T. Mercer, P. E. Morrow and W. Stober, Eds. (Springfield, IL: Charles C. Thomas Publisher, 1972), p. 358.
45. Kotrappa, P. *J. Aerosol Sci.* 2:353 (1971).
46. Kotrappa, P. In: *Assessment of Airborne Particles*, T. T. Mercer, P. E. Morrow and W. Stober, Eds. (Springfield, IL: Charles C. Thomas Publisher, 1972), p. 331.
47. Davies, C. N. "Particle-Fluid Interaction, "Proceedings of Harold Heywood Memorial Symposium, Loughborough, England, Sept. 17-18, 1973.
48. Grimmer, G., H. Böhnke and A. Glaser. *Zbl. Bakt. Hyg., I. Abt. Orig.* B164:218 (1977).
49. Severson, R. F., M. E. Snook, R. F. Arrendale and O. T. Chortyk. *Anal. Chem.* 48:1866 (1976).
50. Lee, M. L., M. Novotuy and K. D. Bartle. *Anal. Chem.* 48:1566 (1976).
51. Tanner, R. L., and L. Newman. *J. Air Poll. Control Assoc.* 26:737 (1976).

CHAPTER 3

INHALATION OF AEROSOLS: PARTICLE DEPOSITION AND RETENTION

J. Heyder, J. Gebhart and W. Stahlhofen

Abteilung für Biophysikalische Strahlenforschung
Gesellschaft für Strahlen- und Umweltforschung m.b.H.
Frankfurt am Main, Federal Republic of Germany

INTRODUCTION

The earth atmosphere is an aerosol. Consequently, inspired air always contains particles which may be potential health hazards when deposited in the respiratory tract. The evaluation of this health hazard for environmental and occupational exposure patterns requires the knowledge of mass of particulate matter deposited in the human respiratory tract, the subsequent removal of this matter from the respiratory tract, its translocation and retention within the body and its biological effectiveness. Therefore, joint efforts of research in aerosol science and environmental and occupational hygiene are needed.

Because of the existence of several detailed reviews on deposition and retention with numerous literature references [1-8], this chapter besides some basic considerations, will be mainly focused on the discussion of experimentally determined human deposition and retention data from this laboratory and their comparison with data found in other laboratories.

GENERAL CONSIDERATIONS

To avoid further confusion in the literature on aerosol inhalation it is necessary to define the words deposition and retention. Although definitions have already been proposed [9] in 1961, these words are still used in different ways even in very recent publications.

Definitions

Inhalability IN is the probability of an airborne particle being inhaled from the ambient air.

Total deposition DE is the average probability of an inspired particle touching a surface of the respiratory tract and thereby being deposited. Deposition in region k of the respiratory tract DE_k is the average probability of an inspired particle touching a surface of the respiratory tract in this region and thereby being deposited. Total deposition is, therefore, the sum of the regional depositions. Usually, it is confined to steady-state breathing and it will be used in this sense in this report.

Total retention RE is the probability of mass from particles initially deposited in the respiratory tract being retained in the body. Retention in region n of the body RE_n is the probability of mass from particles initially deposited in the respiratory tract being retained in this region. Total retention is, therefore, the sum of the regional retentions.

Total and regional retention functions RE (t) and RE_n (t) describe as a function of time t the mass of material present in the total body or in region n of the body and retained from the mass of particles initially deposited in the respiratory tract. The initial condition of the total retention function is RE (t=0) = 1 and of the retention functions of the k regions of the respiratory tract RE_k (t=0) $\leqslant$ 1. whereas the initial conditions of the remaining (n-k) regional retention functions are RE_{n^1} (t=0) = 0 with $K + 1 \leqslant n^1 \leqslant n$. Regional deposition and regional retention function are linked by $DE_k = DE \cdot RE_k$ (t=0).

Particle Transport

Transport of Aerosol Particles Into and Within the Respiratory Tract

It has long been known that in an aerosol flow particles can mechanically or electrically be transported across streamlines by a number of physical mechanisms. These transport phenomena lead to a decreasing inhalability with increasing particle diameter d and increasing particle density ρ so that sufficiently large particles fail to be inhaled. Moreover, the inhalability will be different for mouth- and nose-breathing, is also influenced by the mean inspiratory volumetric flowrate Q and by wind direction and velocity u: IN = IN (d,ρ,Q,u). Very little is known on the dependence of the inhalability on these latter two quantities [10].

On the other hand, when an aerosol is inhaled these transport phenomena lead to the movement of particles towards the surfaces of the respiratory tract where they will be deposited. The mean volumetric flowrate at the entry of the respiratory tract is given by

$$Q = \frac{1}{T} \int_{0}^{T} |\dot{V}| \, dt \tag{1}$$

where $\dot{V}$ is the instantaneous volumetric flowrate and T the duration of a single breath.

Particles inhaled at the beginning of a breath stay longer in the respiratory tract than those inhaled at the end of a breath. A mean residence time can be defined as

$$\tau = \frac{1}{2Q} \int_{0}^{T} |\dot{V}| \, dt \tag{2}$$

It is generally accepted that diffusion, gravitational sedimentation and inertial impaction are the major mechanical transport mechanisms, that interception is significant only with fibrous particles, that the major electrical transport mechanism is due to image forces and that deposition is enhanced by mixing. The contribution of minor mechanisms, such as turbulent diffusion, diffusiophoresis and thermophoresis to deposition is still controversial.

Consequently, deposition of uncharged spherical particles will be dependent on particle diameter d, particle density ρ, mean residence time (MRT) τ and mean respiratory volumetric flowrate (MFR) Q: DE = DE(d,ρ,τ,Q). Usually, deposition is given as a function of the particle diameter with ρ,τ, and Q as parameters. The expected influence of these parameters on total and alveolar deposition is illustrated in Figure 1. So far, no general concept has been developed to describe the influence on deposition of electrical charges carried by aerosol particles.

For particles of different densities there is the possibility to describe their transport and, consequently, deposition as a function of an equivalent diameter, the aerodynamic diameter, which takes into account that a small heavy particle and a large light particle of equal aerodynamic diameter exhibit the

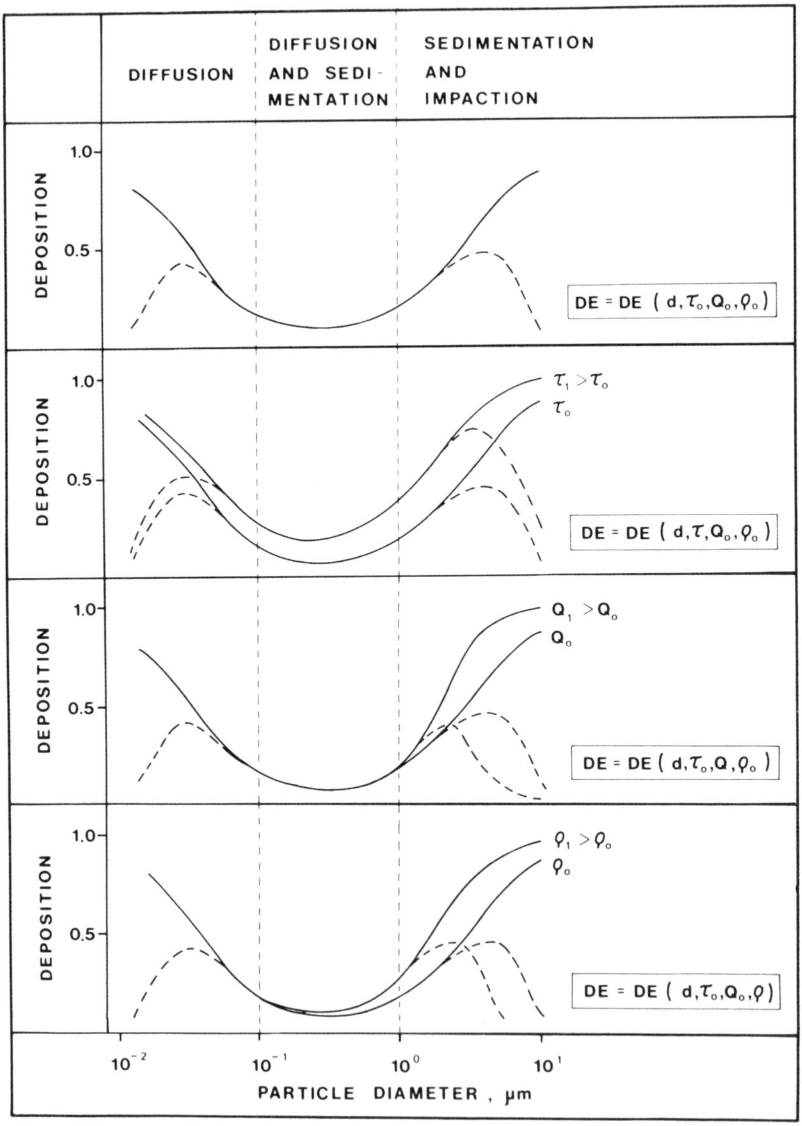

Figure 1. The influence of mean residence time τ, mean volumetric flowrate Q and particle density ρ on the particle diameter dependence of total deposition (solid lines) and alveolar deposition (dashed lines).

same aerodynamic behavior. This concept has also the advantage of taking care of the effect of particle shape on the aerodynamic behavior of aerosol particles. It is, however, only valid in the size range where particle transport is due to sedimentation and impaction but not valid in the size range where diffusional transport occurs. Therefore, deposition cannot be expressed as a function of the aerodynamic diameter for the total particle size range of interest as is often done in the literature, but only for particles larger than about 0.5 μm in geometric diameter.

Certainly, deposition depends not only on the properties of the aerosol particles (d,ρ) and on the characteristics of the breathing pattern (τ,Q) but also on the geometry of the respiratory tract and on the flow fields in the respiratory tract. The dependence on these quantities is not fully understood because of the inaccessibility of the human respiratory tract to direct studies and because of the limited value of investigations on animals or on models of the human respiratory tract. Their influence on deposition, however, will partly be reflected by the intersubject variation of deposition for d,ρ,τ and Q = const.

Transport of Particles Deposited in the Respiratory Tract

When deposited particles are not rapidly dissolved in the fluids covering the surfaces of the respiratory tract there are a number of transport mechanisms by which particles are removed from the respiratory tract. Although some mechanisms are only partly understood and are still the subject of current fundamental research it is generally accepted that the major processes are:

1. the mechanical removal of particles from the unciliated parts of the nose by nose wiping, nose blowing or sneezing;
2. particle removal via mechanical transport of secretions in the extrathoracic and tracheobronchial airways (mucociliary escalator);
3. phagocytosis by alveolar macrophages and transport of bare particles and particle-laden alveolar macrophages either by movement of alveolar fluids along the alveolar-bronchial surfaces onto the mucociliary escalator of the bronchial airways; or
4. by penetration into lymphatic pathways and connective tissue.

The contribution of other processes to the overall removal of particles deposition in the human respiratory tract, especially those effective in the unciliated airspaces of the lower respiratory tract, is still controversial. There appears, however, to be a consensus that dissolution of material from particles deposited in the respiratory tract is an important removal mechanism since even particles considered insoluble in vitro may have a finite in vivo solubility.

Regions of the Respiratory Tract

In environmental medicine it has become customary to define the regions of the respiratory tract on the basis of retention functions for particles consisting of inert material scarcely soluble in the fluids of the respiratory tract:

- extrathoracic airways from which deposited particles are removed within minutes either by mechanical transport of particles or secretions;
- tracheobronchial airways from which deposited particles are removed within hours by mechanical transport of secretions (mucociliary escalator);
- alveolar airspaces from which deposited particles are removed within months or even years mainly by phagocytosis and subsequent cell transport onto the mucociliary escalator or into the lymphatic system.

Conceptual Outline for the Evaluation of Aerosol Inhalation Hazards

Let the particle number concentration of an aerosol be c, let its diameter probability density function be f(d) and let all particles have the same chemical composition so that all can be characterized by the same density ρ. (In the case that the particles of an aerosol are distributed with respect to both size and chemical composition a size-composition probability density function must be considered [11].) Let this aerosol be inhaled in a single breath with mean volumetric flowrate Q via nose or mouth and let the mean residence time be τ. Then, the mass of material retained in region n of the body at time t after inhalation is given by

$$M_n\,(t) = c\,\rho\,\tau\,Q\,\frac{\pi}{6}\int_0^\infty f(d')\,IN(d',\rho,Q,u)\,DE(d',\rho,\tau,Q)\,RE_n(t,d',\rho,\tau,Q)\,d'^3\,dd' \qquad (3)$$

where d' denotes particle diameter. For nonspherical particles $f(d')$ must be replaced by the volume equivalent diameter probability density function and the influence of particle shape on the probabilities IN, DE and RE must be taken into account. However, a sphere has a higher mobility than a nonspherical particle with a volume equivalent diameter equal to the diameter of the sphere and, consequently, the probability of its being deposited will be higher. Therefore, if deposition values determined with spherical particles are used to evaluate inhalation hazards of real aerosols which usually are composed of nonspherical particles, deposition will not be underestimated. Finally, considerations should also be given to changes in particle size due to water vapor absorption by particles in the respiratory tract.

As a first approximation, it is reasonable to assume that the response of each region n to retained material will be dependent on the integral

$$I_n(t) = \int_0^t M_n(t)\, dt \qquad (4)$$

It can be evaluated for environmental and occupational exposure patterns, either experimentally through human or animal inhalation studies or mathematically [12], assuming models such as the deposition and retention model of the ICRP Task Group on Lung Dynamics [2].

Let the probability of region n responding to unit mass of retained material per unit time by an effect i be E_{ni}. A harm, hazard or risk factor r due to the inhalation of an aerosol may then be expressed as

$$r = \sum_n \sum_i p_{ni} E_{ni} I_n(t) \qquad (5)$$

where p_{ni} is a weight factor which takes the harmfulness of each individual effect into account.

In human inhalation experiments with physically well-defined aerosols deposition for nose- and mouth-breathing can be measured as a function of particle size, particle density, mean volumetric flowrate (MFR) and mean residence time (MRT). The retention function in humans, however, can only be evaluated for nontoxic particles. The retention of all other aerosols must be evaluated in animal experiments assuming identical behavior of deposited material in humans and in experimental animals.

The biological response E_{ni} can either be determined by animal experiments with physically and physicochemically well-defined aerosols under environmental or occupational exposure conditions, from epidemiological studies or from in vitro tests using cell or tissue cultures. Although each method has certain disadvantages there is a consensus that animal experiments are most reliable because they can be performed under circumstances close to real exposures. Finally, the question on the magnitude of the factor p_{ni} must be answered by pathologists.

When the material deposited in the respiratory tract is rapidly dissolved in the fluids of the respiratory tract, it can diffuse into the tissues and in the circulating blood even in the short residence time of the deposited material in the extrathoracic and tracheobronchial airways. Consequently, all particles deposited will potentially be harmful.

When, however, the particulate matter is not or only slowly dissolved in the fluids of the respiratory tract, particles deposited in the extrathoracic and tracheobronchial airways are harmless because of their relatively fast removal from the respiratory tract and from the body. Only particles deposited on the surfaces of the alveolar airspaces are retained long enough to be considered potentially harmful. Such particles are especially released by many industrial activities.

MATERIAL AND METHODS

Aerosol Generation and Characterization

Monodisperse spherical aerosol particles were produced either by condensation of di-2-ethylhexyl sebacate vapor upon sodium or silver chloride nuclei [13] in the diameter range 0.1-10 μm or by atomization of an aqueous suspension of iron-oxide with a spinning-top generator [14] in the diameter range 0.6-5 μm. To produce monodisperse insoluble nonleaching radioactive particles for the determination of regional deposition a mixture of aqueous suspensions of iron-oxide and radiogold (^{198}Au) was atomized. Disintegration of the ferrous particles in the respiratory tract could be avoided by heating the ferrous aerosols to about 470°K while they passed through an electrical furnace.

The particle diameter distributions were analyzed by measuring the terminal settling velocity of a large number of individual particles prior to and during each inhalation experiment taking the particle density into account [15]. The sebacate droplets had a density of 0.91 g/cm^3 and the iron-oxide spheres had an average density of 3.2 g/cm^3. The latter value was determined by comparing the geometrical diameter distributions found by electronmicroscopy and the aerodynamic diameter distributions found by measuring the terminal settling velocity of the particles in a sedimentation channel. The diameter distributions of sebacate droplets smaller than 1 μm were analyzed by light scattering on single particles [16]. The relative standard deviations of the diameter distributions were less than 0.12, often less than 0.06.

To ensure that the iron-oxide particles did not contain any water-soluble residues from the conversion of iron-chloride solutions into iron-oxide suspensions, each ferrous aerosol was saturated with water vapor at room temperature and the terminal settling velocity of the particles was measured. Only aerosols of equal terminal settling velocity in dry and humid air were used for the experimental deposition studies. At the beginning of these studies it was often found that particles grew considerably in size due to water vapor absorption. The iron-oxide suspensions had to be thoroughly dialyzed before the particles retained their size in a humid atmosphere. In the past, this effect has not been taken into account so that the size of iron-oxide particles in the respiratory tract could have been underestimated.

The sebacate droplets were electrically neutral whereas the iron-oxide spheres were brought to charge equilibrium prior to inhalation.

Determination of Total Deposition

The determination of total deposition during steady breathing as a function of the aerosol properties and the characteristics of the breathing pattern requires a periodic respiratory volumetric flowrate $\dot{V}$ and a constant particle number concentration c of the inspired aerosol during each inspiration and throughout the entire inhalation experiment. It also requires that wash-in is completed and that the particle properties do not change during the experiment and while the aerosol particles are within the respiratory tract. Total deposition can then simply be calculated from the number of inspired and expired particles per breath by

$$DE = 1 - \frac{N_e}{N_i} = 1 - \frac{\int_{t_e} c|\dot{V}|dt}{\int_{t_i} c|\dot{V}|dt} \tag{6}$$

where t_i and t_e are the periods required for inspiration and expiration respectively. Consequently, deposition can either be determined directly by counting the number of inspired and expired particles (digital method) or indirectly by continuous monitoring of the particle number concentration and the volumetric flowrate (analog methods). The analog method has already been used in two previous studies [17, 18].

A novel photometer has been developed so that the digital method could also be used in the present study. For this purpose the cylindrical beam of an Argon ion laser with an output of 1 W and a wavelength of 514.5 nm is optically shaped into a sheet of light of 0.1 mm thickness and 15 mm height and passes perpendicular to the aerosol flow through the photometer, with its major axis inclined at an angle of 60° to the direction of the aerosol flow. The configuration is schematically shown in Figure 2.

When a particle traversed the light sheet, scattered light was received by a photomultiplier in which this optical signal was transformed into an electrical signal. This signal was then analyzed according to its height and stored in a multichannel device. The signals were stored separately during inspiration and expiration. The time signals for the onset of inspiration and expiration were supplied from a differential pressure transducer connected to a Fleisch tube in the inhalation-exhalation-channel (Figure 2). The photometer could, therefore, be used to compare the diameter distributions of the inspired and expired

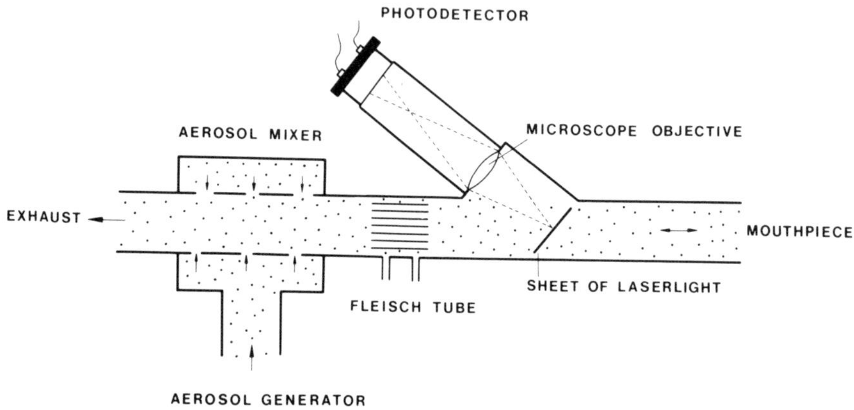

Figure 2. Schematic representation of the photometer for counting the number of inspired and expired particles per breath.

aerosols. Simultaneously, a computer counted the number of signals during inspiration and expiration and calculated the deposition for each breath.

In order to avoid coincidence counting losses in the photometer and also to avoid deadtime corrections, low particle number concentrations were required. Therefore, the present inhalation experiments were performed with particle number concentrations less than $10/cm^3$ when the photometer was operated in the digital mode. At higher concentrations it could, however, also be operated in the usual analog mode. The anode current of the photomultiplier was then proportional to the particle number concentration, and the evaluation of total deposition was based on the continuous monitoring of the particle number concentration and the aerosol flowrate.

Since not all particles inspired or expired passed through the viewing field of the photomultiplier, the number of detected particles during inspiration and expiration must have been representative of the total number of particles inspired and expired. The analog method, therefore, required a uniform particle concentration over the cross section of the inhalation-exhalation-channel in the photometer. This was achieved by an aerosol mixer, as schematically illustrated in Figure 2. The mixer consisted of two concentric cylinders of which the inner one had the same diameter as the inhalation-exhalation-channel and the mouthpiece and was perforated with holes. By introducing the aerosol into the hollow space between the cylinders it passed through these holes into the inhalation-exhalation-channel and was thereby thoroughly mixed.

The digital method, however, required in addition that the velocity of the particles passing the viewing field of the photomultiplier was equal to the mean particle velocity over the cross section of the inhalation-exhalation-channel. To confirm that this additional requirement was met by the novel photometer the particle collection efficiencies of a glass bead filter evaluated with the photometer operating in analog and digital mode were compared. The breathing patterns used in the present human inhalation studies were simulated by ventilating the filter at different mean flowrates with a sinoidal pump. The agreement between the analog- and digital-derived data was taken as evidence that the photometer met the requirements for determining deposition. The agreement between human deposition data evaluated with these two methods was also very good and deposition values presented in this report were obtained both with the analog and with the digital method.

The method for determining total deposition was straightforward. No calibration, no correction or any unproven assumptions were required so that reliable quantitative deposition data could be obtained.

Aerosol Administration

The aerosols were diluted with air until the desired particle concentration was obtained. Before it entered the aerosol mixer it was humidified and warmed to body temperature as already described elsewhere [18]. During inspiration it then passed through a Fleisch tube, the photometer and, finally, the mouthpiece, as illustrated in Figure 2. To prevent condensation of water vapor, all parts of the inhalation apparatus shown in Figure 2 were electrically heated to body temperature. Separation of inspired and expired aerosol occurred automatically in the aerosol mixer so that no valve was needed for this purpose. The instrumental dead space had a volume of 35 cm^3. This relatively large dead space, however, had no effect on the determination of deposition as indicated by the excellent agreement with deposition values obtained previously with a 6-cm^3 dead space design [18].

The aerosols were administered during controlled breathing as described earlier [17]. Sitting in a relaxed position the subject wore a nose clip and breathed, without pauses, through the cylindrical mouthpiece of 16-mm inner diameter at a predetermined and carefully monitored pattern corresponding to an audible and a visible signal. Inspirations were initiated from the subject's functional residual capacity and the subject was asked to maintain a constant flowrate during inspiration and expiration and equal periods for inspiration and expiration.

Determination of Regional Deposition and Retention

Particles deposited in the respiratory tract can either be deposited in the extrathoracic airways, the tracheobronchial airways or the alveolar airspaces so that total deposition can be subdivided into extrathoracic deposition (DE_E), tracheobronchial deposition (DE_B) and alveolar deposition (DE_A): $DE = DE_E + DE_B + DE_A$. Since the removal of insoluble particulate matter deposited in the tracheobronchial airways is much faster than the removal of particulate matter deposited in the alveolar airspaces the determination of regional depositions can be based upon inspiration of radioaerosols and subsequent measurements of the activity retained in the respiratory tract. This radiotracer method was used in the present study and is described in detail elsewhere [19]. Its principles, however, will be briefly summarized.

First of all, total deposition of iron-oxide particles labeled with radiogold was measured as described above. Immediately after inspiration at time zero and at intervals for several days the deposited activity was determined with a novel gamma-ray spectrometer for external in vivo measurements (body counter) [20]. At time zero the activity is found in the extrathoracic airways A_E (t=0), in the intrathoracic airways A_I (t=0) and in the stomach A_S(t=0). The activity deposited in the intrathoracic airways can be deposited onto the surfaces of the tracheobronchial airways A_B (t=0) and onto the surfaces of the alveolar airspaces, A_A (t=0). Thus,

$$A_I\,(t=0) = A_B\,(t=0) + A_A\,(t=0) \tag{7}$$

Later, activity can be removed from the alveolar airspaces into lymphoid tissue, although the body counter is not able to identify this migration of activity within the chest. Consequently, at any time $t > 0$ only activity in the chest, A_C (t), can be evaluated and the alveolar airspaces and the lymphatic system must be considered as a single region. Therefore,

$$A_C\,(t) = A_B\,(t) + A_{AL}\,(t) \tag{8}$$

A_{AL} is the activity found in the alveolar-lymphatic region which initially is A_{AL} (t=0) = A_A (t=0).

From Equation 8 it can be seen that A_A (t=0) could be found by extrapolating that part of the function A_C (t) to t=0 which was measured after completion of particle removal from the tracheobronchial airways or, in other words, when A_B (t) = 0.

Regional depositions were then calculated according to the following set of equations:

$$DE_E = DE \cdot RE_E\ (t=0) = DE \left(\frac{A_E\ (t=0)}{A\ (t=0)} + \frac{A_S\ (t=0)}{A\ (t=0)} \right) \tag{9}$$

$$DE_B = DE \cdot RE_B\ (t=0) = DE \left(\frac{A_I\ (t=0)}{A\ (t=0)} - \frac{A_A\ (t=0)}{A\ (t=0)} \right) \tag{10}$$

$$DE_A = DE \cdot RE_A\ (t=0) = DE\ \frac{A_A\ (t=0)}{A\ (t=0)} \tag{11}$$

where $A\ (t=0) = A_E\ (t=0) + A_I\ (t=0) + A_S\ (t=0)$

Inherent in this method of regional deposition evaluation are several assumptions:

1. The removal of particulate matter from the alveolar region is negligible so long as the removal of particulate matter from the tracheobronchial region is not completed or the alveolar-lymphatic retention function is characterized by a constant removal rate over the period of observation, usually several days.

The validity of this assumption was checked on three subjects [21]. The subjects inhaled an air volume equal to their vital capacity. Particles of 0.6-μm iron oxide were suspended in the first 1500 cm^3 of air so that the aerosol particles were pushed entirely into the alveolar airspaces by particle-free air. Subsequent breath-holding of 30-sec duration ensured that all particles were deposited in the alveolar region. No particles were recovered in the exhaled air. Subsequent measurements of head, chest and stomach with the body counter showed no activity in the head and in the stomach and showed a slow decrease of the chest activity at a constant rate. A rapid removal of particles deposited in the alveolar region as suggested in the literature [2] was never observed. Approximating the decrease in the chest activity by a single exponential function a biological half-time of about 85 days could be estimated.

Therefore, the removal of particulate matter from the alveolar region cannot be neglected during the period of removal of particulate matter from the tracheobronchial airways. However, whatever the mechanism may be for the slow removal from the chest of particulate matter deposited in the alveolar region—whether bare particles or particle-laden alveolar macrophages are removed via the mucociliary escalator or removal occurs by solubilization—the alveolar-lymphatic retention function was proven to be determined by a constant removal rate so that alveolar deposition could be evaluated from the chest retention function by the extrapolation method described above.

2. Particles deposited in the tracheobronchial airways are completely removed from the chest. Inspiration of large particles allows the study of particle removal from the large tracheobronchial airways. This removal

seems to be completed in the healthy subjects within a few hours. However, it cannot be investigated if particles are also completely removed from the lower bronchial airways because deposition in these airways is impossible without simultaneous particle deposition in the alveolar region.

3. Particles removed by the mucociliary escalator are not retained in the esophagus. Therefore, prior to each measurement of chest activity in the body counter particles possibly retained in the esophagus were transported into the stomach by food.

Determination of Activity Deposited in the Respiratory Tract

The activity deposited in the respiratory or gastrointestinal tract was determined by external detection of photons leaving the body with four cylindrical NaJ(T1) detectors. Figure 3 shows a cross-sectional view of the body counter [20]. In supine position the subject was positioned in such a way that the detectors were spatially arranged around the head, thorax or abdomen. Phantom measurements were used to calibrate the counter for the evaluation of activity in the extra- and intrathoracic airways and in the stomach and to determine the contribution to the count rate of photons leaving regions other than the region viewed by the detectors.

Unfortunately, it has been found that when the detectors were positioned around the thorax, the contribution of photons leaving the abdomen to the count rate was time-dependent as the radionuclide moved through the gastrointestinal tract [22]. Therefore, it was determined in vivo in the present experiments. Each subject swallowed several easily digestible capsules filled with a sugar-adsorbed suspension of radiogold, thereby avoiding contamination of the esophagus.

The retention function of the chest activity was found by measuring the chest activity as a function of time after the aerosol was administered. This evaluation is, however, based on the assumption that the total photon-counting efficiency of the four detectors is independent of the distribution of deposited radioactively labeled particles in the respiratory tract. Since simulation experiments revealed little changes in the total photon-counting efficiency when a radiogold point source was translocated within a head, thorax or stomach phantom [23], the counter was considered suitable for determining the chest retention function and, consequently, regional depositions.

RESULTS AND DISCUSSION

Subjects 1, 3 and 4 of a previous investigation [18] volunteered also for all present inhalation experiments. Each individual performed about 300

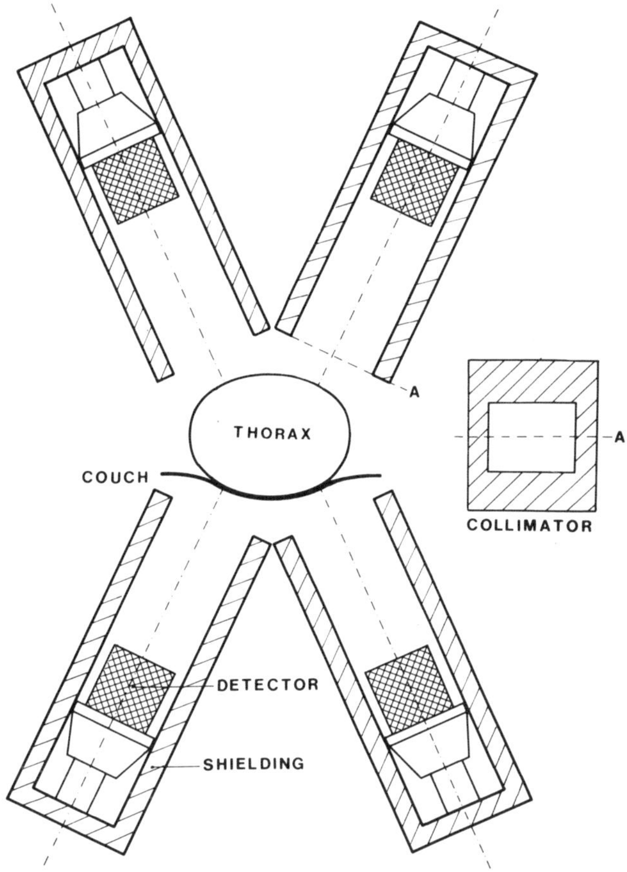

Figure 3. Cross-sectional view of the body counter for the determination of activity deposited in extra- and intrathoracic airways and in the stomach.

inhalation experiments. Their deposition and chest retention was quite similar for all particle sizes and breathing patterns so that, because of the limitation of space here, only the data of subject 3 are given as a representation of the vast amount of data accumulated in this laboratory during recent years. With these data the influence of the properties of the aerosol particles and the characteristics of the breathing patterns on deposition and retention is discussed.

On the other hand, this report is focused on total and alveolar deposition, because of their special importance. The results of the detailed deposition study in extrathoracic and tracheobronchial airways are described elsewhere [19].

Total Deposition

Total deposition of sebacate droplets and iron-oxide spheres for mouth-breathing was evaluated as a function of particle diameter for a variety of breathing patterns to study the influence of the mean residence time (MRT), the mean volumetric flowrate (MFR), and the particle density on the diameter dependence of total deposition. This discussion, however, is focused on the behavior of particles larger than 3 μm since the behavior of smaller particles has already been discussed [18] and since total deposition data of smaller particles which were presented in the previous report were confirmed by the present investigations. Each total deposition value given is the result of at least three experiments. The intrasubject reproducibility of the values was very good, typically DE ± 0.02.

Total deposition was studied for nine different breathing patterns. It is believed that total deposition for any other pattern can be estimated quite accurately from these data. It should be emphasized that deposition depends strongly on the physiological respiration parameters: mean residence time and mean flowrate. Any reference to deposition data is, therefore, incomplete without referring also to these parameters.

Dependence of Total Deposition on Mean Residence Time MRT

The influence of MRT on the diameter dependence of total deposition is shown in Figures 4 and 5. The flowrates are 250 and 750 cm^3/sec. For all particle diameters, deposition increased with increasing MRT since more time was available for depositing particles in the respiratory tract: There is more time for the diffusion of particles smaller than 0.5 μm in diameter and for the sedimentation of larger particles. The dependence of total deposition on MRT was less pronounced at a flowrate of 750 cm^3/sec than at a flowrate of 250 cm^3/sec because more particles were deposited by inertial impaction and, consequently, fewer particles were available for deposition by gravitational transport.

Dependence of Total Deposition on Mean Flowrate MFR

The influence of MFR on the diameter dependence of total deposition is shown in Figures 6 and 7. The mean residence times are 1 and 2 sec. For most particle sizes deposition increased with increasing MFR. However, for particles smaller than a certain size deposition was independent of MFR suggesting that inertial impaction—which is the flow-dependent deposition

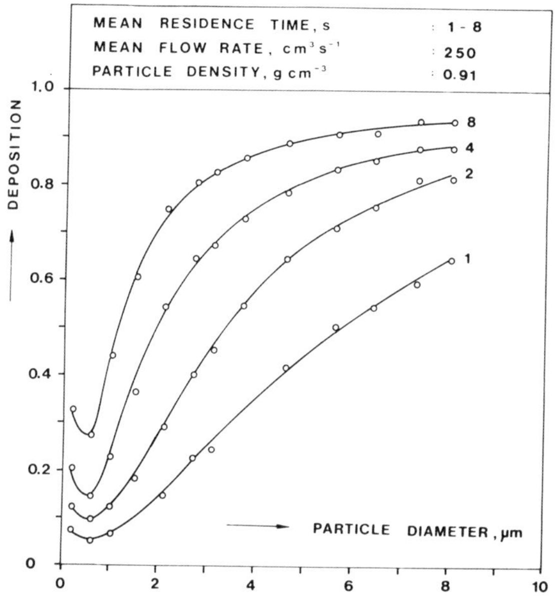

Figure 4. Effect of particle size and mean residence time on total deposition for mouth-breathing at a mean flowrate of 250 cm^3/sec.

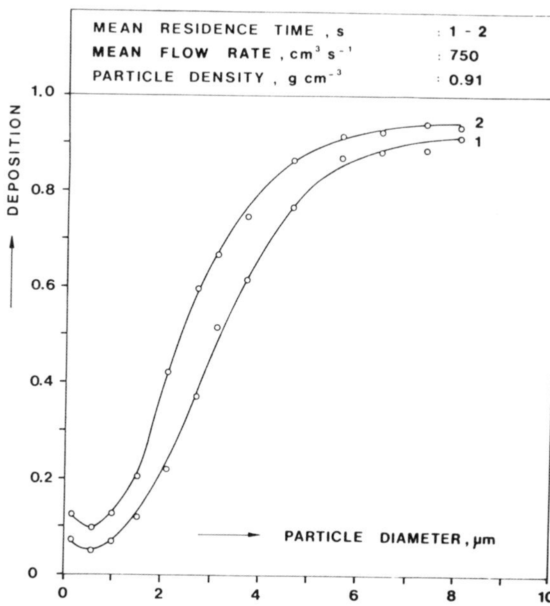

Figure 5. Effect of particle size and mean residence time on total deposition for mouth-breathing at a mean flowrate of 750 cm^3/sec.

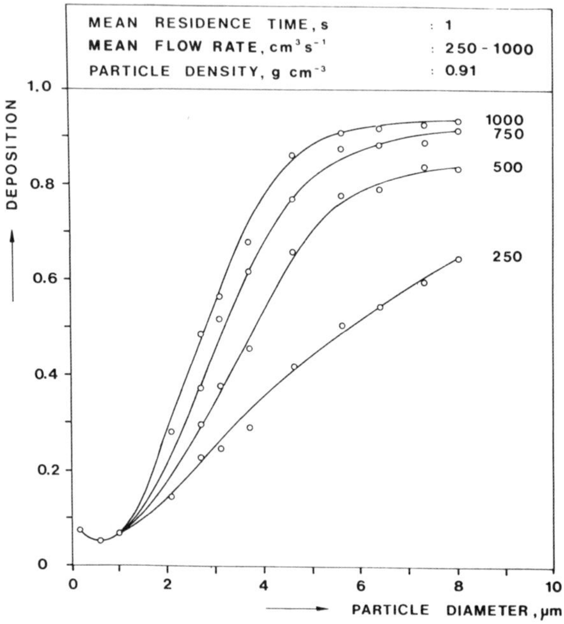

Figure 6. Effect of particle size and mean flowrate on total deposition for mouth-breathing at a mean residence time of 1 sec.

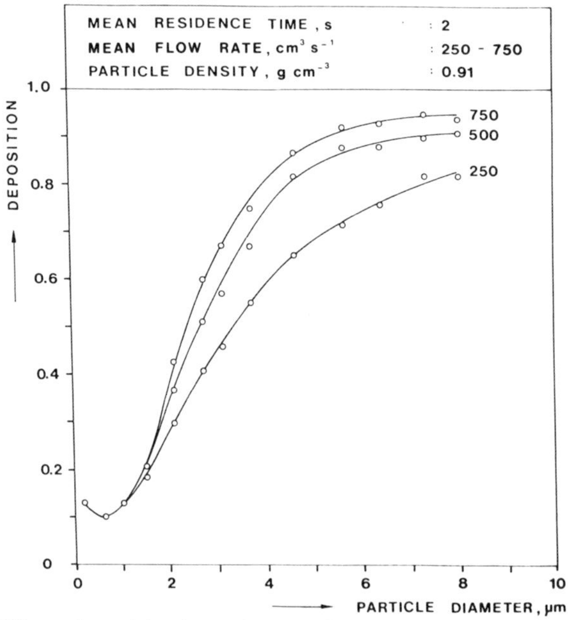

Figure 7. Effect of particle size and mean flowrate on total deposition for mouth-breathing at a mean residence time of 2 sec.

mechanism—was negligible for particles of this size. When aerosols remained in the respiratory tract for a mean time of 1 sec particles smaller than 1 μm were not deposited by impaction even when the aerosols were inspired at MFR = 1000 cm^3/sec (Figure 6). When the aerosols remained in the respiratory tract for a mean time of 2 sec this phenomenon was observed for particles with diameters up to 1.5 μm (Figure 7).

Sedimentation and impaction are competing deposition mechanisms. Sedimentation is governed by MRT and impaction by MFR. Consequently, impaction will be the dominant mechanism at high flowrate and short residence time and most particles will be deposited by sedimentation at low flowrate and long residence time. Figure 8 shows the diameter dependence of total deposition for τQ = const. It can be seen that, for 1-μm particles, deposition increased with increasing MRT so that these particles were mainly deposited by sedimentation, whereas for 8-μm particles deposition increased with increasing MFR so that these particles were mainly deposited by impaction.

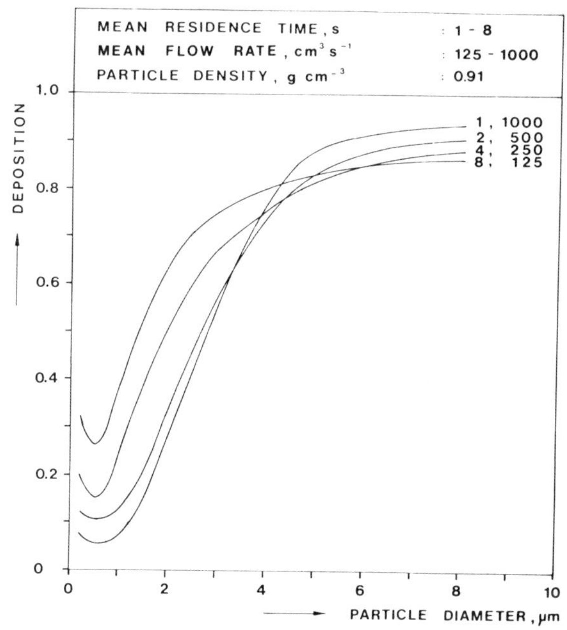

Figure 8. Effect of particle size on total deposition for mouth-breathing at τQ = 1000 cm^3.

A transition region was observed for particles of about 4 μm in diameter. The results, therefore, indicate that, when a subject is mouth-breathing, sedimentation is dominant for unit-density spheres smaller than 4 μm in diameter and impaction is dominant for larger unit-density spheres. Consequently, the increase of total deposition for particles larger than 4 μm will be attributed to deposition in the extrathoracic and tracheobronchial airways and the increase of total deposition will be accompanied by a decrease in alveolar deposition. Hence, there will be a particle size for which alveolar deposition is maximum.

For nose-breathing the transition region was shifted towards smaller particle diameters [18] (Figure 9), suggesting that unit-density spheres smaller than about 1 μm in diameter will mainly be deposited by sedimentation whereas impaction prevails for larger unit-density spheres.

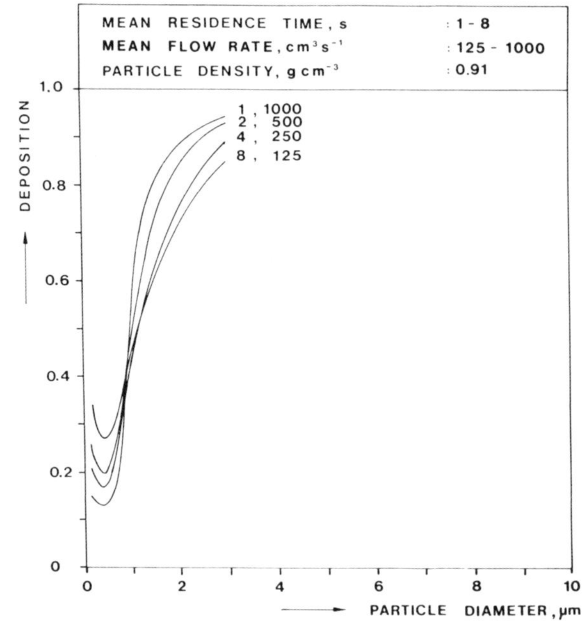

Figure 9. Effect of particle size on total deposition for nose-breathing at $\tau Q = 1000\ cm^3$.

Dependence of Total Deposition on Particle Density

The total deposition for two different breathing patterns was evaluated as a function of particle density. Studies with sebacate and iron-oxide spheres showed that total deposition is a unique function of the aerodynamic diameter (Figures 10 and 11). It should, however, be emphasized here that, for iron-oxide particles, the aerodynamic diameter concept is restricted to particles with an aerodynamic diameter larger than about 1 μm. Therefore, in Figures 10 and 11 the abscissa is not defined for aerodynamic diameters below this value.

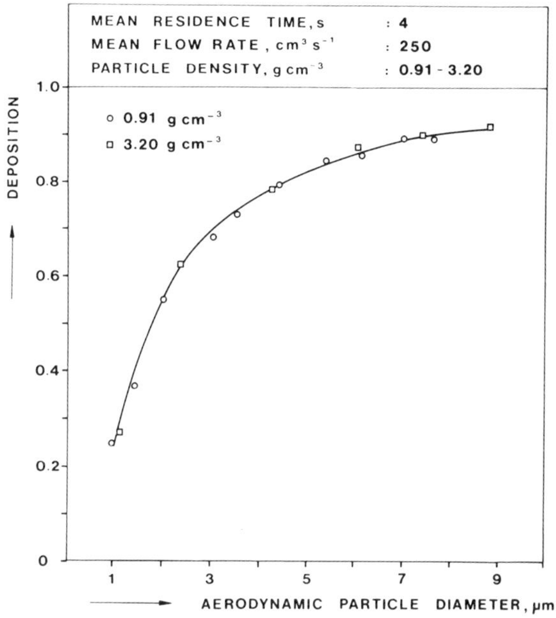

Figure 10. Effect of aerodynamic particle size on total deposition for mouth-breathing at a mean residence time of 4 sec and a mean flowrate of 250 cm^3/sec.

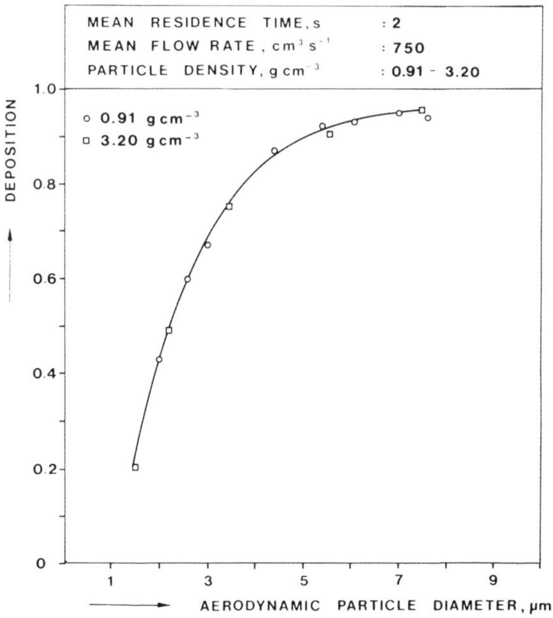

Figure 11. Effect of aerodynamic particle size on total deposition for mouth-breathing at a mean residence time of 2 sec and a mean flowrate of 750 cm^3/sec.

Since the respiratory tract was found to collect airborne particles according to their aerodynamic size, deposition as a function of the geometric diameter can be calculated for spherical particles of different densities and for the breathing patterns considered. Figure 12 demonstrates the importance of particle density for the behavior of aerosol particles in the human respiratory tract. It can be seen that 50% of all 1-μm iron-oxide particles will be deposited, whereas only 23% of 1-μm sebacate droplets will be lost in the respiratory tract when a mixture of 1-μm sebacate and iron-oxide aerosols is inhaled through the mouth at MFR = 250 cm^3/sec and MRT = 4 sec.

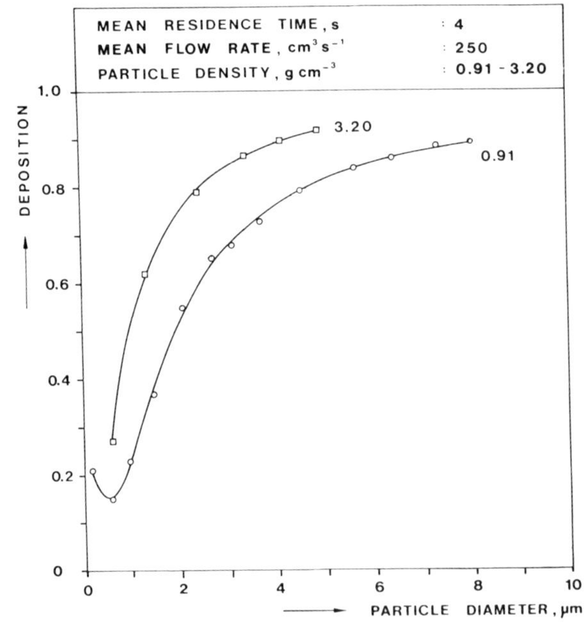

Figure 12. Effect of particle size and particle density on total deposition for mouth-breathing at a mean residence time of 4 sec and a mean flowrate of 250 cm^3/sec.

Dependence of Total Deposition on Electrical Particle Mobility

There is relatively little information available on the effect of the electrical mobility of aerosol particles on deposition. It has, however, recently been shown that total deposition increases with increasing number of electrostatic charges on the particles due to image forces between the unipolar charged particles and the surfaces of the respiratory tract [24]. It has also been shown that electrostatic precipitation of charged particles in the human respiratory tract is a time-dependent mechanism [25].

Particles generated by atomization of liquids have substantial electric mobility. Preliminary experiments in this laboratory with iron-oxide aerosols produced with a spinning top generator showed that for particles as large as 4 μm in aerodynamic diameter, appreciably higher total deposition values were obtained when the aerosols were not brought to charge equilibrium. It is, however, still not known how effective electrostatic precipitation is compared to diffusion, mixing and sedimentation, and whether electrostatic precipitation of negatively charged particles is as effective as for positively charged particles. Nevertheless, deposition will be overestimated when dispersion aerosols are used, unless their charge was neutralized prior to inhalation.

Intersubject Variation of Total Deposition

It has recently been claimed [8] that "with more natural aerosol and respiratory parameters, higher deposition efficiencies would be expected" than have been reported by this laboratory.

All deposition studies of the past, in this laboratory and elsewhere, have been performed with artificially generated test aerosols. If deposition data found by use of more natural aerosols differ from deposition data found with test aerosols, then all deposition studies of the past are erroneous. It is, however, expected that deposition of natural particles will be lower than deposition of spherical test particles because spherical particles have a higher mobility than natural particles of equal volume, equivalent size.

It is apparent that the respiratory parameters chosen in most laboratories were not natural, but strictly controlled with respect to mean residence time and mean volumetric flowrate. To study the effect of controlled breathing on deposition 14 male and female subjects were studied breathing spontaneously, i.e., as normally as possible. These volunteers were normal and healthy with respect to parameters studied in lung function tests (lung volumes, airway resistance and oxygen concentration in the circulating blood). Their tidal volumes varied from 430 to 1360 cm^3 (800 ± 240 cm^3) and their respiratory frequency from 10 to 22/min (15 ± 3/min). Hence, the mean respiratory parameters of this population of 14 subjects relevant to deposition were: MRT = 2 sec and MFR = 400 cm^3/sec. Based on the mean deposition data of the three subjects breathing controlled, a mean deposition between 0.38 (MRT = 2 sec and MFR = 250 cm^3/sec) and 0.51 (MRT = 2 sec and MFR = 500 cm^3/sec) was expected. For the 14 spontaneous breathers, the deposition was found to be 0.40 ± 0.10, which is in good agreement with the value for the three controlled breathers. It can thus be concluded that controlled breathing does not lower deposition.

Controlled breathing, however, reduces the subject-to-subject variation of total deposition. This can be seen by comparing total deposition during spontaneous breathing and total deposition during controlled breathing at MRT = 2 sec and MFR = 500 cm^3/sec (Figure 13). It can also be concluded from Figure 13 that the higher the deposition during controlled breathing, the higher is the deposition during spontaneous breathing. Therefore, controlled breathing does not introduce an artifact into the determination of total deposition.

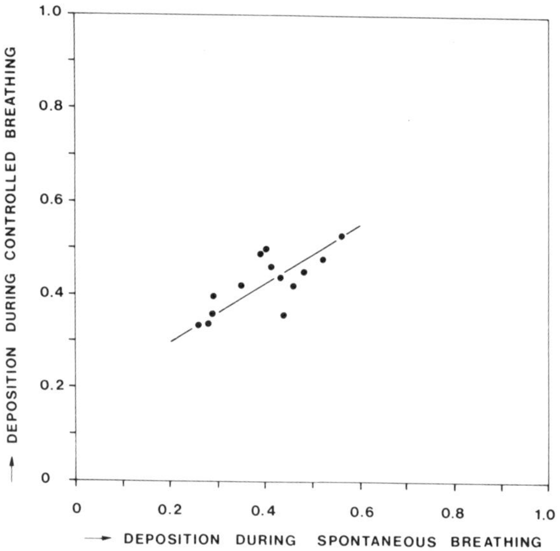

Figure 13. Total deposition of 14 subjects breathing 2.7-μm sebacate droplets spontaneously or under controlled conditions at a mean residence time of 2 sec and a mean flowrate of 500 cm^3/sec through the mouth.

Nevertheless, there is considerable intersubject variability of total deposition. Since all aerosol and respiratory parameters were carefully controlled during the reported experiments, this variability can only be attributed to anatomical differences. It was, therefore, of interest to see whether total deposition can be correlated with a parameter which is easily measured by lung function tests.

Deposition of the population of 14 individuals breathing sebacate droplets of different sizes at different controlled patterns was determined and the following deposition data were obtained: DE = 0.13 ± 0.02 (0.43 ± 0.07) breathing 0.7 μm (2.7 μm) droplets at MRT = 2 sec and MFR = 500 cm^3/sec, and DE = 0.71 ± 0.04 breathing 2.7 μm droplets at MRT = 6 sec and MFR = 250 cm^3/sec. These results support earlier conclusions [25] that deposition values obtained in this laboratory may be used as average data for the normal public.

Plotting these data versus the subjects' functional residual capacities (FRC) revealed that the deposition of 0.7 μm particles was not correlated with FRC (Figure 14). However, for 2.7 μm particles a small FRC resulted in a higher deposition than a large FRC. Under these breathing conditions 0.7 μm particles were mainly deposited by mixing and 2.7 μm particles mainly by gravitational sedimentation. Since sedimentation and mixing lead to particle deposition in the alveolar airspaces the results suggest that the larger the FRC the larger are the distances over which the particles must be gravitationally transported before they come into contact with a surface of the respiratory tract.

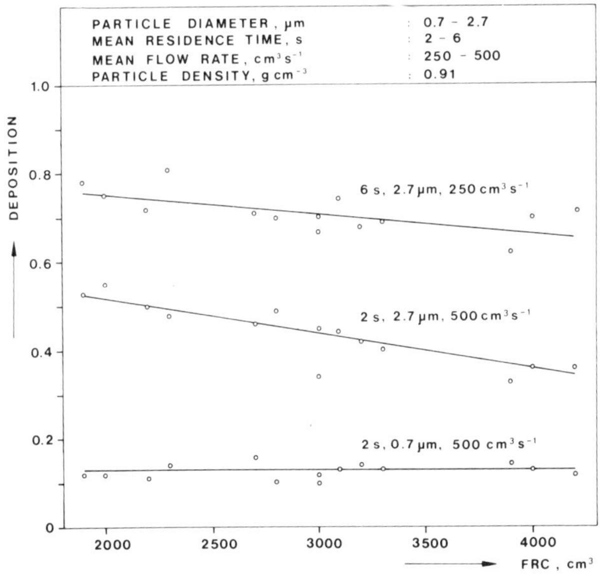

Figure 14. Effect of functional residual capacity (FRC) on total deposition for mouth-breathing.

In a previous study [26], a comparison was made of experimental and theoretical data on the recovery of particles in expired air after single-breath inhalation of aerosols and it was concluded that mixing is flow-induced and coupled with nonsymmetrical expansion of alveolar structures. The present data support this conclusion. Finally, the data also suggest that FRC is not a unique respiratory parameter which can describe intersubject variability of total deposition.

Comparison with Published Data

At present, few deposition data are available for particles larger than 3 μm in diameter. Most of them have been accumulated at the New York University Medical Center [6, 27] where monodisperse iron-oxide particles labeled either with ^{198}Au or ^{99m}Tc were used. The determination of total deposition was based on external in vivo measurements of particle retention and inspired activity. Twenty-two subjects breathed the aerosols at MRT = 2.1 sec and MFR values between 370 and 680 cm^3/sec through the mouth [6]. There is considerable scatter in the deposition data. This may be due partly to intersubject variations and partly to the relatively large variation of the mean volumetric flowrates. Average deposition values of the New York laboratory (NY) are summarized in Table I together with average deposition values of three subjects from this laboratory (FFM). The New York values are slightly higher than the Frankfurt values. However, the agreement is satisfactory if one considers that different approaches were used for the measurement of deposition in the two laboratories.

Recently, total deposition data have been published [28] for mouth-breathing studies performed at the Atomic Energy Research Establishment at Harwell (HAR). The data were determined from the inspired and expired activity of monodisperse charged polystyrene spheres labeled with ^{99m}Tc. These activities were obtained by measuring the activity in the inhalation chamber and on an exhalation filter taking particle losses on a valve assembly and on tube connections into account. Further corrections had to be made on the volume of the expired aerosol because some of it did not pass through the exhalation filter but returned to the inhalation chamber. Because of this complex method of determining total deposition it is not surprising that the Harwell data do not agree as well with the data from this laboratory. The Harwell data obtained for breathing conditions comparable to those used in this study are also summarized in Table I. They are consistently higher than the Frankfurt values which may partly be due to electrical precipitation in the Harwell studies. At Harwell the subjects breathed through a mouthpiece of 10-mm inner diameter, while a 16-mm mouthpiece was used in the Frankfurt experiments. At equal flowrates the inspiratory velocities were, therefore, quite different and inertial losses in extrathoracic airways were enhanced in the Harwell experiments.

For τQ = const., deposition of 3.5-μm particles was found to increase with increasing MRT in this laboratory whereas it decreased in the Harwell ex-

Table I. Total Deposition as a Function of the Aerodynamic Particle Diameter for Mouth-breathing at Several Mean Flowrates Q and Mean Residence Times τ as Evaluated at Different Aerosol Inhalation Laboratories

Laboratory	FFM	HAR	HAR	NY	FFM	HAR	FFM	FFM	HAR	FFM
τ (sec)	1	1.2	2	2.1	2	3	4	8	2	2
Q (cm^3/sec)	1000	833	500	370-680	500	333	250	125	250	250
1.0 μm	0.09				0.15		0.25	0.42		0.15
1.5 μm	0.16				0.24		0.39	0.54		0.21
2.0 μm	0.26				0.37		0.52	0.64		0.28
2.5 μm	0.37				0.47	0.71	0.61	0.70		0.36
3.0 μm	0.49			0.61	0.56		0.67	0.74		0.44
3.5 μm	0.61	0.84	0.81		0.65	0.79	0.72	0.77	0.67	0.50
4.0 μm	0.72			0.75	0.73		0.75	0.79		0.56
4.5 μm	0.80				0.79		0.79	0.81		0.61
5.0 μm	0.86		0.88	0.87	0.82	0.92	0.81	0.83		0.65
5.5 μm	0.89				0.85		0.83	0.85		0.68
6.0 μm	0.91			0.93	0.87		0.85	0.86		0.72
6.5 μm	0.92	0.97	0.94		0.88	0.91	0.87	0.87	0.83	0.75
7.0 μm	0.93			0.96	0.90		0.89	0.88		0.78
7.5 μm	0.94				0.92	0.96	0.90	0.88		0.80
8.0 μm	0.95			0.97	0.93		0.91	0.89		0.83

periments. It is felt that the differences cannot be explained by intersubject variations since the total deposition of a subject who had volunteered for the Harwell experiments was found to be in good agreement with values usually found in this laboratory when that subject was tested in the Frankfurt aerosol inhalation laboratory.

There is now a sufficient amount of evidence that breathing 2-μm spheres of unit density leads to particle deposition by gravitational sedimentation. Inertial losses are almost negligible. Therefore, no or very little deposition is expected at weightlessness. The interpretation of the results of an ingenious study on particle deposition in the human respiratory tract at different levels of gravity-acceleration [29] must, therefore, be revised. It appears that the appreciable value of deposition found at weightlessness was not due to impaction, but most probably due to mixing.

Alveolar Deposition

Alveolar deposition of iron-oxide spheres for mouth-breathing was evaluated as a function of their aerodynamic diameter for two breathing patterns. There are still not enough data available for a detailed discussion of the influence of MRT, MFR and particle density on the diameter dependence of alveolar deposition. However, some preliminary conclusions can be drawn.

Intersubject variability of alveolar deposition is still not known. The present data show little scatter for three subjects breathing aerosols under controlled conditions [19]. The scatter of the vast amount of data accumulated at the New York University Medical Center since 1955 [30] is not exclusively due to intersubject variations, but also due to the large variation in the flowrate [6]. The influence of electrical precipitation on the particle diameter dependence of alveolar deposition is also not known.

Dependence of Alveolar Deposition on Mean Residence Time

Based on physical considerations, there will be a particle size for maximum alveolar deposition, as discussed before. As shown in Figure 15, maximum alveolar deposition was obtained for particles of about 3.5 μm in aerodynamic diameter when breathing at MFR = 250 cm^3/sec and MRT = 4 sec. Particles smaller than about 2.3 μm in aerodynamic diameter were exclusively deposited in the alveolar region of the respiratory tract. Their inertia was not sufficiently high for impaction losses at the mean flowrate of 250 cm^3/sec.

At a constant MFR, the residence time and flowrate are constant in the extrathoracic and tracheobronchial airways when the tidal volume exceeds the volume of these airways. Hence, the rise in total deposition for longer MRT will be due to an increase in alveolar deposition. This was also predicted

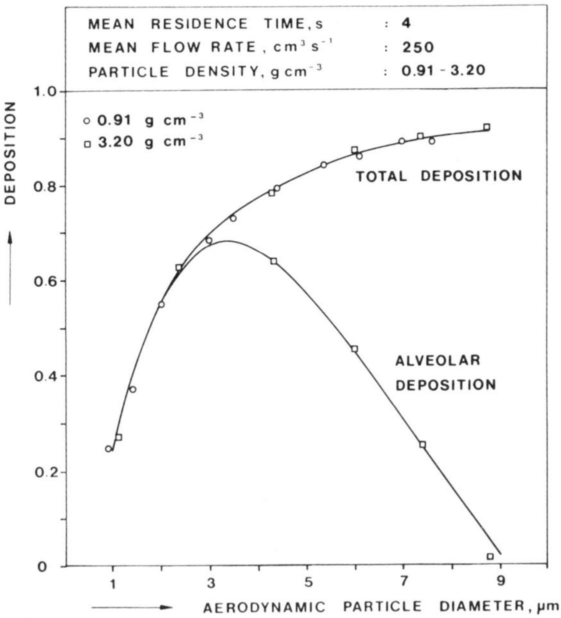

Figure 15. Effect of aerodynamic particle size on total and alveolar deposition for mouth-breating at a mean residence time of 4 sec and a mean flowrate of 250 cm^3/sec.

by a mathematical deposition model [31]. Assuming, therefore, constant deposition in the extrathoracic and tracheobronchial airways at a constant MFR of 250 cm^3/sec the dependence of alveolar deposition on geometric diameter was calculated for different MRT values. The results are shown in Figure 16. They indicate a decrease of the particle size for maximum alveolar deposition as the mean residence time increases.

Dependence of Alveolar Deposition on Mean Flowrate

Taking into account the behavior of aerosol particles in the human respiratory tract, as discussed so far, it is expected that alveolar deposition and the particle size for maximum deposition decrease with increasing MFR. This was experimentally verified and can be seen by comparing alveolar deposition data of Figures 15 and 17. When the MRT was lowered from 4 to 2 sec, the size for maximum alveolar deposition was increased. At the same time the higher flowrate caused a decrease of the size for maximum alveolar deposition. As can be seen, the increase in MFR was more effective with respect to changes of the particle size for maximum alveolar deposition than the

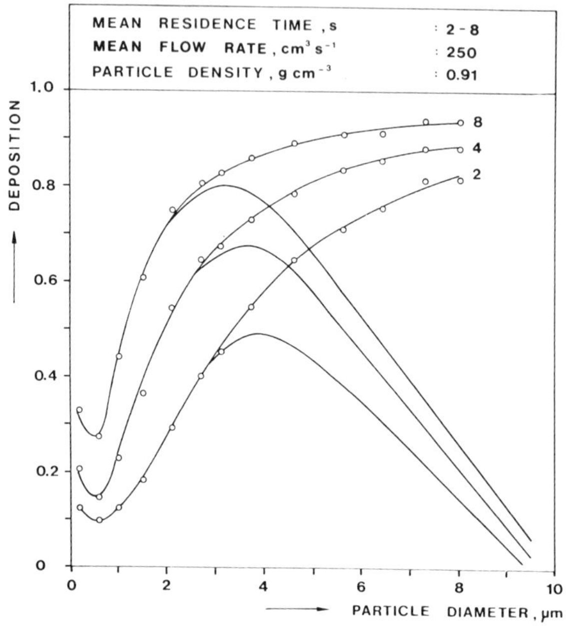

Figure 16. Effect of particle size and mean residence time on total and alveolar deposition for mouth-breathing at a mean flowrate of 250 cm^3/sec.

decrease in MRT. The maximum of alveolar deposition was shifted from 3.5 μm to 3 μm in aerodynamic diameter.

When breathing fast at MFR = 750 cm^3/sec, particles with an aerodynamic diameter smaller than about 1.5 μm were exclusively deposited in the alveolar region of the respiratory tract (Figure 17). This is in good agreement with the conclusion drawn from the total deposition data shown in Figure 7 that inertial impaction is negligible for unit-density spheres smaller than about 1.5 μm when the aerosol remains in the respiratory tract for a mean time of 2 sec. As long as particles are not deposited by impaction, deposition in the extrathoracic and tracheobronchial airways is negligible.

Dependence of Alveolar Deposition on Particle Density

Since the entire respiratory tract was found to collect airborne particles according to their aerodynamic size (Figures 10 and 11), it seems reasonable to assume that each region of the respiratory tract has similar features. Alveolar deposition can then be calculated for spherical particles of different densities as a function of their geometric diameter. Figure 18 shows the

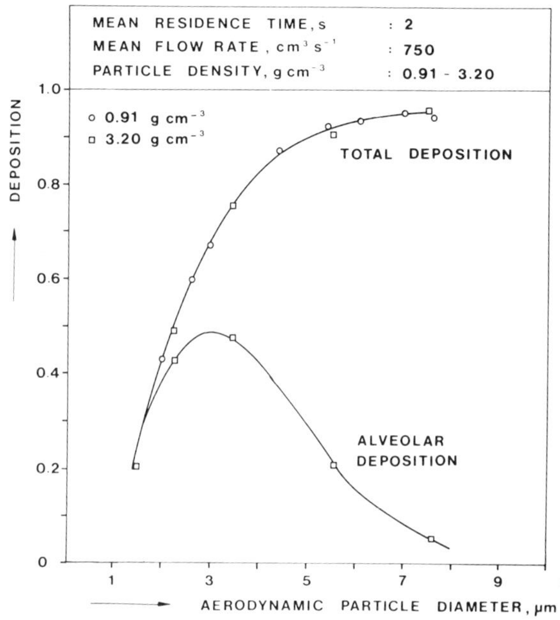

Figure 17. Effect of aerodynamic particle size on total and alveolar deposition for mouth-breathing at a mean residence time of 2 sec and a mean flowrate of 750 cm^3/sec.

difference in alveolar deposition between sebacate droplets and iron-oxide spheres of the same diameter. Whereas a 5-μm ferrous sphere will hardly be deposited in the alveolar airspaces, a 5-μm sebacate droplet will be deposited there with a probability of 0.60.

Comparison with Published Data

Recently, the activity deposited in the alveolar region after inhalation of radioaerosols has been related to the activity deposited in the intrathoracic airways rather than to the inspired activity [28,32]. Therefore, the radiotracer technique has been used in these studies for the determination of the average probability of a particle deposited in the intrathoracic airways being deposited in the alveolar region rather than for the determination of alveolar deposition which is the average probability of an inspired particle being deposited in the alveolar region of the respiratory tract. Consequently, these data are not comparable with the data of the present study.

Average alveolar deposition values found at the New York laboratory [6] are summarized in Table II. Also shown are the average values for the three

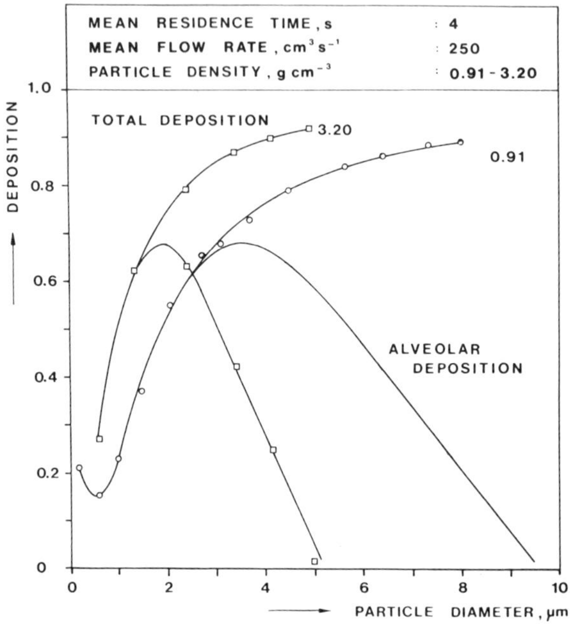

Figure 18. Effect of particle size and particle density on total and alveolar deposition for mouth-breathing at a mean residence time of 4 sec and a mean flowrate of 250 cm^3/sec.

Table II. Alveolar Deposition as a Function of the Aerodynamic Particle Diameter for Mouth-breathing at Several Mean Flowrates Q and Mean Residence Times τ as Evaluated at Different Aerosol Inhalation Laboratories

Laboratory τ (sec) Q (cm^3/sec)	FFM 4 250	FFM 2 750	HAR 2 500	NY 2.1 370-680
1.0 µm	0.26	0.15		
2.0 µm	0.52	0.38		
3.0 µm	0.62	0.50		0.47
4.0 µm	0.61	0.45		0.43
5.0 µm	0.52	0.29	0.27	0.28
6.0 µm	0.40	0.18		0.14
7.0 µm	0.29	0.10		0.06
8.0 µm	0.20			0.04
9.0 µm	0.12			0.02
10.0 µm				0.01

subjects of this laboratory (FFM). The data are in good agreement. Considering the differences in MFR, however, the Frankfurt data seem to be higher than the New York data.

In a single experiment at the Harwell laboratory (HAR), alveolar deposition of 5-μm ^{51}Cr-labeled polystyrene spheres was determined for four subjects. The deposition was found from the activity inspired, the activity deposited in the mouth and in the respiratory tract after mouth-washing and the activity removed from the respiratory tract via feces [28]. The Harwell result is also given in Table II.

Retention

As outlined above, the radiotracer technique allowed only the chest retention functions to be determined. In the present experiments, these functions were evaluated after mouth-breathing of insoluble iron-oxide particles of different sizes under various breathing conditions. Little is known on how retention depends on the chemical composition of the inspired particles. Usually, insoluble material is removed from the alveolar-lymphatic region in a few weeks. However, when soluble substances are deposited, it may take only a few minutes before they are completely removed from the respiratory tract [33].

Chest Retention

It can be seen from Figures 19 and 20 that the chest retention functions found after inhalation of small particles may be approximated quite well by linear functions which suggests that the particulate matter was removed from the respiratory tract at a constant rate. This removal rate was very similar to the rate found when particles were deposited entirely on the surfaces of the alveolar airspaces after inspiration of an aerosol bolus with subsequent breathholding (see the section on "Determination of Regional Deposition and Retention"). Therefore, it can be concluded that these chest retention functions represent alveolar-lymphatic retention functions.

The chest retention functions for the inspiration of larger particles showed two phases: The first phase corresponds to the removal of particulate matter deposited initially onto the ciliary mucosa in the tracheobronchial airways. The removal rate increased with incrasing particle diameter. Up to 40 hours were required for complete particle removal from these airways. The second phase of retention is characterized by the same removal rate as the removal rate of smaller particles which are known to be deposited initially in the alveolar region. This second phase corresponds, therefore, to the removal of particulate matter deposited initially in this region. Consequently, the first phase represents tracheobronchial retention and the second phase alveolar-lymphatic retention.

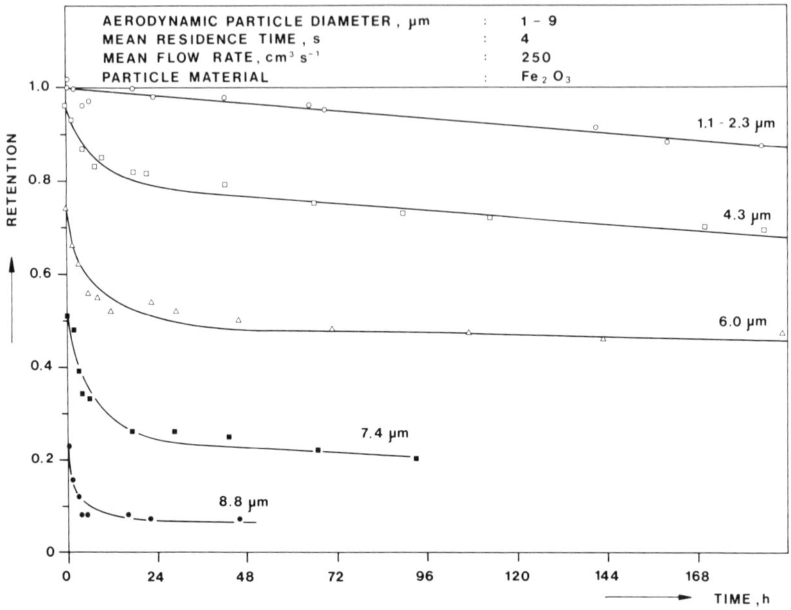

Figure 19. Chest retention after mouth-breathing of monodisperse iron-oxide aerosols of different aerodynamic particle sizes at a mean residence time of 4 sec and a mean flowrate of 250 cm^3/sec.

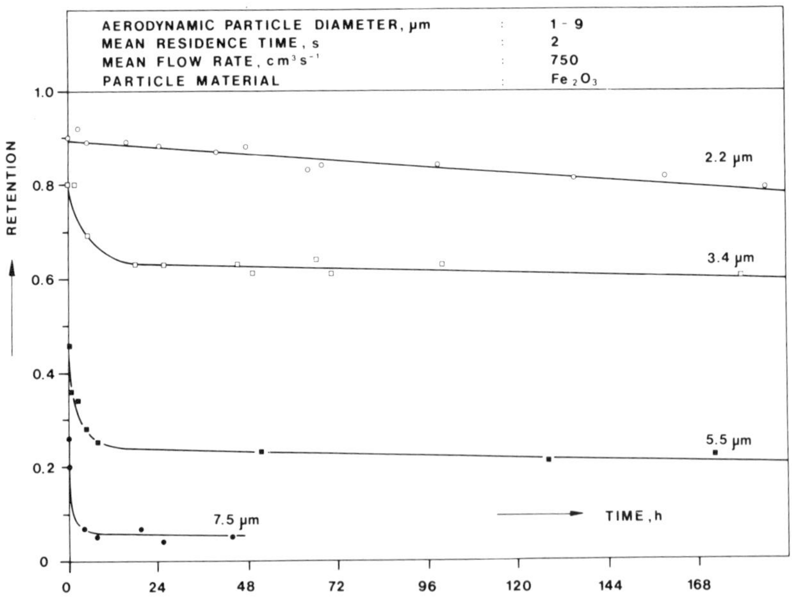

Figure 20. Chest retention after mouth-breathing of monodisperse iron-oxide aerosols of different aerodynamic particle sizes at a mean residence time of 2 sec and a mean flowrate of 750 cm^3/sec.

The ^{198}Au activity of about 0.5 μCi deposited initially in the intrathoracic airways allowed the evaluation of the chest retention for about 12 days. This was not a sufficiently long time for a detailed mathematical analysis of the second phase of the chest retention functions. However, by approximating the second phase of each chest retention function by a single exponential function it was estimated that the biological half-time is about 60 days for alveolar-lymphatic removal processes of iron-oxide particulate matter.

By comparing tracheobronchial retentions after aerosol inhalation with two different breathing patterns (Figures 19 and 20), it may be concluded that the higher the flowrate is the higher is the removal rate. This indicates enhanced inertial particle losses in the larger airways. Alveolar-lymphatic retention seems to be unaffected by the breathing pattern during aerosol administration and appears to be independent of particle size.

Comparison with Published Data

Removal of particulate matter from the tracheobronchial airways has been studied extensively in several laboratories [34-37]. However, in most studies the chest retention functions were evaluated only for a few hours after administration of radioaerosols. The distribution of matter deposited in the intrathoracic airways changes considerably during tracheobronchial removal so that evaluations of tracheobronchial retention functions by the radiotracer technique are influenced by nonuniform photon-counting efficiencies of the body counters used. This may explain at least some of the discrepancies in removal rates reported by different laboratories. Different particle size distributions of the radioaerosols administered under different breathing conditions may also have contributed to these discrepancies. However, all investigators agree with each other in having observed a high intersubject variability and a low intrasubject variability for tracheobronchial retention.

In the extensive studies performed at the New York University, most of the tracheobronchial removal occurred in two phases of 1-2 hours and 4-10 hours duration [38,39]. After inspiration of 5-μm iron-oxide particles, the average time for completion of both phases was about 12 hours. In later studies, in which 16 subjects were tested under unspecified flow conditions, 90% of tracheobronchial removal of iron-oxide particles was completed in 7.6 ± 2.3 hours [40].

The time required for tracheobronchial removal found in these studies seems to be higher than the time found by other investigations. The present experimental removal times are, however, in good agreement with the predicted value of about 30 hours which was found from a mathematical retention model [41]. This model assumes that the mucous transport rate in each generation of airways of an asymmetrical airway model is inversely proportional to the total circumference of these airways. The model assumes further that the deposition pattern in the tracheobronchial region is uniform.

Alveolar-lymphatic retention in humans has been studied most extensively at the University of Rochester Medical School since 1962 [42,43]. By approximating the experimentally determined alveolar-lymphatic retention functions by single exponential functions, the biological half-time for insoluble polydisperse particulate matter in this region was found to be 65 days for manganese dioxide particles labeled with ^{54}Mn, 62 days for ^{51}Cr-labeled iron-oxide particles and 35 days for ^{51}Cr-labeled polystyrene particles. In the present study a half-time of about 60 days was found for ^{198}Au labeled monodisperse iron-oxide particles. An exponential decrease in alveolar-lymphatic activity was also observed at the New York laboratory with similar biological half-times [38]. At the Harwell laboratory, however, half-times of 150-300 days were observed after a single inhalation of 5-μm ^{51}Cr-labeled polystyrene particles followed by a 10-sec breath-holding period [44].

Long-term alveolar-lymphatic removal of particulate matter will change only slightly the distribution pattern of particulate matter in the chest. Hence, evaluations of alveolar-lymphatic retention functions are less sensitive to nonuniform photon-counting efficiencies of body counters than evaluations of tracheobronchial retention functions. There must have been other reasons for the differences in reported half-times. For instance, inspiration of mono- versus polydisperse aerosols, and the different periods over which retention functions have been observed may have influenced the mathematical evaluation of half-times.

Recently, a new technique of measuring long-term alveolar-lymphatic retention in humans has been introduced [45]. The amount of Fe_3O_4 remaining in the respiratory tract at different times after inspiration of a ferromagnetic Fe_3O_4 dust is evaluated with a magnetic detector by measuring the field produced by the magnetized particles around the thorax. For each measurement, the particles in the respiratory tract are first magnetized with an external magnetic field. Average half-times of about 70 days have been observed, but the alveolar-lymphatic retention functions evaluated over a period of one year after aerosol administration were not exponential [46].

ACKNOWLEDGMENTS

The authors gratefully acknowledge the skillful assistance of G. Heigwer (electronics), B. Stuck (technical assistance), F. Haas (mechanics), E. Künzel (drawings) and G. Heise (photography).

REFERENCES

1. Hatch, T. F., and P. Gross. *Pulmonary Deposition and Retention of Inhaled Aerosols* (New York: Academic Press, Inc., 1964), 192 pp.
2. Task Group on Lung Dynamics of ICRP. "Deposition and Retention Models for Internal Dosimetry of the Human Respiratory Tract," *Health Physics* 12:173-207 (1966).
3. R. V. Lourenco, Ed. "Inhaled Aerosol Symposium," *Arch. Int. Med.* 131:21-166 (1973).
4. Davies, C. N. "Deposition of Inhaled Particles in Man," *Chem. Ind.* 1: 441-444 (1974).
5. Mercer, T. T. "The Deposition Model of the Task Group on Lung Dynamics: A Comparison with Recent Experimental Data," *Health Physics* 29:673-680 (1975).
6. Lippmann, M. In: *Handbook of Physiology, Section 9, Reactions to Environmental Agents,* D. H. K. Lee, H. L. Falk, S. D. Murphy and S. R. Geiger, Eds. (Bethesda, MD: American Physiological Society, 1977), 213-232.
7. Brain, J. D. "Anesthesia and Respiratory Defense Mechanisms," *Int. Anest. Clinics* 15:169-198 (1977).
8. Subcommittee on Airborne Particles of National Research Council, *Airborne Particles* (Baltimore, MD: University Park Press, 1979), 343 pp.
9. National Academy of Sciences–National Research Council Publ. No. 848 (1961).
10. Odgen, T. L., and J. L. Birkett. In: *Inhaled Particles IV*, W. H. Walton, Ed. (Oxford: Pergamon Press, 1977), pp. 93-104.
11. Friedlander, S. K. "The Characterization of Aerosols Distributed with Respect to Size and Chemical Composition," *J. Aerosol Sci.* 1:295-307 (1970).
12. Brain, J. D., and P. A. Valberg. "Models of Lung Retention Based on the Report of the ICRP Task Group," *Arch. Environ. Health* 28:1-11 (1974).
13. Stahlhofen, W., J. Gebhart, J. Heyder and C. Roth. "Generation and Properties of a Condensation Aerosol of di-2-Ethylhexyl Sebacate– Description of the Generator," *J. Aerosol Sci.* 6:161-167 (1975).
14. Stahlhofen, W., J. Gebhart, J. Heyder and B. Stuck. "Herstellung von monodispersen Fe_2O_3-Testaerosolen mit Hilfe der Zentrifugal Zerstaubung," *Staub-Reinhaltung der Luft.* 39:73-77 (1979).
15. Stahlhofen, W., L. Armbruster, J. Gebhart and E. Grein. "Particle Sizing of Aerosols by Single Particle Observation in a Sedimentation Cell," *Atmos. Environ.* 9:851-857 (1975).
16. Gebhart, J., J. Heyder, C. Roth and W. Stahlhofen. In: *Fine Particles*, B. Y. H. Liu, Ed. (New York: Academic Press, Inc., 1976), pp. 794-815.
17. Heyder, J., J. Gebhart, G. Heigwer, C. Roth and W. Stahlhofen. "Experimental Studies of the Total Deposition of Aerosol Particles in the Human Respiratory Tract," *J. Aerosol Sci.* 4:191-208 (1973).

18. Hyder, J., L. Armbruster, J. Gebhart, E. Grein and W. Stahlhofen. "Total Deposition of Aerosol Particles in the Human Respiratory Tract for Nose- and Mouth-Breathing," *J. Aerosol Sci.* 6:311-328 (1975).
19. Stahlhofen, W., J. Gebhart and J. Heyder. "Experimental Determination of Regional Deposition of Aerosol Particles in the Human Respiratory Tract," *Am. Ind. Hyg. Assoc. J.* (submitted).
20. Morsy, S. M., E. Werner, W. Stahlhofen and W. Pohlit. "A Detector of Adjustable Response for the Study of Lung Clearance," *Health Physics* 32:243-251 (1977).
21. Stahlhofen, W., B. Eckhardt, J. Gebhart, J. Heyder and B. Stuck. "Experimental Determination of the Alveolar Deposition of Aerosol Particles in the Human Lung," *J. Aerosol Sci.* 10:222-224 (1979).
22. Foord, N., A. Black and M. Walsh. In: *Inhaled Particles IV*, W. H. Walton, Ed. (Oxford: Pergamon Press, 1977), pp. 137-148.
23. Morsy, S. M., B. Eckhardt, W. Stahlhofen and W. Pohlit. "Optimization Studies of a Whole-Body Counter to Cope with Possible Dynamic Transfer between Neighbouring Compartments," *Health Physics* 35:325-332 (1978).
24. Melandri, C., V. Prodi, G. Tarroni, M. Formignani, T. DeZaiacomo, F. G. Bompane and G. Maestri. In: *Inhaled Particles, IV*, W. H. Walton, Ed. (Oxford: Pergamon Press, 1977), 193-203.
25. Heyder, J., J. Gebhart, C. Roth, W. Stahlhofen, B. Stuck, G. Tarroni, T. DeZaiacomo, M. Formignani, C. Melandri and V. Prodi. "Intercomparison of Lung Deposition Data for Aerosol Particles," *J. Aerosol Sci.* 9:147-155 (1978).
26. Taulbee, D. B., C. P. Yu and J. Heyder. "Aerosol Transport in the Human Lung from Analysis of Single Breaths," *J. Appl. Physiol.* 44: 803-812 (1978).
27. Lippmann, M., and R. E. Albert. "The Effect of Particle Size on the Regional Deposition of Inhaled Aerosols in the Human Respiratory Tract," *Am. Ind. Hyg. Assoc. J.* 30:257-275 (1969).
28. Foord, N., A. Black and M. Walsh. "Regional Deposition of 2.5–7.5 μm Diameter Inhaled Particles in Healthy Male Non-Smokers," *J. Aerosol Sci.* 9:343-357 (1978).
29. Hoffman, R. A., and J. Billingham. "Effect of Altered G Levels on Deposition of Particulates in the Human Respiratory Tract," *J. Appl. Physiol.* 38:955-960 (1975).
30. Albert, R. E., and L. C. Arnett. "Clearance of Radioactive Dust from the Lung," *Arch. Ind. Health* 12:99-106 (1955).
31. Yu, C. P., and D. B. Taulbee. In: *Inhaled Particles IV*, W. H. Walton, Ed. (Oxford: Pergamon Press, 1977), pp. 35-46.
32. Camner, P., and K. Philipson. "Human Alveolar Deposition of 4 μm Teflon Particles," *Arch. Env. Health* 33:181-185 (1978).
33. Yeates, D. B., N. Aspin, A. C. Bryan and H. Levison. "Regional Clearance of Ions from the Airways of the Lung," *Am. Rev. Resp. Dis.* 107: 602-608 (1973).

34. Thomson, M. L., and M. D. Short. "Mucociliary Function in Health, Chronic Obstructive Airways Disease and Asbestosis," *J. Appl. Physiol.* 26:535-539 (1969).
35. Lourenco, R. V., M. F. Klimek and C. J. Borowsky. "Deposition and Clearance of 2 μm Particles in the Tracheobronchial Tree of Normal Subjects–Smokers and Nonsmokers," *J. Clin. Invest.* 50:1411-1420 (1971).
36. Camner, P. "The Production and Use of Test Aerosols for Studies of Human Tracheobronchial Clearance," *Env. Physiol.* 1:1-18 (1972).
37. Sanchis, J., M. Dolovich, R. Chalmers and M. Newhouse. "Quantitation of Regional Aerosol Clearance in the Normal Human Lung," *J. Appl. Physiol.* 36:757-762 (1972).
38. Albert, R. E., M. Lippmann, J. Spiegelman, C. Strehlow, W. Briscoe, P. Wolfson and N. Nelson. In: *Inhaled Particles and Vapours II*, C. N. Davies, Ed. (Oxford: Pergamon Press, 1967), pp. 361-377.
39. Albert, R. E., M. Lippmann and W. Briscoe. "The Characteristics of Bronchial Clearance in Humans and the Effect of Cigarette Smoking," *Arch. Env. Health* 18:738-755 (1969).
40. Albert, R. E., M. Lippmann, H. T. Peterson, J. Berger, K. Sanborn and D. Bohning. "Bronchial Deposition and Clearance of Aerosols," *Arch. Int. Med.* 131:115-131 (1973).
41. Yeates, D. B., and N. Aspin. "A Mathematical Description of the Airways of the Human Lungs," *Resp. Physiol.* 32:91-104 (1978).
42. Morrow, P. E., F. R. Gibb and K. Gazioglu. In: *Inhaled Particles and Vapours II*, C. N. Davies, Ed. (Oxford: Pergamon Press, 1967), pp. 351-358.
43. Morrow, P. E., F. R. Gibb and K. M. Gazioglu. "A Study of Particulate Clearance from the Human Lungs," *Am. Rev. Resp. Dis.* 96:1209-1221 (1967).
44. Booker, D. V., A. C. Chamberlain, J. Rundo, D. C. F. Muir and M. L. Thomson. "Elimination of 5 μm Particles from the Human Lung," *Nature* 215(5096):30-33 (1967).
45. Cohen, D. "Ferromagnetic Contamination in the Lungs and Other Organs of the Human Body," *Science* 180:745-748 (1973).
46. Cohen, D., S. F. Arai and J. D. Brain. "Smoking Impairs Long-Term Clearance from the Lung," *Science* 204:514-517 (1979).

CHAPTER 4

DISSOLUTION OF INHALED AEROSOLS

Owen R. Moss

Battelle–Pacific Northwest Laboratory
Richland, Washington 99352

George M. Kanapilly

Lovelace Biomedical and
Environmental Research Institute Inc.
Inhalation Toxicology Research Institute
Albuquerque, New Mexico 87115

INTRODUCTION

The clearance of aerosols deposited in the lung is accomplished by a combination of particle transport and particle dissolution, with subsequent movement of dissolved material into the bloodstream, thus increasing the systemic burden. In contrast with the highly complex variables associated with in vivo testing, in vitro dissolution studies present a relatively simple approach. In this chapter we hope to aid the toxicologist interested in investigating the factors influencing lung clearance by presenting in vitro techniques used to measure the dissolution rates of selected materials.

CLEARANCE MODELS

Inhaled particles may consist of a single component (X), or of several distinct elements combined with one another, or within a base material. In our model, shown in Figure 1, particles are deposited in the lung at some periodic rate K_1. Particles are cleared either by physical removal (indicated by K_2), or by dissolution of component X (K_{31} and K_{32}), with subsequent removal of the dissolved fraction (K_{41} and K_{42}). Only the distinction

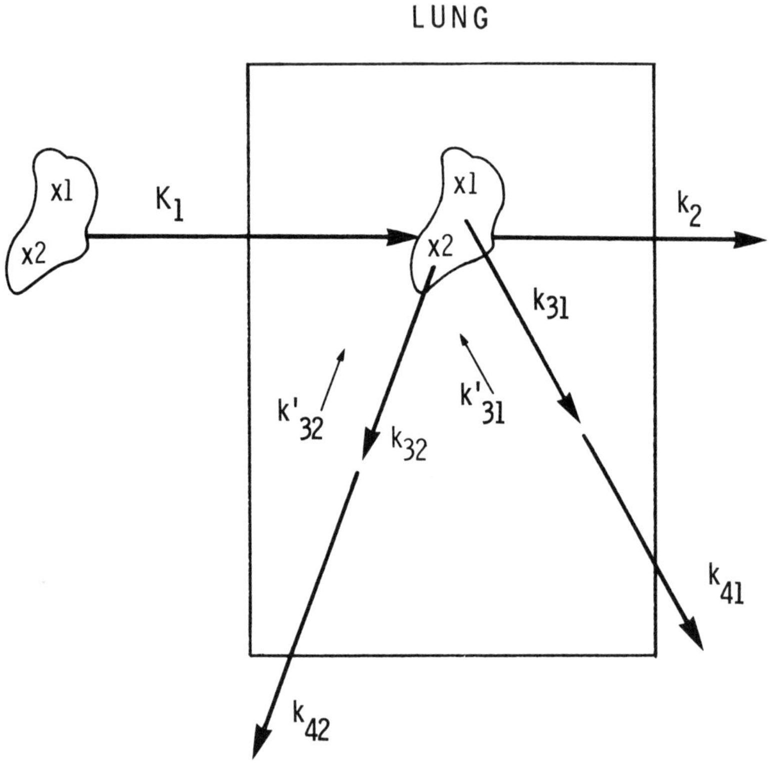

Figure 1. Rate constants associated with deposition and clearance of inhaled particles from the lung.

between particle clearance (K_2) and particle dissolution (K_{31} and K_{32}) is addressed here.

Data collected from animal experiments have historically provided most, but not all, of the information necessary to separate clearly the clearance paths for both single- and multiple-component particles deposited in the lungs. For example, clearance via excreta may result not only from particles cleared from the lungs as a result of swallowing, but also from secretion of the dissolved component into the lower intestinal tract. By following the dissolution and clearance of multicomponent particles from the lungs, distinctions can be made between particle clearance (K_2, Figure 1) and particle components that are dissolved, then cleared, or those that are fixed (K_4=0).

In vivo measurements that can be made to determine the clearance rate constants shown in Figure 1 are restricted to changes (as a function of time) in particle components contained in whole animals, or in organs, urine and feces. This information can be supplemented by measuring particle dissolution rates in solutions that simulate solvent conditions in the lung. In vitro measurements of single-component particles give dissolution rates which can be used to distinguish between dissolution and particle clearance (if the dissolved fraction is rapidly cleared from the lungs). In vitro measurements of multicomponent particles in simulated lung fluid can give the dissolution rate per component or, inversely, the fraction of the component remaining in the particle as a function of time. The ratio of these fractions when compared to the ratio of each component fraction remaining in the lungs, can be used to understand the rate-limiting steps that cause the observed lung clearance. In Figure 1, (fraction XI/fraction X2) is the ratio of those components remaining in a two-component particle.

DISSOLUTION KINETICS

The dissolution rate of a particle depends on: (1) the surface area of the particle exposed to solvent; (2) the diffusion of atoms or molecules through the stagnant layer of solvent around the particle; or (3) the diffusion to the surface of the more soluble component of a multicomponent particle. Dissolution experiments can provide data in the form, mass fraction dissolved $(1 - M/M_0)$ as a function of time t, where M is the remaining mass, and M_0 is the initial mass of the particle sample. The dissolution curves produced from these data will differ in shape, depending on which of the above cases dominates the dissolution rate. The shape of the curve can then be used to describe the type of dissolution taking place and to verify size properties of the sample, in addition to giving the dissolution rate at some time t. In the following paragraphs and in Figure 2, we present the equations used to describe the dissolution $[(1 - M/M_0)$ vs t] curves for each of the above cases. Only five variations of the influence of the rate-limiting step are presented; three deal with surface area and particle size distribution, and two deal with diffusion within the solvent or particle matrix.

Case 1a: Constant Surface, Monodisperse Sample

Consider the dissolution of monodisperse, single-component particles having constant surface during T, the total time of experimental observation. The rate of change in the remaining particle mass M is:

$$-\frac{dM}{dt} = KA \tag{1}$$

for surface area A exposed to solvent, and K, the dissolution rate constant. This relation is true when the solvent is well-stirred or when the solvent is

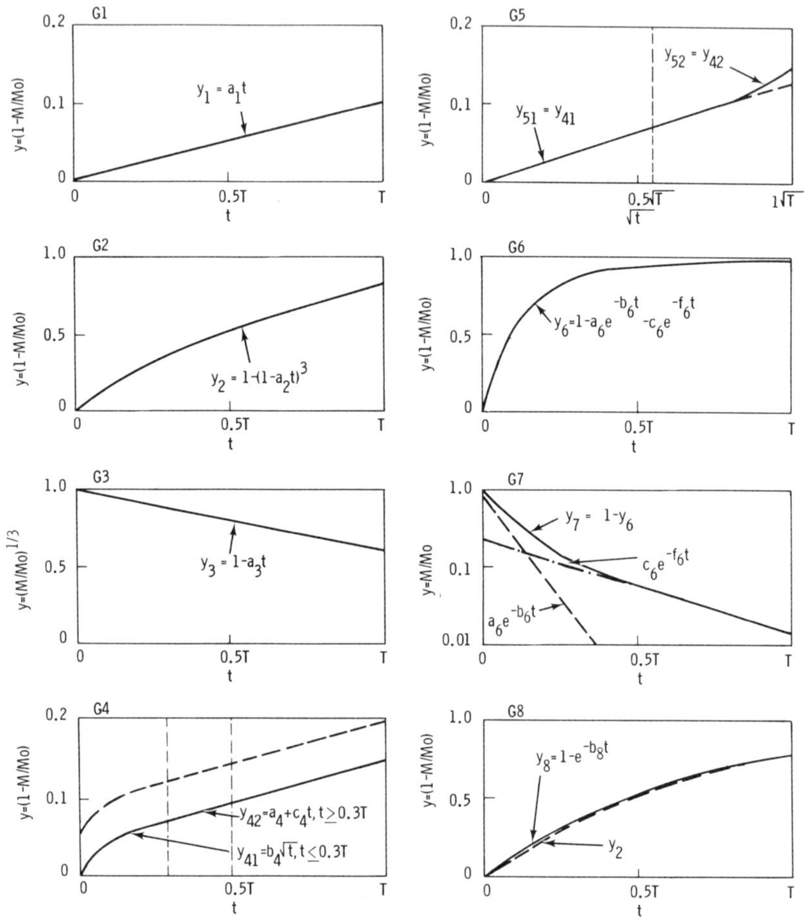

Figure 2. Data analysis of dissolution measurements. The different relationships between undissolved mass fraction (M/M_O) and time (total experimental observation period, T) are shown for various systems: slow dissolution of monodisperse distribution of single-component particles (G1); much greater time or dissolution rate (G2, G3); slow dissolution of monodisperse distribution of multicomponent particles (G4 and G5); log-normal distribution of particles (G6 and G7); or concentration gradient in the solvent (G8).

moving so that solute concentration at the surface of the particle is much lower than the equilibrium concentration.

The mass fraction remaining is:

$$\frac{M}{M_O} = 1 - KS_p t \tag{2}$$

where S_p = specific surface of the sample, cm^2/g. The concentration C in a closed system of volume V is:

$$C = \frac{M_o - M}{V} = \frac{KS_p}{V} t \tag{3}$$

Graph G1, Figure 2, is a plot of the mass fraction dissolved $(1 - M/M_o)$ vs time (t). The dissolution rate is slow enough that less than 10% of the material goes into solution during the total experimental observation period T. If the specific surface of the sample is known, the dissolution rate constant, K_3 (fraction dissolved/cm^2/sec), can be calculated.

Case 1b: Decreasing Surface Monodisperse Sample

Consider the dissolution of monodisperse single-component particles having a decreasing total surface area during time T. The calculation for Case 1a was made assuming that dissolution is slow enough so that over the observation period the surface area of the particles does not change. For relatively insoluble particles, this is certainly true. However, for rapidly dissolving single-component particles of the same size, in a solution where diffusion through the solvent does not affect the dissolution rate, the surface area A of one particle (or of a sample of particles of the same size) is not constant, but is approximated as a function of the remaining particle mass:

$$A = aM^{2/3} \tag{4}$$

where a is a function of the particle density, the surface area and volume shape factors, and is assumed to be constant with respect to particle size. Therefore,

$$\frac{dM}{dt} = -KaM^{2/3} \tag{5}$$

When integrated, the commonly used [1] form of the dissolution equation is:

$$(M/M_o)^{1/3} = 1 - S_{po} Kt/3 \tag{6}$$

where S_{po} is the initial specific surface area of the sample.

Mass fraction dissolved $(1 - M/M_o)$ vs time is shown in Graph G2, Figure 2. This curve differs from G1 by the fraction of material dissolved in time T. Equation 6 indicates that a plot of $(M/M_o)^{1/3}$ vs time t will yield a straight line (Graph G3, Figure 2), from which the dissolution rate constant can be calculated, provided the initial specific surface of the sample is known.

Case 1c: Decreasing Surface, Polydisperse Sample

Consider the dissolution of polydisperse, single-component particles having a decreasing total surface area during time T.

Mercer [2] found that the undissolved mass fraction of a log-normally distributed particle sample could be described by a two-exponential function of the geometric standard deviation GSD and a dimensionless quantity β:

$$\frac{M}{M_o} = f(\mathrm{GSD}, \beta) = f_1 e^{-\lambda_1 \beta} + f_2 e^{-\lambda_2 \beta} \tag{7}$$

and

$$\beta = S_p K t = \frac{\alpha_s}{\alpha_v \rho} \cdot \frac{1}{D_m} \cdot K t \tag{8}$$

where f and λ are functions of GSD

S_p = specific surface

K is the dissolution rate constant

α_s and α_v = surface area and volume shape factors, respectively

D_m is the geometric mass median diameter of each particle

ρ = density

The plot of mass fraction dissolved vs time will be a two-exponential curve (Graph G6, Figure 2), containing the functions described in Equation 7. This curve can be plotted on semi-log paper (Graph G7), and the two exponential components extracted. The half-time for the observed dissolution $t_{1/2}$ can be calculated, along with the dissolution rate constant.

$$t_{1/2} = 0.6 \frac{\alpha_v}{\alpha_s} \frac{\rho D_m}{K} \tag{9}$$

for particles of density ρ and mass median diameter D_m. For a sample described by its aerodynamic diameter D_{ae}, the half-time can be uniquely described by

$$t_{1/2} = \frac{0.6}{6} \frac{\sqrt{\pi}\, \rho_o D_{ae}}{K \sqrt{\alpha_{sae}}} = \frac{0.1772\, \rho_o D_{ae}}{K \sqrt{\alpha_{sae}}} \tag{10}$$

Here, ρ_o = 1 g/cm^3 and α_{sae} = the surface shape factor for the particles described by the mass median aerodynamic diameter, D_{ae} (the surface shape factor is obtained by measuring the specific surface of a sample having a known D_{ae} distribution). The quantity $K\sqrt{\alpha_{sae}}$ can be measured for each polydisperse sample of single-component particles.

Case 2: Constant Surface, Monodisperse Sample, Stagnant Layer of Solvent

Consider again the dissolution of monodisperse single-component particles having constant surface during T, the total time of experimental observation and sampling (as in Case 1a). A stagnant layer of fluid always exists around particles. If the surface reaction is rapid and the diffusion of particle components in the solvent is the rate-controlling step, the equilibrium concentration C_e is maintained at the particle surface.

$$0 = \frac{dC}{dt} = -\frac{1}{v}\frac{dM}{dt} = \frac{KA}{v} - K'AC_e \qquad (11)$$

where K' is the rate constant for molecules returning to the particle surface
v is the volume of solvent in that portion of stagnant layer maintained at C_e

The buildup of concentration in this small volume v about each particle is:

$$\frac{dC}{dt} = \frac{KA}{v} - K'AC = K'A\,(Ce - C) \qquad (12)$$

giving [3]

$$C = Ce\,(1 - e^{-K'At}) \qquad (13)$$

The buildup of concentration in the total solvent volume V is due to diffusion of molecules from the small volume elements of concentration C_e, and is described by Equation 13 when K_D is substituted for K'.

$$K_D = \frac{\Delta}{V\delta_n} \qquad (14)$$

where Δ is the diffusion coefficient and δ_n the thickness of the Nernst diffusion layer (which decreases with increased stirring). The influence of a concentration gradient in the solvent on the observed dissolution of a monodisperse distribution of single-component particles is shown in Graph G8, Figure 2, as a single exponential curve containing the function described in Equation 13. This dissolution, controlled by diffusion within the solvent, is distinguished from rapid dissolution (Case 1b, dashed curve, Graph G8) only by the sensitivity of the curve to changes in mixing agitation and solvent flow.

Case 3: Constant Surface, Monodisperse Sample, Diffusion within Particle

Consider the dissolution of monodisperse multicomponent particles having constant surface during T, the total time of experimental observation. Extraction of molecules due to diffusion within the particle or in its leached layer (nonsteady-state diffusion in mineral or quasisteady-state diffusion through a leached layer) is implied if the dissolution curves are of the form [4]:

$$(1 - M/M_0) = a + b\sqrt{t} \tag{15}$$

where a is the extracted fraction due to initial exchange of surface molecules
b is the parabolic dissolution constant

This type of dissolution is also seen in pharmaceutical applications, such as release of chemical from vaginal devices [5]. At some point, the dissolution rate $b/(2\sqrt{t})$ must become less than the dissolution rate of the particle surface. When this occurs, diffusion within the particle ceases to dominate the dissolution process.

Graph G4, Figure 2, contains a plot of mass fraction vs time for a sample essentially identical to that shown in Graph G1 (constant area, monodisperse), except that the particle is multicomponent, and only one component is being monitored. This case, described in Equation 15, can include an initial concentration caused by surface reactions (dashed line, Graph G4, Figure 2). The diffusion of the trace element within the particle can be distinguished from the surface dissolution of the base material of the particle by plotting mass fraction vs the square root of time, as suggested by Equation 15 (Graph G5, Figure 2). The rate constant from the linear dissolution phase can be calculated from the straight-line portion of Graph G4, and the dissolution constant for diffusion within the particle can be calculated from the straight-line portion of Graph G5.

Figure 2 contains the eight curves describing the five dissolution cases above. Graph G1, the curve for dissolution of monodisperse distribution of single-component particles, becomes G2, for increased time of observation T or increased fraction dissolved. Without additional information, the curve in G2 (decreasing surface area of monodisperse sample) is difficult to distinguish from that shown in G8 (constant surface of monodisperse sample in a stagnant layer of solvent). The family of curves would be complete if one includes the effect of solvent diffusion layer and polydisperse log-normal particle size on observed dissolution rates. This has not been worked out for time periods involving all phases of dissolution. Such a calculation would be useful to relate the first half-time period to both parabolic and linear dissolution constants. One could then estimate the minimum time required to measure in vitro dissolution on polydisperse particle samples in order to obtain both constants.

IN VITRO DISSOLUTION MEASUREMENTS

Limitations in experimental modeling of the lung environment have been discussed previously [6,7]. Composition, temperature and pH of the experimental model must be matched to those of the lung to insure that dissolution rates observed in vitro and in vivo are comparable. Batch, flowthrough, and flowpast are the names given to the three basic in vitro techniques currently used in estimating the dissolution component of inhaled particle clearance.

Simulated Lung Solutions

The environment of both the lung and the solvent in contact with inhaled particles has been discussed [6-13]. These reports attempt to describe the lung microenvironment and related models, and to draw conclusions about lung clearance rates from measuring the dissolution of material in simulated lung fluid. The solvents used in these reports have been summarized by Kanapilly [7] and, more recently, by Moss [8]. The components of one such solution [6,8-10] are:

1. magnesium chloride, hexahydrate, 0.203 g/l;
2. sodium chloride, 6.019 g/l;
3. potassium chloride, 0.298 g/l;
4. sodium phosphate, dibasic, anhydrous, 0.142 g/l;
5. sodium sulfate, anhydrous, 0.071 g/l;
6. calcium chloride, dihydrate, 0.368 g/l;
7. sodium acetate, trihydrate, 0.953 g/l;
8. sodium bicarbonate, 2.604 g/l; and
9. sodium citrate, dihydrate, 0.097 g/l.

Using a starting volume of water that is at least 95% of the desired final volume, the compounds can be added sequentially in the order given without precipitation, if they are added very slowly, with rapid stirring. Under sterile conditions, the solution can be maintained indefinitely at pH 7.4 by slowly bubbling 5% CO_2 in O_2 or air through the solution. Protein has been substituted for citrate [6,9], but, for simplicity, the general assumption has been made that citrate, and similar compounds and proteins, are alike in forming chelate with dissolved metal ions.

Dissolution Measuring Techniques

Of the three commonly used techniques, batch (static) system, flowthrough and flowpast, the batch system is the simplest to use and results in data that are easiest to analyze. The batch system consists of a container of solvent (volume V_0) in which is suspended a mass M_0 of the sample (Figure 3). The mass M_0 is suspended uniformly in the solvent (B1) or separated from the solvent by filter barriers (B2). Samples are drawn from system B1 and filtered to differing degrees [6,7,9,10]. After samples are drawn from system B2, the volume remaining in the container is recorded. The data from both series of samples can be represented by a graph of concentration C, or mass fraction dissolved ($1 - M/M_0$) against time t. In sampling system B1, we assume that (1) the sample contains particles and dissolved materials such that $m/M_0 = v/V_0$, where m is the mass of material (particulate and dissolved) removed in the sample, and v, the sample volume; and (2) no precipitation takes place before the particles and solvent are separated. In sampling system B2, the assumption is made that no particles are drawn into the solvent sample. A disadvantage of the batch system is that it does not easily lend itself to automation. Furthermore, it is particularly dependent on the requirement that the concentration C of dissolved material in the solvent be much less than the equilibrium solubility; chemical detection techniques ultimately determine how much less C must be. Usually C_{max} is less than Ce/10, as a first approach.

In the flowthrough system, the filter used to collect the airborne sample is placed between finer filters (0.1 μm pore size; Figure 3). Fluid is pumped past the sample so that the observed dissolution rate does not vary, even when flow is increased up to 10 times. Flowrate and concentration of solvent passing through the filter and sample are monitored as a function of time. Analysis of the fluid sample gives mass fraction dissolved ($1 - M/M_0$) as a function of time, provided that volume of flow through the system is also monitored. The system gives accurate values of the dissolution rate constant, assuming that no particles pass through the filter into the sampled solvent, and that the flowrate past the filter is known for the entire experiment. Although more complex than the batch system, this method allows measurement of rapid dissolution rates and lends itself to automation and continuous sampling. On the other hand, it is less complicated to set up than the flowpast system, and there are no air bubbles or stagnant zones in the system. The main limitation of this approach is that flowrate does not remain constant over a run lasting several days [6,7]: filter resistance changes and driving force must be increased to keep flow constant. Furthermore, as in any flow system, large volumes of solvent are needed.

In the flowpast system (Figure 3), a filter containing a particle sample is sandwiched between two filters of finer pore size (0.1 μm or less) and clamped in a flow cell that causes solvent to pass parallel to one or both outer filter surfaces. The flow along the filter surface is from an inlet point at the edge of the filter to an exit point diagonally opposite [7], or from

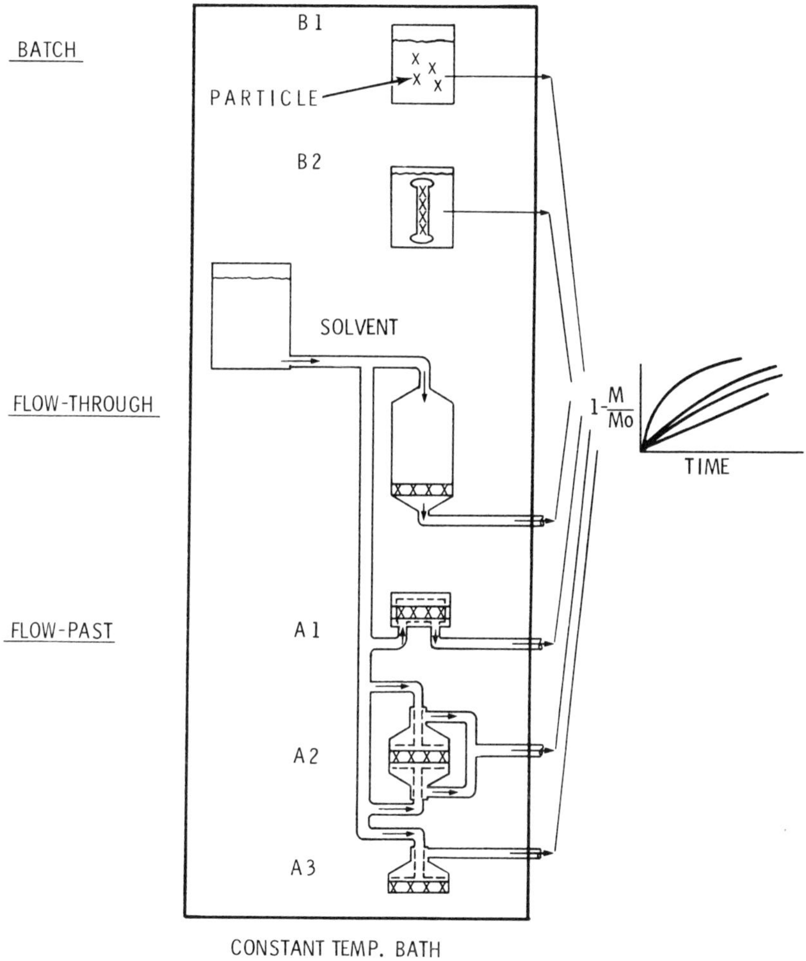

Figure 3. Dissolution systems: particles are either dispersed in solvent (batch B1) or held in a filter sandwich. The filter is then either suspended in solvent (batch B2), held in a filter holder with solvent passing through (flowthrough) or held in a flow cell with solvent passing parallel to only one of the filter surfaces (flowpast, A1 and A3) or parallel to both outer filter surfaces (flowpast, A2).

a center inlet point radially to the outer edge of the filter (described below and in Figure 4). In each case, care must be taken to prevent formation of stagnant zones of solvent that may remain even during increased flow. We assume, then, that the dissolution rate constant (K_{31} and K_{32}, Figure 1) is the correct dissolution rate constant for the sample. The advantages of this system over the batch system are the same as for the flowthrough system. In addition, in the flowpast system, there is no change in flow during the course of the experiment. Disadvantages are the possibility of stagnant solvent areas in flow (the degree of influence on the measured dissolution rate constants is unknown); the need for large volumes of solvent; and (as

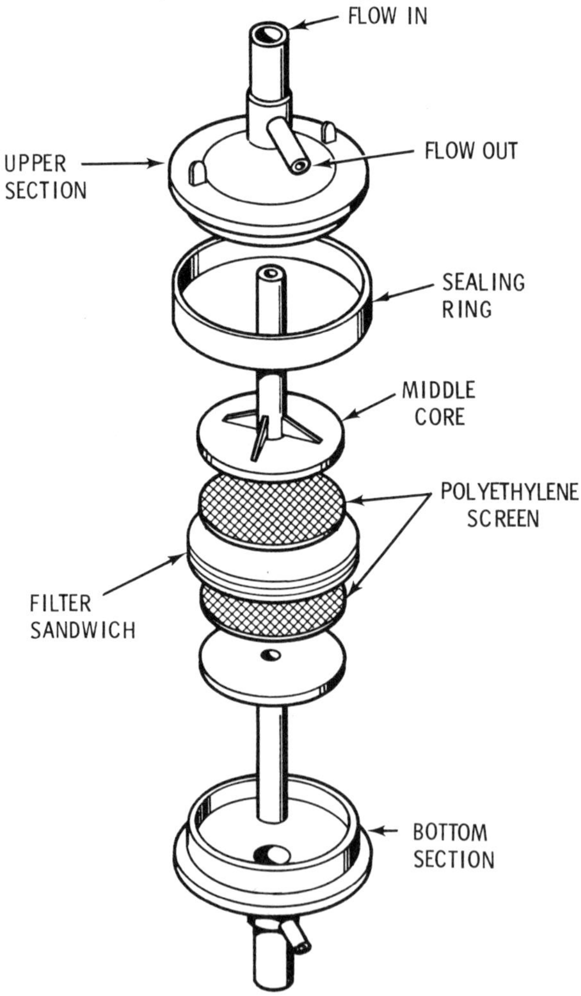

Figure 4. A miniature flowpast cell.

in the flowthrough system) for data analysis, flowrate and concentration as a function of time must be known if the total solvent is not collected and analyzed.

The radial flowpast system mentioned above and shown in Figure 3, A2 and A3, was developed recently by the first author to gain a more uniform fluid flow, such as that in commercially available ultrafiltration devices (e.g., the Pellicon system from Millipore®). The main goal was to design a unit in which the solvent would flow radially from the center towards the outer edge of the filter disk. In addition, other design constraints included were:

1. to collect solvent as soon as possible after it clears the edge of the filter surface disk;
2. to minimize the volume of fluid, both in the collection section and above each filter surface;
3. to prevent the wet filter from sealing against the smooth surface of the plate separating inlet and exit flow;
4. to allow the user to seal the filter holder unit permanently, or to clamp it shut so that it can be opened later for inspection; and
5. to be able to adjust the holder for either filter paper samples or "cakes" of samples for dissolution testing.

Figure 4 shows an exploded view of such a device; Figure 3 (A2 and A3) shows the application. In Figure 4, solvent enters through the middle core and passes radially between the filter sandwich surface and the bottom plate of the middle core by flowing within the woven mesh of a polyethylene screen. At the outer edge of the bottom plate the solvent flow turns back, traveling between the middle core and the upper section to the outlet line. Each 25- or 47-mm diameter unit can be made by injection molding (Meyer Plastic, 2001 Saint Street, Richland, WA 99352), using forms built and owned by Pacific Northwest Laboratory (PNL).

Allen et al. [14] have used these disposable flow cells in experiments on the dissolution characteristics of reactor fuel-sodium aerosols. They used the half-configuration indicated by A3, Figure 3 by substituting a polyethylene disk for the bottom filter and clamping the sandwich between both halves of the flow unit (A2, Figure 3). Solvent was thereby restricted to only one-half the holder. The volume of solvent contained in half the 25-mm flow cell is 1.1 ml, with an open volume of approximately 0.05 to 0.1 ml within the screen separating the bottom plate of the middle core and the outer surface of the filter sandwich. Allen et al. [14] found that the measured dissolution rate was independent of flow for flow equal to or greater than 0.78 ml/min. Besides being able to measure the dissolution rates of samples of mixed fuel-sodium aerosols, they were able to observe the rapid increase in the fraction dissolved when the solvent was switched from distilled water to a saturated carbonate-bicarbonate water mixture. Such experiments allowed them to conclude that sodium-uranium mixed oxide dissolved after air oxidation in the presence of carbonate in the solvent, becoming uranyl (VI) tricarbonate complex.

Kalkwarf [15] tested a different application of the 25-mm flowpast cell as part of his work on solubility classification of airborne products from uranium ores and tailings piles. The filter sandwich used in the half-configuration of the flow cell (A3, Figure 3) consisted of two 0.25-μm diameter pore size Millipore filters on either side of a Teflon® ring (0.085 cm thick, 2.3 cm inside diameter), containing approximately 0.35 cm^3 of ore. He compared the dissolution of uranium from yellow cake in lung fluid simulant [8], as measured in the flowpast or batch systems. At pH = 7.4 and 37°C, the batch system gave varying results, depending on whether the bottles were shaken (the first half-time of dissolution, $T_{1,1/2}$ = 0.6 days) or stirred internally ($T_{1,1/2}$ = 0.2 days). The first half-times of dissolution, using the flowpast cell, were 8.5 days or 5.0 days, depending on whether the solvent flow was 0.33 or 0.66 ml/min. Solvent flows higher than 0.7 ml/min were impractical in his experimental system, thus he did not use the flow cell for the majority of his work. The diffusion problems associated with using a packed sample disk that filled a Teflon ring might be reduced if the full-configuration flow cell (A2, Figure 3 and Figure 4) were used, even for total flow as low as 0.66 ml/min; this remains to be tested.

The dissolution of benzoic acid in PO_4^- and Cl^- buffers was recently measured in the first author's laboratory using both the static system and the 25-mm flowpast cell in half-configuration, with a sample ring in the filter sandwich (A2, Figure 3 and Figure 4) and flows of 3-5 ml/min. The dissolution half-times measured with either system were approximately the same, in spite of problems in continuously detecting the compound. Further work is planned on the use of the miniature flow cells in measuring the dissolution of soluble material and the influence of solvent changes on dissolution rates.

APPLICATION OF IN VITRO DISSOLUTION MEASUREMENTS

Two applications of the dissolution measurements previously discussed are given below: The dissolution of uranium and vanadium from ore particles in the first example, and the dissolution of yttrium-91 from fused-clay spheres in the second. Both examples deal with the dissolution of a trace material from a base particle, and both relate in vitro dissolution rates to in vivo observations.

Dissolution of Uranium and Vanadium from Uranium Ore Particles

Specific reports [6,16-19] describing in vivo clearance of two or more components of particles deposited in the lungs have posed questions that cannot be answered from animal data alone. In particular, Wilson [16] studied the clearance of uranium and vanadium from the lungs of rats in an effort to determine the synergistic effects of trace amounts of vanadium in the uranium ore. He saw no such effect, but reported that over the 30-day exposure period, the molar ratio of vanadium to uranium in the lungs doubled that observed for the original material, which he showed to be homogeneous with a 1:1 ratio in the respirable size range. Two possible explanations are: (1) uranium may dissolve from the vanadium-containing particles fixed in the lungs; or (2) vanadium may rapidly dissolve from the ore particles and become fixed in the lungs. The dissolution of uranium and vanadium from high-grade ore particles in simulated lung fluid was subsequently measured by Moss [6] to gain insight into the lung clearance observed by Wilson [16]. The experiment focused on measuring the early (less than 10-day) dissolution rate of both elements (U and V) contained within the base matrix of tyuyamunite particles, since this early time period promised the most information.

The measurements were made on aerodynamically size-separated particles [20,21], which were dispersed in a batch dissolution system (Figure 3). Measurements taken were: the buildup of uranium and vanadium in simulated lung fluid; the specific surface of the material before aerodynamic separation (used to normalize all the data); and projected area diameter measurements for selected aerodynamic diameter-size fractions, since this diameter was expected to be a good estimator of particle surface and volume.

Each size of ore fraction was placed in a bottle containing simulated lung fluid, without protein, at 37°C [8]. Samples of the uniformly mixed suspension were withdrawn by syringe and immediately filtered through 0.8- and 0.2-μm filters. The solvent was analyzed for uranium (by fluorometric analysis of fused NaF) and vanadium (by carbon-rod atomic absorption spectroscopy).

Data from the dissolution measurements showed that parabolic dissolution (Graphs G4 and G5, Figure 2) ended after the first four days. The U and V parabolic dissolution constants for the samples are shown in Table I. Surface area for all the aerodynamically size-separated particle fractions was assigned as that measured for the < 10-μm sieve fraction (5.0 ± 0.2 m^2/g), in order to normalize the data. If surface area varies as $1/D_p$ (projected area diameter), the dissolution constants, as calculated, should do likewise. The U and V dissolution constants, independent of particle size, then, equal the dissolution

Table I. Dissolution of Uranium and Vanadium from Tyuyamunite Ore Particles in Simulated Lung Fluid; T = 37°C

Aerodynamic Diameter (μm)	pH of Solvent	Dissolution During the First 4 Days[a] Parabolic Dissolution Constants[a] (μmoles extracted/m^2/$(day)^{1/2}$)	
		Uranium	Vanadium
1.3 ± 0.3	7.5	11.0 ± 3.6	29.6 ± 1.6
2.0 ± 0.2	7.5	16.3 ± 3.2	32.3 ± 2.1
2.3 ± 0.3	7.45	12.1 ± 0.7	26.6 ± 7.4
3.1 ± 0.3	7.4	7.3 ± 1.0	25.3 ± 1.8
3.7 ± 0.3	7.45	7.4 ± 0.7	25.2 ± 2.0
4.4 ± 0.4	7.4	5.2 ± 0.3	18.8 ± 0.3
5.4 ± 0.6	7.5	7.0 ± 1.0	18.5 ± 2.4
7.2 ± 1.2	7.2	4.4 ± 0.4	18.5 ± 0.8
Sieve Size 0 - 10 μm	7.7	5.7 ± 0.5	15.0 ± 1.8

[a]Surface area was assigned as that measured for the <10 μm sieve fraction (5.0 ± 0.2 m^2/g).

rate measured for the smallest-size fraction used in making the specific surface measurements: 15.0 ± 1.8 and 5.7 ± 0.5 μmol/m^2/$(day)^{1/2}$, respectively (Table I). Vanadium dissolves out of the particles during this early phase of dissolution 2.6 times faster than uranium. After four days, 80% of the vanadium and 30% of the uranium have been removed from the particles.

The higher dissolution rate of vanadium suggests that the vanadium was separating from the carnotite by dissolution, that it was precipitating, or becoming fixed in the lungs of the animals [6,22].

Yttrium-91 Dissolution from Clay Spheres

A study [23] on the dissolution of monodisperse samples of spherical particles in simulated lung fluids was initiated in 1970. Its purposes were to evaluate experimental techniques used in measuring particle dissolution rates and to compare in vitro and in vivo dissolution of fused-clay spheres. Measurements were made on the changes in dissolution rate constants as a function of particle size, solvent temperature and method of measurement.

Size-separated samples of ^{91}Y fused-clay spheres with GSD < 1.08 were placed in static, flowthrough and flowpast dissolution systems (Figure 3). The solvent [7] was maintained at either 23 or 37°C. The ^{91}Y fused-clay spheres were produced by heat-treating a nebulized clay suspension at 1100°C [24] and separated into monodisperse size groups in an aerosol centrifuge [25].

The dissolution rates of ^{91}Y from fused-clay spheres measured with the static system were lower than those measured with the flowpast system, which gave lower dissolution rates than those measured with the flowthrough system (Figure 3); higher temperatures increased dissolution rates.

The curves of the fraction of ^{91}Y dissolved vs time were identical to Graphs G4 and G5 in Figure 2; two distinct dissolution processes are shown. The initial phase of dissolution of ^{91}Y from fused-clay spheres follows parabolic dissolution, while the later phase follows linear dissolution of the clay-particle surface. The measured parabolic dissolution constant and the linear dissolution rate constant are given in Table II for each particle size.

Table II. Dissolution of ^{91}Y Fused-Clay Spheres in Simulated Lung Fluid; T = 23°C, pH = 7.3–7.4

Particle Diameter (μm)	Dissolution During First 2 Days (Parabolic Dissolution Constant Mass Fraction/(day)$^{1/2}$) x 10^{+4}	Dissolution After 2 Days (Linear Dissolution Rate Constant Mass Fraction/day) x 10^{+4}
1.42	5.3	3.2
1.28	6.7	3.9
0.83	–	4.5
0.70	15.8	6.6
0.56	36.3	7.7

Both the parabolic dissolution constant and the linear dissolution rate constant are inversely proportional to particle size. The dissolution constants should be independent of particle size when normalized for surface area, which can be done by dividing them by the specific surface, S_p ($S_p = 6/\rho D$, where D = diameter in cm and ρ = 2.25 g/cm^3 for the fused-clay spheres). The average values from columns 2 and 3, Table II are $(4.4 \pm 2.2) \times 10^{-8}$ g/cm^2/(day)$^{1/2}$ and $(1.7 \pm 0.2) \times 10^{-8}$ g/cm^2/day, respectively, for the parabolic dissolution constant and the linear dissolution rate constant. Approximately 0.16% of the ^{91}Y activity in 1.1-μm diameter particles dissolved during the first two days. The remaining activity dissolved at a constant rate having a dissolution half-time of 1213 days.

The calculations described above are for a monodisperse sample of 1.1-μm ^{91}Y fused-clay spheres at 23°C. At 37°C, the linear dissolution rate constant was measured as $(3.4 \pm 0.3) \times 10^{-8}$ g/cm^2/day, which gives a half-time dissolution of 606 days. A polydisperse distribution of clay spheres having a mass median diameter of 1.1 μm and a geometric standard deviation between 1.5 and 2.0, was reported to clear from the lungs with a half-time of 354 ± 100 days [23]. Such a polydisperse sample is predicted by Equation 9 to have a half-time dissolution of 727 days [$T_{1/2}$ = 0.6 (ρD/6K) for the clay spheres].

These last two half-times, the biological half-time, T_B = 354 days, and the dissolution half-time, T_D = 727 days, can be used to calculate the particle clearance half-time from Beagle dog lungs, $T_P = T_D T_B/(T_D - T_B) = 689$ days. This estimation of the particle clearance (transport) rate is normally not calculated and can be useful in evaluating toxic effects provided the dissolution taking place in the lungs is similar to that observed in the in vitro measurements.

In the above example for ^{91}Y in fused-clay spheres and in the example for uranium ore particles, information on the in vitro dissolution of the components of the particles was used to obtain valuable insight into the observed biological clearance of these materials. Similar observations of other inhaled dusts should lead to greater understanding of the effects of toxic substances on the lung.

ACKNOWLEDGMENTS

We would like to thank Tom Mercer at the University of Rochester, Department of Radiation Biology and Biophysics, for sharing so many of his ideas on particle dissolution with the first author and for advising on the uranium-vanadium dissolution work. We would also like to thank George Newton, Bruce Boecker and Richard Cuddihy at the Inhalation Toxicology Research Institute and Roy Thompson at Battelle for reviewing the document.

Portions of this work were performed in Richland, Washington, at Pacific Northwest Laboratory which is operated by Battelle for the Department of Energy under contract EY-76-C-06-1830; at the Inhalation Toxicology Research Institute, Albuquerque, New Mexico by Department of Energy Contract EY-76-C-04-1013; and at the University of Rochester, Department of Radiation Biology and Biophysics by a National Institutes of Health Toxicology Trainee Grant (01781) and Department of Energy contract with the office of Health and Environmental Research Project.

REFERENCES

1. Withey, R. J. "The Kinetics of Dissolution for a Non-Disintegrating Standard Substrate," *J. Pharm. Pharmacol.* 23:573-582 (1971).
2. Mercer, T. T. "On the Role of Particle Size in the Dissolution of Lung Burdens," *Health Physics* 13(11):1211-1221 (1967).
3. Noyes, A. A., and W. R. Whitney. "The Rate of Solution of Solid Substances in Their Own Solutions," *J. Am. Chem. Soc.* 19:930-934 (1897).

4. Luce, R. W., R. W. Bartlett, and G. A. Parks."Dissolution Kinetics of Magnesium Silicates," *Geochim. Cosmochim. Acta* 36:35-50 (1972).
5. Chien, Y. W., and E. P. K. Lau. "Controlled Drug Release from Polymeric Delivery Devices. IV: In Vitro–In Vivo Correlation of Subcutaneous Release of Norgestomet from Hydrophilic Implants," *J. Pharm. Sci.* 65(4):488-492 (1976).
6. Moss, O. R. "Dissolution of Uranium and Vanadium from Aerodynamically Size-Separated Ore Particles in a Simulated Lung Fluid," Ph.D. Thesis, The University of Rochester, Rochester, NY (1976).
7. Kanapilly, G. M., O. G. Raabe, C. H. T. Goh and R. A. Chimenti. "Measurement of In Vitro Dissolution of Aerosol Particles for Comparison to In Vivo Dissolution in the Lower Respiratory Tract after Inhalation," *Health Physics* 24(5):497-507 (1973).
8. Moss, O. R. "Simulants of Lung Interstitial Fluid," *Health Physics* 36(3):447-448 (1979).
9. Morrow, P. E., F. R. Gibb, H. Davis, and M. Fisher. "Dust Removal from the Lung Parenchyma: An Investigation of Clearance Simulants," *Toxicol. Appl. Pharmacol.* 12:372-396 (1968).
10. Steckel, L. M., and C. M. West. "Characterization of Y-12 Uranium Process Materials Correlated With In Vivo Experience," Y-1544-A, Health and Safety, Union Carbide Corporation, Nuclear Division, Oak Ridge Y-12 Plant, Tennessee, p. 56 (Springfield, VA: National Technical Information Service, 1966).
11. Kanapilly, G. M. "Alveolar Microenvironment and its Relationship to the Retention and Transport into Blood of Aerosols Deposited in the Alveoli," *Health Physics* 32(2):89-100 (1977).
12. Morrow, P. E. "Alveolar Clearance of Aerosols," *Arch. Intern. Med.* 131:101-108 (1973).
13. Cooke, N., and F. B. Holt. "The Solubility of Some Uranium Compounds in Simulated Lung Fluid," *Health Physics* 27(1):69-77 (1974).
14. Allen, M. D., J. K. Briant, O. R. Moss, E. J. Rossignol, D. D. Mahlum, L. G. Morgan, J. L. Ryan and R. P. Turcotte. "Dissolution Characteristics of LMFBR Fuel-Sodium Aerosols," *Health Physics* (in press).
15. Kalkwarf, D. R. "Solubility Classification of Airborne Products from Uranium Ores and Tailings Piles," U.S. Nuclear Regulatory Commission, NURG/CR-0530, PNL-2870 (Springfield, VA: National Technical Information Service, 1979), p. 59.
16. Wilson, H. B., H. E. Stokinger and G. E. Sylvester. "Acute Toxicity of Carnotite Ore Dust," *Arch. Ind. Hyg. Occup. Med.* 7(4):301-309 (1953).
17. Stuart, B. O., and T. M. Beasley. "Non-Equilibrium Tissue Distributions of Uranium and Thorium Following Inhalation of Uranium Ore by Rats," In: *Inhaled Particles and Vapors II*, C. N. Davies, Ed. (New York: Pergamon Press, 1971), pp. 291-298.
18. Rundo, J. "A Case of Accidental Inhalation of Irradiated Uranium," *Brit. J. Radiol.* 38(445):39-50 (1965).

19. Cuddihy, R. G., S. R. Gomez and R. C. Pfleger. "Inhalation Exposures of Beagle Dogs to Cerium Aerosols: Physical, Chemical and Mathematical Analysis," *Health Physics* 29(2):257-265 (1975).
20. Stober, W., and H. Flachsbart. "Size-Separating Precipitation of Aerosols in a Spinning Spiral Duct," *Environ. Sci. Technol.* 3(12):1280-1296 (1969).
21. Moss, O. R. "Prepared Comments During a Panel Discussion on Aerosol Centrifuges," Aerosol Measurement Workshop, 24-26 March 1976, University of Florida. In: *Aerosol Measurement*, D. A. Lungren, Ed. (Gainesville, FL: University of Florida Press, 1979), pp. 24-28.
22. Carlberg, J. R., J. V. Crable, L. P. Limtiaca, H. B. Norris, J. L. Holtz, P. Mauer and F. R. Wolowicz. "Total Dust, Coal, Free Silica, and Trace Metal Concentrations in Bituminous Coal Miners' Lungs," *Am. Ind. Hyg. Assoc. J.* 32(7):432-440 (1971).
23. Kanapilly, G. M., and C. H. T. Goh. "Dissolution of ^{91}Y Fused Clay and ^{90}Sr Fused Clay Particles," Fission Product Inhalation Program Annual Report, 1970-1971, LF-44. Lovelace Foundation. (Springfield, VA: National Technical Information Service, 1971).
24. Raabe, O. G., G. M. Kanapilly and G. J. Newton. "New Methods for the Generation of Aerosols of Insoluble Particles for Use in Inhalation Studies," In: *Proceedings of an International Symposium on the British Occupational Hygiene Association* (Surrey, England: Unwin Brothers, 1970), pp. 3-17.
25. Kotrappa, P., and O. R. Moss. "Production of Relatively Monodisperse Aerosols for Inhalation Experiments by Aerosol Centrifugation," *Health Physics* 21(4):531-535 (1971).

CHAPTER 5

RESPONSE OF THE RESPIRATORY TRACT TO INHALED POLLUTANTS

Robert F. Phalen and Peter Reischl

Air Pollution Health Effects Laboratory
Community and Environmental Medicine
University of California
Irvine, California 92717

Edward J. Faeder

Research and Development
Southern California Edison Company
Rosemead, California 91770

Finis L. Cavender

International Research and Development Corporation
Mattawan, Michigan 49071

INTRODUCTION

The respiratory system is a major area in which mammals contact their external environment. During a typical breath about 600 cm^3 of air enters the nose. A fraction of this fills the "deadspace"–about 150 cm^3 including the nose, trachea and bronchial tubes that have relatively thick walls and do not participate in oxygen and carbon dioxide exchange. The remainder, about 450 cm^3 in the moderately active adult enters the respiratory portion of the lung where it is diluted by the 2400 cm^3 of air that remained unexhaled from the previous breath. Each day approximately 10,000 liters of air is inspired by the average adult. Since the respiratory system is designed to efficiently deliver this air to 70 m^2 of delicate surfaces, the potential for disease from air pollutants is understandably great. The respiratory tract has varied and elaborate systems of defense, including rapid healing of epithelial injury. Although the list of agents that cause or promote lung disease is a long and growing one, these defense systems may explain why many of the changes seen in laboratory investigations after short-term inhalation exposure are transient. Such brief changes cannot be appropriately classified as disease but do show that the material is toxic enough to cause some injury and might lead to progressive damage and disease on repeated exposure.

Broadly, the lung can respond to pollutants in as many ways as its total number of properties and functions. Thus, if one can define and measure a property or function, this measurable factor can be made the biologic endpoint in an inhalation study. Furthermore, it appears that no matter what is selected for measurement, there will be a great number of inhalable materials that, in sufficient concentration and over a suitably long exposure time, will produce a statistically significant change. But until the role of the measured property is understood in terms of its relationship to respiratory tract disease, we do not know just how to treat the result. From a toxicologic point of view we have either identified a toxin or generated basic scientific information, but we may not have the ability to distinguish between the two. Although new scientific findings are of great, even fundamental, importance to all scientific disciplines, they generally require further development before they can be readily justified in toxicologic investigations. Although many of us–the authors of this chapter included–engage in basic studies on the respiratory system, we will direct our attention here to toxicologically meaningful perturbations of basic processes in the lung with emphasis on materials inhaled in polluted air.

Although the science of inhalation toxicology is still in an early stage of development, a great number of meaningful measures of injury to the respiratory tract have been developed. For simplicity these measures can be categorized into three broad types: anatomic, physiologic and biochemical. These categories represent the specific scientific disciplines that have primarily been applied to assessing toxic injury in the respiratory tract. Each discipline has its own way of viewing the respiratory system, its own tools and animal models and its own state of sophistication. For this reason, each of the three main areas of investigation will be presented separately, though in reality the lines of separation are often blurred.

TESTING STRATEGIES

The increase in public awareness of toxicological phenomena in environmental, industrial and occupational settings appears to have led to a desire to test all materials in every possible testing system. Such activity has helped to produce increased costs for these tests and has helped to point out the need for regulations and standards of performance of tests. One problem that results from this greater demand for toxicological testing involves attempts to select inexpensive or short-term tests over longer-term or traditional testing procedures. Another attempt to reduce the costs of testing involves the utilization of a single species for testing and making judgments as to the hazards of the test material in relation to the response of this single species. While it is true that certain tests, using a given species, can provide information concerning the relative hazard of two or more test materials, a single test using a single species is inadequate for toxicological characterization of any test material.

The respiratory tract will respond in one way or another to any airborne pollutant that is inhaled. A variety of toxicity tests, or specific endpoints, have been applied to detect and quantitate these responses. The eventual goal is to interpret for mankind the hazard of the inhaled pollutant, in view of the responses seen in one or more of these tests. Two items that plague an attempt to make precision measurements for statistical interpretation are biological variability and defense mechanisms. If measurements are so fine-tuned as to detect miniscule physiologic or biochemical excursions from control values, we are often confronted with the dilemma that such excursions might be considered normal in the next batch of animals tested. Even when we obtain a statistically significant result for a given material, lifetime exposures to this material may not indicate any toxicity or potential hazard. This may be due to our initial measurement of a pulmonary defense mechanism or other transient physiological or biochemical response that is not seen in longer-term exposures. Often "adaptation" or some other phenomenon is initiated to counter the response.

Alarie [1] has developed a simple rapid method for evaluating the degree of upper airway irritation in rodents resulting from exposures to gases and aerosols. The mechanism underlying the method is that irritant chemicals stimulate the free nerve endings in the nasal mucosa; and through a reflex pathway, mediated by the trigeminal nerve, the respiratory rate is depressed as a function of irritant concentration [2]. The test involves four female mice per exposure group and has been of considerable value in evaluating the irritancy of personal and household aerosol products. Perhaps the greatest value of the test comes in discriminating between two or more formulations and when used in this way, i.e., the ranking of upper airway irritants, it is a very useful tool. For test materials that are not upper airway irritants, including many toxic dusts, this test alone is not adequate.

The irritant response in guiena pigs has been characterized by Amdur and associates [3, 4]. The respiratory system of the guinea pig is capable of dramatic bronchoconstriction in response to inhaled pollutants. Bronchoconstriction results in increased pulmonary resistance and in decreased pulmonary compliance. Slight changes in pulmonary resistance and compliance can be measured by placing the guinea pig in a plethysmograph. Results from this testing system are highly significant if one uses each guinea pig as his own control. However, it is quite difficult using this test to get statistically significant results by randomly choosing a group of control animals to be tested against an identical exposure group. Also, it is possible to obtain statistically significant results from otherwise nontoxic aerosols. Finally, long-term exposures to pollutants that give positive results in the plethysmograph system have, at times, failed to show any toxicity in either rats or guinea pigs, even in lifetime studies at levels 10- to 100-fold higher than those detected in the plethysmograph system [5, 6]. Thus, neither this test system alone nor the use of a single species is appropriate as the basis for regulation of the test material.

The changing of the breathing rate and/or bronchoconstriction are physiological responses that serve to reduce exposure to inhaled pollutants by limiting their penetration into the lungs. Other defense mechanisms can be activated once pollutants deposit in the deep lung. Particulates that deposit in the gas exchange region of the lung are taken up by macrophages in the lung which have been recruited by these deposition stimuli. Particle-laden macrophages are then swept up the mucociliary escalator or are transported via the lymphatics to bronchus-associated lymphoid tissue. When the particulate burden is greater than the recruitment process can handle, granulomas form in the lung and lymphoid tissue [7, 8]. The transient recruitment of neutrophils and macrophages to the lung may not indicate a toxic response per se. It is only when recruitment cannot handle the magnitude of the insult or repeated insults or when the macrophages cannot ingest the entire particle, e.g., asbestos fibers, that toxic effects are seen, and only then after a long latent period or after years of repeated insult.

Interestingly, recruitment stimulus and the ability to clear the recruited cells from the lung varies with each species. In granuloma formation in the rat, more neutrophils (3:1) than macrophages are present in the fluid that can be lavaged, or rinsed from the lungs, while, in guinea pigs, more macrophages (4:1) than neutrophils are obtained. This may relate to the inflammatory response seen in oxygen toxicity in these species. When activated in the lung neutrophils produce superoxide, O_2^-, perhaps as an antimicrobial agent. The presence of increased levels of superoxide induces the enzyme, superoxide dismutase in lung tissue [9]. A similar response occurs when adult rats are placed in 85% oxygen for five days. Once the induction of superoxide dismutase occurs, these rats can survive in 100% oxygen. On the other hand, guinea pigs seem unable to induce the enzyme superoxide dismutase in lung tissue and are unable to survive in 85% or 100% oxygen. Perhaps, the recruitment of much larger numbers of macrophages over neutrophils is a teleological explanation of why guinea pigs are sensitive to high oxygen concentrations.

They do not have the need to induce superoxide dismutase since relatively few neutrophils are recruited to the lung. Instead, they rely on macrophages to clear the lung of deposited materials.

The traditional method of histopathological examination of target organs does not always tell us which pollutants are toxic. For instance, it is possible to see histopathological evidence of toxicity that is, in actuality, the result of "super" stimulation of bronchoconstriction. The guinea pig bronchi are capable of constriction to the point of asphyxiation. Laryngeal or bronchial spasm can completely eliminate the supply of oxygen resulting in histopathological manifestations in lung tissue. Thus, using the guinea pig we may identify the potential hazards of a given pollutant but we cannot determine if these hazards are due to direct toxic effects or to "superstimulated" defense mechanisms.

The conclusion is that multiple tests with multiple species are necessary for hazard evaluation of airborne pollutants. It is not likely we will ever eliminate the hazards of airborne pollutants, but through proper evaluation, we can reduce or minimize the health hazards in the environment and the occupational workplace.

ANATOMICAL CONSIDERATIONS

Because the respiratory system is not one organ, but rather an organ system, it is useful to simplify by identifying major compartments, i.e., regions that exhibit some internal anatomic similarity. The compartmental scheme proposed by the Task Group on Lung Dynamics of the International Commission on Radiologic Protection (ICRP) that was devised in order to describe the deposition and retention of inhaled particles is a very useful one in inhalation toxicology [10].

The model divides the respiratory tract into three regions based on anatomical features and deposition and clearance phenomena. The regions, called the nasopharynx (N), the tracheobronchial region (TB) and the pulmonary or parenchymal region (P) also are to some extent characterized by their unique responses to air pollutants.

The nasopharynx begins at the anterior nares and includes the respiratory airway down to the level of the larynx, i.e., "voice box." The nasal cavity, which is roughly triangular in cross section, contains a central septum and turbinates that form narrow channels through which air flows. The forward portion of the nose is lined by cells not unlike those of the body skin. The remainder being mostly lined with ciliated epithelium. The cilia, i.e., small numerous motile hairlike projections, produce movement of overlying mucus rearward to a point where it is swallowed. The most common disease state, rhinitis, is characterized by inflammation; nasal membranes are swollen and excessive secretion or excessive dryness are both possible states. Rhinitis, a term which includes the common cold, is commonly associated with infection by inhaled viruses, bacteria or fungi. Rhinitis can also be produced by allergens and by irritating gases such as chlorine or formalin vapor or by prolonged exposure to dusts like those which occur in milling and stone cutting [11]. Rhinitis often precedes the formation of benign growths or polyps. Other responses of the nasopharynx include ulceration, which can be produced by very irritating gases or particles. Cancer is seen in nickel refinery workers who chronically inhale nickel carbonyl [12]. Although inhalation toxicologic studies often neglect tissue changes in the NP region, the potential for disease in this compartment is real and should not be overlooked.

The tracheobronchial region begins at the larynx and includes the trachea and the 16 to 20 generations of cilia-covered bronchial airways down to and including the terminal bronchioles. Interspersed in the ciliated cells are goblet cells that, along with mucous glands deep in the bronchial walls, secrete the mucus. Various mammalian species differ in the number and location of mucous glands. Muscle bundles that control airway caliber are present throughout the bronchial tree [13].

The responses of the tracheobronchial region to inhaled materials are varied and include excessive constriction of muscles leading to increased resistance to airflow, hypersecretion of mucus, swelling or edema of bronchial walls, bronchial infection and cancer.

Bronchoconstriction can be induced in sensitive asthmatics and in normal human or animal subjects by a great variety of inhaled substances. The most potent are usually organic materials that first establish a state of allergic sensitization and later induce an allergic response; typical agents include gases or vapors and dusts. Representative causative agents are pollens, mold spores, animal danders, feathers, insecticides, fabric lint and grain dust. Direct stimulation of bronchoconstriction in nonsensitized subjects occurs with sulfur dioxide, ammonia, cigarette smoke, sulfuric acid mists, a number of sulfate salts and so-called "inert dusts" [14-17]. The severity of bronchospasm can range from mild—detectable only be sensitive pulmonary function tests—to life-threatening bronchospasm where death due to asphyxiation may occur as in status asthmaticus.

Bronchitis with inflammation and excessive secretion of thick tenacious mucus that may or may not be effectively removed by cough can be produced by many agents. Most commonly viruses or mycoplasma, and less commonly bacteria are causative factors with common air-pollutants probably playing a secondary, potentiating role [14]. In sufficient concentrations, sulfur dioxide, chlorine and nitrogen dioxide can lead to this condition [14]. In the previously healthy person bronchitis is not commonly fatal; cough and treatment with antibiotics usually results in complete resolution.

A serious, often fatal form, *bronchiolitis obliterans* (silo-fillers' disease). progresses to a condition where nodules of scar tissue completely occlude the bronchial tubes. Most commonly, inhalation of nitrogen dioxide from nitric acid spills, burning of nitrocellulose or working in silos is the precipitating factor. This progressive fatal disease is characterized by shortness of breath, unremitting cough and cyanosis, i.e., poor oxygenation of tissues [14].

The most common tumor of the respiratory tract of man, bronchiogenic (usually squamous) carcinoma, arises from the cells lining the tracheobronchial tree [18]. Such growths usually occur in the central bronchi involving the mainstem (lobar) bronchus or its largest branches and upon enlargement completely obstruct bronchi and lead to collapse and infection of tissues supplied by the affected airway. Although as long as 20 years may lapse between exposure to a causative agent and the development of disease, bronchial carcinoma, once established, invades extensively and can produce metastases, i.e., daughter growths, throughout the body [18]. The prime causative agent is tobacco smoke, but asbestos, nickel, chromium, arsenic, beryllium, coal tar, radioactive particles, vinyl chloride and polyurethane are among the airborne agents that appear to produce the disease in man and experimental animals [14, 18, 19]. Sulfur dioxide appears to act as an irritant cofactor in the production of experimental squamous carcinoma by benzo[a] pyrene in rat lung [19].

A less common class of bronchial tumors of man, the bronchial adenomas and adenocarcinomas, appear to derive from glands of the bronchi rather than epithelia of central bronchi per se [18]. The causative agents of this form are less certain than those of bronchial carcinoma [19]. There are also a number of rare forms of bronchial tumors, including lipomas, papillomas, cystadenomas and melanomas [19].

The third compartment, the pulmonary or parenchymal region represents the functional gas-exchange sites of the lung. It includes respiratory bronchioles–small bronchial tubes with some air-sacs, i.e., alveoli, opening into their lumens, alveolar ducts–ducts whose walls are composed totally of alveoli and alveolar sacs, i.e., terminating alveolar ducts. The basic unit of this compartment, the alveolus is a thin-walled polyhedral sac with one face open to the air. The interior walls are covered by flat,thin pulmonary epithelial cells. A dense network of small capillary blood vessels intimately surrounds the alveoli where exchange of oxygen and carbon dioxide between the air and blood occurs. Alveoli in man are about 300 micrometers in diameter, being much larger than the inhaled particles that penetrate to the deep lung. In addition to the alveolar epithelial (Type I) cells, alveolar macrophages (mobile cells that engulf and digest foreign bodies) and alveolar septal (Type II) cells (which secrete surfactant) are also present. Alveoli are lined by a thin layer of surfactant that prevents their collapse [20].

Responses of the parenchymal region that are commonly induced by inhaled substances include inflammation, edema, fibrosis and cancer. The term "pneumoconioses" refers to parenchymal diseases caused by dusts. It is perhaps best characterized by a failure of the self-cleansing mechanisms of the lung that results in the development of specific responses depending on the type of material that is deposited and retained [21]. As the variety and number of materials that produce parenchymal disease are large, no attempt will be made to completely explore this topic.

Gases that are relatively insoluble in water are not well absorbed in the NP and TB regions and penetrate into the deep lung. Ozone, or triatomic oxygen, a component of photochemical air pollution, is a good example of a gas that is sufficiently insoluble to pass beyond the NP and TB compartments to produce damage in the lung parenchyma. In concentrations of about 1 part per million (ppm), ozone has many effects, including killing of pulmonary alveolar epithelial cells and damaging alveolar capillary endothelium which produces hemmorhage and edema, i.e., filling of airspaces and interstitial tissues with plasma ultrafiltrate [22]. Even in concentrations below 1 ppm, ozone destroys cilia, interferes with the ability of phagocytic cells to inactivate microorganisms, produces edema with thickening of alveolar walls and causes various biochemical and breathing-pattern changes [22]. A major feature of this injury pattern is that it points to the vulnerability of the thin alveolar epithelial cell and the closely neighboring capillary endothelial cells.

Tissue destruction followed by inflammatory changes in the pulmonary region is produced by the very irritant gases ammonia, chlorine and phosgene. Inhalation of these types of gases can lead to acute pulmonary edema, a condition that may resolve or in severe cases may progress to fibrosis of the deep lung. Similar responses have been caused by the inhalation of a variety of dusts including cadmium, osmium tetroxide, bauxite fumes, beryllium and a variety of organic materials including molds from hay, sugar cane, sawdust, bark and cheese [18]. In many of the organic dust cases a hypersensitivity reaction apparently plays a role in the lung response which is described as extrinsic allergic alveolitis. In such reactions an allergic response involving an antibody against the toxin, that appears after repeated exposure, participates in the formation of products that injure cells of the lung [18].

Several dusts lead, on chronic inhalation, to fibrosis without an apparent inflammatory phase. Silicon dioxide, asbestos, aluminum and possibly talc are notable examples of important fibrogenic dusts [14, 18].

The epithelial cells of the bronchiolar-alveolar zone also give rise to neoplasms in man and animals. Alveolar adenomas occur spontaneously in certain mouse strains and can be induced by many agents of which some but not all are chemical carcinogens. Bronchoalveolar adenomas are relatively rare in humans, probably equally prevalent in both sexes and not apparently related to smoking or occupation [18].

The foregoing has served to illustrate, by example, the variety of anatomical responses of the lung; it is by no means a complete treatment in any sense of the term.

Because the respiratory system is composed of a large and varied set of tissues and organs, one must devise careful morphologic sampling techniques in inhalation studies. The problem of obtaining an adequate number of representative samples from tissues that have been properly preserved is discussed in a review by Dungworth et al. [23].

PHYSIOLOGIC CONSIDERATIONS

Physiologic measures of lung injury are important because they assess life-sustaining functions. The measures are largely nondestructive and many are therefore applicable to both human and laboratory animals. A major challenge in assessing physiological lung injury results from the substantial functional reserve that is present in the respiratory system. This functional reserve permits the slightly injured lung to often appear near normal in clinical tests.

Gas exchange of the lung is one example which shows a large functional reserve. This prime function of the lung is to allow oxygen to move from the ambient air into the venous blood and carbon dioxide to move out. During rest, gas exchange across the lung is at a rate only 1/10 of that possible during exercise. Should any toxic agent impede gas exchange, then the lung reserve capacity is often more than adequate to supply the needed oxygen and remove the metabolic CO_2 during rest. Only after the reserve is exhausted does one see dramatic changes of arterial O_2 and CO_2.

Surely, one must seek measurements which show clinical degradation before such reserve capacity is exhausted. One way is to expend the reserve capacity through exercise while making measurements that detect a decrement in the lung function. Such an approach is used in several laboratories [22, 24], but is complicated in that measurements must be made in a moving experimental preparation.

A somewhat different approach is to look at the respiratory mechanical function of the lung, which reflects the cellular and tissue properties of the airway. If these properties are modified by experimental intervention, then the mechanical function of the lung may be altered.

The mechanical function of the lung is commonly described by pulmonary compliance and resistance. Compliance is the index of the functional stiffness of the lung; compliance will decrease if the lung becomes stiffer through such mechanisms as constriction of alveolar duct smooth muscle, cellular infiltration or alveolar edema [25]. Airway closure, which is thought to occur in small airways, will also make the lung functionally stiffer by reducing the amount of parenchyma available to accept inspired gas.

Pulmonary compliance is the ratio of the pressure differences between the mouth and the pleural space and lung volume. Pulmonary compliance may be determined from the shape of a pressure-volume curve near the lung volume of interest at zero airflow.

Lung resistance is more difficult to measure than lung compliance. It is the relationship between pulmonary airflow and transpulmonary pressure and is complicated in that this transpulmonary pressure not only must overcome the resistance to airflow but compliance of the lung as well. To compute resistance, transpulmonary pressure and airflow are measured once during inspiration and again at the same lung volume during expiration.

Resistance to airflow is influenced by airway size, flow direction and lung tissue resistance. Although all patent airways have an influence on pulmonary resistance, the large airways account for the majority of the resistance to airflow because the aggregate cross section of small airways is much greater. The caliber of the upper and central airways can be altered by changing bronchomotor tone, by bronchial edema, or by the presence of increased amounts of mucus. Low doses of 0.1-1 ppm ozone have been shown to increase airway resistance in humans exposed for one hour over 75 days [26]. Resistance has been used in toxicological studies as an index of pulmonary response.

The frequency of respiration is under central nervous system control but may be influenced by mechanical factors such as resistance and compliance, or respiratory reflexes. Depending on the nature of the interaction between a given irritant and the central nervous system, a given stimulus may increase or decrease respiratory frequency [27].

Tests which measure the distribution of inspired gas to the alveoli require the use of gases such as N_2, H_2 or helium, which are so poorly soluble that they do not pass in great quantity from the alveolar gas into the blood or pulmonary tissue. The simplest of these tests, the "pulmonary N_2 emptying rate," requires only that the subject breathe O_2, usually for about 7 minutes. If the inspired O_2 is distributed evenly to all alveoli during the 7-min period, the N_2 will be washed out evenly and the final alveolar gas sample will contain less than 2.5% N_2. More detailed analysis may involve use of the entire washout curve, as by exponential analysis or moment analysis [28]. However, if some areas are markedly hypoventilated during normal breathing, they will still have a high N_2 concentration. When there is uneven ventilation in the lung, there may be different washout rates in different parts of the lung.

The cause of uneven ventilation may be a regional reduction of the number of elastic fibers in the lung such as occurs in advanced pulmonary emphysema; regional obstruction such as occurs in asthma; or regional changes in expansion caused by fluid or exudate in the alveoli, pulmonary congestion or restrictive disease.

BIOCHEMICAL CONSIDERATIONS

Homeostatic mechanisms within an organism responsible for maintaining its integrity and well-being rely upon complex networks of biochemical reactions for rapid response to pollutant exposures. Thus, biochemical response to different classes of pollutants might be expected to be general and nonspecific. Recent reviews have dealt at length with this subject [29, 30]. We will consider here only general highlights of biochemical response to sulfur oxide and oxidant pollutants.

Sulfur Oxides

Sulfur dioxide is a highly soluble gas capable of rapid equilibration in an aqueous environment, producing sulfite:

$$SO_2 + H_2O \rightleftharpoons SO_3^{2-} + 2H^+ \qquad (1)$$

Sulfite, in turn can react with free protein (R-SS-R) or amino acid disulfide groups, Reaction 2, and by exchange lytic processes produce S-sulfonate compounds capable of circulating in blood [31-33]:

$$\text{R-S-S-R}' + SO_3^{2-} \rightleftharpoons \text{R-S-}SO_3^- + \text{R}'\text{-S} \qquad (2)$$

These circulating S-sulfonates have been measured in rabbit and dog plasma after exposure to high (10-25 ppm) levels of sulfur dioxide [31-33]. Sulfite can also be oxidized enzymatically to sulfate. This process, catalyzed by the enzyme sulfite oxidase can occur in many tissues, including lung, liver and kidney. It has been calculated [34] that even 100-500 times the maximum estimated average daily sulfite intake of humans could not saturate the capacity for oxidation to sulfate and excretion of that end product. More recent studies [35] point out the lack of correlation in different species between endogenous plasma S-sulfonate levels and sulfite oxidase activity, leaving the conclusion that other factors may be important in the partitioning between the competing reactions to form S-sulfonates and sulfates. However, within a single species, S-sulfonate formation correlates with sulfite dose [36] and may still serve as a useful indicator of exposure. Similar detailed studies need to be performed with inhalation exposures to sulfur dioxide, since much of the previous work has relied upon sulfate supplied in the diet, or by injection.

In vitro guinea pig lung slice exposure to various sulfate salts [37] has indicated that associated ammonium cations can cause histamine release. The physiological significance of the observation for humans has been questioned [38] in light of the high in vivo ammonia levels measured in exhaled breath [39].

Ozone and Nitrogen Dioxide

Ozone and nitrogen dioxide, strong oxidizing agents, are both believed to be harmful to biological tissue because of that particular property. And because of the physiological similarity of lung lesions produced by exposure to these gases and those caused by ionizing radiation, radiomimetic properties are ascribed to them [30].

Ozone has been studied both in vitro and in vivo. When exposure of organs, tissue homogenates or pieces occurs, lipid peroxidation, sulfhydryl group oxidation and loss of enzymatic activity can all occur [29]. However, these same changes, although postulated to occur in many models of ozone action, have not always been observed in the whole animal exposures to the gas. Evidence for malonaldehyde production (a product of lipid peroxidation) in lung tissue has been reported by some groups, but not found by others [29].

Nitrogen dioxide, more soluble than ozone and less reactive, can penetrate more deeply into the lung and react with endothelial tissue. On the basis of deep lung injury and a hypothesis expecting "neurohumoral factor" release [40], experiments using continuous low level exposure of guinea pigs to nitrogen dioxide were conducted [41]. Increases in urinary protein excretion were noted in exposed animals when compared to controls. Quantitating the changes or ascribing them to kidney damage response to released chemical factors from lung was not possible.

Since the lung is not only the central portal by which oxygen enters and carbon dioxide leaves animals, but also an important biosynthetic organ providing a wide range of essential "chemical services," it has been logical to study the impact of acute and chronic oxidant gas exposure on enzyme activities and metabolite levels. Details of these studies have been recently reviewed [29] and will be generally summarized here. Glucose metabolism, lipid metabolism, nucleic acid and protein synthesis, all essential for highly metabolically active tissue, all show significant decreases when lung homogenates or lung slices from animals are exposed to high levels of either nitrogen dioxide (greater than 5-10 ppm) or ozone (greater than 1-2 ppm) for short periods of time. However, in animals exposed to lower levels of oxidant gas for longer periods of time, an augmentation in enzyme activities and metabolite levels was observed. This suggests some type of repair and/or adaptive processes occurring in the chronically exposed animals where massive initial tissue destruction had not occurred. Other in vitro testing in acutely and chronically exposed rats has shown similar results for hexose monophosphate shunt enzymes (key in producing NADPH, an essential intracellular reducing agent), monamine oxidase and glutathione.

Also, it has been shown that the nutritional status of an animal is important in reducing susceptibility to these oxidant-induced biochemical changes. Both vitamin C and vitamin E can protect against lethality from ozone and nitrogen dioxide. The exact extent and mechanism of this protection is not yet known.

Finally, although early indication of oxidant change can be observed in biochemical changes, specific initial responses need still be observed and understood.

ACKNOWLEDGMENTS

Lisa Lieberman provided valuable editorial and typing services in the preparation of this chapter.

REFERENCES

1. Alarie, Y. "Irritating Properties of Airborne Materials to the Upper Respiratory Tract," *Arch. Environ. Health* 13:433-449 (1966).
2. Ulrich, C. E., M. P. Haddock and Y. Alarie. "Airborne Chemical Irritants, Role of the Trigeminal Nerve," *Arch. Environ. Health* 24:37-42 (1972).
3. Amdur, M. O., R. Z. Schulz and P. Drinker. "Toxicity of Sulfuric Acid Mist to Guinea Pigs," *Arch. Ind. Hyg.* 5:318-326 (1952).
4. Amdur, M. O. "The Respiratory Response of Guinea Pigs to Sulfuric Acid Mist," *Arch. Ind. Health* 18:407-414 (1958).
5. Cavender, F. L., H. W. Sugg, B. Y. Cockrell and W. M. Busey. "Health Effects of Sulfuric Acid Mist," presented at the 71st Annual Meeting of the Air Pollution Control Association, Houston, TX, June, 1978.
6. Cavender, F. L. "Respiratory Defense Mechanisms and Inhalation Toxicology," presented at the Eighth Conference on Environmental Toxicology, Dayton, OH, October, 1977.
7. Steinhagen, W. H., F. L. Cavender and B. Y. Cockrell. "Respiratory Tract Lesions in Guinea Pigs Exposed to Sulfuric Acid Mist," *J. Environ. Path. and Tox.* 1:267-277 (1978).
8. Cavender, F. L., J. Campbell and B. Y. Cockrell. "Functional Changes in the Free-Cell Population Lavaged from Lungs of Rats and Guinea Pigs During Chronic Inhalation Exposures," presented at EPA Symposium, Williamsburg, VA, February, 1978.
9. McCord, J. and J. Crapo. Personal communication.

10. Task Group on Lung Dynamics. "Deposition and Retention Models for Internal Dosimetry of the Human Respiratory Tract," *Health Physics* 12:173-207 (1966).
11. Bell, E. T. In: *A Textbook of Pathology*, (Philadelphia, PA: Lea and Febiger, 1956), pp. 551-554.
12. Nelson, N. "Inhalation Carcinogenesis in Man: Environmental and Occupational Hazards," in *Inhalation Carcinogenesis*, M. G. Hanna, Jr., P. Nettesheim and J. R. Gilbert, Eds. (Oak Ridge, TN: U.S. Atomic Energy Commission, 1970), pp. 3-12.
13. Bloom, W., and D. W. Fawcett. *A Textbook of Histology* (Philadelphia: W. B. Saunders Company, 1968), pp. 629-651.
14. Beeson, P. B., and W. McDermott. *Textbook of Medicine* (Philadelphia: W. B. Saunders Company, 1967), pp. 492-551.
15. Altman, P. L., and D. S. Dittmer, Eds. *Environmental Biology* (Bethesda, MD: Federation of American Societies for Experimental Biology, 1966), pp. 269-310.
16. Alarie, Y. "Sensory Irritation by Airborne Chemicals," *Crit. Rev. Toxicol.* 2:299-363 (1973).
17. Amdur, M. O., J. Bayles, V. Ugro and D. W. Underhill. "Comparative Irritant Potency of Sulfate Salts," *Environ. Res.* 16:1-8 (1978).
18. Crofton, J. W., and A. Douglas. *Respiratory Diseases* (Oxford: Blackwell Scientific Publications, 1975), pp. 562-611.
19. Laskin, S., M. Kuschner, A. Sellakumar and G. V. Katz. "Combined Carcinogen-Irritant Animal Inhalation Studies," in *Air Pollution and the Lung*, E. F. Aharonson, A. Ben-David and M. A. Klingberg, Eds. (New York: John Wiley and Sons, Inc., 1976), pp. 190-218.
20. Macklin, C. C. "The Pulmonary Alveolar Mucoid Film and the Pneumocytes," *Lancet* 266:1099 (1954).
21. Committee on Pneumoconioses. "The Pneumoconioses," *Arch. Environ. Health* 7:130-171 (1963).
22. Committee on Medical and Biologic Effects of Environmental Pollutants. *Ozone and Other Photochemical Oxidants* (Washington, DC: National Academy of Sciences, 1977).
23. Dungworth, D. L., L. W. Schwartz, W. S. Tyler and R. F. Phalen. "Morphological Methods for Evaluation of Pulmonary Toxicity in Animals," *Ann. Rev. Pharm. Tox.* 16:381-399 (1976).
24. Folinsbee, L. J., F. Silverman and R. J. Shephard. "Exercise Response Following Ozone Exposure," *J. Appl. Physiol.* 38:996-1001 (1975).
25. Drazen, J. M. "Physiologic Basis and Interpretation of Common Indices of Respiratory Mechanical Function," *Environ. Health Pers.* 16:11-16 (1976).
26. Goldsmith, J. R., and J. A. Nadel. "Experimental Exposure of Human Subjects to Ozone," *J. Air Poll. Control Assoc.* 19:329-330 (1969).
27. Alarie, Y. "Sensory Irritation by Airborne Chemicals," *Crit. Rev. Tox.* 2:299-263 (1973).
28. Saidel, G. M., R. B. Salmon and E. H. Chester. "Moment Analysis of Multibreath Lung Washout," *J. Appl. Physiol.* 38:328-334 (1975).
29. Mustafa, M. G., and D. F. Tierney. "Biochemical and Metabolic Changes in the Lung with Oxygen, Ozone and Nitrogen Dioxide Toxicity," *Amer. Rev. Respir. Dis.* 118:1061 (1978).

30. Coffin, D. L. and H. E. Stokinger. "Biochemical Effects of Air Pollutnats," in *Air Pollution*, A. C. Stern, Ed. (New York: Academic Press, Inc., 1977), Vol. 2, pp. 232-360.
31. Gunnison, A. F., and A. W. Benton. "Sulfur Dioxide: Sulfite, Interaction with Mammalian Serum and Plasma," *Arch. Environ. Health* 22:381 (1971).
32. Gunnison, A. F., and E. D. Palmes. "A Model for Metabolism of Sulfite in Mammals," *Toxicol. Appl. Pharmacol.* 38:111 (1976).
33. Gunnison, A. F., and E. D. Palmes. "Persistence of Plasma S-Sulfonates Following Exposure of Rabbits to Sulfite and Sulfur Dioxide, Biological Origin and Metabolism of SO_2," *Toxicol. Appl. Pharamacol.* 24:206 (1973).
34. Rajagopalan, K. V., and J. L. Johnson. In: *Biochemical Effects of Environmental Pollutants*, S. D. Lee, Ed. (Ann Arbor, MI: Ann Arbor Science Publishers, Inc., 1977), pp. 307-314.
35. Gunnison, A. F., and E. D. Palmes. "Species Variability in Plasma S-Sulfonate Levels During and After Sulfite Administration," *Chem.-Biol. Interact.* 21:315 (1978).
36. Gunnison, A. F., C. A. Bresnahan and E. D. Palmes. "Comparative Sulfite Metabolism in the Rat, Rabbit and Rhesus Monkey," *Toxicol. Appl. Pharmacol.* 42:99 (1977).
37. Charles, J. M., and D. B. Menzel. "Ammonium and Sulfate Ion Release of Histamine from Lung Fragments," *Arch. Environ. Health* 30:314 (1975).
38. Faeder, E. J., and A. V. Colucci. "The Effects of Sulfate Aerosols on Human Health," Annual Meeting American Society of Civil Engineers, Chicago, Illinois, 1978.
39. Larson, T. V., D. S. Covert, R. Frank and R. J. Charlsen. "Ammonia in the Human Airways: Neutralization of Inspired Acid Sulfate Aerosols," *Science* 197:161 (1977).
40. Fairchild, E. J. "Nuralhumoral Factors in Injury from Inhaled Irritants," *Arch. Environ. Health* 6:79 (1963).
41. Sherwin, R. P., and L. J. Layfield. "Proteinuria in Guinea Pigs Exposed to 0.5 ppm Nitrogen Dioxide," *Arch. Environ. Health* 26:336 (1974).

CHAPTER 6

PARTICLE CHARGE EFFECTS

Walter John

Air and Industrial Hygiene Laboratory
California Department of Health Services
Berkeley, California 94704

INTRODUCTION

Particles in ambient air come into charge equilibrium with ions created by natural radioactivity and cosmic rays. Consequently, the particles carry very little charge. However, in health effects studies, artificially generated aerosols are used within a short time after generation. These particles can carry appreciable charge which will increase particle deposition in the respiratory tract, introducing an uncontrolled variable into the experiments.

This chapter will discuss the effect of electrical charge on deposition in the respiratory tract, charging phenomena in aerosol generation and methods for measuring and neutralizing particle charge.

DEPOSITION OF CHARGED PARTICLES IN THE RESPIRATORY TRACT

Physics of Charged Particle Deposition

There are two forces which cause charged particles to deposit on surfaces, one due to the image charge in the surface and one due to the mutual repulsion between like charges. In Figure 1, a particle with charge q, located at distance x from a conducting surface, is attracted to the surface as though there were an image charge as shown. The velocity of the particle towards the surface is given by

$$v = \frac{Bq^2}{16\pi\epsilon_o x^2} \tag{1}$$

where B is the mobility
ϵ_o is the permittivity of free space (rationalized MKS units are used here)

The time to reach the surface is given by:

$$t = \frac{16\pi\epsilon_o x^3}{3Bq^2} \tag{2}$$

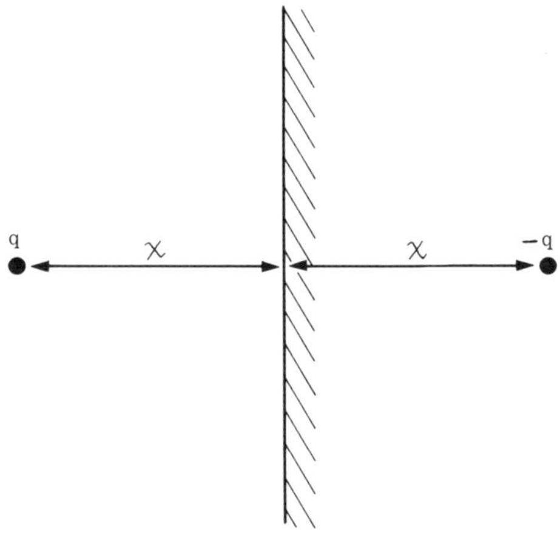

Figure 1. Image of a charge in a conducting surface.

Equations 1 and 2 show that the image force is most effective near the surface. Image charge deposition is independent of the sign of the charge and of the particle concentration; it is a single particle effect. It depends on the size and shape of the container. Yu [1] has derived expressions (see comments and corrections in references 2 and 3) for the image charge deposition efficiency P_I for slug flow in a circular tube:

$$P_I = 1 - \left(\frac{r_c}{R}\right)^2 \tag{3}$$

where r_c is the critical radius outside which all particles are deposited
R is the tube radius

The critical radius r_c is related to the dimensionless residence time τ by:

$$\tau = 4\left(\frac{R}{r_c} + 2\ln\frac{r_c}{R} - \frac{r_c}{R}\right) \tag{4}$$

with

$$\tau = \frac{Bq^2L}{4\pi\epsilon_o R^3 U_o} \qquad (5)$$

where L is the tube length
U_o is the flow velocity

If the wall material is a dielectric rather than a conductor, the image charge -q is replaced by $[(\epsilon_r\text{-}1)/(\epsilon_r+1)](\text{-}q)$, where ϵ_r is the relative permittivity. The reduction of the image charge force depends on the magnitude of ϵ_r; for example, for $\epsilon_r = 3$, the image charge becomes -2/3 q. For calculating deposition in the respiratory tract, conducting walls are probably a good approximation.

The other charge deposition mechanism arises from the mutual repulsion of the like charges which causes a unipolar aerosol cloud to expand into the walls of the container. It can be shown that the aerosol concentration decreases uniformly over all portions of the cloud. The deposition depends on the particle concentration and the residence time but is independent of the size and shape of the container, although Yu [3] has pointed out that particles deposited on the walls could modify the fields except for a sphere or infinite cylinder. Fuchs [4] and others [5, 6] have shown that the mutual repulsion deposition efficiency in a cylinder P_{MR} is given by

$$P_{MR} = \frac{4\pi n_o R^3 \tau}{1 + 4\pi n_o R^3 \tau} \qquad (6)$$

where n_o is the number concentration of particles. (Unfortunately the factor 4π is missing in Equation 25 of reference 1. The figures and tables in references 1 and 3 based on this equation are also in error, affecting the comparison which was made between P_I and P_{MR}).

The image force and mutual repulsion force deposition efficiencies become comparable for $1 \lesssim n_o R^3 \lesssim 10$. The physical interpretation is that the forces are comparable when the mean distance between particles is of the order of the radius of the tube. When $n_o R^3 \gg 10$, mutual repulsion predominates and when $n_o R^3 \ll 1$, the image force predominates. No theoretical treatment is available for the difficult intermediate case.

Deposition in the Respiratory System

Under normal conditions $n_o R^3 \ll 1$ and thus deposition in the human respiratory tract will be mainly due to the image charge force. Yu and Chandra [7] have calculated the image force deposition for flow with a parabolic velocity profile which gives significantly lower deposition than slug flow. Deposition in each generation of bifurcation according to the Weibel model was evaluated for 1-μm aerodynamic diameter particles carrying 100 elementary charges/particle, breathed at 12 respirations/min with 1 liter tidal volume. Total image force deposition was 0.31, most of it in the last 6 of the 23 generations. For comparison, the total gravitational deposition was estimated to be 0.28. Thus 100 charges/particle resulted in deposition about equal to that due to gravitation. This result is in agreement with the earlier rough estimate by Fuchs [8]. While the theoretical calculations can only be approximate, they are useful estimates.

Experimental measurements of charged particle deposition in humans by Melandri et al. [9] are listed in Table I (column 4 is the average of the values in reference 9 for four subjects). The result for 1.1-μm particles is 60% of the deposition calculated by Yu and Chandra for 1-μm diameter particles.

Table I. Deposition of Charged Particles in Humans. (From Melandri et al., [9])

Particle Diameter (μm)	Concentration (Number/cm^3)	Elementary Charges	% Increase in Deposition
0.33	1×10^5	29	22 ± 5
0.65	4×10^4	45	27 ± 6
1.10	2×10^4	95	19 ± 5

In exposure experiments one normally desires to eliminate particle charge as a variable. The above discussion implies that the allowable charge on 1-μm particles is less than about 10 elementary charges and even less for smaller particles, depending on the mobility. On the other hand, for research on deposition in the respiratory tract and particularly for aerosol therapy it may be useful to employ charged particles [5]. The total deposition can be increased and, furthermore, deposition can be shifted towards the upper portion of the respiratory tract. Theoretically, 1-μm diameter particles will be completely deposited in the first two generations by the image force if their charge exceeds 10^4 units. The mutual repulsion efficiency would also approach 100% if the concentration were greater than $10^4/\text{cm}^3$.

Particles in ambient air are usually assumed to have been suspended for several hours or more and hence to be essentially uncharged. However, exposure to fresh aerosol can occur near sources. These particles could carry sufficient charge to affect lung deposition. Such exposure is even more likely in the workplace. This problem deserves investigation.

ELECTRICAL CHARGE ON EXPERIMENTAL AEROSOL

Aerosol formed by spraying solutions, mechanically suspending dusts, or from electrical discharges is generally charged. An exception is a condensation aerosol created at temperatures low enough to avoid thermionic emission. In most cases, the charging mechanisms are poorly understood.

Separating dust particles acquire high charges by triboelectric effects [10]. Several mechanisms have been proposed for spray charging. Statistical fluctuations in the number of ions in the vicinity of the forming drop are predicted to produce symmetric charging. Measured charging by Natanson and by Chow and Mercer [11] is about twice that predicted by theory. Solutions of 10% concentration, moreover, yielded asymmetric charging. Shearing of double layers at the liquid-orifice interface has been proposed to account for asymmetric charging. However, the simple theory has been found inadequate, and charging much larger than predicted has been reported [12].

Whitby and Liu [13] found the median charge on polystyrene latex from atomized suspensions to be of the order of 10-100 units with some 1.3-μm diameter particles carrying more than 600 units of charge. High charges on latex particles were also observed by John and Davis [14].

Spray charge from a vibrating-orifice generator is exceptionally large. Table II lists the measurements of Reischl et al. [15]. Spray charge depends

Table II. Spray Charge on Approximately 3-μm Diameter Particles from a Vibrating Orifice Aerosol Generator

Substance	Elementary Charges
Methylene blue	-1640
Potassium biphthalate	-5540
Sodium chloride	-2640
Dioctylphthalate (DOP)	-10000
DOP + 10 ppm methylene blue	-1400
Distilled water (residue)	1300

on the solution concentration. Figure 2 shows the charge on NaCl particles as a function of solution concentration, as measured in the author's laboratory. The sign reversal and asymptotic trend at high concentration is typical of many substances. However, some observers report negative charge on pure water whereas Figure 2 starts at positive charge, probably due to the nature of the impurities in the distilled water.

The present brief discussion is intended to emphasize that most aerosol is charged as it is generated, that the charging mechanisms are poorly

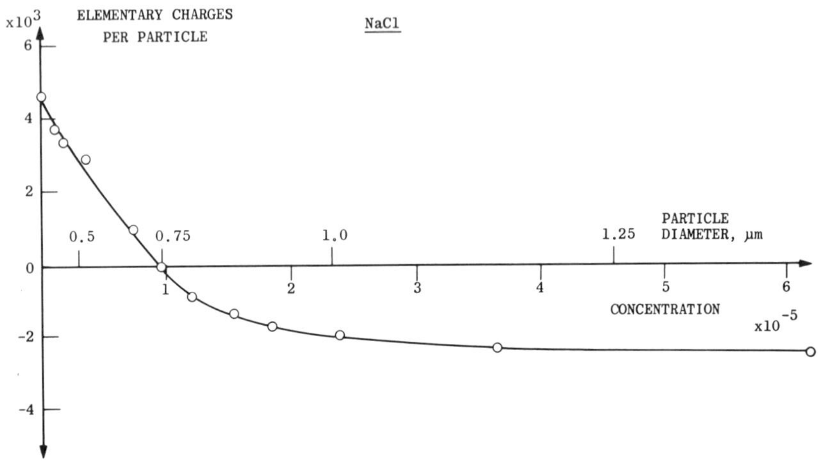

Figure 2. Spray charge on NaCl particles vs solution concentration.

understood and that the charging is affected by subtle chemical effects. Therefore, in most instances, it is prudent to neutralize experimental aerosol.

Subsequent to generation and neutralization, the aerosol is transported through tubing or ducts, sometimes through some apparatus, before use in exposure studies. In order to avoid possible static charging effects, it is advisable to employ grounded metal for the surfaces exposed to aerosols. However, even metals can impart charge to particles. The transfer of charge between metal surfaces and particles has been investigated by John et al. [16]. A charge transfer curve is shown in Figure 3; the y-intercept gives the contact charge, in this case 500 units, arising from the contact potential between the particle and surface. Contact charging was found to vary as the cube of the particle diameter and directly with the impact velocity. The fraction of precharge transferred to the metal (lost by the particle) was found to be determined by the fraction of the particle area in contact with the surface, typically only about 4% for impact velocities of 75 m/sec. Thus a particle of insulating material impacting on a metal surface will, in general, rebound with a charge whether or not it was initially charged.

MEASUREMENT AND NEUTRALIZATION OF PARTICLE CHARGE

Charge Measurements

The total charge on an aerosol can be measured by collecting it in a Faraday cup. By dividing the collected charge by the number of particles

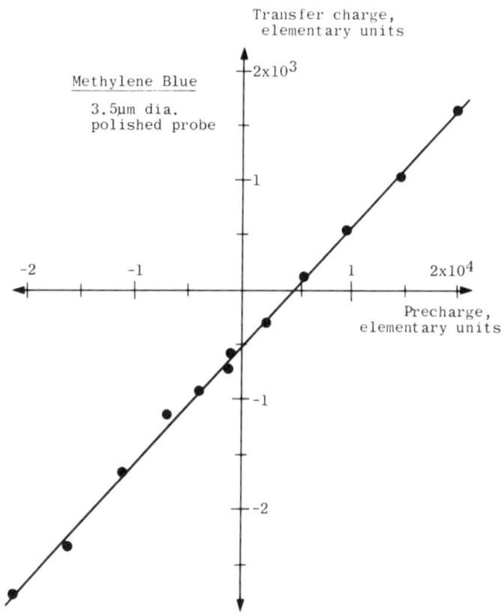

Figure 3. Transfer of charge to a stainless steel surface by impacting methylene blue particles.

sampled, the average particle charge can be obtained. An example of a Faraday cup, developed by John et al. [16], is shown in Figure 4. The aerosol is pumped through a high-efficiency filter surrounded by a metal can insulated from ground. The instant a charge enters the cavity, according to Gauss' law, an equal induced charge appears on the outside of the metal, to be conducted to ground via the electrometer which records the current. It is, of course, immaterial how long it may take the charge on the filter to be neutralized by the induced charge of opposite sign left on the inside of the can, i.e., the filter medium can be an insulator.

The Faraday cup is surrounded by a grounded metal shield to isolate it from the surroundings. The entrance is kept small to reduce leakage of external fields *into* the cup. The cup is allowed to "see" only grounded metal surfaces outside of the entrance. Leakage of flux from the collected charge *out* of the cup is minimized by backing the filter with a metal screen. The cup is designed to be mechanically rigid to avoid induction of currents by vibrations. Exposure of insulating surfaces to the aerosol flow is kept small to avoid static charging effects. The amount of aerosol collected can be determined after removing the filter. The interior of the cup is shaped for negligible wall losses.

Modern electrometers have a sensitivity (offset current) of the order of 10^{-15} A with only 10^6 input impedance because of the feedback. Thus,

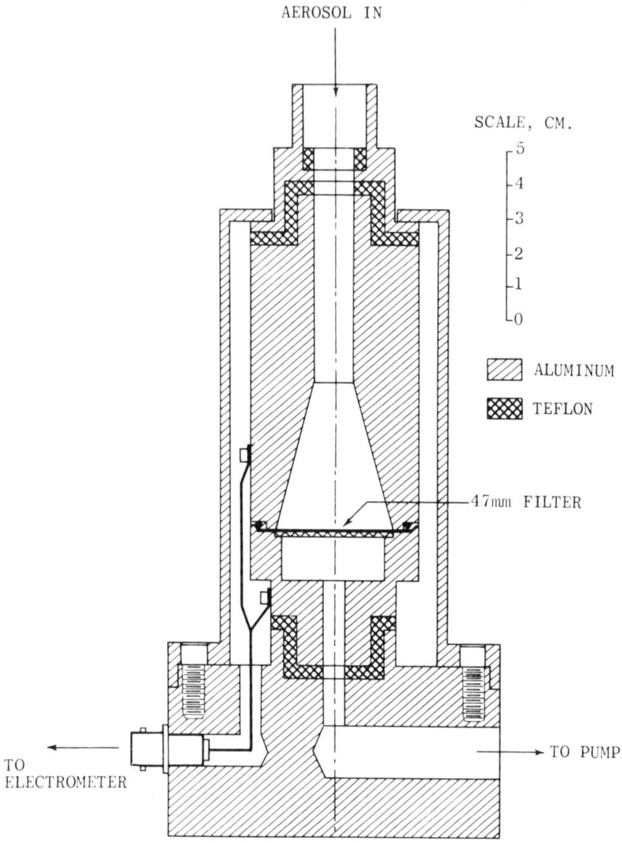

Figure 4. Faraday cup for measurement of charge on aerosols.

reduction of leakage to ground to 1% of the current signal requires a resistance of the cup to ground of 10^8 ohms; a value easily achieved. The Faraday cup in Figure 4 has a resistance $>10^{14}$ ohms to ground.

Measurement of the charge distribution requires the use of an electrical mobility analyzer [17]. The aerosol is passed through a transverse electrical field and a measurement made on either the current collected on the electrodes or that transmitted through the analyzer. Alternatively, the transmitted particles can be monitored, thus avoiding a current measurement. Furthermore, the aerosol detector can also be used to determine the particle size distribution which is needed for a complete analysis of a polydisperse aerosol. The required setup is shown schematically in Figure 5. Examples of this arrangement are described by Whitby and Liu [13] and by John and

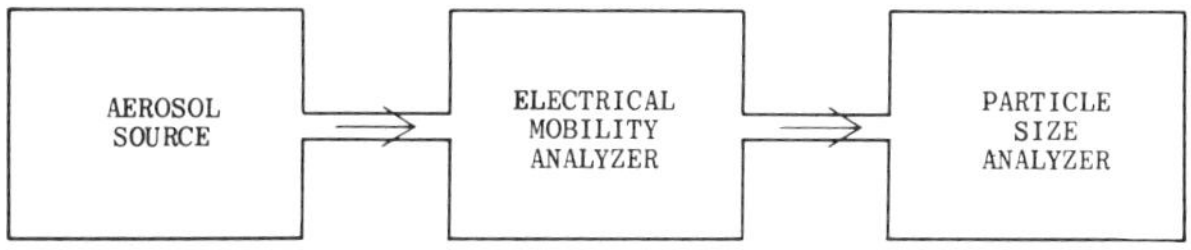

Figure 5. Arrangement for measurement of the charge distribution of an aerosol.

Davis [14]. Determination of the sign of the charge requires an asymmetry in the electrical mobility analyzer. For example, the aerosol can be introduced near one electrode and collected near the other, oppositely charged, electrode.

Charge Neutralization

Radioactive sources are commonly employed to "neutralize" aerosol charge. The charge distribution established as a result of particle ion interactions is described by the Boltzmann distribution. It should be recognized that the reduced magnitude of the charge is just adequate for exposure studies in view of the foregoing discussion. It points up the need for care in ensuring the establishment of the Boltzmann equilibrium. The necessary Nt product (ion pair concentration X time) has been determined by Liu and Pui [18].

Radioactive sources must be handled with respect for the radiation hazard. They are cumbersome and expensive. A different method of neutralization has been devised by Reischl et al. [15] for use with a vibrating-orifice generator. An electrical field is applied to the breaking jet, resulting in induction charging of the droplets. Data from operation of the charger is plotted in Figure 6. The induced charge is linear with induction voltage but opposite in sign. The y-intercept is the spray charge on the aerosol (methylene blue). The x-intercept corresponds to zero particle charge, i.e., the induced charge is just equal and opposite to the spray charge.

In practice there are some limitations to this method. The induced charge depends on droplet diameter, hence the droplets must be reasonably monodisperse. The spray charge was observed to change after 10-20 minutes of operation, possibly due to changing jet breakup conditions. Spray charge is also highly dependent on the composition and concentration of the solution. Therefore maintenance of neutralization requires frequent adjustment of the induction voltage. This could be automatic by monitoring the current drawn from the spray solution to provide a signal for control of the induction voltage.

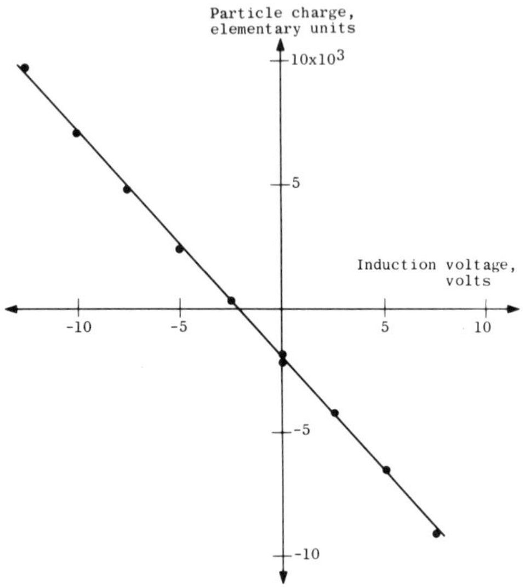

Figure 6. Induction charging and neutralization of aerosol by the method of Reischl et al. [15].

REFERENCES

1. Yu, C. P. "Precipitation of Unipolarly Charged Particles in Cylindrical and Spherical Vessels," *J. Aerosol Sci.* 8:237-241 (1977).
2. Pich, J. "Comments on the Paper: C. P. Yu: Precipitation of Unipolarly Charged Particles in Cylindrical and Spherical Vessels," *J. Aerosol Sci.* 9:275-277 (1978).
3. Yu, C. P. "Author's Reply," *J. Aerosol Sci.* 9:279-280 (1978).
4. Fuchs, N. A. *The Mechanics of Aerosols* (New York: Macmillan Publishing Company, 1964), pp. 102-106.
5. Wilson, I. B. "The Deposition of Charged Particles in Tubes, with Reference to the Retention of Therapeutic Aerosols in the Human Lung," *J. Colloid Sci.* 2:271-276 (1947).
6. Faith, L. E., S. N. Bustany, D. N. Hanson and C. R. Wilke. "Particle Precipitation by Space Charge in Tubular Flow," *Ind. Eng. Chem. Fundam.* 6:519-526 (1967).
7. Yu, C. P., and K. Chandra. "Deposition of Charged Particles from Laminar Flows in Rectangular and Cylindrical Channels by Image Force," *J. Aerosol Sci.* 9:175-180 (1978).
8. Fuchs, N. A. *The Mechanics of Aerosols* (New York: Macmillan Publishing Company, 1964), pp. 233-240.
9. Melandri, C., V. Prodi, G. Tarroni, M. Formignani, T. DeZaiacomo, G. F. Bompane and G. Maestri. In: *Inhaled Particles IV*, W. H. Walton, Ed. (Oxford: Pergamon Press, 1977), pp. 193-201.

10. Kunkel, W. B. "The Static Electrification of Dust Particles on Dispersion into a Cloud," *J. Appl. Phys.* 21:820-832 (1950).
11. Chow, H. Y., and T. T. Mercer. "Charges on Droplets Produced by Atomization of Solutions," *Am. Ind. Hyg. Assoc. J.* 32:247-255 (1971).
12. Matteson, M. J. "The Separation of Charge at the Gas-Liquid Interface by Dispersion of Various Electrolyte Solutions," *J. Colloid Interface Sci.* 37:879-890 (1971).
13. Whitby, K. T., and B. Y. H. Liu. "Polystyrene Aerosols–Electrical Charge and Residue Size Distribution," *Atmos. Environ.* 2:103-116 (1968).
14. John, W., and J. W. Davis. "The Measurement of the Charge Distribution on Monodisperse Aerosols," *Atmos. Environ.* 8:1029-1034 (1974).
15. Reischl, G., W. John and W. Devor. "Uniform Electrical Charging of Monodisperse Aerosols," *J. Aerosol Sci.* 8:55-65 (1977).
16. John, W., G. Reischl, W. Devor and J. J. Wesolowski. "Investigation of Particulate Matter Monitoring Using Contact Electricity," U.S. Environmental Protection Agency Report No. EPA-600/2-78-212, Research Triangle Park, NC, October 1978.
17. Whitby, K. T. In: *Fine Particles,* B. Y. H. Liu, Ed. (New York: Academic Press, Inc., 1976), pp. 581-624.
18. Liu, B. Y. H., and D. Y. H. Pui. "Electrical Neutralization of Aerosols," *J. Aerosol Sci.* 5:465-472 (1974).

CHAPTER 7

DELIQUESCENCE PROPERTIES AND PARTICLE SIZE CHANGE OF HYGROSCOPIC AEROSOLS

Ignatius N. Tang
Department of Energy and Environment
Brookhaven National Laboratory
Upton, New York 11973

INTRODUCTION

Ambient aerosols frequently contain large proportions of hygroscopic inorganic salts such as chlorides, sulfates and nitrates, which may induce adverse health effects when inhaled. Thus, in exposure studies employing animal or human subjects, test aerosols are routinely generated from saline solutions of known composition and the observed physiological effects are often correlated not only with the chemical nature, the dosage, but with the particle size [1, 2] of the test aerosol as well. Since in most cases the nebulized test aerosol is mixed with a carrier gas before entering the test chamber where the relative humidity may or may not be regulated, the particle size is subject to change as a result of water vapor condensation or evaporation. A knowledge of the hygroscopic properties of salt aerosols, therefore, is essential in estimating particle size variations under either dry or humid conditions.

More important, however, is the fact that inhaled salt particles are always exposed to an environment of high humidity in the respiratory tract, where the particles often deposit. Milburn et al. [3] have considered the effect of particle rehydration on the retention of hygroscopic dusts in the human lungs. They concluded that, for NaCl particles of initial dry diameter less than 2 μm, rehydration would be most likely to occur in the lungs, resulting in a significant increase in retention.

This paper discusses the deliquescence properties of pure and mixed salt aerosols and the particle size change as a function of relative humidity. Experimental results of the particle growth characteristics are presented for some mixed-salt aerosols of chlorides, sulfates and nitrates which are of physiological or environmental importance.

THERMODYNAMIC CONSIDERATIONS

Deliquescence Properties

Most of the inorganic salts exhibit the properties of deliquescence on exposure to moisture. The phase transition from a solid particle to a saline droplet occurs rather suddenly when the relative humidity reaches a specific value corresponding to the water activity of the saturated solution. Water activity data [4] necessary for determining aerosol deliquescence properties are usually available in the literature for aqueous solutions of single electrolytes. The theory of droplet growth for single-component salt aerosols [5-8] has been well established on the basis of thermodynamics. It is relatively simple, therefore, to calculate the size of a saline droplet in equilibrium with the water vapor of the surrounding air. The calculation, however, requires extensive thermodynamic properties such as water vapor activity, solution density and surface tension as a function of droplet composition and temperature. Such information is usually available only for dilute electrolyte solutions [4, 9].

On the other hand, the growth of a mixed-salt aerosol is complicated by the fact that, dependent upon its initial composition, a salt particle may pass through several regions of multiphase equilibria before its complete dissolution to form a homogeneous solution droplet. Recent thermodynamic considerations [10] have shown that the phase transformation and droplet growth of a mixed-salt aerosol in a moist atmosphere can be understood with reference to the phase diagram of the multicomponent system of interest. Stage-wise phase transformation during growth is often expected and the deliquescence humidity can be predicted from the thermodynamic properties in the solid-liquid coexistence regions.

Consider, for example, the growth characteristics of letovicite aerosols. Letovicite, $(NH_4)_3H(SO_4)_2$, is a mixed salt composed of 0.75 mole fraction of ammonium sulfate and 0.25 mole fraction of sulfuric acid. Its presence in atmospheric particulates has occasionally been observed and identified by Brosset et al. [11] using X-ray diffraction techniques. Referring to the solubility diagram shown in Figure 1 for the H_2O-$(NH_4)_2SO_4$-H_2SO_4 system, letovicite in its crystalline state has its chemical composition represented by point B on the diagonal line.

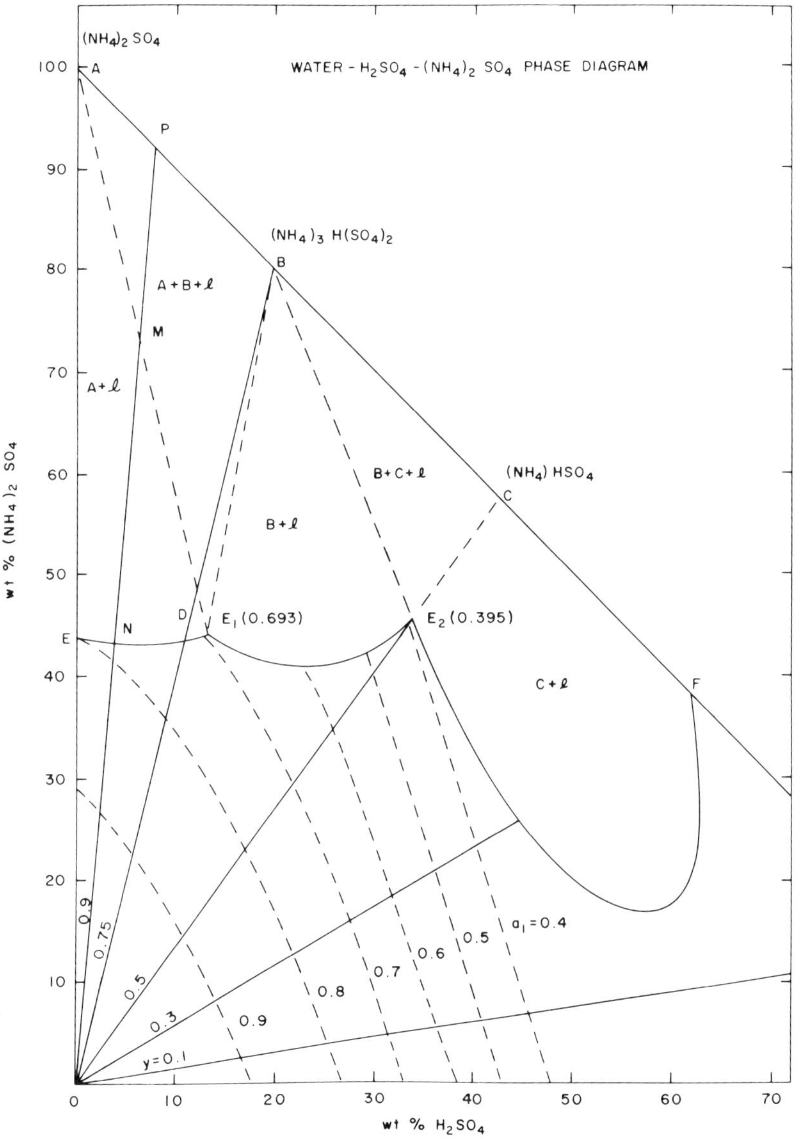

Figure 1. Solubility diagram for the H_2O-$(NH_4)_2SO_4$-H_2SO_4 system at 30°C.

Upon exposure to an increasing relative humidity, a solid letovicite particle would take up water with its composition changing along the line BO. However, since the particle must first pass through the three-phase coexistence region indicated by A+B+*l*, in which only one degree of freedom is allowed by the phase rule, nothing will then happen until the relative humidity (r.h.) reaches a value of 69.3% corresponding to the water activity of the eutonic point E_1 as shown in the diagram. This is the first deliquescence point for letovicite and all other mixed $(NH_4)_2SO_4$-H_2SO_4 aerosols having initial composition between pure ammonium sulfate and letovicite as represented by the line segment AB. At this deliquescence humidity, the dry salt particle suddenly transforms into a partially dissolved solution droplet, thus marking the onset of growth. A second marked growth will occur at 72% r.h. when the line BO crosses the solubility line EE_1, and all of the remaining solid dissolves in the solution at this time.

Figure 2 shows the growth curves calculated for sulfate aerosols $(NH_4)_2SO_4$, NH_4HSO_4, $(NH_4)_3H(SO_4)_2$ and H_2SO_4 at 25°C. While the H_2SO_4 aerosol grows continuously and smoothly with increasing humidity, the $(NH_4)_2SO_4$ aerosol displays the deliquescence property at 79.5% r.h. with a distinct onset of growth. The $(NH_4)_3H(SO_4)_2$ aerosol shows a two-stage growth as dictated by the phase diagram. The NH_4HSO_4 aerosol, although being a mixed salt, is expected to behave like a single-component salt aerosol since its growth path goes through the second eutonic point E_2 as shown in the phase diagram in Figure 1. The growth curve for the NH_4HSO_4 has been verified experimentally [12]. The deliquescence properties for the mixed sulfate aerosols have been determined [13].

Equilibrium Size of Saline Droplets

It has been shown [11] that for a solution droplet containing nonvolatile multicomponent electrolytes, the equation

$$ln \frac{p_1}{p_1^o} = ln\gamma_1\chi_1 + \frac{2\bar{v}_1\sigma}{RTr_e} \tag{1}$$

is quite general and applies to both single and multicomponent systems as long as the solution properties are determined for the system. Equation 1 relates the equilibrium radius r_e of a droplet of composition χ_1 (mole fraction) to the relative humidity, $H = p_1/p_1^o$, and the solution properties such as the activity coefficient γ_1, partial molar volume $\bar{v}_1$ and surface tension σ. Here the subscript 1 refers to the solvent, i.e., water in the present case. For droplets of radius larger than ~0.05 μm, the contribution of the second term on the right-hand side of Equation 1 is small and one may relate the composition of the droplet directly to that of a bulk solution in equilibrium with its water vapor pressure at temperature T. The calculation, however, requires vapor pressure data for the entire concentration range extending from the solid-liquid coexistence regions to infinite dilution. Such information is

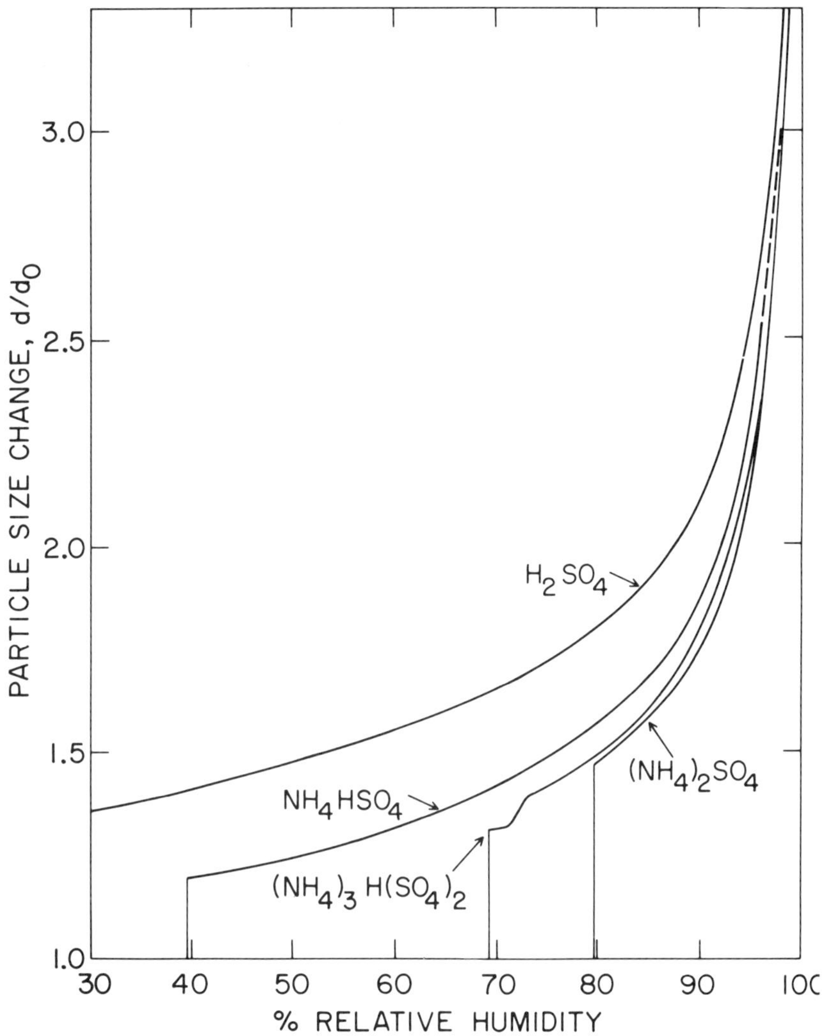

Figure 2. Growth curves for sulfate aerosols at 25°C.

usually not available in the literature for multicomponent electrolyte solutions.

The change in particle size at a given relative humidity can readily be deduced from a material balance on salt content before and after droplet growth to its equilibrium size. The following equation is obtained:

$$\frac{d}{d_o} = \left(\frac{100\rho_o}{W_2\,\rho}\right)^{1/3} \tag{2}$$

Here, d and ρ are, respectively, the diameter and density of a droplet containing W_2% by weight of salt, and d_o and ρ_o are, respectively, the diameter and density of the dry salt particle. The calculation, of course, requires density data as a function of composition. The following relationship [14] may be used to estimate the density of a solution droplet at composition W_2:

$$\rho = \rho_1 + \left(1 - \frac{\rho_1}{\rho_2}\right)\rho\, W_2 \tag{3}$$

where ρ_1 is the density of pure water and ρ_2 is the hypothetical density of the solutes. The factor $(1 - \rho_1/\rho_2)$ may be evaluated from the known density and composition of the saturated solution and taken to be independent of composition. Such a procedure has been demonstrated [12] to be quite adequate for practical droplet growth calculations.

In the case of a multicomponent salt particle which is only partially dissolved under certain humidity conditions, the fraction of undissolved salt may be estimated from the solubility diagram. Referring again to Figure 1 for the H_2O-$(NH_4)_2SO_4$-H_2SO_4 system, one observes that a mixed sulfate particle having initial composition denoted by point P will deliquesce at 69.3% r.h., at which time the particle becomes partially dissolved and the fraction of undissolved solid is given by the ratio of the line segment ME_1 to AE_1. The liquid phase should have a composition denoted by the eutonic point E_1, and the solid phase is pure $(NH_4)_2SO_4$ as dictated by the two-phase coexistence A + *l*. The solid-to-liquid ratio should be taken into consideration when making growth calculations.

EXPERIMENTAL PROCEDURE

The deliquescence properties of mixed salt aerosols and subsequent droplet growth by water vapor condensation were investigated in a continuous-flow apparatus at 25°C as a function of relative humidity. The design, construction and characterization of the apparatus have been described in detail elsewhere [15]. Briefly, salt aerosols were generated by bubbling nitrogen gas through a coarse glass frit immersed in an aqueous solution of known composition. The aerosol stream was dried and passed through an electrostatic separator where charged particles were sorted according to their electrical mobility. A distinct size fraction, usually exhibiting a log-normal particle-size distribution with a geometric standard deviation ranging from 1.07 to 1.14 was selected and extracted from the separator.

Figure 3 shows the electron micrograph of a size-selected letovicite aerosol sample collected on a 0.4-μm Fluoropore® filter. The size distribution shown with the electron micrograph was obtained with a Carl Zeiss Particle Size Analyzer. Since the particle concentration was typically about 10 particles/cm^3 in the growth chamber, a long sampling time was required to collect enough particles on the filter for electron microscopic examination. In this particular case, a 6-hr sampling time was used, during which period some fluctuation of the operating conditions for the aerosol generation system was unavoidable. This might have contributed to the fact that the collected aerosol sample showed a rather large σ_g (1.14). Under normal experimental conditions, the extracted aerosol is quite monodisperse with good size resolution.

The size-selected aerosol stream was mixed in a thermostated growth chamber with N_2 carrier gas containing a measured amount of moisture. The particle-size distribution after equilibration at each given relative humidity was continuously monitored with an optical particle counter. The output signal of the particle counter was preamplified, shaped and sorted in a 100-MHz pulse-height analyzer and subsquently stored in a 1024 multichannel analyzer. The data was displayed on the screen of an oscilloscope and later retrieved by a high-speed printer for processing.

A composit of three oscilloscope tracings for a letovicite aerosol grown at different humidities is shown in Figure 4. The dry salt aerosol has a median diameter of 0.67 μm. As the humidity was gradually raised to the deliquescence point of 69.3% r.h., there was a sudden shift of the peak position from channel No. 100 to 137 with only a small increment of relative humidity, thus marking the onset of particle growth. From the particle size change before and after exposure to moisture, aerosol growth by water vapor condensation was obtained as a function of relative humidity.

RESULTS AND DISCUSSION

Deliquescence Properties and Droplet Growth

Inorganic salt aerosols composed of either single component or combinations of NaCl, KCl, $(NH_4)_2SO_4$, H_2SO_4 and NH_4NO_3 were investigated at 25°C in the continuous-flow apparatus. The size of the initial dry aerosols used in the investigation ranged from 0.4 to 0.9 μm in median diameter. The relative humidity at which phase transformation took place was determined for each aerosol, and the subsequent droplet growth was measured as a function of relative humidity. In the case of mixed sulfate aerosols, the observed deliquescence properties were further substantiated by vapor pressure measurements made in a separate study [16].

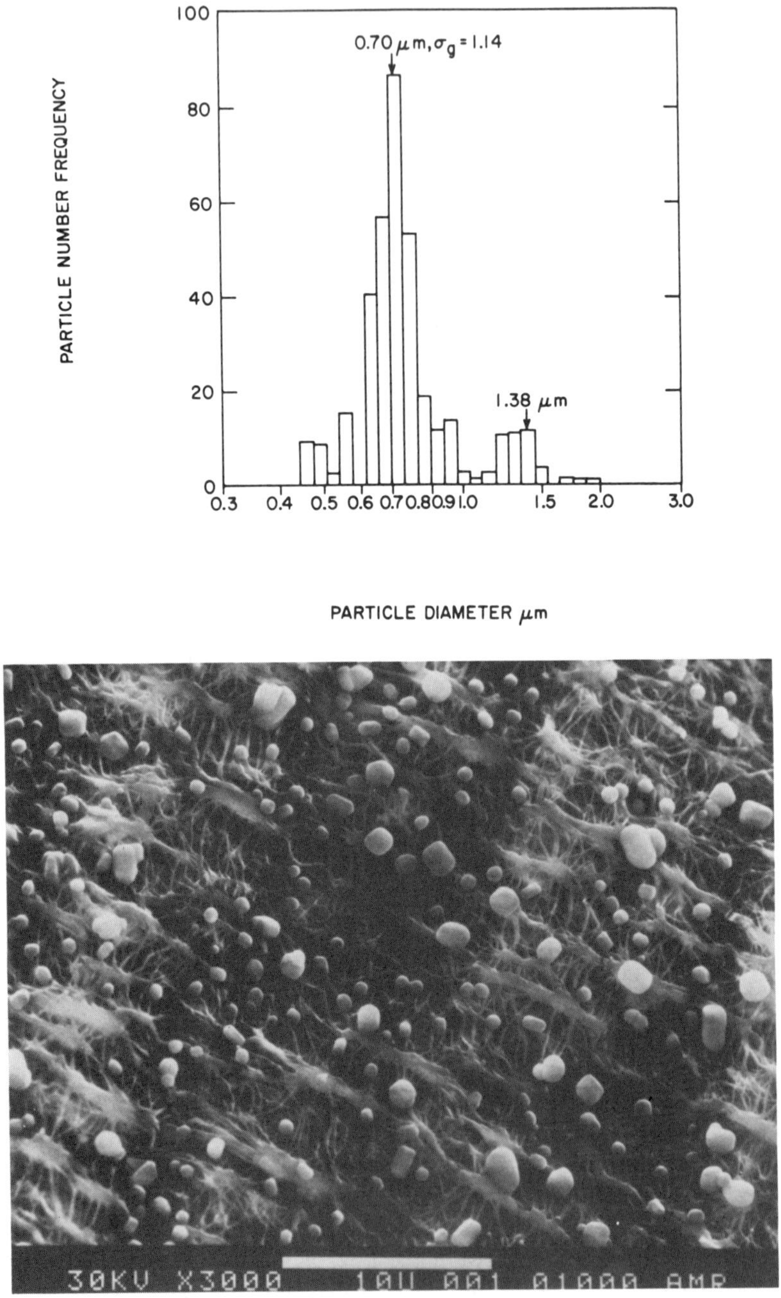

Figure 3. Electron micrograph and particle size distribution of a letovicite aerosol.

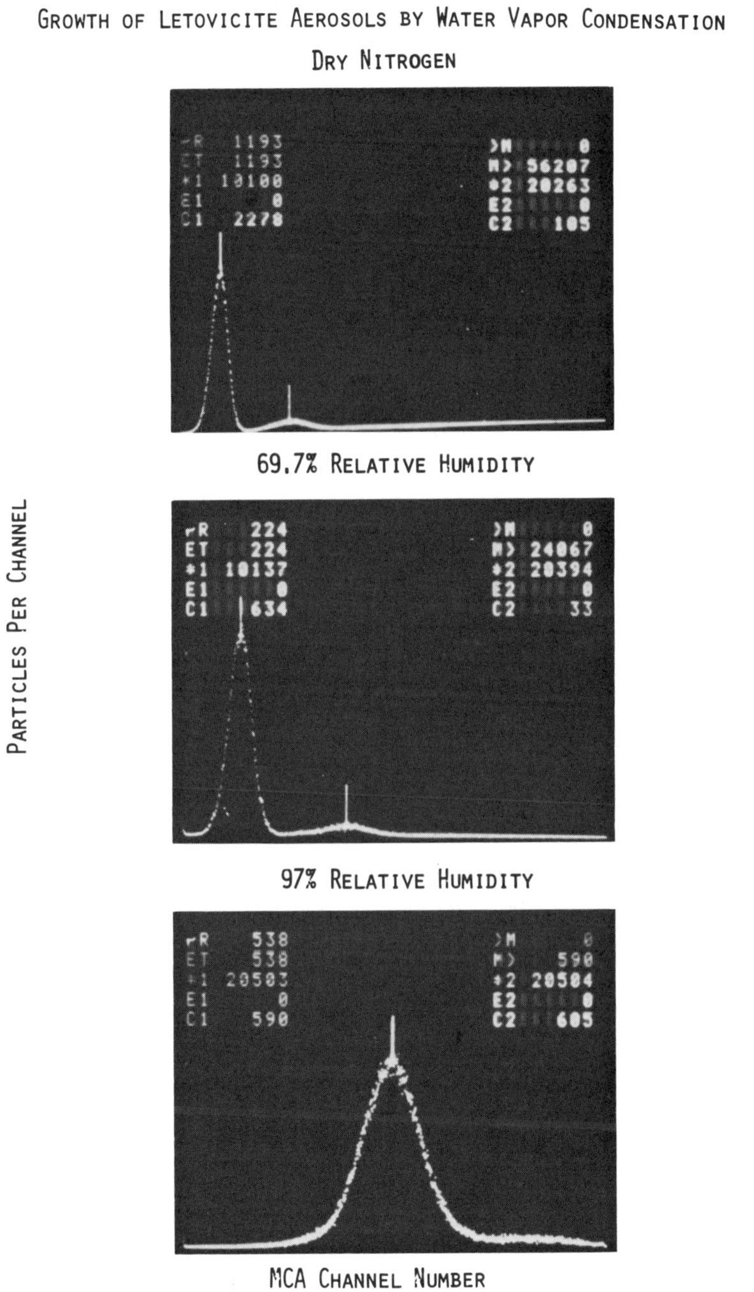

Figure 4. Oscilloscope tracings showing the growth of a letovicite aerosol as monitored with the optical particle counter.

The growth curves for two mixed sulfate aerosols showing the typical stage-wise growth characteristics predicted from thermodynamic considerations are given in Figure 5. The observed deliquescence point at 69.0 ± 0.5% r.h. is in good agreement with the water activity at the eutonic composition.

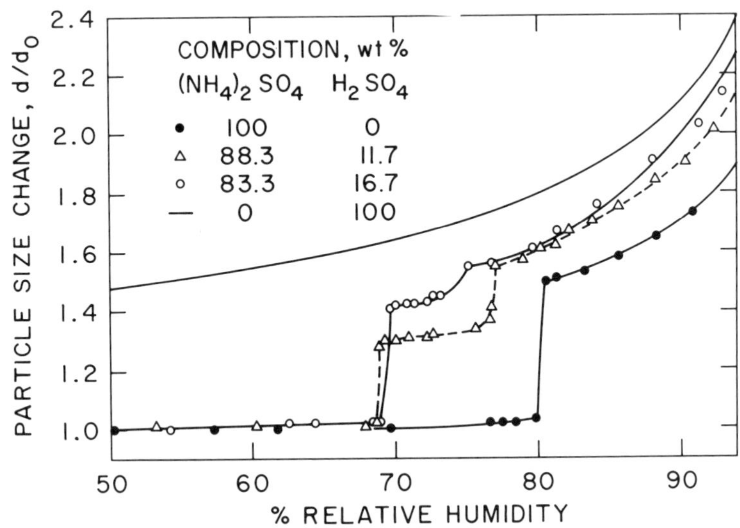

Figure 5. Growth of sulfate aerosols as a function of relative humidity.

The deliquescence humidities of some environmentally important aerosols investigated in our laboratory are given in Table I. The growth of the NH_4NO_3 aerosol was also investigated, however, no deliquescence point was experimentally established for this salt aerosol. As indicated by the data

Table I. Deliquescence Properties of Some Salt Aerosols at 25°C

Aerosol Composition	Deliquescence Humidity
NaCl	75.7
KCl	84.3
NaCl-KCl	73.8
$(NH_4)_2SO_4$	79.5
$(NH_4)_3H(SO_4)_2$	69.0
NH_4HSO_4	39.0
$2NH_4NO_3 \cdot (NH_4)_2SO_4$	56.4

shown in Figure 6, the NH_4NO_3 aerosol already became quite hygroscopic at ~30% r.h., and the phase transition at the predicted deliquescence point of 62% r.h. was not observed with certainty. In contrast, a definite phase transition was observed to occur at 56.4% r.h. for a mixed salt aerosol containing 45% by weight NH_4NO_3 and 55% by weight $(NH_4)_2SO_4$, in agreement with the behavior predicted from the phase diagram of the system. The measured droplet growth for the NH_4NO_3 aerosol, as shown in Figure 6, follows the theoretical curve rather closely at relative humidities greater than the expected deliquescence point.

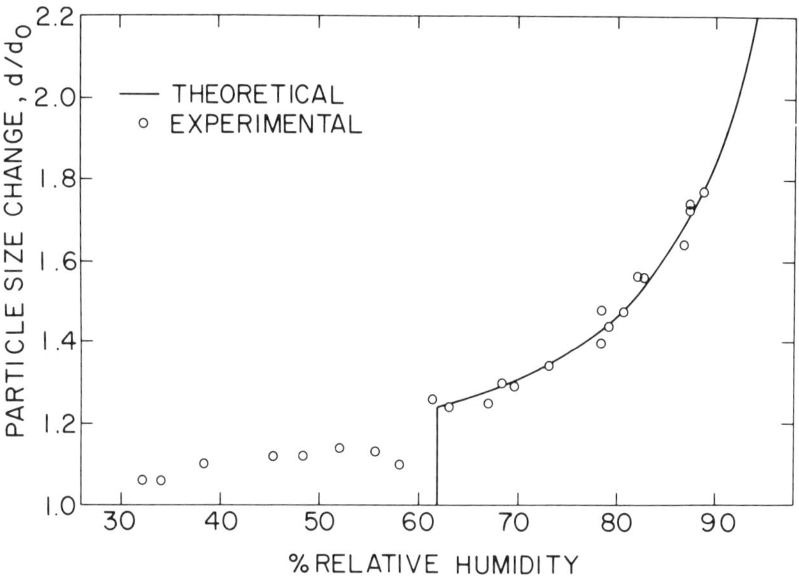

Figure 6. Growth of NH_4NO_3 aerosols in a moist atmosphere at 25°C.

Droplet Evaporation and Crystallization

Evaporation of solvent water will take place when a wet saline aerosol is mixed with dry carrier gas, a situation often encountered within many exposure facilities. The particle-size distribution of the nebulized aerosol will depend upon the transit time and the relative humidity finally reached in the mixing process. Equations 1 and 2 are equally applicable to the case of decreasing relative humidity for calculating the ultimate size of a saline droplet, provided that equilibrium prevails at the final relative humidity.

In reality, however, it is very common for a saline aerosol to exhibit the so-called "hysteresis" phenomenon, which is typically illustrated by the $(NH_4)_2SO_4$ aerosol as shown in Figure 7.

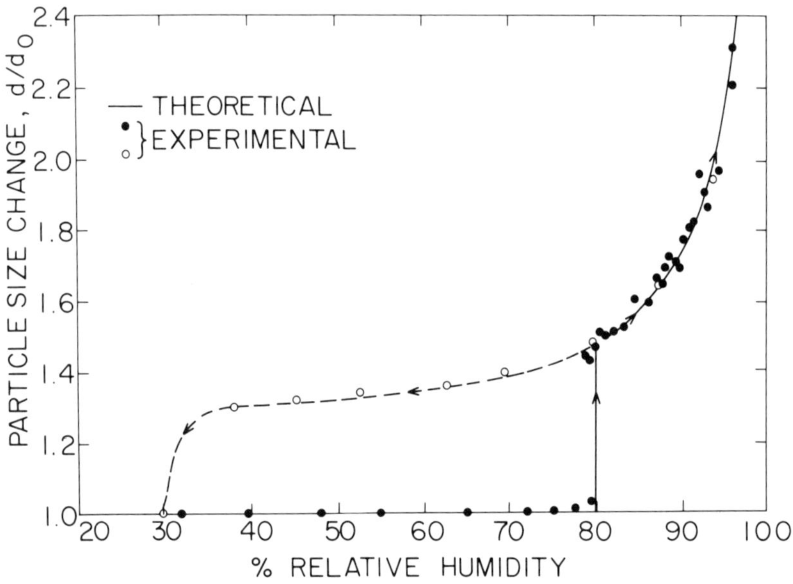

Figure 7. Condensational growth and evaporation of $(NH_4)_2SO_4$ aerosols in a moist atmosphere at 25°C.

The measured growth of $(NH_4)_2SO_4$ aerosols with increasing relative humidity is shown by the filled points in Figure 7. The deliquescence properties at 79.5% r.h. and subsequent droplet growth are seen to agree with theoretical predictions. The open circles represent experimentally determined size changes when the relative humidity was gradually decreased. The droplet size follows the growth curve very closely until the deliquescence point is reached. A further decrease in humidity does not usually result in an abrupt reduction in particle size by crystallization of the salt content as might be expected. Instead, a continuous and gradual reduction in size takes place as a result of the droplet becoming supersaturated. These "hysteresis" phenomena are well known for many inorganic salt solutions and have been observed by other investigators [6, 17]. In fact, "hysteresis" is responsible for the often noted lingering of smogs [18] below the humidities at which they first appeared.

When crystallization does occur, it will occur spontaneously and the particle size will suddenly shift to its initial dry size. Although the lingering time, in general, depends upon the extent of humidity lowering, there appears for each saline aerosol a specific humidity below which the lingering time is order-of-magnitude shorter than at slightly higher humidities. To demonstrate this fact with NaCl, KCl and mixed NaCl-KCl aerosols the reciprocal of lingering time, called transition rate, is plotted in Figure 8 vs the relative humidity at which the crystallization occurs. In each case, a sharp rise in transition rate is noted to occur about a specific humidity. These crystallization humidities are quite reproducible in our experiments and compare fairly well with the observations by Orr et al. [6], who employed salt aerosols in the size range of 0.01 to 0.1 μm in radius. Table II gives the crystallization humidities for a few salt aerosols.

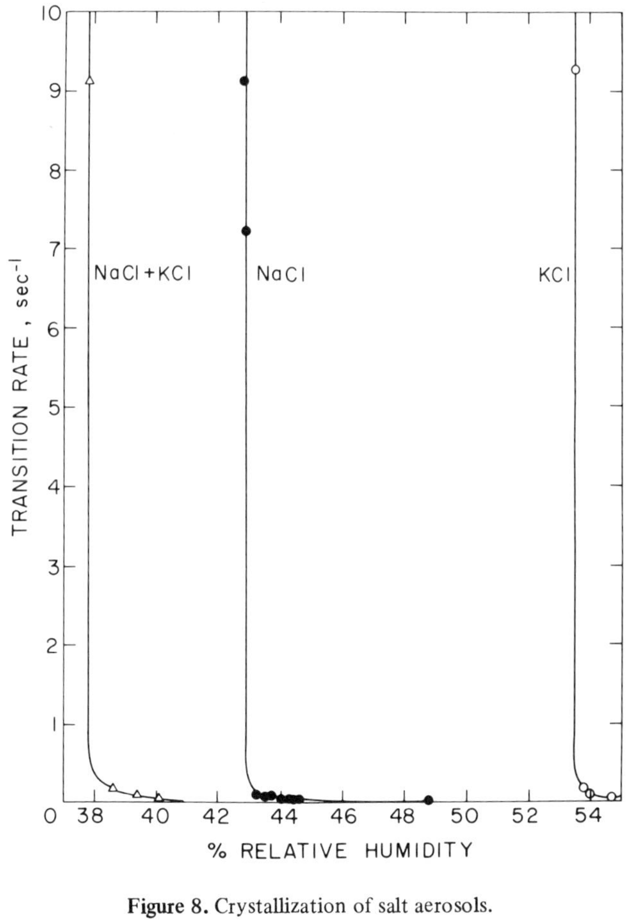

Figure 8. Crystallization of salt aerosols.

Table II. Crystallization Humidities of Salt Aerosols

Aerosol Composition	This Work	Reference 6
NaCl	43	42
KCl	53	48
NaCl-KCl	38	–
$(NH_4)_2SO_4$	~36	40

CONCLUSIONS

Theoretical and experimental results are presented to show that the growth of a mixed salt aerosol by water vapor condensation may go through several stages of multiphase equilibria before its complete dissolution to form a homogeneous solution droplet. The deliquescence properties are predictable from the phase diagram of the system of interest. The size of a saline droplet in equilibrium with the relative humidity in the surrounding air may be calculated from the thermodynamic properties of the bulk solutions. The increase in particle size as a result of rehydration and condensational growth may significantly affect the retention of inhaled particulates in the respiratory tract.

Experiments with $(NH_4)_2SO_4$ and chlorides show that for each deliquescent salt aerosol there is a threshold humidity below which the solution droplets will quickly evaporate to become crystalline particles. The threshold crystallization humidity is generally much lower than the deliquescence humidity of the given salt aerosol. The information should provide a basis for selecting a suitable humidification procedure to generate test aerosols for exposure studies.

ACKNOWLEDGMENTS

The author wishes to thank his colleagues H. R. Munkelwitz, J. G. Davis and W. L. Kalinowski for their invaluable contributions in performing the experiments. He is indebted to Dr. L. Newman for constant encouragement and helpful discussions. This research was performed under the auspices of the United States Department of Energy under Contract No. EY-76-C-02-0016.

REFERENCES

1. Frank, R. "Sulfur Oxides and Particles: Effects on Pulmonary Physiology in Man and Animals," *Conference on Health Effects of Atmospheric Salts and Gases of Sulfur and Nitrogen in Association with Photochemical Oxidant*, Newport Beach, CA, January 7-8, 1974.

2. Crocker, T. T. "Effects of Sulfur Oxides in Animal and Man," *Conference on Health Effects of Atmospheric Salts and Gases of Sulfur and Nitrogen in Association with Photochemical Oxidant*, Newport Beach, CA, January 7-8, 1974.
3. Milburn, R. H., W. L. Crider and S. D. Morton. "The Retention of Hygroscopic Dusts in the Human Lungs," *AMA Arch. Ind. Health* 15:59-62 (1957).
4. Robinson, R. A., and R. H. Stokes. *Electrolyte Solutions* (London: Butterworths, 1970).
5. Kohler, H. "The Nucleus in and the Growth of Hygroscopic Droplets," *Trans. Faraday Soc.* 32:1152-1161 (1936).
6. Orr, C., Jr., F. K. Hurd and W. J. Corbett. "Aerosol Size and Relative Humidity," *J. Colloid Sci.* 13:472-482 (1958).
7. Dufour, L. and R. Defay. In: *Thermodynamics of Clouds* (New York: Academic Press, Inc., 1963), Chapter 8.
8. Sedunov, Yu. S. In: *Physics of Drop Formation in the Atmosphere* (New York: John Wiley & Sons, 1974), Chapter 2.
9. West, C. J. and C. Hull. *International Critical Tables* (New York: McGraw-Hill Book Company, 1933).
10. Tang, I. N. "Phase Transformation and Growth of Aerosol Particles Composed of Mixed Salts," *J. Aerosol Sci.* 8:361-371 (1976).
11. Brosset, C., K. Andreasson and M. Ferm. "The Nature and Possible Origin of Acid Particles Observed at the Swedish West Coast," *Atm. Environ.* 9:631-642 (1975).
12. Tang, I. N., and H. R. Munkelwitz. "Aerosol Growth Studies–III. Ammonium Bisulfate Aerosols in a Moist Atmosphere," *J. Aerosol Sci.* 8:321-330 (1977).
13. Tang, I. N., H. R. Munkelwitz and J. G. Davis. "Aerosol Growth Studies –IV. Phase Transformation of Mixed Salt Aerosols in a Moist Atmosphere," *J. Aerosol Sci.* 9:505-511 (1978).
14. Moelwyn-Hughes, E. A. *Physical Chemistry* (Oxford: Pergamon Press, 1961), p. 808.
15. Tang, I. N., H. R. Munkelwitz and J. G. Davis. "Aerosol Growth Studies –II. Preparation and Growth Measurements of Monodisperse Salt Aerosols," *J. Aerosol Sci.* 8:149-159 (1977).
16. Tang, I. N., and H. R. Munkelwitz. "The Optical and Thermodynamic Properties of $(NH_4)_3H(SO_4)_2$ (Letovicite) Aerosols," presented at the May 1978 Industrial Hygiene Conference, Los Angeles, CA.
17. Winkler, P., and C. Junge. "The Growth of Atmospheric Aerosol Particles as a Function of the Relative Humidity," *J. Rech. Atm.* 4: 617-638 (1972).
18. La Mer, V. K., and S. Cotson. "The Growth and Shrinkage of Aerosols," *Science* 118:516-7 (1953).

SECTION II

METHODS OF AEROSOL GENERATION AND CHARACTERIZATION

CHAPTER 8

DRY-DISPERSION AEROSOL GENERATORS

William C. Hinds
Harvard School of Public Health
Department of Environmental Health Sciences
Boston, Massachusetts 02115

INTRODUCTION

The most widely used technique for generating solid particle test aerosols is dry dispersion, defined here as the pneumatic redispersion of a dry powder to form an aerosol. Dry-dispersion methods can accommodate a wide range of powdered materials and have the advantage over nebulization methods of producing aerosols that do not contain excessive moisture or solvent vapors. Dry-dispersion methods can be used at feedrates ranging from less than a milligram to more than a kilogram per minute to give dust concentrations prior to dilution from mg/m^3 to greater than 100 g/m^3. This paper describes the characteristics of dry-dispersion aerosol generators suitable for inhalation toxicology studies and discusses recent developments in generator designs. The operation and performance of selected commercially available generators are described.

Inhalation studies, which may last years, require constant generator output during exposure because inhaled dose is frequently estimated from the product of aerosol concentration, exposure duration and breathing rate. But more important is the need to avoid large fluctuations in concentration that are likely to affect animal response, e.g., a brief acute exposure during a planned chronic study. Furthermore, it is necessary for the particle size distribution to remain constant because the toxicity of inhaled particulates depends on particle size, density and shape, which govern the site of deposition within the respiratory system. Because the failure to achieve complete dispersion leads to variations in size distribution, it is essential that a powder be thoroughly dispersed into unitary particles, rather than agglomerates of two or more particles. This problem of fully dispersing a dust to unitary particles represents an important consideration in the use of dry-dispersion aerosol generators that is not a consideration for liquid aerosols produced by nebulization or condensation.

Although the general subject of dry-dispersion aerosol generators has been the subject of prior reviews [1-5], this paper is concerned exclusively with generators suitable for inhalation toxicology research, i.e., those that produce aerosols primarily in the respirable size range and are able to maintain a reasonably constant mass concentration and size distribution for at least one hour without recharging. Many of the dust generators described here evolved from the development of preexisting dry-dispersion aerosol generators used in other areas of research. These include filter and respirator testing, air cleaning studies, meteorological tracer studies, metalizing processes and calibration of aerosol measuring instruments. In general, dust generators for inhalation toxicology research require greater stability, lower concentrations and longer running times than those for other areas of research.

Ideally, dry-dispersion aerosol generators produce suspended particulates having the same size distribution and physical characteristics as the unitary powder particles. Typically, the particles have an irregular shape and sizes that cover a broad range, but powders that have been size-classified can be dispersed to produce near monodisperse aerosols. If there is incomplete dispersion or size-selective losses in the system (as by sedimentation of the largest sizes) the aerosol particle size distribution will differ from that in the original powder. In theory any material that can be prepared in powder form could be used; in practice, hygroscopic materials or plastics that develop high electrostatic charges present many problems and require special generators. Methods for preparing test aerosols of fibrous materials are covered by Spurny in Chapter 13 of this volume.

OPERATING MECHANISMS

Basic requirements for all dry-dispersion dust generators for animal studies are: (1) a means of continuously metering a powder into the generator at a constant rate and (2) a means of dispersing the powder to form an aerosol. An essential element of the dispersion mechanism is a means to break up agglomerates or to remove them from the aerosol stream by size selective classification.

Dust Metering

The simplest dust metering systems use gravity feed of loose powder into an air stream, usually assisted by vibrators or agitators. They give uneven delivery and a variable bulk powder density, both of which lead to variations in aerosol concentration. An improvement on the gravity feed is the turntable dust feed described below, where the powder is gently and continuously packed into a groove of fixed dimensions in a rotating disk and subsequently removed from the groove at a constant rate by an air ejector. Other systems use discrete volume elements to meter the dust. These devices provide a constant feed rate over a long time period but deliver the dust in pulses. Generators using such feed systems are designed to minimize concentration pulses by providing adequate air mixing volume or gradual delivery from several volume elements. There are physical limitations on the size of grooves or volume elements that provide satisfactory metering. If the element is too small, the powder bridges the gap and fails to completely fill the element; if too large, the dust does not fill uniformly and it resembles a loose powder feed system. A more stable metering method is the use of a cylinder of compressed powder that is eroded or scraped away at a constant rate. This method, used by Wright [6] and Dimmock [7], is reliable for long-term studies because the bulk density of the powder can be controlled during packing. It is, however, unsuitable for dusts containing a significant fraction of particles larger than 10 μm because of the poor packing characteristics of large particles.

Dust Dispersion

Dispersibility of a powder depends on powder material, particle size and size range, particle shape, and moisture content. Incomplete dispersion results in an aerosol particle size distribution that is larger than the size distribution of the unitary powder particles. Hydrophobic materials such as talc are easier to disperse than are hydrophilic materials such as quartz and limestone and dry powders are easier to disperse than humidified powders. To fully disperse a powder as an aerosol it is necessary to supply sufficient energy to a small volume of the bulk powder to separate the particles by overcoming the attractive forces between them. These forces, which depend on the surface properties of the material, are proportional to the first power of particle diameter d but, because the separating forces are proportional to d^3 for vibration and d^2 for drag, dispersibility increases rapidly with particle size and there is a size below which particles cannot be satisfactorily dispersed with a given generator. Fuchs [8] gives results of Andreason on the relative dispersibility of 21 different powders. *Lycopodium* spores, the easiest to disperse, had 70 times greater dispersion than the most difficult material, finely powdered chalk dust.

A metered powder may be dispersed and agglomerates broken up using the energy of a turbulent air jet directly, or by passing the dust-laden air stream through an impactor or fluidized bed. The powder itself can be fluidized to disperse it or the powder can be introduced into a fluidized bed of 100- to 200-μm beads to break up agglomerates, suspend particles in the air stream and, by elutriation, prevent large agglomerates from escaping the fluidized bed. It is essential for good results that the compressed air used to disperse the dust be clean, dry and free from compressor oil mist. Extremely dry air, i.e., relative humidity less than 5%, can cause strong electrostatic forces between particles which reduces dispersibility.

Fluidized-Bed Aerosol Generators

Studies by Guichard and Magne [9], Guichard [10] and Willeke et al. [11] characterized the use of fluidized beds as an aerosol dispersion mechanism. This led to the development of laboratory prototypes [12] and a commercial dust generator [13] using this principle. These devices generate aerosols with a very stable output of particles in the 0.5 to 30 micrometer range. Output concentration ranges from mg/m^3 to a maximum of about 4 g/m^3. These generators use a fluidized bed of glass or metallic beads in the 100- to 200-μ-m size range, into which the powder is introduced and emerges as an aerosol. Although the exact dispersion mechanisms of a fluidized bed are very complicated, they involve the deposition of particles onto the fluidized beads and their subsequent reentrainment into the air stream by the combined action of impaction as beads collide with one another and intense aerodynamic turbulence. In addition, the whole system acts as an elutriator, preventing large particles and agglomerates from leaving the system. This type of generator can be used only with dry nonsticky powders. One characteristic of a fluidized bed is that it always produces electrically charged aerosols. Consequently, the aerosol must be discharged as it leaves the fluidized bed. The charging phenomenon is so pronounced that Guichard [14] used fluidized beds to study triboelectrification charging mechanisms. In addition, care must be taken to ensure that the beads that are used do not oxidize or abrade and thereby generate an additional metallic or glass aerosol. Air elutriation velocities that provide good fluidization for 100- to 200-μm beads range from 5 to 15 cm/sec, which limits maximum output particle size to 40 to 70 μm. Because there is a significant particle holdup time in the fluidized bed, it acts as an averaging volume, or capacitor, to damp out fluctuations in output dust concentration.

Charging and Discharging of Dusts

A problem common to all dry-dispersion aerosol generators is buildup of particle charge during dispersion which gives variations in output concentration due to variable loss to the walls of the system. This is a result of triboelectrification or contact charging of the dust particles as they touch and separate from the surfaces of the generator and each other. The charging effect is greatest for powders with low moisture content commonly used to prevent clumping and improve flowability. The sign of the charge is determined by the relative discharge potential of the powder and generator surfaces, a function of their position in the triboelectric series. In practice, this is usually complicated by the presence of impurities and surface oxide layers.

Although charging can be reduced by careful selection of generator materials, it is more common to discharge the aerosol to Boltzmann equilibrium charge distribution as it leaves the generator by mixing the aerosol with high concentrations of bipolar ions, produced by corona discharge or a radioactive source. Whitby [15] described a sonic jet ion generator capable of producing high concentration of ions ($10^9/cm^3$ at exit of jet) for charge neutralization of aerosols. The device uses air discharged at sonic velocity through a 1.6-mm jet to strip away bipolar ions from a high-voltage corona discharge formed at the tip of a needle. The bipolar ion stream is mixed with the aerosol stream immediately upon exiting from the aerosol generator. In our laboratory we have found that the output from this ion generator changes with time because the needle tip radius changes with use and the high air flowrate (68 liter/min) necessitates undesirable dilution of the aerosol. Zamorani [16] found that under most conditions, neutralization by ac corona discharge yields aerosols with a charge greater than Boltzmann's equilibrium charge. These difficulties are overcome by the alternative approach of passing the aerosol past a radioactive source which produces large numbers of air ions. A 10-mCi Krypton-85 source in a 50-cm by 9-cm diameter cylinder is manufactured for this purpose by TSI, Inc., (St. Paul, MN). These sources have a half-life of 10.3 years and produce 0.695-MeV beta radiation that is contained by the 1.6 mm thick aluminum shell. Liu and Piu [17] calculated the equilibrium ion concentration inside this device to be 8.8×10^6 ions per cm^3. The time constant for aerosol neutralization by a bipolar ion concentration of N ions per cm^3 is $(4\pi NeZ)^{-1}$, where e is one electron charge and Z is ion mobility.

Wear

Component wear is an important consideration for long-term stability and one that is frequently overlooked. Many dusts, such as quartz, are very abrasive and cause wear wherever they strike metal surfaces, particularly at high velocity or at high pressure such as the scraper blade of the Wright Dust Feed. This wear can modify dust feedrates or air flowrates, either of which will change output dust concentration and the latter may affect particle size. Experience in our laboratory has been that the impactor orifice in the Wright Dust Feed enlarged 10% in diameter in about 100 hours of operation with an abrasive dust. The material that is worn away may introduce minute amounts of nickel, chromium, copper or other toxic materials into the generated aerosols, an important consideration for animal exposure studies.

SPECIAL DUST GENERATORS

This section describes selected laboratory-type dry-dispersion aerosol generators which are either recent developments or modifications of current designs that provide a stable dust output for periods long enough to be useful for animal exposure studies.

Laboratory Dust Generators

Fuchs and Marashkevich [18] described a dust generator with a continuous stable output in the range of 1 mg to 10 g/min. The dust is continuously metered through a coarse sieve to an aspirating nozzle under the action of pneumatic pulsations. The height of the powder layer above the screen is maintained constant by a photoelectric device connected to a dust supply hopper. The metered dust is aspirated to a dispersing nozzle consisting of a single-turn spiral made of narrow tubing that achieves nearly complete dispersion of particles larger than 1 μm. The deviation of the instantaneous concentration from the mean is reported to be less than 5-10%. The air entering the generator is ionized by means of an ac corona discharge to reduce triboelectrification.

A dust generator suitable for dispersing powders with mass median diameters from 2 to 80 μm has been described by Zahradnicek and Loffler [19]. A homogenous column of powder is pushed up a cylinder at a constant rate to a rotating brush where it is dispersed into the air stream of a dispersing nozzle. The dust feedrate can be adjusted to give output concentrations of 0.025-25 g/m^3 at 6.6 liter/min for the 5-mm diameter cylinder and 1-100 g/m^3 at 17 liter/min for the 80-mm cylinder. The device can operate without recharging from a few minutes to 75 hours depending on the feedrate. At feedrates greater than 80 mg/min, concentration variations are reported to be less than 5% of the mean. The device responds in one or two seconds to changes in dust feedrate.

Aerosol Generators with Size Classification Capability

It is frequently necessary to introduce a size-selective classifier on the output stream of a dust feeder to ensure that agglomerates or large particles are removed from the exiting aerosol. Elutriation columns [20], cyclones [12], impactors [6] and centrifugal separators [7] have been used for this purpose. Cyclones operated to match the ACGIH respirable size cut off curve [21] have been used to eliminate nonrespirable particles from aerosols used for animal exposure studies [22].

All the size separation techniques described above have the disadvantage that the oversize material is lost from the final aerosol stream. Moss and Millman [23] overcame this problem by comminuting the output of the Wright dust feed by passing it through a Trost air impact pulverizer [24]. The unit, shown in Figure 1, uses the collision of particles in opposed jets of compressed air at total flow of 5.5 cfm to reduce particle size. Particles pass from the impaction zone through a centrifugal flow region from which oversized particles are recycled back through the impaction zone and acceptable particles exit through the center outlet part. The degree of grinding is controlled by the size and flowrates of the jets and is limited to a minimum size of about 1 μm. For constant input conditions the output is stable with respect to particle size and concentration. The unit can be used with any type of positive pressure dust feed.

In Moss and Millman's experiments [23] the addition of the air mill increased by a factor of three the output of respirable fly ash produced by their system. Significant wear of stainless steel jets and liners was observed by Moss and Millman [23]; it was reduced by using urethane liners and nearly eliminated with tungsten carbide liners. The air mill is not suitable for sticky or damp materials. Like dust generators, the air mill produces a highly charged aerosol that must be neutralized to obtain a stable output concentration.

Turntable Dust Feed

The turntable dust feed [1, 25] is a relatively simple device that provides feed rates from 0.1 to about 100 g/min. Powder flows from a hopper, usually assisted by vibration or agitation, to flood a turntable, where a scraper bar forms a ribbon of powder of constant width and height on the turntable and deflects excess to waste. As the turntable rotates at constant speed, the dust ribbon passes under an air ejector, and is aspirated into the compressed air stream. Output concentration is varied by controlling the turntable speed and by using dust ribbons of different dimensions. This device can accommodate powders having a mass median diameter from 2 to 100 μm. Primary

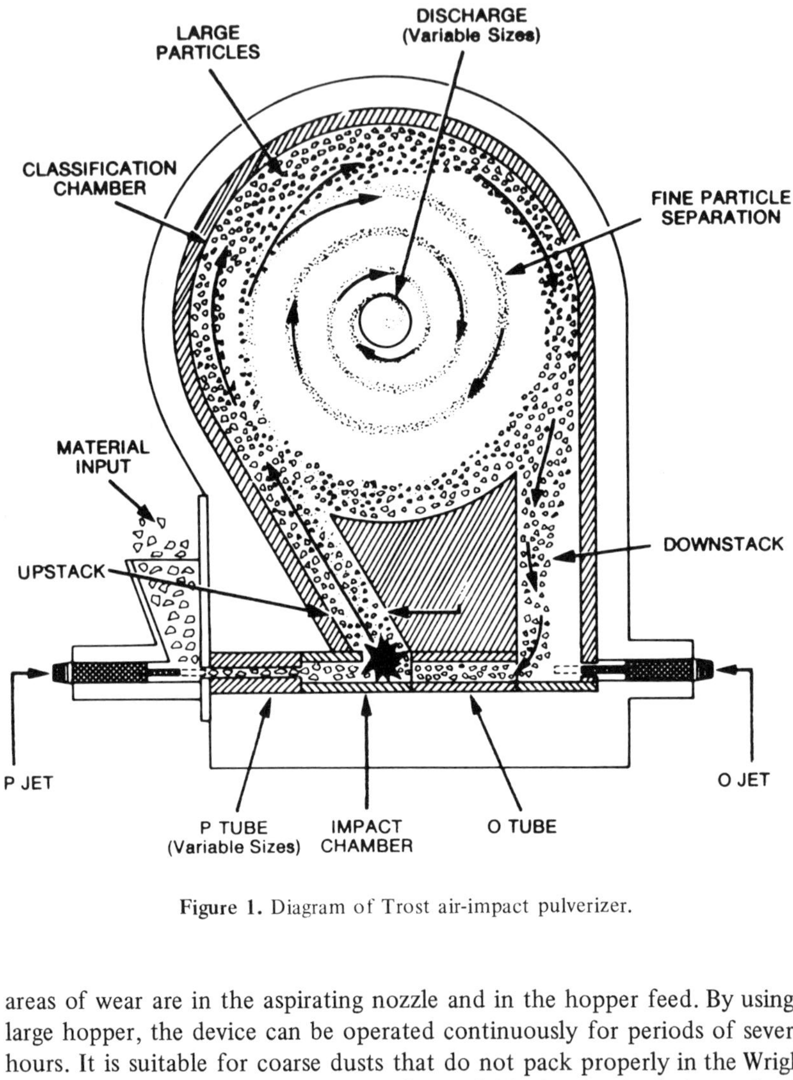

Figure 1. Diagram of Trost air-impact pulverizer.

areas of wear are in the aspirating nozzle and in the hopper feed. By using a large hopper, the device can be operated continuously for periods of several hours. It is suitable for coarse dusts that do not pack properly in the Wright dust feeder. Although not commercially available, its simple design and relatively low cost enable it to be fabricated readily. Because its output concentration stability depends on the flow properties of bulk powder, its short-term and long-term stability are less than other more precise metering systems discussed in this section.

The turntable dust feeder has been modified in our laboratory to improve its output stability. The turntable has four concentric grooves machined in its surface, 2.5, 5, 7.5 and 10 cm in radius. Each groove is 0.8 cm in width at the top, 0.3 cm wide at the bottom, and 0.4 cm deep. The dust is packed gently into one of the grooves by two hard rubber wiper blades. The first blade, along the turntable radius, wipes across two-thirds of the groove top. The second wiper is placed at a 45° angle to the radius and sweeps the balance of the dust from the groove top over the edge of the turntable and to waste. This scraper design permits minimum dust waste and prevents incomplete filling of the groove caused by aggregates trapped on the scraper blade. The gentle packing of the dust into the groove eliminates some of the variability associated with the loose powder feed system described previously.

At a turntable speed of 1 rpm, this device feeds dust at 3.5 to 14 cm^3/min, depending on which dust groove is used. Although the speed of the turntable can be varied to vary dust output rate, it is convenient to change grooves to provide rapid and accurate changes in output concentrations over a range of 4:1. This device has been used routinely for a number of years in our laboratory to disperse fly ash having a CMD of 0.3 μm. Scanning electron micrographs of the aerosol show good dispersion of particles as small as 0.1 μm.

Dust Generators for Problem Dusts

A number of laboratory-type generators have been built to handle difficult to disperse dusts. Laskin et al. [26] described a generator designed to disperse plastic powder that develops high electrostatic charges. A generator described by Dimmick [7] uses air erosion of a compacted plug to generate aerosols of bacteria and spores. Generators for the production of fibrous aerosols have been described by Holt and Young [27] and Timbrell et al. [28] and are reviewed in this volume by Spurny. Brown et al. [29] have described a generator for deliquescent dusts.

COMMERCIALLY AVAILABLE DUST GENERATORS

Although a great many dry-dispersion aerosol generators have been described [1-5], most are special-purpose laboratory devices that are not available commercially. Three commercial generators suitable for inhalation studies are described below.

Wright Dust Feed

The most widely used dry-dispersion aerosol generator for inhalation toxicology research is the Wright dust feed [6], shown in Figures 2 and 3. By packing the dust in a feed cylinder under controlled pressure and scraping the surface of the dust away and dispersing it at a constant rate, this device has eliminated the stability problems characteristic of loose dust feeding systems. The dust cylinder, containing compressed dust, rotates about the

Figure 2. Wright Dust Feed

stationary scraper head. Filtered dry compressed air passes up through an annular space in the spindle to the scraper head and passes through a groove at the outer edge of the scraper and flows radially along the scraper blade entraining the dust that is being cut away from the dust cake. Dust and air exit at high velocity through an axial hole to an impactor jet that breaks up any remaining agglomerated particles. Differential gearing propels the dust cylinder down onto the scraper blade at a constant rate, insuring a constant feedrate of dust.

The operating parameters for this generator are given in Table I. By changing the drive gear ratios the rate at which the dust is metered, and consequently output concentration, can be varied by a factor of 96 in 14 increments. Dust cylinders come in two diameters, 12.7 and 38 mm, to give a ninefold variation in dust feedrate. The long term maximum output variations are ± 15% for averaging periods of 3 minutes or longer [30, 31]. Short-term variations (periods of a few seconds) are much larger as the dust output tends to go in spurts. This necessitates a large averaging volume, ca. 100 liters, to

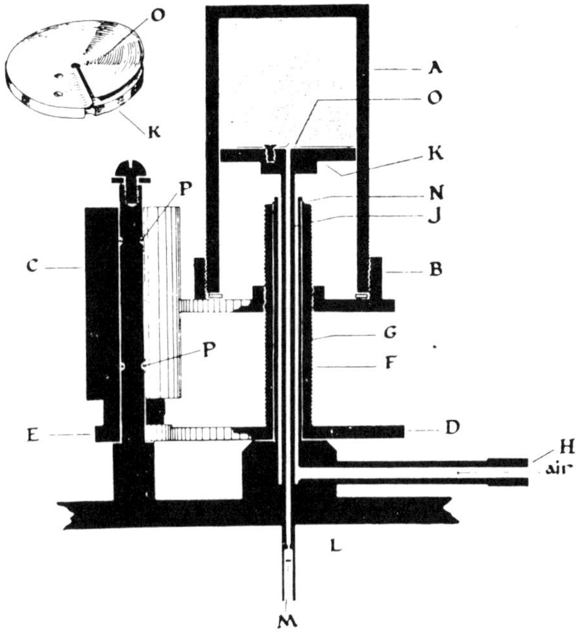

Figure 3. Schematic diagram of Wright Dust Feed: (A) dust cylinder (B) (C) (D) (E) differential gear train, (F) threaded spindle, (H) compressed air inlet, (K) scraper head, (L) outlet jet, (M) impaction plate, (O) scraper blade.

reduce these variations to the long period level. The source of long period output variation is the variations in density of the packet powder layers in the dust feed cylinder. The device works best with hard dry material when 90% of the particles are less than 10 μm, such as silica, uranium dioxide or other mineral dusts. It has been found to be unsatisfactory for soft sticky dusts such as coal and carbon black and for fine iron oxide and zinc oxide powders. An operating air flow that is too great tends to erode the dust column at a rate greater than it is metered and too low a flow does not fully break up agglomerates in the impactor nozzle outlet and can cause clogging in that section. Some initial experimentation is usually needed to determine proper packing pressure for loading the dust into the feed cylinder. With hard dry powder materials such as silica, the device is capable of providing good dispersion for particles as small as 0.1 or 0.2 μm [32]. Principle problems with this dust feed are wear at the scraper blade and outlet tube, enlargement of impactor orifice, and secondary wear along the threaded spindle and walls of the dust feed cylinder.

Table I. Comparison of Three Dry-Dispersion Aerosol Generators Commercially Available in the US

	Wright Dust Feed	NBS	TSI Fluidized Bed
Air flowrate[a], liter/min	10-40	50-85	5-20
Volumetric feedrate[a], mm^3/min	0.24-210	1,200-50,000	1.2-36
Range of output aerosol mass conc. g/m^3 (ρ_p=1.0 g/cm^3)	0.012-11.5	15-(100)	0.13-4.0
Recommended particle size range, μm	(0.2)-10	1-100	0.5-40
Variation in output concentration, relative S.D., % (averaging time $>$5 min)	7	10	5
Source and cost, US$ (Feb. 1979)	BGI[b] $1,930	FEW[c] $975	TSI[d] ($5,000)
Maximum time between recharging, hr	300	17[e]	(100)
Equilibration time after startup, min	~5	~1	200 (60 for change in concentration)

[a]Controlled by operator within ranges indicated.
[b]BGI Inc., 50 Guinan St., Waltham, MA 02154.
[c]F. E. Weideman & Sons, Box 197, Ossipee, NH 03864.
[d]TSI, Inc., 500 Cardigan Road, St. Paul, MN 55164.
[e]With auxillary hopper.
Numbers in parentheses are estimates.

NBS Dust Generator

The NBS Dust Generator, Figure 4, was originally developed to give reproducible high concentration test dust for filter testing [33], but is suitable for inhalation toxicology studies of low toxicity dusts or where a high flowrate is required. The dust flows from a hopper (200 cm^3 for the standard hopper or 1200 cm^3 for an auxiliary hopper), assisted by a vibrator, into the spaces between the teeth of a metering gear. A contoured spreader plate removes the excess dust as the gear rotates at constant speed (adjustable over a 40 to 1 range). The dust is sucked up from between the teeth by a compressed air ejector which is positioned diagonally to extract dust from parts of three teeth to minimize pulsations. An impactor plate is fitted to the ejector outlet to remove any large agglomerates. The operating parameters for this generator are summarized in Table I. The output of the generator can be extended to lower feedrates by grinding down the height of the gear teeth, thereby

Figure 4. NBS Dust Feed.

reducing the volume between the teeth or by partially filling the space between the teeth. Dennis et al. [34] found that this dust generator, when operated at 40 psig, produced an aerosol with fewer large agglomerates and a reduced mass median diameter than when operated at the recommended 2 psig [33]. Wear in the ejector may change the flowrate through the device when operated at a fixed pressure and thereby change output mass concentration over time. The position of the spreader plate is adjustable to compensate for wear.

TSI Fluidized-Bed Aerosol Generator

The commercial version of the fluidized-bed aerosol generator, based on the design of Marple et al. [12], has a 5.1 cm diameter fluidized-bed chamber filled with 180-μm bronze beads to a depth of 1.5 cm. As shown in Figure 5, the dust is eluted directly from the fluidized bed of bronze beads into a Krypton 85 radioactive discharging chamber at 9 liter/min. This elutriation velocity restricts output to particles less than 50 μm aerodynamic diameter. The dust is metered into the fluidized bed by a conveyor feed system using a continuous loop of number 3 ball chain originally described by Ebens and Vos [35]. The space between each ball in the ball chain picks up a fixed volume of dust and transports it to the fluidized bed through a close-fitting tube made of resilient material to remove excess dust from the chain. Instead of having pulleys or wheels within the fluidized bed, the ball chain passes over a Teflon skid. By varying the speed of the chain, dust flow into the bed

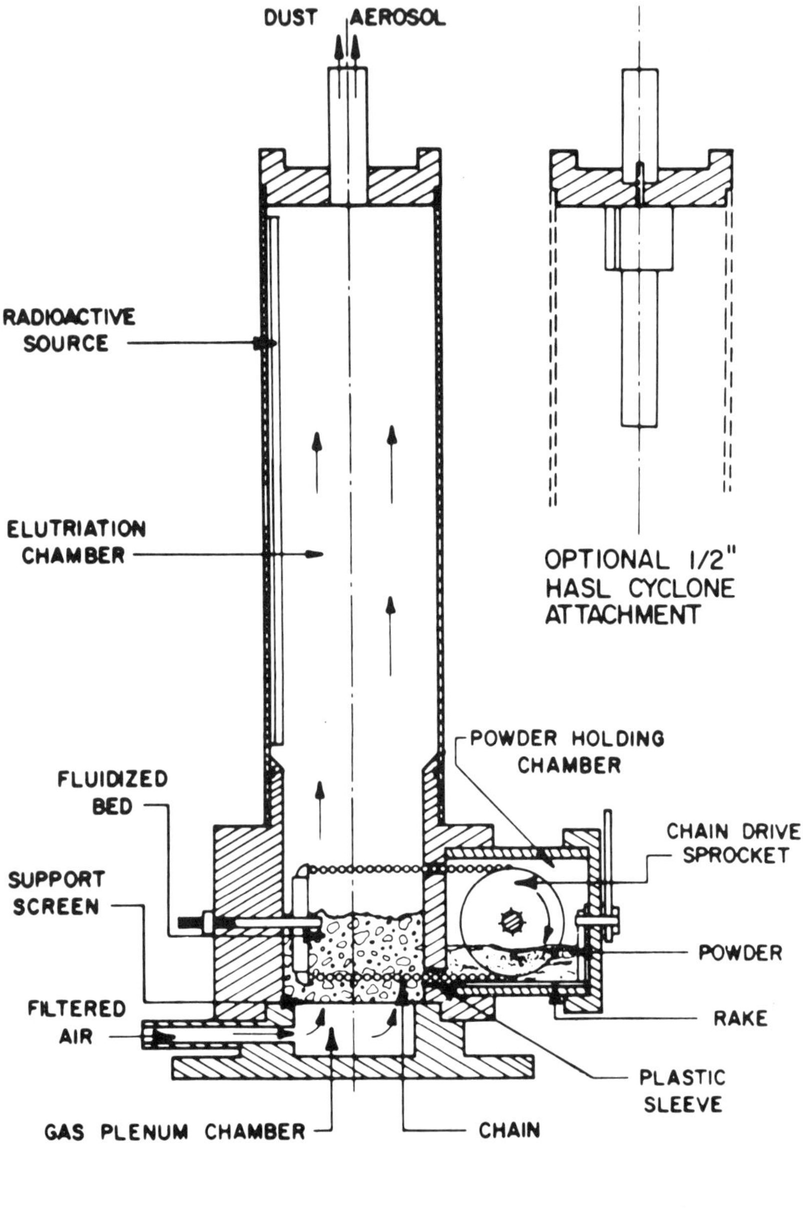

Figure 5. Diagram of fluidized-bed aerosol generator. (From Marple et al. [12].)

and dust generation rate can be varied by a factor of 30 (Table I). Immediately above the fluidized bed, output concentration fluctuates by about 10 to 15% with a frequency that matches the frequency with which beads deliver powder into the bed. This fluctuation can be damped out with a suitable averaging volume of about five liters to give long-term fluctuations of less than ± 10%. Because of the long residence time in the fluidized bed, it takes about 3.5 hr for the bed to achieve constant output following the initial introduction of powder. For a change in dust metering rate, equilibration takes place in about 1 hr. A 0.5-in. respirable mass cyclone can be attached at the outlet to limit the output to respirable dust.

Summary of Commercial Dust Generators

The operating ranges of particle size and concentration for the three commercial aerosol generators, compared in Table I, are shown in Figure 6. The Wright Dust Feed, which uses high-velocity impaction to break up agglomerates, has the capability to disperse the finest particles [36]. The generator with the most constant output concentration is the TSI fluidized-bed generator, which is also the most expensive. The NBS dust feed can achieve the highest output concentration but its minimum output concentration, prior to dilution, may be too high for small-scale studies. The minimum output concentration for the NBS dust feed can be reduced by a factor of 10 by the

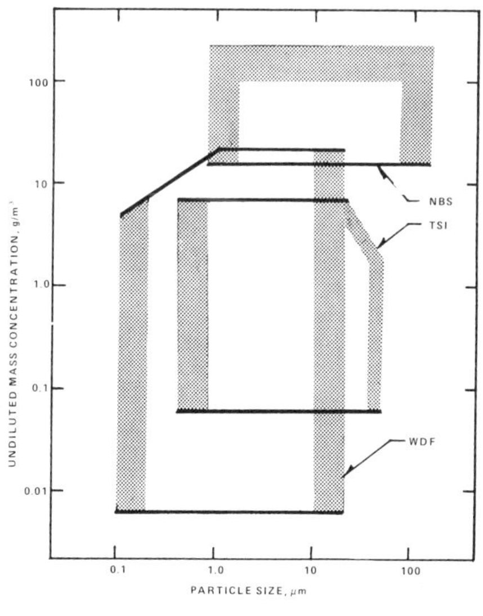

Figure 6. Range of application of commercially available dust feeds.

steps outlined above. Together, these three generators cover the full range of concentration and particle size needed for inhalation toxicology studies.

REFERENCES

1. Silverman, L., and C. E. Billings. "Methods of Generating Solid Aerosols," *J. Air Poll. Control Assoc.* 6:1-8 (1956).
2. Zahradnicke, A. "Methoden zur Aerosolherstellung aus vorgegebenen Feststoffhaufwerken," *Staub* 35:226-231 (1975).
3. Grassel, E. E. "Aerosol Generation for Industrial Research and Product Testing," in *Fine Particles*, B. Y. H. Liu, Ed., (New York: Academic Press, Inc., 1976), pp. 145-172.
4. Corn, M., and N. A. Esmen. "Aerosol Generation," in *Handbook on Aerosols*, R. Dennis, Ed., TID-26608, USERDA (1976), pp. 9-39.
5. Ebens, R. "Results With Some Continuous Dust Metering Devices for Maintaining Constant Concentrations in Test Installations," *Staub Engl.* 29(3):1-5 (1969).
6. Wright, B. M. "A New Dust-Feed Mechanism," *J. Sci. Inst.* 27:12-15 (1950).
7. Dimmick, R. L. "Jet Dispersor for Compacted Powders in the One-to-Ten-Micron Range," *Arch. Ind. Health* 20:8-14 (1959).
8. Fuchs, N. A. *Mechanics of Aerosols* (Oxford: Pergamon Press, Ltd., 1964), p. 373.
9. Guichard, J. C., and J. L. Magne. "Application du lit fluidise en milieu gazeux a la generation des aerosols," Note interieure No. 51 IRCHA, Vert-le-Petit, France.
10. Guichard, J. C. "Aerosol Generation Using Fluidized Beds," in *Fine Particles*, B. Y. H. Liu, Ed. (New York: Academic Press, Inc., 1976), pp. 173-193.
11. Willeke, K., C. S. K. Lo and K. J. Whitby. "Dispersion Characteristics of a Fluidized Bed," *J. Aerosol Sci.* 5:449-455 (1974).
12. Marple, V. A., B. Y. H. Liu and K. L. Rubow. "A Dust Generator for Laboratory Use," *Am. Ind. Hyg. Assoc. J.* 39:26-32 (1978).
13. Thermo Systems, Inc., 2500 Cleveland Ave. N., St. Paul, MN 55113.
14. Guichard, J. C. "The Triboelectrification of Metallic Particulate Aerosols Generated by a Two-Component Fluidized Bed," *Staub Engl.* 33(4):174-179 (1973).
15. Whitby, K. T. "Generator for Producing High Concentrations of Small Ions," *Rev. Sci. Inst.* 32(12):1351-1355 (1961).
16. Zamorani, E., and G. Ottobrini. "Aerosol Particle Neutralization to Boltzmann's Equilibrium by AC Corona Discharge," *J. Aerosol Sci.* 9:31-39 (1978).
17. Liu, B. Y. H., and D. Y. H. Pui. "Electrical Neutralization of Aerosols," *J. Aerosol Sci.* 5:465 (1974).
18. Fuchs, N. A., and F. I. Murashkevich. "Laboratory Powder Disperser (Dust Generator)," *Staub Engl.* 30(11):1-3 (1970).
19. Zahradnicek, A., and F. Löffler. "Eine neue Dosiervorrichtung zur Erzeugung von Aerosolen aus vorgegebenen feinkörnigen Feststoffen," *Staub* 36(11):425-427 (1976).

20. Rudnick, S., Sc.D. Thesis, Harvard University, October, 1978.
21. ACGIH. "Threshold Limit Values for Chemical Substances in the Workroom Environment," (Cincinnati, OH: ACGIH, 1977), p. 34.
22. McFarland, K. D., O. G. Raabe, B. Tarkington and B. Prentice. "Generation of Aerosols of Power Plant Fly Ash for Inhalation Exposure of Rodents and Monkeys," presented at American Industrial Hygiene Conference, May 12, 1978, Los Angeles, CA, paper No. 314.
23. Moss, O., and E. M. Milliman. "Respirable Aerosol from a Combined Trost Air Jet Mill–Wright Dust Feed Aerosol Generator," presented at American Industrial Hygiene Conference, May 12, 1978, Los Angeles, CA, paper No. 313.
24. Garlock Inc., Plastomer Products, Newtown, Pennsylvania 18940.
25. Billings, C. E., M. W. First, R. Dennis and L. Silverman. USAEC Report NYO-1590, Harvard School of Public Health, Air Cleaning Laboratory (1961).
26. Laskin, S., S. Posner and R. Drew. "New Designs for Dust Generating Systems for Inhalation Studies," presented at 1968 AIHA Conference St. Louis, MO, May 1968.
27. Holt, P. F., and D. K. Young. "A Dust Feed Mechanism Suitable for Fibrous Dust," *Ann. Occup. Hyg.* 2:249-256 (1960).
28. Timbrell, V., A. W. Hyett and J. W. Skidmore. "A Simple Dispenser for Generating Dust Clouds from Standard Reference Samples of Asbestos," *Ann. Occup. Hyg.* 11:273-281 (1960).
29. Brown, J. R., J. Harwood and E. Mastromatteo. "A Dust Dispenser for Deliquescent Material," *Ann. Occup. Hyg.* 5:145-147 (1962).
30. Leach, L. J., C. J. Spiegl, R. H. Wilson, G. E. Sylvester and K. E. Lauterbach. "A Multiple Chamber Exposure Unit Designed for Chronic Inhalation Studies," *Am. Ind. Hyg. Assoc. J.* 20:12-22 (1959).
31. Collins, G. F. "Measurement of Variation in Output Rate from Aerosol Generators," *J. Aerosol Sci.* 6:169-172 (1975).
32. Kotrappa, P. "Shape Factors for Aerosols of Coal, UO_2, and ThO_2 in Respirable Size Range," in *Assessment of Airborne Particles*, T. T. Mercer, P. E. Morrow and W. Stober, Eds. (Springfield, IL: Charles C. Thomas, Publisher, 1972), pp. 331-358.
33. Dill, R. S. "A Test Method for Air Filters," *Trans. Am. Soc. Heat. Vent. Eng.* 44:379 (1938).
34. Dennis, R., L. Silverman, C. E. Billings, A. T. Rossano, E. Conners, R. D. Coleman, D. M. Anderson, W. R. Samples and P. Drinker. "Air Cleaning Studies Progress Report July 1, 1953 to June 30, 1954," NYO-4608 (1956) p. 17. Work done under AEC Contract No. AT (30-1)841.
35. Ebens, R., and M. Vos. "A Device for the Continuous Metering of Small Dust Quantities," *Staub Engl.* 28(5):24-26 (1968).
36. McFarland, A. R., C. Ortiz, G. L. Fisher and O. B. Raabe. "Aerosolization of Bulk Powders Utilizing Fluidized-Bed and Wright Dust Feed Techniques," 1977 Annual Report, Radiobiology Laboratory, Univ. of California at Davis, pp. 10-13.

CHAPTER 9

EXPERIMENTAL STUDY ON THE DISPERSION OF FINE PARTICLES INTO AIR

Koichi Iinoya
Department of Chemical Engineering
Kyoto University
Kyoto, Japan

Hiroaki Masuda
Department of Chemical Engineering
Hiroshima University
Hiroshima, Japan

INTRODUCTION

Various kinds of dispersers have been developed in order to generate test aerosols from powdered materials [1, 2]. However, the aerosol produced by dispersing fine powder is usually agglomerated [3] and the size distribution is different from that of a completely dispersed sample. Agglomerated aerosols are not suitable for the examination of industrial dust collectors, for the generation of airborne dust in an exposure chamber or for investigations of problems such as particle deposition or bounce.

On the other hand, quasimonodisperse aerosols may be obtained by some size-classification methods, if the particles are completely dispersed. Then the size-related investigation of a particulate process becomes possible by use of any industrial powder.

This study presents experimental results obtained by use of a modified mixer-type disperser and a fluidized bed which utilizes glass beads as a dispersion aid. Dispersion of powder through a narrow tube is also examined.

EXPERIMENTAL APPARATUS AND PROCEDURES

The experimental arrangement for testing the performance of dispersers is shown in Figure 1. A special microfeeder (Sankyo Dengyo Co. Ltd., Japan*) developed in our laboratory [4] was utilized to achieve constant and continuous feeding of a small amount of fine powder. The particles were airborne in an ejector and were then ducted to the disperser. The size distributions of the test particles before and after the disperser were measured by a cascade impactor (Schimadzu Seisakusho Ltd., Japan) which was also developed in our laboratory and measures particle size distributions irrespective of the particles' degree of agglomeration.

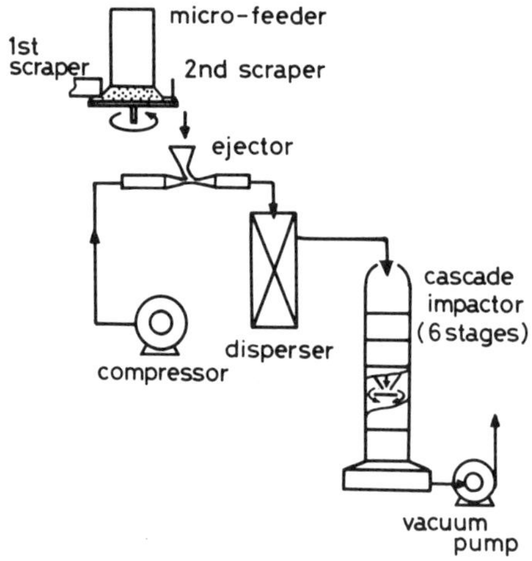

Figure 1. Schematic representation of the experimental arrangement.

On the other hand, the size distribution of completely dispersed particles is obtained by a liquid sedimentation method or by use of a Coulter counter with dispersion agents.

Figure 2 shows the modified mixer-type disperser with two different rotors. Figure 3 shows the fluidized bed where coarse glass beads (0.5 mm) are used as a dispersion aid. Calcium carbonate, metallic silicon and talc are used as test powders. Their characteristics are listed in Table I.

---

*Minimum feeding rate of the microfeeder is about 0.1 cm^3/min.

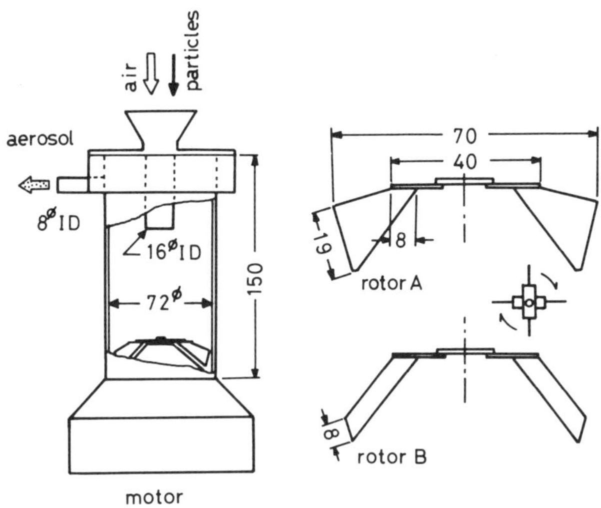

Figure 2. Mixer-type disperser; dimensions are in mm.

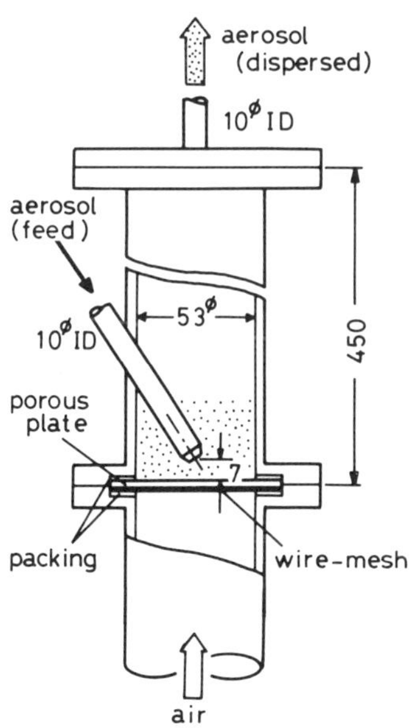

Figure 3. Fluidized-bed disperser; dimensions are in mm.

Table I. Characteristics of Test Powders

Powder	Solid Density ρ_p (g/cm^3)	Mass Median Diameter[a] D_{p50} (μm)	Cohesion Force per Unit Area[b] (dyne/cm^2)
Calcium Carbonate $CaCO_3$	2.6	1.4[c]	4900
Metallic Silicon MeSi	2.15	0.4[d]	8900
Talc	2.7	7.0[d]	2900

[a]Completely dispersed condition.
[b]By a shear tester.
[c]Sedimentation balance.
[d]Coulter counter.

EXPERIMENTAL RESULTS AND DISCUSSIONS

Mixer-Type Disperser

Some of the results are shown in Figure 4. The size distribution of dispersed calcium carbonate particles is almost the same as that of completely dispersed ones, which is measured by a liquid sedimentation method. The performance of dispersion of metallic silicon particles is, however, not as good as that for calcium carbonate. The mass median diameter of dispersed metallic silicon particles is six times as large as that of the completely dispersed ones determined by a liquid sedimentation method. The results show

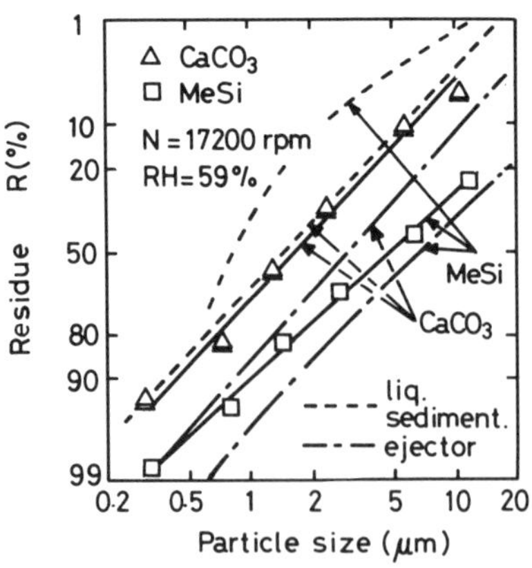

Figure 4. Size distributions by mass of calcium carbonate and metallic silicon particles dispersed by the mixer-type dispenser.

that the dispersion of finer and more cohesive powder (see Table I) is very difficult and that the dispersion of finer powder does not always generate finer test aerosols.

Figure 5 shows the results obtained with talc particles. It seems that the particles are easily broken down by impingement or strong turbulence in the mixer, resulting in a wider size distribution and a smaller mass median size. Such a breakage of particles may always be associated with strong dispersion [5]. It may be necessary in the future to disperse brittle and large particles with less force so that their breakup may be avoided. Reentrainment of collected particles in the cascade impactor might also affect the measured size distribution, because the talc particles are very slippery.

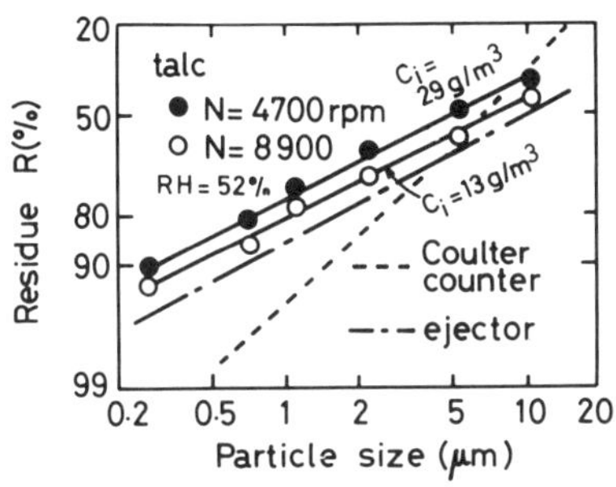

Figure 5. Size distributions of talc particles dispersed by the mixer-type disperser.

The results obtained by use of the two different rotors A and B are virtually the same as shown in Figure 6, if the air flowrates through the mixer are kept constant.

Figure 7 shows the relationship between the mass median diameters D_0 of the dispersed particles and the rotational speed N. The diameters D_0 of calcium carbonate and metallic silicon particles decrease gradually with the rotational speed, but the diameters D_0 of talc particles increase with the rotational speed. This fact shows that higher rotational speed promotes not only dispersion but also agglomeration.

The humidity of air also affects the dispersion somewhat. Lower humidity of air is preferred because it results in a smaller median size, as shown in Figure 7.

Fluidized-Bed Disperser

Size distributions of calcium carbonate particles dispersed by the fluidized bed are shown in Figure 8. Dispersion at the smallest air flowrate of

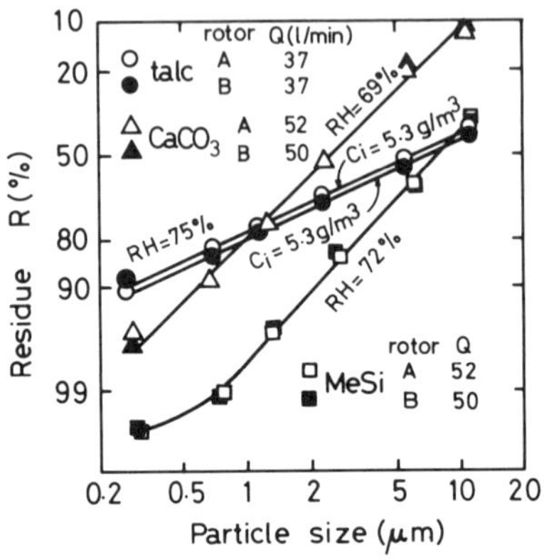

Figure 6. Effect of rotor dimension on the performance of the mixer-type disperser (RH = relative humidity).

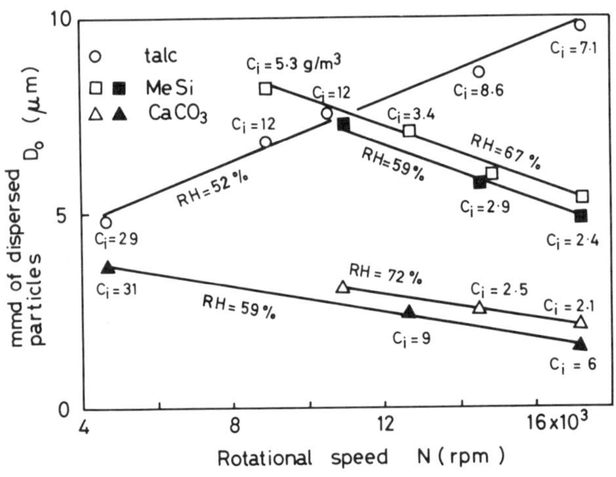

Figure 7. Relationship between mass median diameter and rotational speed (c_i = dust concentration).

47 liter/min is less than the dispersion obtained by the ejector. A higher flow-rate is required to achieve better dispersion, but an upper limit for the flow-rate is rendered when coarse glass beads (dispersion aid) are entrained with the effluent flow. In this sense, heavier particles such as bronze spheres may be preferable as a dispersion aid [6]. Mechanical vibrations may also help the dispersion [7].

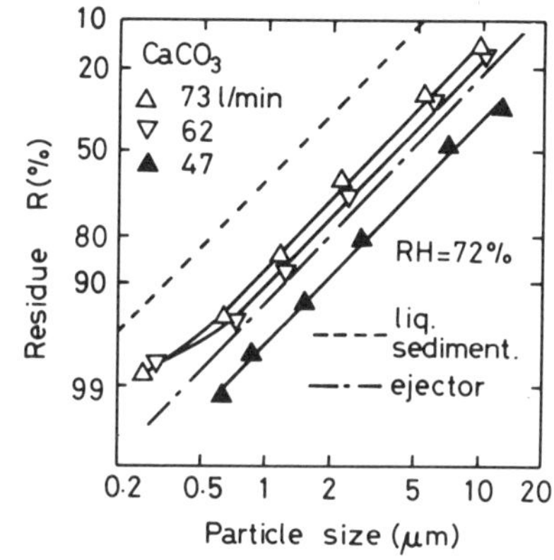

Figure 8. Size distributions by mass of calcium carbonate particles dispersed by the fluidized bed.

The relationship between the mass median diameters D_0 and the air flowrate Q is shown in Figure 9. While the results for talc suggest agglomerations with increasing flowrate, the mass median diameters of calcium carbonate and metallic silicon particles decrease gradually with increase in flowrate.

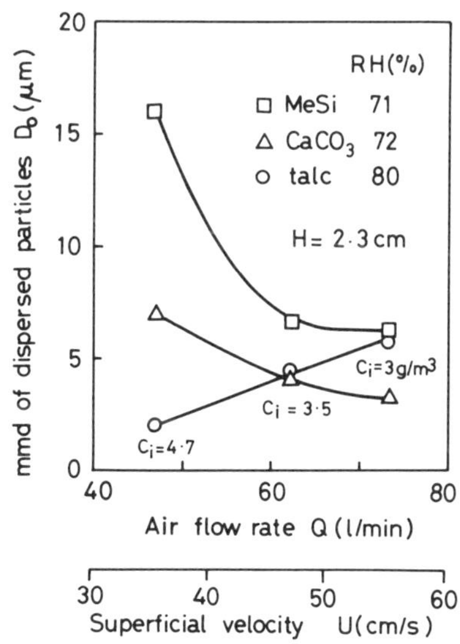

Figure 9. Effect of air flowrate on the mass median diameters of particles dispersed by the fluidized bed.

Another factor affecting the dispersion is the amount of coarse glass beads. Figure 10 shows the relationship between the mass median diameter of the dispersed particles and the bed height of coarse glass beads. The bed heights of beads for both calcium carbonate and metallic silicon are best where the mass median diameters of the dispersed particles have the minimum values. At the optimum bed heights, the performance of dispersion is comparable with that obtained by the mixer-type disperser. However, the bed height has no effect on the dispersion of talc.

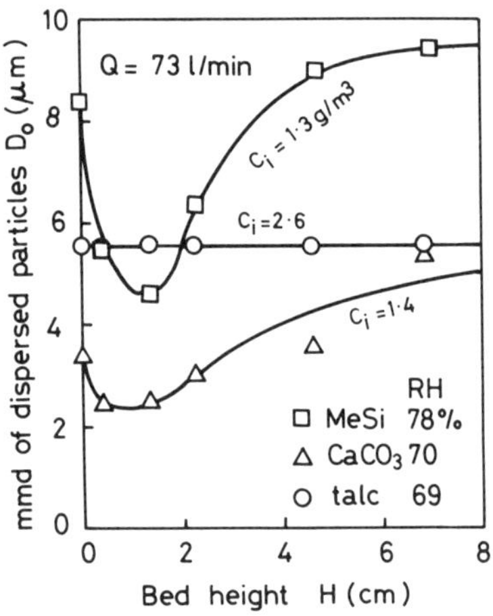

Figure 10. Effect of bed height on mass median diameters of particles dispersed by fluidized bed.

Residence Time in Dispersers

The residence time of particles in the dispersers was measured by an electrostatic method as follows. A brass tube (20 cm in length, 6 mm diameter) was inserted into the disperser outlet and a small amount of powder was fed into the disperser as an impulsive input. The electric current generated by the impact of particles on the tube wall is proportional to the particle flowrate [8]. The current from the tube wall to earth was recorded by an electromagnetic oscillograph. The distributions of residence time τ were represented by log-normal distributions as shown in Figure 11. Table II shows the mean residence time $\overline{\tau}$ obtained by use of the following equation [9].

$$\bar{\tau} = \exp(\mu + \tfrac{1}{2}\sigma^2) \tag{1}$$

where $\mu = \ln \tau_{50}$
$\sigma = \ln \tau_{16} - \ln \tau_{50}$

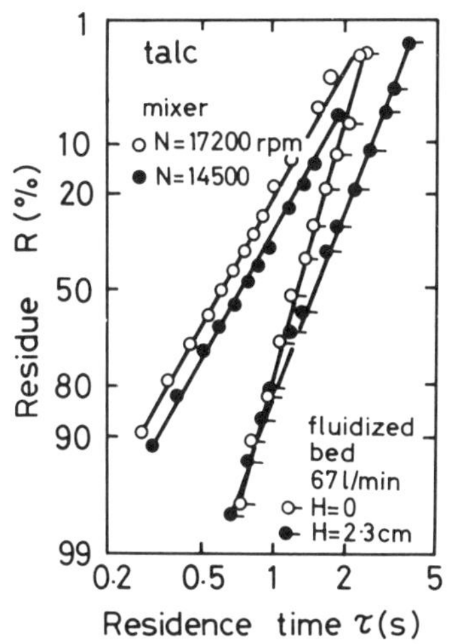

Figure 11. Residence time distributions in dispersers (log-normal).

Because of back mixing in the dispersers, the mean residence time $\bar{\tau}$ is longer than the corresponding apparent residence time τ', which is calculated as the inner volume of a disperser divided by the air flowrate. The residence times of the mixer-type disperser at different air flowrates are nearly constant for the experimental conditions, as shown in Table II. This fact may be due to stronger back mixing in the mixer for higher flowrates. The residence times for the fluidized bed are also almost constant because the experiments were performed over a small range in bed heights.

Effect of Particle Concentration on Performance

The time derivative of the number concentration dn/dt is proportional to the number concentration n in the disperser if the dispersion depends on turbulence and/or shear flow. On the other hand, particle-to-particle collisions may result in agglomeration of particles. Then the number concentration of particles in the disperser changes with dispersing time, which may be predicted by

Table II. Mean Residence Time

Powder	Disperser	Air Flowrate Q (liter/min)	Apparent Residence Time τ' (sec)	Mean Residence Time $\bar{\tau}$ (sec)
$CaCO_3$	Mixer Type[a]	36	1.0	1.0
		54	0.7	0.9
		76	0.5	1.0
$CaCO_3$	Fluidized Bed	67 (H=0cm)	1.0	1.5
		67 (H=2.3)	1.0	1.8
		77 (H=2.3)	0.9	1.6
MeSi	Mixer Type[a]	36	1.0	2.1
		54	0.7	1.6
		76	0.5	1.6
MeS_i	Fluidized Bed	67 (H=0)	1.0	1.7
		67 (H=2.3)	1.0	1.8
		77 (H=2.3)	0.9	1.7
Talc	Mixer Type[a]	36	1.0	1.0
		54	0.7	0.9
		76	0.5	1.0
Talc	Fluidized Bed	67 (H=0)	1.0	1.4
		67 (H=2.3)	1.0	1.7
		77 (H=2.3)	0.9	1.6

[a]Rotor A.

$$\frac{dn}{dt} = \alpha n - \beta n^2 \tag{2}$$

The first term on the right-hand side represents the dispersing action and the second term represents agglomeration [10]. The solution of Equation 2 is given by

$$\frac{n_o}{n_i} = \frac{\beta n_i}{\alpha}[1 - \exp(-\alpha\bar{\tau})] + \exp(-\alpha\bar{\tau}) \tag{3}$$

where n_i and n_o are the initial and final values of the number concentration, respectively. Note that we have the following relationship from material balance considerations:

$$\frac{D_o}{D_i} = \left(\frac{n_i}{n_o}\right)^{1/3} \tag{4}$$

where D_i and D_o are the initial and final mass median diameters of particles in the disperser, respectively.

The hold-up number concentration* n_i is related to the mass-flow concentration* c_i through

$$n_i = \frac{6 c_i \bar{\tau}}{\pi D_i^3 \rho_p \tau'} \qquad (\tau' < \bar{\tau}) \tag{5}$$

where ρ_p is the density of agglomerated particles which may be less than the particles' true density. Substitution of Equations 4 and 5 into Equation 3 gives

$$D_o^3 = \frac{6\beta\bar{\tau}}{\pi\rho_p \alpha \tau'}[1 - \exp(-\alpha\bar{\tau})]\, c_i + D_i^3 \exp(-\alpha\bar{\tau}) \tag{6}$$

Equation 6 shows that the third power of the mass median diameter of dispersed particles D_o is proportional to the mass concentration c_i as long as the dispersing conditions are kept constant throughout the experiments.

Figure 12 shows one of the experimental results. A narrow tube (20 cm in length, 6 mm in diameter) was also examined as a disperser. These results support the above discussions.

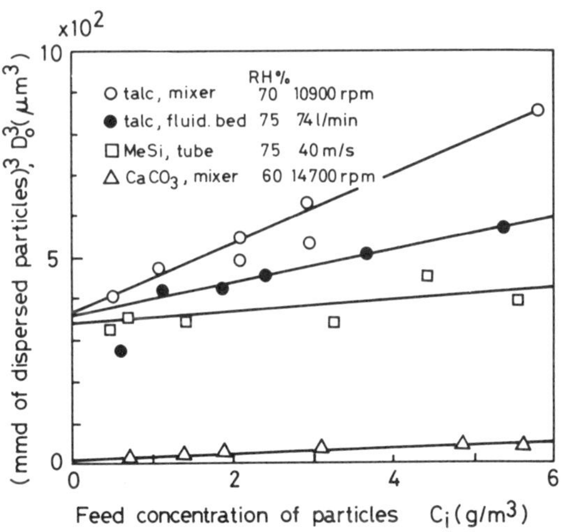

Figure 12. Effect of particle feed concentration on dispersed mass median diameter.

*The mass-flow concentration is equal to the mass flowrate of particles divided by volumetric flowrate of gas. On the other hand, the holdup mass-concentration is equal to the mass of particles in the unit volume of suspension. If the mean velocity of particles is lower than that of the gas the holdup mass-concentration is larger than the mass-flow concentration. The values of these concentrations coincide with each other only when the particle velocity equals the gas velocity.

Dispersion and Agglomeration Constants

The constants α and β in Equation 6 are determined from the experimental results and are listed in Table III. It is assumed that the initial size of agglomerated particles is 20 μm and that the particle density is equal to the true density.

Equation 6 may be simplified as follows

$$\frac{D_o}{D_i} = \sqrt[3]{\frac{\beta n_i}{\alpha} + \exp(-\alpha\bar{\tau})} \qquad (7)$$

because the experimental results showed that

$$1 - \exp(-\alpha\bar{\tau}) \cong 1 \qquad (8)$$

Table III. Dispersion and Agglomeration Constants

Powder	Disperser	Dispersion Constant α (sec^{-1})	Agglomeration Constant $\beta \times 10^5$ (cm^3/sec)	$(\beta/\alpha) \times 10^5$ (cm^3)
	Mixer, 14,700 rpm	7.3	4.9	0.67
$CaCO_3$	Fluidizer, 74 liter/min	3.2	0.7	0.22
	Tube, 40 m/s	10.4	2.6	0.25
	Mixer, 14,700 rpm	1.5	6	4
MeSi	Fluidizer, 74 liter/min	1.4	6	4.3
	Tube, 40 m/s	5.9	8	1.4
	Mixer, 10,900 rpm	3.1	39	13
Talc	Fluidizer, 74 liter/min	1.9	8	4.2
	Tube, 40 m/s	4.7	23	4.9

Note: assuming $D_i = 20\ \mu m$ and ρ_p = true particle density.

The mass median diameter of dispersed particles D_o is determined by Equation 7 as an equilibrium state between dispersion and agglomeration. Equation 7 shows $D_o \cong D_i$ for the special condition $\alpha = \beta n_i$, because of Equation 8. It is also obvious from Equation 7 that $D_o < D_i$ when α is larger than βn_i and $D_o > D_i$ when α is smaller than βn_i. The former is the apparent dispersion and the latter is the apparent agglomeration. Further discussion on the dispersion needs the functional relationship between the constant α, β and the operating conditions. It has been shown that the agglomeration constant β increases with energy dissipation per unit volume [11], which may be proportional to the third power of the rotor speed in a mixer [12].

Data for the dispersion constant are, however, not available at this stage of the work. Further experimental study of the constants α and β may promote a better understanding of the dispersion process.

CONCLUSIONS

Fine powders of calcium carbonate, metallic silicon and talc were dispersed into air by use of a narrow tube, a mixer-type disperser, and a fluidized bed which utilized coarse glass beads as a dispersion aid.

It is difficult to get completely dispersed aerosols from a fine powder such as metallic silicon, which is held together by large cohesion forces. On the other hand, talc particles are easily broken down by strong dispersing actions and will partially agglomerate again in the disperser. As a result, they tend to have a wider size distribution and a smaller mass median diameter than the original particles. Therefore, metallic silicon and talc may not be adequate as test aerosols.

Calcium carbonate powder is almost completely dispersed by a mixer-type disperser, resulting in a good test aerosol. In a mixer-type disperser, higher rotor speed gives better dispersion of particles for both calcium carbonate and metallic silicon. The shape of the rotor does not have much of an effect on dispersion.

In a fluidized-bed disperser, low velocity of fluidization causes agglomeration instead of dispersion. There is an optimum bed height of coarse glass beads, where the mass median diameter of dispersed particles is smallest.

The inlet concentration of feed particles should be as small as possible because the mass median diameter of dispersed particles increases with particle concentration. There is an optimum operating condition for every disperser, because strong dispersing actions induce agglomeration as well.

ACKNOWLEDGMENTS

The authors express their appreciation to Mr. S. Fushiro for carrying out the experimental work.

REFERENCES

1. Iinoya, K. "Feeding Methods of Dust," (in Japanese) in *Shujin Sochi*, K. Iinoya, Ed. (Tokyo: Nikkan Kogyo Shinbunsha, 1963), pp. 84-89.
2. Zahradnicek, A. "Methoden zur Aerosolherstellung aus vorgegebenen Feststoffhaufwerken," *Staub Reinhalt. Luft* 35:226-231 (1975).
3. Mori, Y., A. Suganuma, M. Oka and Y. Kaida. "Influence of Air Humidity on Airborne Dust Generation Phenomena," (in Japanese) *Kagaku Kogaku* 34:198-203 (1970).

4. Masuda, H., H. Kurahashi, M. Hirota and K. Iinoya. "Micro-Feeding by Table Feeder," (in Japanese) *Kagaku Kogaku Ronbunshu* 2:286-290 (1976).
5. Zahradnicek, A., and F. Löffler. "Desagglomeration von feinkörnigen Feststoffen in Gasströmungen," *Verfahrenstechnik* 11:229-232 (1977).
6. Willeke, K., C. S. K. Lo and K. T. Whitby. "Dispersion Characteristics of a Fluidized Bed," *J. Aerosol Sci.* 5:449-455 (1974).
7. Spurny, K., C. Boose, D. Hochrainer and F. J. Mönig. "Ein Schwingbett Aerosolgenerator für fäserige und pulverförmige Stäube," *Zbl. Bakt. Hyg.* B161:362-370 (1976).
8. Masuda, H., T. Komatsu and K. Iinoya. "The Static Electrification of Particles in Gas-Solids Pipe Flow," *Am. Inst. Chem. Eng. J.* 22:558-564 (1976).
9. Chakravarti, I. M., R. G. Laha and J. Roy. *Handbook of Methods of Applied Statistics, Vol. 1* (New York: John Wiley & Sons, Inc., 1967), p. 153.
10. Zebel, G. "Coagulation of Aerosols," in *Aerosol Science*, C. N. Davies, Ed. (London: Academic Press, Inc., 1966), p. 37.
11. Friedlander, S. K. *Smoke, Dust and Haze* (New York: John Wiley & Sons, Inc., 1977), p. 192.
12. Schwartzberg, H. G., and R. E. Treybal. "Fluid and Particle Motion in Turbulent Stirred Tanks," *I.E.C. Fund.* 7:1-6 (1968).

CHAPTER 10

SIZE-RELATED CHEMICAL AND PHYSICAL PROPERTIES OF POWER PLANT FLY ASH

Gerald L. Fisher
Radiobiology Laboratory
University of California
Davis, California 95616

INTRODUCTION

Recently, increasing emphasis has been placed on the utilization of coal as the major fuel in electric power generation. Associated with the combustion of coal is the production of vast quantities of gaseous and particulate by-products. This chapter describes the physical and chemical properties of fly ash, the major primary particulate matter associated with coal combustion. Furthermore, to aid in the elucidation of the chemical properties of fly ash, appropriate results from short-term bioassays will also be briefly reviewed.

To appreciate the magnitude of present and future fly ash production, one need only consider that in 1977 fly ash produced and collected by coal-fired power plants was considered to be the sixth most abundant mineral in the United States [1]. It is estimated that of the 50 million tons of fly ash produced in 1977, only 13% was reused commercially, predominantly as a partial replacement for cement or as fill material in construction. Since most of the in-plant collected fly ash is disposed of by deposition in landfills, power plant fly ash is presently considered to be the most abundant solid waste product in the United States. In general, the in-plant collection efficiency for aerosolized particulate matter varies inversely as a function of aerodynamic particle size.

In a review of particulate abatement technologies, Vandergrift et al. [2] described collection efficiency as a function of particle size for a variety of control technologies including electrostatic precipitators, fabric filters, wet scrubbers and cyclones. Average collection efficiencies for a medium-efficiency electrostatic precipitator (ESP) were 90%, 70% and 35% for 1.0 μm, 0.1 μm and 0.01 μm particles, respectively. Interestingly, the venturi wet scrubber (VWS) was more efficient (99.5%) for 1.0-μm particles and less efficient ($<$ 1%) for 0.01-μm particles. A crossover in the ESP and VWS efficiency curves was observed at 0.35 μm.

In this chapter, toxicologically important physical and chemical properties of stack-collected size-classified coal fly ash are discussed. Because lung deposition and clearance [3] are dependent on the size-distribution of inhaled aerosols and because the collection efficiency of current particulate abatement technology is minimal for respirable particles, the size-dependence of the physical and chemical properties of coal fly ash is emphasized.

FLY ASH COLLECTION

Four size-classified fly ash fractions were collected from the stack breeching of a large, modern power plant burning low-sulfur high-ash pulverized coal [4]. Collection took place at 95°C over a 30-day period, downstream from the power plant's cold-side ESP. During the collection period, unfractionated fly ash samples were also obtained from the ESP collection hopper.

The stack-collected fly ash was size-classified in situ utilizing a unique combination of two cyclone separators in series followed by a 25-jet centripeter, as described by McFarland et al. [4]. After passage through the appropriate classification device, fly ash was collected on high-efficiency fabric filters. The filters were pulsed every two minutes with high-velocity pulses of air which deposited the collected fly ash in stainless steel collection hoppers. Air pulsation provided two important functions: (1) continuous operation of the fractionator without creation of a large pressure differential was possible, and (2) the flue gas was continuously drawn through collected particulate matter deposited on the fabric filter, thus avoiding the potential sampling artifacts reported by Pitts et al. [5]. After collection, the individual samples were homogenized in acid-cleaned double-distilled-deionized (DDD) water-washed containers and weighed aliquots stored in either acid-cleaned, DDD-water-washed flame-sealed glass ampules or in polyethylene containers which were subsequently sealed with paraffin.

The resulting four fractions had volume median diameters (VMDs), determined by Coulter analysis, of 2.2, 3.2, 6.3 and 20 μm with associated geometric standard deviations (σ_g's) of approximately 1.8 for all fractions. The volume distribution data agreed well with mass distributions determined in aqueous dispersion from Stoke's law of settling. Comparison of the relative mass of collected particles of calculated aerodynamic diameters less than 2 μm with the relative mass of such particles in an isokinetically collected sample, indicated a 15-fold enhancement of these fine respirable particles in the finest fly ash fraction [6].

FLY ASH MORPHOLOGY

Light-microscopic (LM) or scanning-electron-microscopic (SEM) analyses have provided useful information on the morphogenesis of fly ash during the coal combustion process. On the basis of LM morphological analyses, a morphogenesis scheme has been developed which relates shape and relative opacity to particle composition and to the extent and duration of exposure within the combustion zone of the boiler [6]. Eleven LM-defined morphological particle classes have been employed in the morphogenesis scheme. If the parent material is a clay mineral, then exposure to boiler temperatures will initially result in the formation of amorphous particles with rounded edges, containing vesicles from mineral decomposition and subsequent outgassing. Further heating will result in sphere formation. A variety of spherical particles are found in coal fly ash:

1. nonopaque solid glass spheres;
2. cenospheres (hollow spheres);
3. plerospheres (spheres packed with spheres); and
4. spherical particles containing quench crystals.

Opaque spheres, which are predominantly magnetite (Fe_3O_4) may also be observed in fly ash as small aggregates due to magnetic attraction. The increased coloration in spherical particles from water-white to yellow, orange and red has been demonstrated by SEM X-ray analysis to be associated with increased iron content within the individual particles [7]. When the parent material is the organic component of coal, incomplete combustion may give rise to soot formation. Particles which are mixtures of the organic and inorganic components have also been identified.

The relative abundances of the morphological classes has been determined by LM analysis of 3000 individual particles for each of the four fractions [6]. In all fractions, spherical particles were most abundant (Figure 1). However, amorphous and vesicular particles, including cenospheres, predominated in the coarsest fraction (66% by number) while solid, nonopaque spheres predominated in the finest fraction (87% by number). Similarly, density determinations indicated an inverse dependence on particle size, which can be explained by the greater abundance of solid spheres and the lesser abundance of vesicular particles in the finer fly ash fractions.

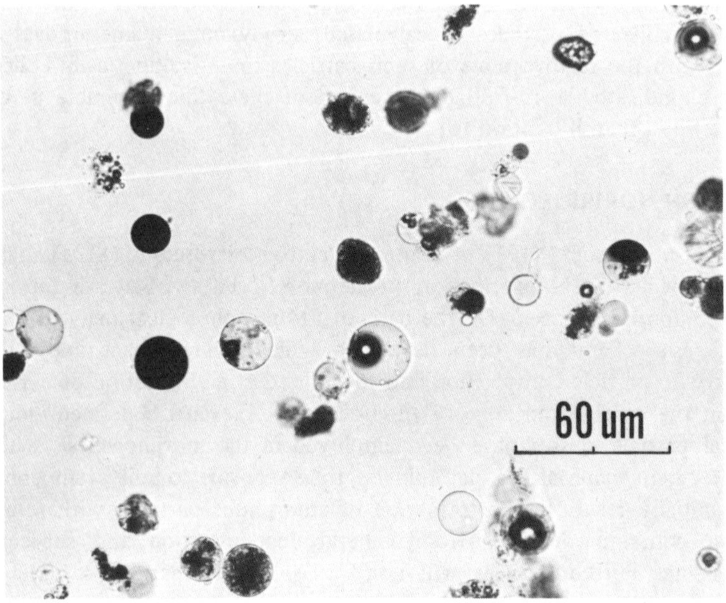

Figure 1. Light micrograph of a general field of fly ash particles indicating the predominant morphological particle classes: Clear, solid spheres, opaque spheres, spheres with crystals, cenospheres and amorphous, nonopaque particles.

Plerosphere, Cenosphere and Surface Crystal Formation

The mechanism of plerosphere and cenosphere formation has been discussed in detail by Fisher and Natusch [8]. It is hypothesized that cenosphere formation is the result of gaseous decomposition, which provides the driving force to inflate the molten parent particle. Gaseous products released by fracturing fly ash in laboratory studies have been identified as H_2O and CO_2, which are thought to be the result of clay mineral and carbonate decomposition, respectively [9]. Plerosphere (Figure 2) formation is hypothesized to result from a process similar to cenosphere formation. As the aluminosilicate parent particle is progressively heated, a molten surface layer develops. Mineral decomposition results in formation of a bubble around the solid core, which remains attached to the molten shell. As the core boils away from the shell, concomitant formation of fine particles may result. The process can be repeated until the plerosphere is full of other plerospheres, cenospheres or solid particles or until the particle freezes. Based on heat transfer and mass transport considerations, plerosphere formation time has been estimated to be approximately 1 msec [9].

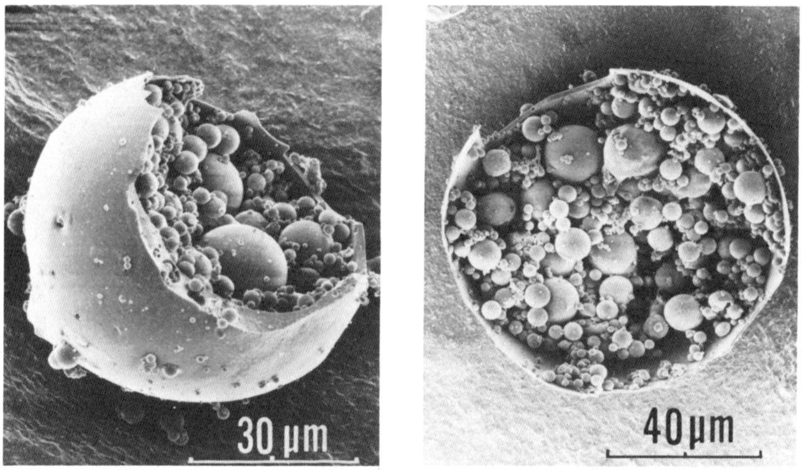

Figure 2. Scanning electron micrographs of fly ash plerospheres; fine particles contained within the plerospheres are thought to be the result of internal fragmentation processes resulting from mineral decomposition and rapid particle growth.

Surface crystal [Figures 3(a) and (b)] formation on fly ash surfaces has been documented by SEM analysis. Electron microprobe analysis of crystalline material in fly ash similar to surface microcrystals has resulted [9] in the tentative identification of $CaSO_4$. It is hypothesized that surface crystal formation results from SO_2 hydration and subsequent oxidation on fly ash surfaces to form H_2SO_4, which then reacts with metal oxides, predominantly CaO, or with ambient NH_3 to form either $CaSO_4$ or $(NH_4)_2SO_4$. Such a mechanism could also result in formation of soluble compounds from relatively insoluble oxides, e.g., conversion of PbO to $PbSO_4$. The rate of atmospheric conversion of power plant emitted SO_2 to H_2SO_4 that occurs on fly ash surfaces, and the potential catalysis of this oxidation process by surface-associated metals deserve careful study.

INDIVIDUAL PARTICLE ELEMENTAL ANALYSIS

Individual particle elemental analysis using a scanning electron microscope with an energy dispersive X-ray analyzer has demonstrated extreme variability in the matrix composition of fly ash particles. Pawley and Fisher [10] showed major differences in the elemental composition of fly ash particles with similar SEM morphological appearance. Interestingly, the matrix heterogeneity of individual particles was found in a NBS standard reference

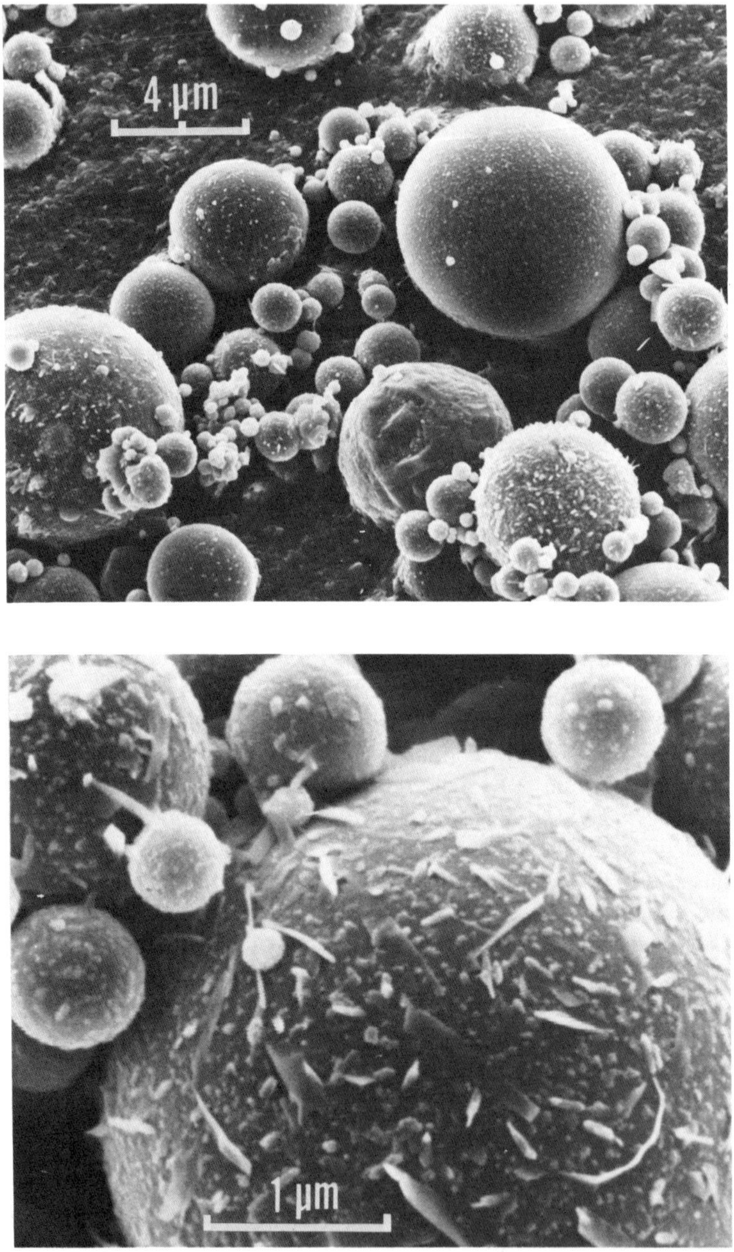

Figure 3. Scanning electron micrographs of coal fly ash indicating marked variation in surface morphology and surface growth of microcrystals.

material fly ash (SRM-1633), which is certified as homogeneous for bulk sample analysis of 12 trace elements.

Individual particle analyses of the LM-defined morphological classes demonstrated the opaque amorphous particles were predominantly unburned coal [11]. Average elemental compositions were found to be similar for uncolored spherical particles (cenospheres, water-white solid spheres and spheres which contain quench crystals [7]). As previously described, the degree of pigmentation from white to yellow, orange, red and, finally, opaque correlated with the iron content. Cenospheres had an apparently greater concentration of silicon than other particle types. The class of amorphous, nonopaque particles, which is considered the parent material for the other nonopaque particle classes, displayed the greatest chemical heterogeneity. Particles containing high concentrations of Ca, Fe and Si, as well as those with elemental concentrations typical of aluminosilicates were observed. The morphological and chemical heterogeneity of fly ash may provide a special toxicological mode [12]. It is possible that uniquely toxic particles are present in fly ash and that efficient public health protection is best afforded by abatement of the atmospheric release of that unique particle type.

BULK ELEMENTAL ANALYSIS

The concentrations of 42 minor and trace elements in the four fly ash fractions have been determined by atomic absorption spectroscopy (AAS) [13] and instrumental neutron activation analysis (INAA) [14]. Elemental concentrations in fly ash can be classified into two categories: elemental concentrations (1) dependent on particle size and (2) independent of particle size [15]. Concentration dependence with particle size was determined qualitatively with the criterion that constant concentration trends beyond experimental uncertainty be observed for each of the four fractions analyzed.

The elements not displaying clearcut concentration dependence on particle size for all fractions analyzed were Al, Fe, Ca, Na, K, Ti, Mg, Sr, Ce, La, Rb, Nd, Th, Ni, Sc, Hf, Co, Sm, Dy, Yb, Cs, Ta, Eu and Tb, although Na, Sr, Ni and Co displayed marked enhancement in the finest fraction relative to the coarsest fraction [15]. In agreement with previous findings of Natusch et al. [16], the most volatile elements (or oxides) displayed the greatest inverse concentration-dependence on particle size. In order of decreasing dependence on particle size, the elements Cd, Zn, As, Sb, W, Mo, Ga, Pb, V, U, Cr, Ba, Cu, Be and Mn displayed increased concentration with decreasing particle size. Silicon was the only element to decrease in concentration with decreasing particle size [15]. These results are consistent with the hypothesis of Natusch et al. [16] that concentration enhancement in fine particles is due to vaporization and subsequent condensation of elements volatilized during the combustion process. It is important to note, however, that refractory elements also displayed concentration trends inversely dependent on particle size. Therefore, processes other than vapor condensation may be involved in the concentration-size relationship. Mineral decomposition and elemental distribution within the organic component of coal may, in part, explain the elemental trends of the high boiling chemical species.

Coles et al. [17] have described the elemental behavior in the four size-classified fractions in terms of elemental enrichment factors relative to the parent coal. The elements were grouped into three classes: Group I elements displayed little or no enrichment in fine particles and were lithophilic; Group II elements displayed marked enrichment and were chalcophilic (sulfur-associated); and Group III consisted of elements with behavior intermediate to Groups I and II. Group I elements included Al, Ca, Cs, Fe, Hf, K, Mg, Mn, Na, Rb, Sc, Ta, Th, Ti, Ce, Dy, Eu, La, Nd, Sm, Tb and Yb; Group II elements were As, Cd, Ga, Mo, Pb, Sb, Se, W and Zn; and Group III consisted of Ba, Be, Co, Cr, Cu, Ni, Sr, U and V.

In a separate report, Coles et al. [18] described enrichment factors (EFs) for ^{228}Th, ^{228}Ra, ^{210}Pb, ^{226}Ra, ^{238}U and ^{235}U relative to ^{40}K in the four size-classified fractions of stack fly ash. Although the EFs for all radionuclides appeared to increase with decreasing particle size, ^{210}Pb, the most volatile radionuclide, showed the greatest size-dependence. Uranium enrichment was thought to be due to decomposition of uranyl carbonates and formation of volatile UO_3. Thorium enrichment was rationalized to be due to coexistence in submicron grains of zircon in the coal.

Filtration Studies

To obtain further information on the dependence of elemental concentrations on particle size, filtration studies were performed with the finest fly ash fraction (VMD = 2.2 μm) [19]. Neutron-activated fly ash samples were dispersed in buffer at pH 7.4 and rapidly filtered through a membrane with pore size of 5, 2, 0.8, 0.4, 0.2, 0.1, 0.05 or 0.03 μm. The elements were classified into four groups based on their behavior:

1. Na, Ca, Co, Se, Mo and Ba were partially soluble and did not display filtrate concentrations that were pore-size dependent;
2. Sb, As, Zn, W, Cr and U displayed a pattern of filtrate concentrations that appeared to be bimodal;
3. K, Si, Fe, Ce, Sm, Eu and Th were only detected in filtrates from membranes >2 μm in pore size; and
4. Zr, Cs, Nd, Rb, Tb, Yb, Hf and Ta were not detected in the filtrates.

For those elements displaying apparent bimodal behavior, a relatively large increase in concentration was observed in filtrates derived from the 0.4-μm membrane. The concentration profile as a function of pore size remained constant thereafter. These data suggest a specific concentration maximum for Sb, As, Zn, W, Cr and U in fine particles less than 0.4 μm in diameter. The submicron mode probably is the result of coagulation of primary particles formed from condensation of volatile oxides and reaction products.

Differential Dissolution Studies

In an effort to further study the distribution of trace elements in fly ash, the differential solubility of fly ash components in varying concentrations of either HCl or HF was studied [20]. The solubility of an elemental component of fly ash depends on the distribution of the element within the matrix, the chemical form and solubility in the extracting medium. Surface-deposited elements would, for the most part, be expected to be dissolved in HCl, while the elements, associated with the aluminosilicate matrix would be minimally soluble in HCl but soluble in HF.

Preliminary analyses of these data indicate that the most volatile elements display the greatest degree of differential solubility in HCl compared to HF. Based on the solubility behavior it appears that 100% of the mass of the As and Se and 80-90% of the Mo, Sb and Zn are not intimately associated with the fly ash matrix. Similarly, relatively high differential solubilities were observed for Cd, V, U, W, Cr, Cu and Ca. These results may be interpreted as quantitative demonstration of either the extent of surface-association or bimodal size-distribution. With the exception of calcium, these results are in good agreement with previously described studies indicating fine particle enhancement for all these elements and possible bimodal size distribution for Sb, As, Zn, W, Cr and U. The behavior for calcium probably reflects enhanced surface concentration through sulfate crystal formation on fly ash surfaces.

MUTAGENICITY STUDIES

Mutagenicity studies are briefly reviewed to provide further description of the size-dependence of the physical and chemical properties of coal fly ash. Chrisp et al. [21] described the mutagenicity of filtrates derived from addition of the finest fly ash fraction to normal horse serum, phosphate-buffered saline (PBS) or cyclohexane. Five strains of histidine-requiring (his^-) auxotrophs of the bacterium *Salmonella typhimurium* were used. With addition of serum filtrate, an increase in his^+ revertants was observed with the two frameshift tester strains, TA-1538 and TA-98. Horse serum was demonstrated to be an efficient extractant of mutagenic components, presumably due to complexation of mutagens with serum protein. The testing methodology has been described [21,22]. Briefly, sufficient histidine was added to agar plates to allow his⁻ cells to replicate several times in the presence of the test substance. After incubation, the his^+ bacterial colonies were counted. In addition, optimal concentrations of liver homogenates from rats treated with polychlorinated biphenyl were added to test systems to determine the requirement of microsomal enzymes for conversion of promutagens to ultimate mutagens.

Further studies on the physical properties of the fly ash mutagens indicated that the two finest fly ash fractions were the most mutagenic [23]. However, the 3.2-μm fraction was more mutagenic than the finest fraction (VMD = 2.2 μm). In contrast to the results with the stack-collected fly ash fractions, samples of size-classified and unclassified fly ash collected from the electrostatic precipitator hopper of the power plant were not found to be mutagenic. Furthermore, heating of mutagenic fly ash fractions prior to extraction resulted in loss of mutagenic activity. The loss of activity appeared to reach a plateau initially at 200° and 250°C and total loss of activity occurred at 350°C. These observations may be explained either by (a) volatilization of two chemical classes of mutagens, (b) decomposition of two classes of mutagens, or (c) a combination of volatilization and decomposition.

These studies indicate that (a) the most respirable stack-collected fly ash samples are the most mutagenic, (b) the temperature of collection may drastically affect the relative mutagenicity, and (c) the mutagenicity of fly ash is completely removed by heating to 350°C. Studies of the chemical composition of mutagens in fly ash are required to confirm these conclusions and aid in development of appropriate control technologies.

ACKNOWLEDGMENTS

This summary reflects the collaborative efforts of a number of fine scientists. In particular, I acknowledge with great pleasure the stimulating collaboration of C. E. Chrisp, O. G. Raabe, B. A. Prentice, D. Silberman, D. P. Y. Chang and M. Brummer from the University of California at Davis; A. R. McFarland, Texas A&M University; T. L. Hayes, University of California at Berkeley; J. M. Ondov, R. E. Heft and D. G. Coles, Lawrence Livermore Laboratory; and J. B. Pawley, University of Wisconsin, Madison. This research was supported by the U.S. Department of Energy.

REFERENCES

1. "Power Plant Ash Disposal–A Growing Problem," *Chem. Eng. News* 56(45):26 (1978).

2. Vandergrift, A. E., L. J. Shannon and P. G. Gorman. "Controlling Fine Particles," *Chem. Eng.* 80:107-114 (1973).
3. Yeh, H. C., R. F. Phalen and O. G. Raabe. "Factors Influencing the Deposition of Inhaled Particles," *Environ. Health Pers.* 15:147-156 (1976).
4. McFarland, A. R., R. W. Bertch, G. L. Fisher and B. A. Prentice. "A Fractionator for Size-Classification of Aerosolized Solid Particulate Matter," *Environ. Sci. Technol.* 11:781-784 (1977).
5. Pitts, J. N., K. A. Van Cauwenberghe, D. Grosjean, J. P. Schmid, D. R. Fitz, W. L. Belser Jr., G. B. Knudson and P. M. Hynds. "Atmospheric Reactions of Polycyclic Aromatic Hydrocarbons: Facile Formation of Mutagenic Nitro Derivatives," *Science* 202:515-519 (1978).
6. Fisher, G. L., B. A. Prentice, D. Silberman, J. M. Ondov, A. H. Biermann, R. C. Ragaini and A. R. McFarland. "Physical and Morphological Studies of Size-classified Coal Fly Ash," *Environ. Sci. Technol.* 12: 447-451 (1978).
7. Fisher, G. L., T. L. Hayes, C. Lai and B. A. Prentice. "Elemental Analysis of Individual Morphological Particle Classes in Coal Fly Ash," in *1978 Radiobiology Laboratory Annual Report* (Springfield, VA: National Technical Information Service, in press).
8. Fisher, G. L., and D. F. S. Natusch. In: *Analytical Methods for Coal and Coal Products, Volume III*, C. Karr, Jr., Ed. (New York: Academic Press, Inc., in press).
9. Fisher, G. L., D. P. Y. Chang and M. Brummer. "Fly Ash Collected from Electrostatic Precipitators: Microcrystalline Structures and the Mystery of the Spheres," *Science* 192:553-557 (1976).
10. Pawley, J. B., and G. L. Fisher. "Using Simultaneous Three Color X-ray Mapping and Digital-scan-stop for Rapid Elemental Characterization of Coal Combustion By-products," *J. Micros.* 110:87-101 (1977).
11. Fisher, G. L., C. E. Chrisp and T. L. Hayes. In: *Proceedings of the Symposium on Carbonaceous Particles in the Atmosphere*, T. Navakov, Ed. (City: Publisher, in press).
12. Hayes, T. L., G. L. Fisher and J. B. Pawley (in preparation).
13. Silberman, D., and G. L. Fisher. "Room-temperature Dissolution of Coal Fly Ash for Trace Metal Analysis by Atomic Absorption Spectrometry," *Anal. Chim. Acta* (in press).
14. Ondov, J. M., R. C. Ragaini, R. E. Heft, G. L. Fisher, D. Silberman and B. A. Prentice. In: *Methods and Standards for Environmental Measurement*, W. K. Kirchoff, Ed. U.S. Government Printing Office (1977).
15. Fisher, G. L. "Size-dependence of the Physical and Chemical Properties of Coal Fly Ash," presented at Division of Fuel Chemistry, Second Joint Chemical Institute of Canada/American Chemical Society Conference, May 1977, Montreal.

16. Natusch, D. F. S., J. R. Wallace and C. A. Evans, Jr. "Toxic Trace Elements: Preferential Concentration in Respirable Particles," *Science* 183:202-204 (1974).
17. Coles, D. G., R. C. Ragaini, J. M. Ondov, G. L. Fisher, D. Silberman and B. A. Prentice. "Chemical Studies of Stack Fly Ash from a Coal-fired Power Plant," *Environ. Sci. Technol.* 13:455-459 (1979).
18. Coles, D. G., R. C. Ragaini and J. M. Ondov. "Behavior of Natural Radionuclides in Western U.S. Power Plants," *Environ. Sci. Technol.* 12:442-446 (1978).
19. Fisher, G. L., D. Silberman, B. A. Prentice, R. E. Heft and J. M. Ondov. "Filtration Studies with Neutron-activated Coal Fly Ash," *Environ. Sci. Technol.* 13:689-693 (1979).
20. Fisher, G. L., D. Silberman and B. A. Prentice. Unpublished data.
21. Chrisp, C. E., G. L. Fisher and J. Lammert. "Mutagenicity of Respirable Coal Fly Ash," *Science* 199:73-75 (1978).
22. Fisher, G. L., and C. E. Chrisp. "Physical and Biological Studies of Coal Fly Ash," in *Application of Short-term Bioassay in the Fractionation and Analysis of Complex Environmental Mixtures,* M. D. Waters, S. Newnow, J. L. Huisingh, S. S. Sandhu and L. Claxton, Eds., EPA-600/9-78-027, Williamsburg, VA (1978), pp. 441-462.
23. Fisher, G. L., C. E. Chrisp and O. G. Raabe. "Physical Factors Affecting the Mutagenicity of Fly Ash from a Coal-Fired Power Plant," *Science* 204:879-881 (1979).

CHAPTER 11

GENERATION AND CHARACTERIZATION OF POWER PLANT FLY ASH

Otto G. Raabe
Radiobiology Laboratory and
Departments of Radiological Sciences
and Civil Engineering
University of California
Davis, California 95616

INTRODUCTION

During the combustion of coal in power plant furnaces, small particles of aluminosilicate and other components formed by fusion or sintering of mineral residues are carried with the exhaust gases as fly ash. In addition, volatile coal constituents including hydrocarbons, polycyclic organic compounds and inorganic compounds can be vaporized and undergo chemical transformations. Fly ash particles are carried with exhaust gases to electrostatic precipitators, scrubbers and/or baghouse filtration systems. Most of the fly ash is collected by these abatement devices. However, up to a few percent of the mass of fly ash particles may negotiate the abatement equipment and be released to the environment via a smoke stack. Usually, these are respirable particles with a high proportion smaller than 5 μm aerodynamic diameter, D_{ar} [1]. Releases totaling thousands of kilograms per day are associated with large modern coal-burning power plants.

Particles released from a power plant to the atmosphere that are greater than about 5 μm D_{ar} have relatively high settling speeds (greater than 5 cm/min), are otherwise relatively unstable, and are not usually carried long distances in the air. Hence, the respirable particles smaller than about 5 μm D_{ar} are the most likely to form durable aerosols in the atmosphere that might be inhaled by people. In addition, smaller particles may be the most hazardous to inhale.

As the effluent cools in the plant exhaust ducts, in the smoke stack and finally upon release to the atmosphere, organic and metallic compounds can diffuse to and collect upon fly ash surfaces. This results in these materials having a higher mass concentration among the smaller, more respirable particles.

Chrisp et al. [2] have found fly ash collected from the stack of a coal-burning power plant to be mutagenic using the Ames *Salmonella* assay system [3]. Further, Fisher et al. [4] found that two groups of size-fractionized fly ash particles with mass median (physical) diameters (MMD) of 2.4 μm and 3.7 μm, respectively, were more mutagenic per unit mass of fly ash than two fractions with MMD values of 6.0 μm and 18.5 μm, respectively. It is not possible to predict the health implications of these results, but the apparent biological activity of fly ash and the fact that the public may regularly inhale small quantities in the environment indicate that the possibility of a potential health hazard associated with these fly ash aerosols needs to be carefully evaluated.

In this review, methods for generation and characterization of aerosols of fly ash are discussed. Examples are based on fly ash collected from coal-burning power plants, but the general information given is also applicable to fly ash produced in burning oil. In both inhalation studies and physico-chemical research, the presence of particles larger than 5 μm D_{ar} may tend to be misleading and should be avoided. Emphasis is therefore upon the simulation in experimental systems of the respirable fraction of fly ash as would provide a durable aerosol in the environment rather than the size distribution at the time of release.

GENERATION OF RESPIRABLE AEROSOLS OF FLY ASH

Fly Ash Collection and Treatment

Various investigators have developed methods for reaerosolizing fly ash suitable for inhalation exposure of experimental animals in the laboratory and for studying fly ash properties. These studies have four parts:

1. collection of suitable samples of fly ash as released from power plants;
2. reaerosolization with appropriate treatment to provide a stable respirable aerosol under laboratory conditions;
3. delivery of the aerosol via the exposure system to the experimental animals permitting them to inhale the fly ash under controlled conditions; and
4. characterization and monitoring of the aerosols as found in the exposure system.

A reasonable goal for a polydisperse respirable aerosol is one with a mass median aerodynamic diameter ($MMAD_{ar}$) of 1-2 μm with geometric standard deviation (σ_g) not exceeding 2.0. The electrostatic charge distribution should be brought to Boltzmann equilibrium.

Ideally, samples of fly ash are collected directly from the effluent stream as it is released from the power plant smoke stack. McFarland et al. [5] have collected size-separated respirable fractions of fly ash directly from the breeching of a smoke stack using a specially designed high-volumetric-flow inertial sampler. Others may use ash collected from hoppers of the plant's electrostatic precipitator (ESP) or baghouse filter systems since large quantities are readily available. However, hopper ash represents the particles not released by the power plant and may be expected to be quite different in both physical and chemical properties from released particles.

Most of the particles in hopper ash are large and nonrespirable. Raabe et al. [6] solved this problem by collecting large quantities of ESP hopper ash and aerodynamically size-separating it prior to attempting aerosolization to obtain a sample containing only particles smaller than 5 μm D_{ar}. Their methods of size-classifying fly ash particles are applicable to other dry dusts. By working with size-classified fine particle samples, many equipment and technique requirements and operational problems encountered with gross samples are avoided. However, dry powders do agglomerate in storage, making deagglomeration an essential aspect of a satisfactory reaerosolization system.

Fluidized-Bed Aerosolization Systems

A fluidized-bed (FB) generator (Figure 1) [7] has a bed of nearly spherical large particles (greater than 100 μm in diameter) of a suitable metal or metallic oxide. The bed particles are fluidized by suspending them in an upward flow of air (air speed at least 8 cm/sec) maintained at a low enough velocity to avoid release of the bed material. Dry fly ash is added and agitated in the bed and flows upward. Particles and aggregates small enough to have lower settling speeds than the average upward air velocity are carried out with the air stream. It might be expected that the agitation of fly ash and material will lead to high electrostatic charging of the bed so that the potential for deagglomeration of the fly ash may be compromised. A suitable electrostatic discharging device must be placed near the bed outflow.

Willeke et al. [8] described the generation of fly ash aerosol using a fluidized-bed aerosol generator with an air-driven feeder to continuously supply fly ash to the bed. Willeke et al. [8] used a ^{85}Kr discharging unit to reduce charge to Boltzmann equilibrium and reported efficient deagglomeration. They generated an aerosol with a count median (physical) diameter

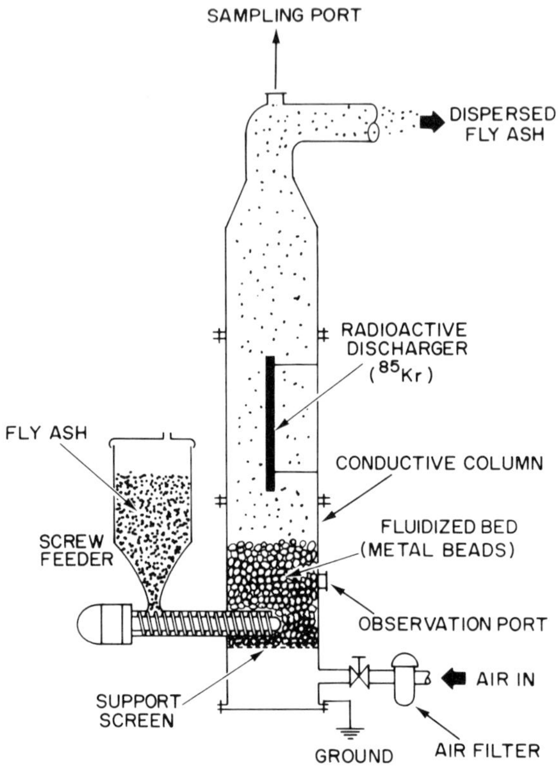

Figure 1. Schematic illustration of a typical fluidized-bed aerosol generator as used to generate laboratory aerosols of fly ash collected from power plants (similar to that described by Hedley et al. [7].

(CMD) of 1.35 μm, with $\sigma_g = 2.7$. This is equivalent to a calculated $MMAD_{ar}$ of 9.3 μm. Better results may have been possible if they had used size-classified fine fly ash. Willeke et al. [8] operated their generator at 15 liter/min, but did not report output concentration.

A similar fluidized-bed system was described by Marple et al. [9]. Their generator used a conveyor chain feeder to supply dust and had a ^{85}Kr discharger in the outlet section. Although they did not aerosolize fly ash, their output conditions with other dusts may have been similar to conditions achievable with fly ash. This system operated with output flowrate from 9 to 30 liter/min. They generated an aluminosilicate dust at concentrations up to 180 mg/m^3 with $MMAD_{ar}$ 6.5 μm and σ_g 2.6. They attempted to obtain a more respirable aerosol by passing the flow through a 1.3-cm diameter cyclone at 9 liter/min and achieved aerosol concentrations up to 60 mg/m^3 with $MMAD_{ar}$ of 4.5 μm and $\sigma_g = 2.3$. Better results may have been possible if size-classified fine particles had been used as the test dust. However, the output concentration of the fluidized-bed system continued to increase and took several hours to stabilize under continuous-feed conditions.

Wright Dust Feed Aerosolization Systems

The Wright Dust Feed (WDF) was invented in 1950 by B. M. Wright [10] and has been widely and successfully used in lengthy stable aerosol generation in inhalation toxicology [11]. The Wright Dust Feed (Figure 2) is more

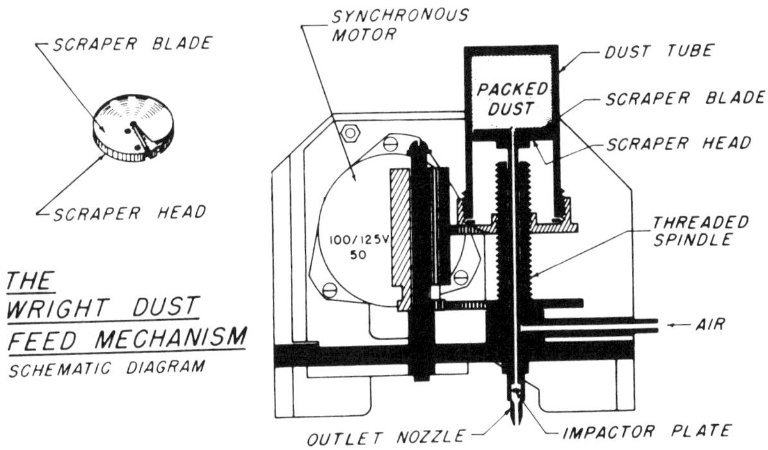

Figure 2. Schematic drawing of the Wright Dust Feed (WDF) aerosol generator invented by B. M. Wright [10] for dispersing dry powders (from Leach et al [11]).

than a feeder, it is an aerosol generator suitable for deagglomerating and dispersing dry dusts. Prior to operations, the fly ash or other test dust is packed under hydraulic pressure into a 4-cm diameter by 6-cm long cylindrical dust cup. A 3.8-cm diameter circular cutting blade is set against the packed dust cake and the dust cup is turned by an electric motor via a replaceable gear-drive system so that a thin (about 30 μm) swath of packed dust half the cake diameter is continually and uniformly removed. A stream of air (10 to 30 liter/min) carries the liberated dust out of the cutting chamber of the dust cup under turbulent conditions via an outlet at the center of the circular cutting blade. The outlet tube is partially obstructed by a steel band upon which the larger particle aggregates impact and consequently are forcefully deagglomerated prior to release from the device.

The Wright Dust Feed was used by Alarie et al. [12], Alarie et al. [13], and MacFarland et al. [14] to generate fly ash aerosols for exposure of monkeys and rodents. They collected fly ash from four power plant stacks by high- volume filtration without size separation and pooled the samples. The Wright Dust Feed was used for reaerosolization. They did not report using a discharger or the extent of deagglomeration. The particle size distributions of fly ash aerosols in various experiments had concentrations between 0.16 mg/m^3 and 0.6 mg/m^3 with MMD between 2.6 and 5.3 μm (σ_g 1.6 to 2.2). If the physical density of the fly ash was 2.45 g/cm^3, the MMD values corresponded to $MMAD_{ar}$ of 4.2 and 8.4 μm, respectively.

Raabe et al. [6] and Raabe et al. [15] developed complete systems using the Wright Dust Feed for generation of fly ash aerosols that simulate respirable particles in effluents from coal-burning power plants for exposure of rodents and nonhuman primates. Two types of size-classified fly ash samples were used: (a) fly ash collected and size classified in situ at the breeching of the smoke stack downstream of the plant's electrostatic precipitators (ESP); and (b) fly ash obtained from the ESP hoppers and subsequently size-classified. The stack ash was collected by McFarland et al. [5] as four size fractions, including one with volume median (physical) diameter (VMD) of 2.2 μm (σ_g = 1.9) and CMD of 0.92 μm (σ_g = 1.5). The hopper ash was obtained from the dry discharge of the same plant's ESP hoppers and subsequently size-classified using a Donaldson Acucut Production Classifier (Majac Division, Donaldson Co. Inc., Minneapolis, MN). A total of 75 kg of size-classified respirable particles of ESP collected hopper fly ash was yielded from 3000 kg of hopper ash (2.5%). A Coulter counter analysis of a grab sample of separated hopper ash was found to have a VMD of 1.77 (0.06 SE) μm, with geometric standard deviation of 1.52 (0.04 SE).

Raabe et al. [6] dispersed the previously size-classified fly ash samples with a Wright Dust Feed and passed the resulting aerosols through a specially built 2.9-cm diameter cyclone separator (Figure 3) to remove agglomerates

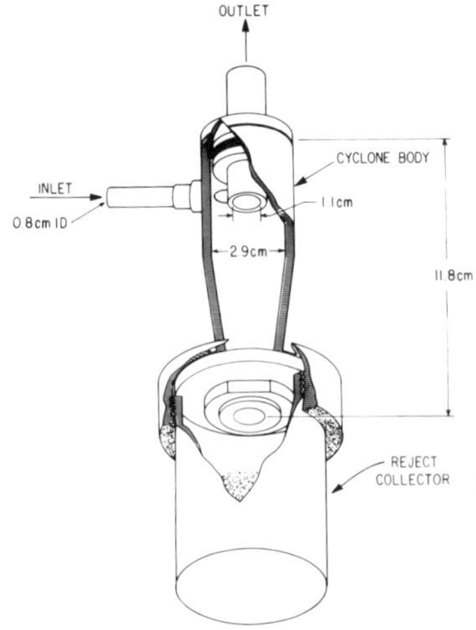

Figure 3. Schematic illustration of cyclone separator used in conjunction with a Wright Dust Feed by Raabe et al. [6] to generate respirable aerosols of power plant fly ash.

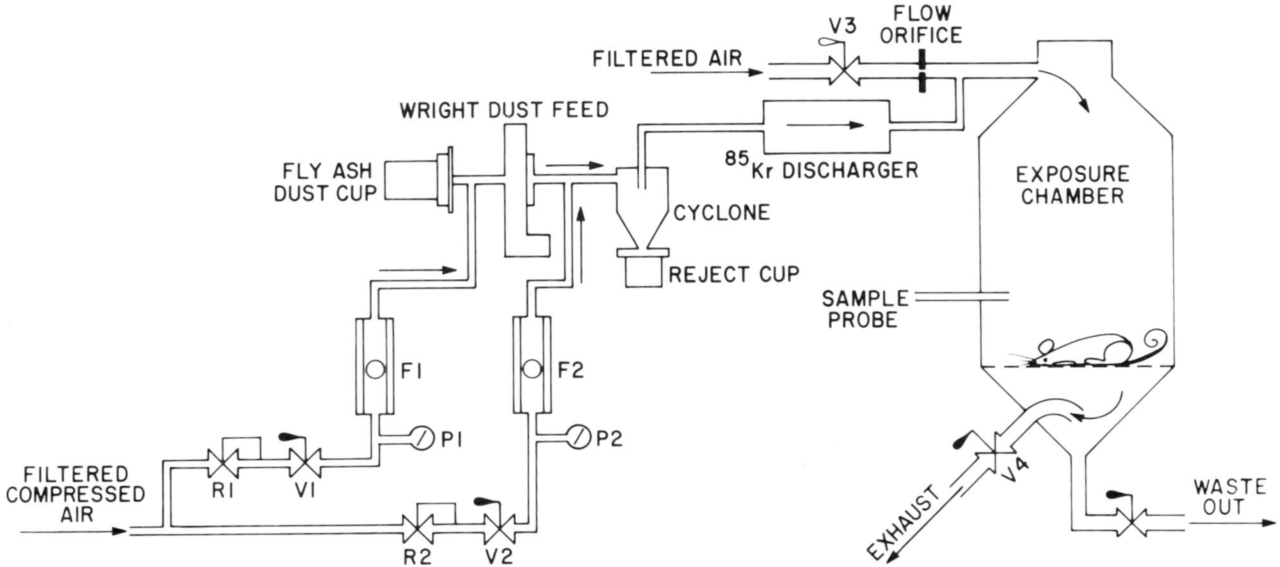

Figure 4. Schematic of the fly ash aerosol generator system developed by Raabe et al. [6] for generation of respirable aerosols of power plant fly ash and exposure of experimental animals.

and large particles. The system operated at 40 liter/min. A ^{85}Kr discharger was used to reduce the aerosol electrostatic charge distribution to Boltzmann equilibrium. The resulting aerosol was introduced into a 3.5-m^3 exposure chamber as shown in Figure 4. During a 180-day exposure of rats, aerosol samples collected by electrostatic precipitation were evaluated by electron microscopy (Figure 5) and found to have an average CMD of 0.68 (0.02 SE) μm, with an average σ_g of 1.54 (0.02 SE). Aerodynamic size distributions determined with a cascade impactor had an average $MMAD_{ar}$ of 1.98 (0.02 SE) μm, with average σ_g of 1.65 (0.02 SE). The mass concentration had a mean of 4.2 (1.4 SD) mg/m^3 with 2000 liter/min of diluting air at the chamber. This is equivalent to 210 mg/m^3 for the 40-liter/min generator system. The methods described by Raabe et al. [6] are suitable for use in other laboratory studies requiring reaerosolization of collected fine particulate matter.

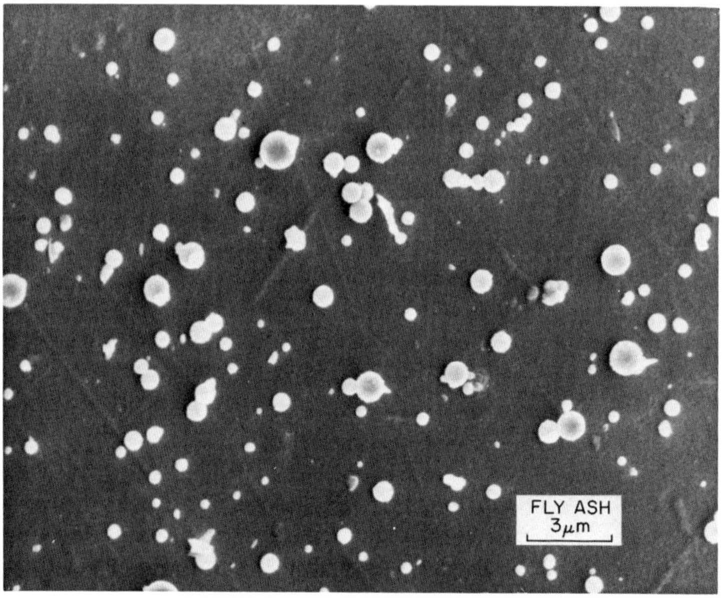

Figure 5. Scanning electron micrograph of sample of laboratory generated aerosol of power plant fly ash from Raabe et al. [15].

Moss and Milliman [16] operated the Wright Dust Feed in series with a Trost airjet mill (Plastomer Products Division of Garlock Corp., Newton, PA) that uses comminution and centrifugal separation to deaggregate the particles (Figure 6). They reported improvement in mass yield of respirable aerosol [17] with this arrangement over operation of the WDF alone. They obtained

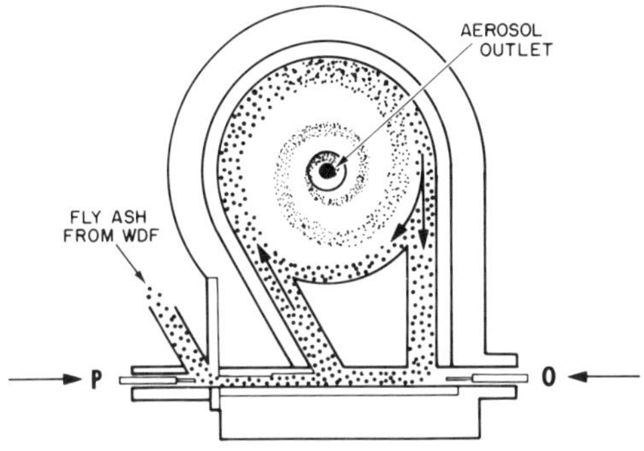

Figure 6. Schematic drawing of Trost airjet mill used by Moss and Milliman [16] to improve the output of respirable fly ash generated by WDF. Compressed air at 40 psig is introduced at P and at 60 psig at 0 for a total throughput of 170 liter/min.

fly ash collected by filtration of effluents from an experimental combustor, some of which had been obtained from coal previously enriched with nickel for special experiments. Unfortunately, the combustion conditions were not optimized, and the samples contained a large fraction of opaque amorphous material identifiable as incompletely burned carbonaceous particles. The presence of aluminosilicate spheres was masked by the presence of the carbonaceous material. The fly ash was not size-classified prior to use. The output of two WDF devices in parallel yielded 600 mg/m^3 of aerosol at 170 liter/min (including diluting air) during a 6-hr exposure period. The concentration of respirable aerosol was determined to be 218 mg/m^3 so that the respirable fraction was about 36%.

When the output of one WDF was introduced into the Trost airjet mill operating at an aerosol output flowrate of 130 liter/min, total concentration of fly ash aerosol was 310 mg/m^3. The concentration of respirable aerosol was 183 mg/m^3 (59%). Apparently the Trost mill served to break up the carbonaceous material in this fly ash, decreasing the particle size distribution. This technique involves violent impaction, comminution and abrasion of the Trost mill so that special tungsten-carbide liners were recommended. Moss and Milliman [16] did not describe the aerodynamic size distribution of the aerosols, but an aerosol with $MMAD_{ar}$ of 3 μm and σ_g 2.0 would have a 60% respirable fraction.

WDF and FB Comparison

McFarland et al. [18] generated fly ash with a fluidized-bed aerosol (FB) generator in 5- to 20-min tests. They compared the resulting aerosols to those produced with a Wright Dust Feed operated separately and in series with the FB generator. They reported that the more stable and better deaggregated aerosols were achieved with the Wright Dust Feed alone than with the fluidized-bed generator alone or with the two in series. They used a FB generator with two types of bed materials, zirconium dioxide beads and nickel beads. For generation of size-classified fly ash (MMD = 2.4 μm) [5] the FB generator was operated at 13.4 liter/min and the aerosol was diluted with clean air. A ^{85}Kr discharge tube was placed above the generator to reduce the electrostatic charge to Boltzmann equilibrium. Samples were collected with filters and a cascade impactor. The undiluted aerosol concentration was 835 mg/m^3 with MMAD$_{ar}$ of 4.8 μm with σ_g 1.9.

For separate short tests, the WDF was operated at 22.4 liter/min in place of the FB generator in the same experimental apparatus with the WDF output introduced at the level of the fluidized bed. The undiluted fly ash aerosol concentration was 510 mg/m^3 with MMAD$_{ar}$ of 1.9 μm and σ_g 1.6. This laboratory aerosol was stable over successive 10-min periods, and the particles contained fewer aggregates than generated by the FB. To further evaluate these results, the WDF was used as the feeder for the FB system. The dispersion characteristics of the aerosol generated with this hybrid system showed poorer deagglomeration than with the WDF alone, suggesting that the FB system was causing the agglomeration of particles. Based on these results, McFarland et al. [18] recommended the WDF in preference to the FB generator.

The various methods that have been used for generating fly ash aerosols are summarized in Table I. The aerosols concentration are all normalized to output flowrate of 200 liter/min for comparison.

CHARACTERIZATION OF FLY ASH

Basic Considerations

The physical and chemical properties of fly ash are extremely complex. The particles are heterogeneous and properties vary with both particle type and size. Whether or not the laboratory-generated aerosols physically and chemically mimic the durable respirable-sized aerosols actually released from power plants depends upon many factors including the source of the fly ash, effects of treatment (such as size-classification), effects of handling and effects of aerosolization and exposure conditions.

Whatever method is used for laboratory aerosolization, detailed physical and chemical characterization of the resulting aerosols is required. Such characterization must be as exhaustive as feasible since many properties may participate in influencing the biological fate of the inhaled particles. Measure-

Table I. Comparison of Methods Used to Generate Laboratory Aerosols of Power Plant Fly Ash with Output Concentration Normalized to a Flowrate of 200 liter/min.[a]

Reference	Method	Size Classified Fine Ash Used	$MMAD_{ar}$ (SE) (μm)	σ_g (SE)	Concentration (mg/m^3 at 200 liter/min)	Efficiency (%)
Willeke et al. [8]	FB	No	9.3 (Est.)	2.7	NA	NA
McFarland et al. [18]	FB	Yes	4.8	1.9	56	NA
Moss and Milliman [16]	WDF (two)	No	6.0 (Est.)	2.0 (Est.)	510	17
Moss and Milliman [16]	WDF-Trost	No	3.0 (Est.)	2.0 (Est.)	200	25
McFarland et al. [18]	WDF	Yes	1.9	1.6	57	NA
Raabe et al. [7]	WDF-Cyclone	Yes	1.98 (0.02)	1.65 (0.02)	42	15
Alarie et al. [12, 14]	WDF	No	8.4	2.0	0.55	NA

[a]NA indicates data was not available; Est. indicates reasonable guess.

ments made to evaluate the laboratory aerosols need to be compared to similar measurements made of fly ash released from power plants. It is possible that some property that may be neglected could be instrumental in determining the biological significance. Hence, an understanding of the biological response from inhaled fly ash requires detailed physical and chemical characterization of the fly ash aerosols.

Physical Characterization of Aerosols

Fisher et al. [19] observed the unique morphological properties of fly ash and Fisher et al. [20] developed a model system for classifying and quantifying the various morphological types of fly ash particles. They studied particles size-classified and collected directly from the smoke stack of a power plant. Many of the larger particles were hollow cenospheres and some were particle-filled plerospheres. Some particles of relatively pure iron oxide were present in all size classes. The relative abundances of morphological particle types were size-dependent. Particles composed of elemental carbon derived from incompletely burned coal were common in the larger fly ash particles, while fused aluminosilicate spheres predominated among the finer, more respirable particles. They found the average physical density of fly ash to vary from 2.45 g/cm^3 for the fine particles to 1.85 g/cm^3 for the larger particles.

Most of the reported uses of laboratory generated aerosols of fly ash included have limited information concerning the aerosol physical properties. Filter samples were used for mass concentration measurements, and cascade impactors for determination of the aerodynamic size distribution.

During exposure of rats to fly ash aerosols Raabe et al. [6] continuously monitored the aerosol concentration using a Royco 225 light-scattering particle counter (Royco Instruments, Menlo Park, CA) providing a strip chart recording of particle number concentration. Comparison measurements were made using a Climet 208 light-scattering particle counter (Climet Instruments, Redlands, CA). Particle size distribution data were collected with these instruments operated in conjunction with a multichannel pulse height analyzer. Samples were collected periodically with a small point-to-plane electrostatic precipitator [21] designed to collect representative samples for evaluation by electron microscopy. These samples were studied using a Zeiss EM 10A transmission electron microscope (Carl Zeiss, Inc., NY) and an ETEC Autoscan VI scanning electron microscope (ETEC Corp., Hayward, CA). Size data from electron micrographs were collected using a Zeiss particle size analyzer (Model TGZ3, Carl Zeiss, Co., NY) and log-normal functions were fit to the size distribution data using the grouped data of projected-area diameters and the maximum likelihood method described by Raabe [22].

Raabe et al. [6] also collected 90-minute samples at 21 liter/min using a Sierra 216 AL Radial slot impactor (Sierra Instrument Co., Carmel Valley, CA) to evaluate the aerodynamic size distributions of the aerosols. The mass distribution data collected with the cascade impactor were analyzed by fitting log-normal functions with respect to D_{ar} using the weighted least-squares method described by Raabe [23].

Mass concentrations were measured by collection of samples on pre-weighed glass fiber and membrane filters by weighing the samples collected for 90 min at 25 liter/min. These results are summarized in Table II. From these data, Raabe et al. [6] calculated an average physical density of 2.43 (0.11 SE) g/cm^3 for the respirable fly ash particles.

Table II. Physical Characteristics of Laboratory Aerosols of Fly Ash Generated by Raabe et al. [6].

Parameter	Number of Sample	Mean (SE)
CMD	12	0.68 (0.02) μm
VMD	12	1.19 (0.05) μm
$MMAD_{ar}$	65	1.98 (0.02) μm
σ_g	65	1.65 (0.02) μm
Mass conc.	65	4.2 (1.4) mg/m^3
Number conc.[a]	105	2.4 (0.4) x 10^9/m^3

[a]Particles larger than about 0.6 μm D determined during 8 hr per sample with Royco Analyzer.

Fisher et al. [24] compared aerosols generated as described by Raabe et al. [6] using size-classified fine particles separated and collected in situ from the power plant stack to aerosols generated from fine particles of size-classified ESP hopper fly ash. The $MMAD_{ar}$ and σ_g values were 1.98 (0.02 SE) μm and 1.6 (0.02 SE) for the ESP hopper-collected ash and 1.65 (0.04 SE) μm and 1.45 (0.05 SE) for the stack fly ash. Electron-micrographs of samples (Figure 7) showed the abundant presence of very small particles less than 0.1 μm D in the stack fly ash aerosol and a sparsity of such ultrafine particles in the hopper fly ash aerosol.

Chemical Characterization

It is not strictly possible to subdivide chemical and physical characterization since the chemical properties of fly ash are a complex function of particle size [25-29]. Also, the chemical properties have been shown to be heterogeneous even for morphologically identical particles [30]. Some chemical species may be found in relatively pure particles. The surface

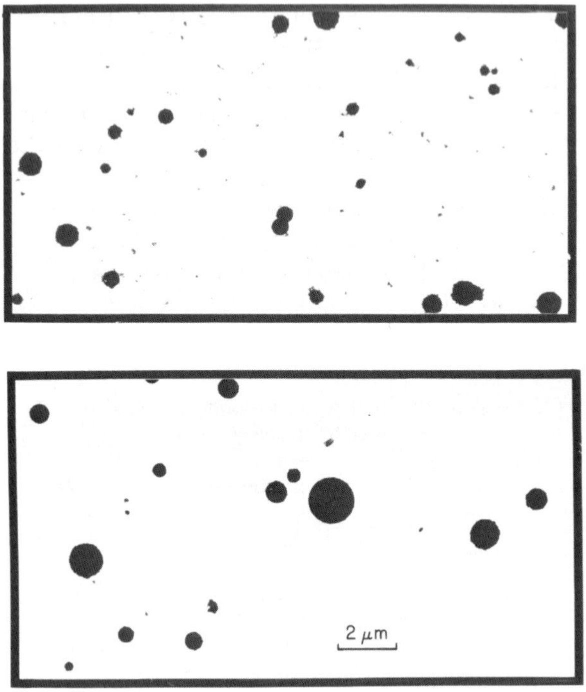

Figure 7. Comparison of transmission electron micrographs of samples of laboratory generated aerosols of power plant fly ash as described by Raabe et al. [6] showing (a) ESP hopper-collected size-classified ash and (b) stack-collected size-classified fly ash (from Fisher et al [24]).

properties are of particular interest in inhalation toxicology, since the chemical constituents on the surface of the particles are the most readily available for biochemical interactions upon deposition in the lung. Volatile trace metal compounds and organic compounds condense on fly ash particles during the cooling of the effluent stream in the power plant smoke stack and during emission into the atmosphere. The condensed materials coat the particles resulting in a relative mass concentration that increases with decreasing particle size. Important elements such as Se, Cd, As, V, Zn, Sb and Be have been found to exhibit this size dependence [25,27-29].

In the transport of condensing materials to particle surfaces by diffusion under ideal tranquil conditions the flux to a spherical particle is given by [31]:

$$\phi_t = 2\pi D \Delta C_o \tag{1}$$

where ϕ_t = the total flux (molecules or particulate clusters hitting per unit time assuming total sticking)

D = the particle geometric diameter

Δ = the diffusion coefficient of the condensing material

C_o = its particulate or molecular number concentration far from the particle

For an extremely thin surface coating, the flux per unit mass of particles is given by:

$$\phi_t/m = 12\Delta C_o/\rho D^2 \tag{2}$$

where the particle mass m is taken as $\rho\pi D^3/6$ with ρ the particle physical density. The resultant mass concentration of condensing material would therefore be inversely proportional to the squared particle diameter.

On the other hand, ideal fluid conditions do not apply to very small particles (< 0.3 μm D) nor do tranquil conditions occur in power plant smoke stacks or plumes. Under extreme conditions of turbulent convective mixing, the flux to a spherical particle is given from kinetic theory by [31]:

$$\phi_t = \pi D^2 \bar{v} C_o/4 \tag{3}$$

where $\bar{v}$ is the average diffusional velocity of the diffusing molecules or particulate clusters hitting the particle. Then, for a thin surface coating the flux per unit mass of particle is given by

$$\phi_t/m = 3\bar{v}C_o/2\rho D \tag{4}$$

so that the resultant mass concentration of condensed material would be inversely proportional to the particle diameter.

The data [25-28] for fly ash collected from smoke stacks of coal-burning power plants suggest the actual conditions for particles bigger than 0.3 μm D are somewhere between the two extremes given by Equations 2 and 4.

Methods of chemical analysis include both wet chemical and instrumental techniques. Dissolution studies help in identifying mobilizable components. Particularly useful methods for identifying elemental concentrations include atomic absorption spectroscopy (AAS), instrumental neutron activation analysis (INAA) and particle-induced X-ray emission (PIXE) analysis. Raabe et al. [32] compared the elemental concentrations of laboratory-generated ESP hopper fly ash measured by AAS to similar measurements of stack ash made by Fisher et al. [28]. This comparison is summarized in Table III. The observed differences reflect the source and treatment of the laboratory ash. Such factors may be important in experimental evaluations. However, the full details concerning the chemical composition of fly ash are not yet avail-

Table III. Comparison of Selected Elements Measured by Atomic Absorption Spectroscopy in Fly Ash Samples[a]

Element	Fly Ash Laboratory Aerosol [percent (range)]	Fly Ash Stack-Collected [percent (range)]
Si	25.4 (0.7)	26.8 (0.1)
Al	12.6 (0.1)	14.1 (0.3)
Fe	4.53 (0.08)	3.44 (0.31)
Ca	2.28 (0.02)	2.38 (0.09)
Mg	0.82 (0.01)	0.63 (0.01)
Na	1.61 (0.07)	1.85 (0.03)
K	0.89 (0.01)	0.81 (0.03)
Ti	0.61 (0.02)	0.78 (0.06)
C	1.9 (0.9)	0.27 (0.04)
	[μg/g (range)]	[μg/g (range)]
Ba	6350 (0.3)	4090 (180)
Be	13.1 (0.3)	10 (0.5)
Cd	5.1 (0.1)	4.6 (0.2)
Co	36 (4)	22 (0.5)
Cr	127 (8)	71 (4)
Cu	136 (8)	137 (1)
Mn	299 (8)	309 (3)
Ni	61 (3)	40 (2)
Pb	221 (46)	278 (3)
Sr	672 (43)	714 (10)
Zn	428 (20)	746 (218)

[a]Raabe et al. [32].

able. Chemical speciation of fly ash surfaces is in its infancy, and interesting new developments will be forthcoming.

DISCUSSION

Some techniques are probably unacceptable for generating aerosols of fly ash. For example, nebulization of a suspension of particles could result in significant alterations of the surface chemical properties and may involve dissolution or chemical reactions with the solvent. Also, aggregates may be formed in droplets produced during nebulization.

Losses and delivery efficiency for laboratory-produced aerosols of fly ash are sometimes not addressed in the published reports. When the chosen generator system does not remove the larger particles or aggregates, they may settle in various other parts of the experimental system or may settle on the experimental animals being exposed. These settling losses will, of course, effect a reduction in the reported MMAD, but systems like those described by Raabe et al. [1] have the advantage of removing the larger particles in a controllable manner and avoiding overly dusty conditions. Likewise, removal of the nonrespirable particles and aggregates as part of the generator system leads to greater system stability. Raabe et al. [6] operated their system regularly and reproducibly for eight hours per day for up to 180 days.

The ultimate test of the respirability of the laboratory fly ash aerosol is the deposition observed in the pulmonary region of the lung. Yeh et al. [33] described the expected deposition of inhaled airborne particles in humans. Particles larger than about 5 μm D_{ar} are mostly deposited in the nasopharyngeal region with less than 10% deposited in the pulmonary region. Raabe et al. [34] found that particles 3.15 μm D_{ar} had high nasal deposition in rats and hamsters suggestive of the undesirability of using larger particles where pulmonary deposition is under study. On the other hand, very small particles carry little mass. A reasonable balance between pulmonary deposition and required mass concentration is an aerosol with $MMAD_{ar}$ between 1 μm and 2 μm.

Since electrostatic charge may lead to anomalous behavior and particle coagulation, laboratory aerosols of fly ash should be reduced to Boltzmann equilibrium by mixing with ionized air or use of radioactive discharge systems. Most if not all of the reported generator systems use some form of charge reduction. Although the exact state of electrostatic charge of fly ash released from power plants has not been reported, it is likely that it is not highly charged. Also, electrostatic charge in laboratory aerosols may affect the deposition of particles inhaled by experimental animals.

CONCLUSIONS

The generation and characterization of laboratory aerosols of power plant fly ash requires appropriate consideration of the complex physical and chemical properties of fly ash and the alterations that may be introduced by experimental procedures. If fly ash aerosols are used in inhalation toxicology studies, an understanding of the observed biological effects is predicated on a detailed understanding of the biologically relevant physical and chemical properties of the particles.

Each of the various methods for generating laboratory aerosols of fly ash has succeeded in part in producing useful experimental conditions. Raabe et al. [6] have apparently generated the best characterized and reproducible laboratory aerosols of fly ash. The degree to which these are representative of fly ash as released by power plants depends on an understanding of the complex chemical mixtures, including surface coatings. Further research in this area is indicated.

ACKNOWLEDGMENTS

Research supported by the Office of Health Effects Research (OHER) of the U.S. Department of Energy (DOE) under Contract EY-76-C-03-0472 with the University of California, Davis.

REFERENCES

1. Raabe, O. G. "Aerosol Aerodynamic Size Conventions for an Inertial Sampler Calibration," *J. Air Poll. Control Assoc.* 26:856 (1976).
2. Chrisp, C. E., G. L. Fisher and J. E. Lammert. "Mutagenicity of Filtrates from Respirable Coal Fly Ash," *Science* 199:73-75 (1978).
3. Ames, B. N., J. McCann and E. Yamasaki. "Methods for Detecting Carcinogens and Mutagens with the Salmonella/Mammalian-Microsome Mutagenicity Test," *Mutation Res.* 31:347-364 (1975).
4. Fisher, G. L., C. Chrisp and O. G. Raabe. "Physical Factors Affecting the Mutagenicity of Fly Ash," *Science* 204:879-881 (1979).
5. McFarland, A. R., R. W. Bertch, G. L. Fisher and B. A. Prentice. "Fractionator for Size-Classification of Aerosolized Solid Particulate Matter," *Environ. Sci. Technol.* 11:781-784 (1977).
6. Raabe, O. G., K. D. McFarland and B. K. Tarkington. "Generation of Respirable Aerosols of Power Plant Fly Ash for Inhalation Studies with Experimental Animals," *Environ. Sci. Technol.* 13:836-840 (1979).
7. Hedley, W. H., S. M. Mehta, C. M. Moscowitz, A. D. Snyder, H. H. S. Yu and D. L. Zunders. "Sources and Characterization of Fine Particulate Test Dusts," EPA-650/2-74-117, U.S. Environmental Protection Agency, Washington, D.C. (1974).
8. Willeke, K., C. S. K. Lo and K. T. Whitby. "Dispersion Characteristics of a Fluidized Bed," *Aerosol Sci.* 5:449-455 (1974).
9. Marple, V. A., B. Y. H. Liu and K. L. Rubow. "A Dust Generator for Laboratory Use," *Am. Ind. Hyg. Assoc. J.* 39:26-32 (1978).
10. Wright, B. M. "A New Dust Feed Mechanism," *J. Sci. Instrum.* 27:12-15 (1950).
11. Leach, L. F., C. J. Spiegl, R. H. Wilson, G. E. Sylvester and K. E. Lanterbach. "A Multiple Chamber Exposure Unit Designed for Chronic Inhalation Studies," *Amer. Ind. Hyg. Assoc. J.* 20:13-22 (1959).
12. Alarie, Y. C., R. J. Kantz II and C. E. Ulrich. "Long-Term Continuous Exposure to Sulfur Dioxide and Fly Ash Mixtures," *Arch. Environ. Health* 27:251-253 (1973).
13. Alarie, Y. C., A. A. Krumm, W. M. Busey, C. E. Ulrich and R. J. Kantz II. "Long-Term Exposure to Sulfur Dioxide, Sulfuric Acid Mists, Fly Ash, and Their Mixtures," *Arch. Environ. Health* 30:254-262 (1975).

14. MacFarland, H. N., C. E. Ulrich, A. Martin, A. Krumm, W. M. Busey and Y. Alarie. "Chronic Exposure of Cynamolgus Monkeys to Fly Ash," in *Inhaled Particles III* (Surrey, England: Unwin Brothers Limited, 1971), pp. 313-26.
15. Raabe, O. G., K. D. McFarland, B. Tarkington, S. Teague and B. Prentice. "Generation of Respirable Aerosols of Power Plant Fly Ash for Inhalation Exposure of Experimental Animals," in *Radiobiology Laboratory 1977 Annual Report*, UCD 472-124, University of California, Davis. (1977) pp. 2-5.
16. Moss, O., and E. Milliman. "Respirable Aerosol from a Combined Trost Air Jet Mill-Wright Dust Feed Aerosol Generator," American Industrial Hygiene Conference, Los Angeles, CA, May 1978.
17. Lippmann, M. "Respirable Dust Sampling," *Amer. Ind. Hyg. Assoc. J.* 31:138-159 (1970).
18. McFarland, A. R., C. A. Ortiz, G. L. Fisher and O. G. Raabe. "Aerosolization of Bulk Powders Utilizing Fluidized Bed and Wright Dust Feed Techniques," *J. Aerosol Sci.* (submitted).
19. Fisher, G. L., D. P. Y. Chang and M. Brummer. "Fly Ash Collected for Electrostatic Precipitators: Microcrystalline Structures and the Mystery of the Spheres," *Science* 192:553-555 (1976).
20. Fisher, G. L., B. A. Prentice, D. Silberman, J. M. Ondov, A. H. Biermann, R. C. Ragaini and A. R. McFarland. "Physical and Morphological Studies of Size-Classified Coal Fly Ash," *Environ. Sci. Technol.* 12:447-451 (1978).
21. Morrow, P. E., and T. T. Mercer. "A Point-to-Plane Electrostatic Precipitator for Particle Size Sampling," *Am. Ind. Hyg. Assoc. J.* 25:8 (1964).
22. Raabe, O. G. "Particle Size Analysis Utilizing Grouped Data and the Log-Normal Distribution," *J. Aerosol Sci.* 2:289-303 (1971).
23. Raabe, O. G. "A General Method for Fitting Size Distributions to Multicomponent Aerosol Data Using Weighted Least-Squares," *Environ. Sci. Technol.* 12:1162-1167 (1978).
24. Fisher, G. L., O. G. Raabe, B. A. Prentice and D. Silberman. "Comparison of Fractionated Electrostatic Precipitator Hopper Ash and Stack-Collected Fly Ash," *Radiobiology Laboratory 1978 Annual Report*, University of California, Davis (1978).
25. Davison, R. L., D. F. S. Natusch and J. R. Wallace. "Trace Elements in Fly Ash," *Environ. Sci. Technol.* 8(13):1107-1113 (1974).
26. Ondov, J. M., R. C. Ragaini, R. E. Heft, G. L. Fisher, D. Silberman and B. A. Prentice. In: *Methods and Standards for Environmental Measurement*, Publication 464, National Bureau of Standards, pp. 565-572 (1977).
27. Natusch, D. F. S., J. R. Wallace and C. A. Evans, Jr. "Toxic Trace Elements: Preferential Concentration in Respirable Particles," *Science* 183:202-204 (1974).
28. Fisher, G. L., B. A. Prentice, D. Silberman, J. M. Ondov, R. C. Ragaini,

A. H. Biermann, A. R. McFarland and J. B. Pawley. "Size Dependence of the Physical and Chemical Properties of Coal Fly Ash," Second Joint Chemical Institute of Canada/American Chemical Society Conference, Montreal, Canada (1977).

29. Gladney, E. S., J. A. Small, G. E. Gordon and W. H. Zoller. "Composition and Size Distribution of In-Stack Particulate Material at a Coal-Fired Power Plant," *Atmos. Environ.* 10:1071-1077 (1976).
30. Pawley, J. B., and Fisher, G. L. "Using Simultaneous Three Colour X-Ray Mapping and Digital-Scan-Stop for Rapid Elemental Characterization of Coal Combustion By-Products," *J. Microscopy* 110(2):87-101 (1977).
31. Raabe, O. G. "The Adsorption of Radon Daughters to Some Polydisperse Submicron Polystyrene Aerosols," *Health Physics* 14:397-416 (1968).
32. Raabe, O. G., W. Tyler, J. Last, L. Schwartz, G. L. Fisher and F. D. Wilson. "Studies of the Chronic Inhalation of Fly Ash by Rats," *Radiobiology Laboratory 1978 Annual Report*, University of California, Davis (1978).
33. Yeh, H. C., R. F. Phalen and O. G. Raabe. "Factors Influencing the Deposition of Inhaled Particles," *Environ. Health Pers.* 15:147-156 (1976).
34. Raabe, O. G., H. C. Yeh, G. J. Newton, R. F. Phalen and D. J. Velasquez. "Deposition of Inhaled Monodisperse Aerosols in Small Rodents," in *Inhaled Particles IV*, W. H. Walton, Ed. (New York: Pergamon Press, 1977), pp. 3-21.

CHAPTER 12

PHYSICAL AND CHEMICAL CHARACTERISTICS OF AIRBORNE FIBERS

Howard E. Ayer
Institute of Environmental Health
University of Cincinnati
Cincinnati, Ohio 45267

Ralph D. Zumwalde
National Institute for Occupational Safety & Health
U.S. Public Health Service
Cincinnati, Ohio 45226

INTRODUCTION

In this chapter, chemical and physical properties of fibers commonly present in the air are reviewed. For purposes of classification, a reasonable definition of a fiber is presented. The presence of organic fibers is noted, and their predominance in some circumstances is pointed out. The principal emphasis, however, is on inorganic mineral fibers, because of their commercial importance, their persistence in the air, and their potential health hazards.

DEFINITION

A common definition of a fiber, e.g., "a slender and greatly elongated natural or synthetic filament (as of wool, cotton, asbestos, gold, glass or rayon) typically capable of being spun into yarn" refers to an object too large to remain airborne for any length of time. Airborne fibers are objects which are slender and greatly elongated, but of microscopic dimensions. Thin flat crystals can appear to meet the definition if seen on their side in a two-dimensional view, but would not be considered fibers. Thus a regular cross section would appear to be implicit in the definition of a fiber; a cross-sectional width-to-height ratio greatly exceeding 3 would probably exclude a particle from definition as a "fiber." The definition of a "fiber" in terms of length versus average diameter might most reasonably be made by considering departure from the terminal falling velocity of regular particles which approximate spheres. Timbrell's empirical equation for falling speed of glass fibers [1]

$$D_e/D_f = 66 \left(\frac{L_f/D_f}{2 + 4\ L_f/D_f} \right)^{2.2}$$

where D_e = equivalent sphere diameter
D_f = fiber diameter
L_f = fiber length

demonstrates that beyond a length to diameter ratio of 5, length as a factor in falling speed becomes relatively unimportant, resulting in only a one-sixth decrease in falling speed, which then becomes a function of diameter only. Beyond an aspect ratio of 10 there is only a 7% decrease in falling speed for glass fibers, with similar values for the major types of asbestos. An aspect ratio of 5 would eliminate most nonfibrous particles such as quartz crystals, some of which meet a criterion of 3, though they do not appear "fibrous."

The foregoing operative definition will distinguish those cylindrical particles which behave aerodynamically and biologically as do spheres, from those which behave differently because of their extreme length-to-diameter ratios.

ORGANIC FIBERS

The organic fibers most commonly found in indoor air are of animal and vegetable origin. Examination of a home furnace filter, or a space where relatively quiescent air permits dust to preferentially settle, e.g., under a bed, reveals a significant proportion of fibers. Air sampling throughout a home shows that these fibers occur in highest concentration where clothing or household linen is handled, suggesting their origin [2]. Concentrations of these macroscopic fibers are low by number, so that they do not significantly contribute to a number count as determined by optical microscopy of filter samples. The monofilamentous synthetic fibers are more resistant to wear, and thus less frequently found. Athough of far greater economic importance than inorganic fibers, in domestic air their size tends to retard their becoming airborne and to cause their removal by settling, so that the importance of organic fibers in the air is primarily as a nuisance, usually of concern only in clean rooms.

Industrially, airborne organic fibers are only of importance in mills processing these materials, where they may contribute to the fire hazard if allowed to accumulate.

Microscopic organic fibers may be observed in very low concentrations in samples from ambient air, where they apparently are primarily of vegetable origin, with occasional insect fragments [3]. They have not been reported to be of significance as an air pollutant from either a nuisance aspect or as a potential health hazard.

Organic fibers in the air have a density of approximately 1 g/cm^3. Few are found with a diameter of less than 10 μm, and the shape of the fiber is usually ribbon-like or irregular, rather than that of a smooth elongated cylinder.

BIOLOGICAL EFFECTS OF FIBERS

This review will not deal with the potential health hazards of fibrous aerosols. Nevertheless, if it were not for the excesses of lung cancer, mesothelioma of the pleura and peritoneum, and cancer of the gastrointestinal tract among some groups of asbestos workers there would be little or no interest in airborne fibers. Almost all the sources in this review are papers dealing in one way or another with real or potential health hazards. The review will cover physical and chemical properties which may be related to potential health hazards. A great deal of emphasis will be placed on size and shape of airborne fibers. Not only do these properties determine which fibers might reach the air spaces in the lung, but they may determine the degree of cancer hazard as well. A frequently quoted paper by Stanton [4] suggests that on the basis of implantation experiments in animals, small diameter long fibers present a greater cancer hazard than other fibers. If this should prove to be so, then the proportion of a fibrous aerosol in this biologically significant fraction should have to be determined to estimate health hazard. Such information is presented for some of the materials covered in this review.

AMORPHOUS MINERAL FIBERS

Mineral Wool

Mineral wools are made from the slag byproduct of iron blast furnaces, or from rock such as argillaceous limestone by passing a stream of molten rock or slag in front of a steam jet, by throwing a molten material off a cylindrical rotor at high speed or by throwing molten material off a spinning concave rotor at high speed [5]. In either case an organic binder may be added to cause the fiber to adhere. In none of these processes is all the rock or slag fiberized, so that the fibers are accompanied by "shot," unfiberized or partially fiberized mineral which must be removed. Currently about 70% of the mineral wool sold in the United States is produced from blast furnace slag, with most of the remainder from copper, lead and iron smelter slag. Although average fiber diameters are in the range of 3.5-7.0 μm, there is a broad distribution of fiber diameters. In a study of two mineral wool manufacturing facilities Corn et al. [6] found that 78% and 66% of the fibers had diameters <3 μm, respectively, although only 16% and 15% of fibers had diameters <1 μm. Lengths of airborne fibers in the two facilities were longer than 20 μm 41% and 39% of the time, but 79% and 71%, respectively, were longer than 10 μm.

A larger survey of mineral wool producers and users was reported by Fowler [7]. Both bulk and airborne fibers were measured, and results are summarized in Table I. Typical bulk fibers had geometric mean diameters of 3-4 μm, but airborne fibers typically had geometric mean fiber diameters of 2 μm, with geometric standard deviations near 2. Fiber lengths were quite variable by operation, but geometric mean lengths of 15-20 μm were often encountered, with geometric standard deviations near 3.

Inasmuch as slag wool is a byproduct, with additions to balance properties, its chemical composition can be expected to vary between manufacturers, depending on the source of slag. Four analyses over a 2-yr period by one

Table I. Mineral Wool Fiber Diameters and Lengths[a]

	Fiber Diameter (μm)				Airborne Fiber Length (μm)	
	Bulk		Airborne			
Facility	M_g[b]	S_g[c]	M_g	S_g	M_g	S_g
Mfg A	3.9	1.8	1.2–2.4	1.8–2.7	6.7–20	2.3–4.3
Mfg B	3.7	2.0	1.0–2.5	1.6–2.0	8.9–26	2.5–3.3
Mfg C	3.4	2.7	1.6–3.0	1.6–2.2	7.4–36	2.2–4.4
Mfg D	3.6	2.3	1.5–3.2	1.6–2.9	7.5–33	1.6–3.3
Mfg E	2.2	2.1	1.6–2.6	1.4–3.1	8.3–21	2.3–3.3
User F	4.8	1.7	1.6	1.8	12	3.3
User G	3.2	2.7	–	–	–	–
User H	–	–	2.3–2.4	1.7	17–28	2.8–3.1
User I	–	–	1.2–2.8	1.3–2.1	4.2–26	1.5–3.1
User J	2.3	2.2	2.2–2.7	1.8–1.9	28–45	3.9
User K	3.4	2.0	1.9	1.7	12	2.8

[a]Adapted from Fowler [7].
[b]M_g = geometric mean, μm.
[c]S_g = geometric standard deviation.

producer reported by Fowler [10] showed relatively consistent composition: Silica, 35.8-41.4%; alumina, 9.3-11.6%; ferric oxide, 0.8-1.8%; CaO, 34.7-37.2%; MgO 7.0-9.8%; and sulfur trioxide, 0.5-1.9%. Electron microprobe analyses of fiber show the same elements to be the major components of the fibers as well. Trace metals were in relatively low concentration with the exception of one facility which used lead smelter slag and thus had a relatively high lead content.

Fibrous glass

Although a dress fabricated from glass fibers for the Empress Eugenie was exhibited at the St. Louis World's Fair in 1893, textile fiber was not made in the U.S. until 1934. Preceding the large-scale development of the textile fibers was the development of air filters and thermal insulation [8] starting in 1930. A wide range of fibers of different compositions and fiber sizes are now available. Some of the chemical and physical properties of the various fibers are summarized in Table II adapted from Shand [9]. It may be noted that only the "fine wool" and "ultrafine wool" in this table have average diameters that would lead to fibers which could be airborne for any significant length of time. However, in addition to the perhaps 1% of production going into the "ultrafine" fibers averaging about 1 μm in diameter, there is a wide range of fiber diameters produced at many of the manufacturing operations. Further, the smaller diameter particles are separated by elutriation, so that at a construction site [10], mean fiber diameters in parent insulating material of 4.0-10.2 μm can result in mean airborne fiber diameters of 2.3-8.4 μm. Likewise, Dement [11] found median airborne fiber diameters of 1.1-2.1 μm in three facilities manufacturing large diameter glass fibers, with a fourth plant having median diameters from 2.1-4.3 μm in four operations. These results may be contrasted with those of Balzer [12], who found geometric mean glass fiber diameters of 2.2 μm in ambient air, 1.3 μm from fibrous glass-lined ventilation ducts, and 8.4 μm at occupational exposures. There are undoubtedly sampler entry biases on particle sizes as determined in some of the foregoing studies. None of the biases compare with that of Gross et al. [13] who found mean fiber diameters determined by optical microscopy in lungs at autopsy to range from 1.4-2.8 μm, with most mean fiber diameters around 2 μm. As noted previously, the aerodynamic diameter of fibers is primarily a function of fiber diameter, so that fiber lengths tended to reflect the operation and any sampler size selection bias. Dement [11], for example, reported median lengths of airborne fibers at wool insulation manufacturing operations to range from 19 to 70 μm, while Gross et al. [13] found mean lengths of fibers in human lungs at autopsy to range from 17-28 μm with the majority of the fibers between 5 and 25 μm

Table II. Chemical and Physical Properties of Some Commerical Fibrous Glass[a]

Glass Type	Component (% by weight)							Fiber Diameter range (μm)	Density (g/cm^3)	Refractive Index
	SiO_2	Al_2O_3	CaO	MgO	B_2O_3	Na_2O	Other			
Low alkali, lime-alumina borosilicate	54.5	14.5	22.0	–	8.5	0.5		6-9.5	2.60	1.548
Soda-lime borosilicate (textile)	65.0	4.0	14.0	3.0	5.5	8.0	K_2O 0.5	6-9.5	2.54	1.541
Soda-lime borosilicate (wool)	50.0	4.5	16.0	5.5	3.5	11.0	0.5	7.5-15	2.61	1.549
Soda-lime	73.0	2.0	5.5	3.5	–	16.0		115-250	2.47	1.512
Lime-free borosilicate (fine wool)	59.5	5.0	–	–	7.0	14.5	ZrO_2, TiO_2, F 16.0	0.75-5	2.57	1.537
Lime-free borosilicate (ultrafine wool)	59.5	5.0	–	–	7.0	14.5	ZrO_2, TiO_2, F 16.0	0.25-0.75	2.57	1.537
High lead silicate	34.0	3.0	–	–	–	0.5	K_2O 3.5 PbO 59.0	6-9.5	4.30	–

[a]Adapted from Shand [9].

in length. These figures are not unlike those reported for airborne slag wool fibers.

Phytoliths

An observation by Gross et al. [14], confirmed by Langer et al. [15] was that in the lungs at autopsy of persons not occupationally exposed to fibrous dust were numerous fibers which resembled those obtained from burning paper or leaves. These fiber diameters were most frequent at 1.5 μm, with lengths most frequently between 15 and 20 μm. The amorphous nature and numerous potential sources for such fibers does not make them susceptible to ready identification. The tendency of the lung to concentrate fibers far above the proportion in which they exist in air breathed suggests that these fibers would not be seen in high concentrations in ambient air. Their presumed vegetable origin, however, would tend to make their existence perhaps as much a factor in rural as in urban air.

Fragments of diatoms, which may be considered to be of animal origin are also seen in human lungs; small fragments are sometimes difficult to distinguish from fibers. These particles are composed entirely of amorphous silica.

Summary: Amorphous fibers

These materials can be found in urban air samples, and in the lungs of humans at autopsy. Both fibrous glass and slag wool products usually have mean fiber diameters on the order of 3-7 μm. However, the airborne fibers are smaller in diameter, averaging about 2 μm. The lengths of airborne fibers sampled from the air directly or from lungs will have a greater variance than the diameter. If a fiber is defined by a 5 to 1 ratio as suggested here, the minimum length will be about 6 μm, with maximum lengths greater than 50 μm.

CRYSTALLINE MINERAL FIBERS

The definitions of fibers as related to minerals can be somewhat circular; a fiber being defined in one dictionary [16] as "threadlike," while "mineral fiber" is defined by Campbell et al. [17] as "the smallest elongated crystalline unit. . . that exhibits a resemblance to organic fibers," "fibril" as "a single fiber, which cannot be separated into smaller components without losing its fibrous properties or appearances," being distinguished from "prismatic" and "acicular" by their fiber-like appearance. The crystalline shapes and patterns as shown by Campbell et al. [17] are presented in Figure 1. Although such distinctions may be useful mineralogically, they will not

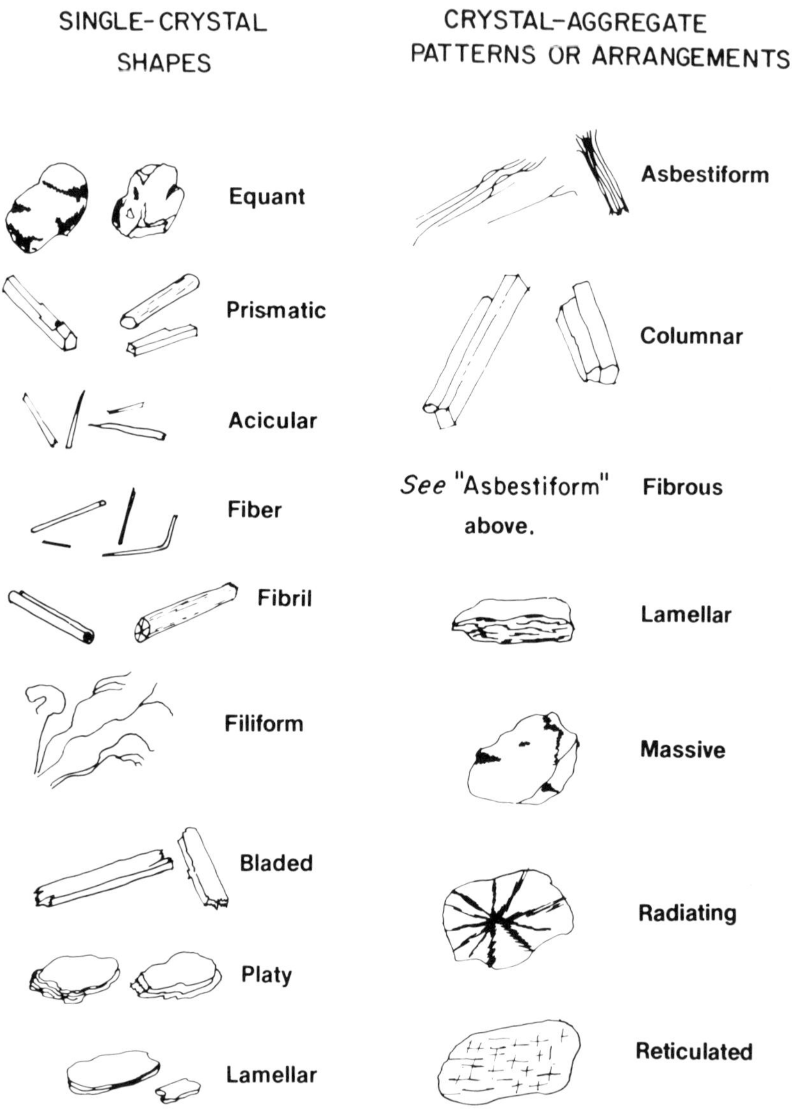

Figure 1. Various mineral crystal and aggregate shapes (from Campbell et al. [17]).

necessarily be observed in this review, for the properties of aerosols may be only incidentally related to such distinctions. In defining asbestos, however, this resemblance to organic fibers was explicit in the classification, and the minerals included as asbestos are recognized by this definition.

Asbestos

Asbestos is classified into two main classes, serpentine, of which the only member is chrysotile, and amphiboles of which the five recognized varieties are crocidolite, amosite,* anthophyllite, tremolite and actinolite. Approximate chemical formulae for these minerals, as presented by Gilson [18] after Hodgson [19] and idealized chemical formulae indicating their variable composition from Speil and Leineweber [20] are shown in Table III. The most abundant cation is shown first in the latter categorization. Chrysotile is by far the more common of the two asbestiform mineral classes and accounts for more than 95% of current commercial asbestos production [20].

Table III. Idealized Chemical Formulae for Asbestos Minerals

Type	Approximate Formulae	
	Average [18,19]	Indicating Variability [20]
Chrysotile	$3\ MgO, 2\ SiO_2, 2\ H_2O$	
Amphiboles		
Crocidolite	$Na_2O, Fe_2O_3, 3\ FeO, 8\ SiO_2, H_2O$	$(Na_2Fe_2^{2+}Fe_2^{3+})\ Si_8O_{22}\ (OH)_2$
Amosite	$5.5\ FeO, 1.5\ MgO, 8\ SiO_2, H_2O$	$(Fe^{2+}, Mg)_7\ Si_8O_{22}\ (OH)_2$
Anthophylite	$7\ MgO, 8\ SiO_2, H_2O$	$(Mg, Fe^{2+})_7\ Si_8O_{22}\ (OH)_2$
Tremolite	$2\ CaO, 5\ MgO, 8\ SiO_2, H_2O$	$Ca_2Mg_5\ Si_8O_{22}\ (OH)_2$
Actinolite	$2\ CaO, 4\ MgO, FeO, 8\ SiO_2, H_2O$	$Ca_2\ (Mg, Fe^{2+})_5\ Si_8O_{22}\ (OH)_2$

Being crystalline, asbestos can be determined chemically by X-ray diffraction in bulk samples. Quantitative analysis has been achieved in laboratory samples by enhancing the X-ray diffraction pattern using electrostatic alignment of the fibers [21], but for routine analysis the limit of detection is on the order of 5% [22]. As reported by Timbrell [23], asbestos fibers of the sizes which become airborne can be aligned magnetically when suspended in air or liquid. Differing modes of alignment are observed, with samples from some asbestos mining areas aligning parallel to the applied magnetic field, others normal to the field, and still others showing a mixture of normal and parallel alignment. Individual fibers can be identified by electron microscopy using either selected area electron diffraction or electron microprobe analyses using energy-dispersive X-ray detectors [22, 24, 25]. Larger crystalline fibers can be distinguished optically from amorphous fibers by polarizing

*Although perhaps more correctly categorized mineralogically as cummingtonite-grunerite [17], the common name amosite will be used here.

microscopy, but many, perhaps most, airborne asbestos fibers are too small in diameter for them to be so distinguished.

A principal distinction between the amorphous glass and slag fibers on the one hand, and the crystalline fibers including asbestos on the other, is that the method of production leads to inherent size differences. The drawing or blowing of the molten mineral most easily produces large fibers of several μm diameter, while the formation of acicular or fibrous crystals can start with fibrils of a few nm diameter.

Chrysotile

Chemically, chrysotile from different sources varies [20] about the theoretical composition of 43.4% SiO_2, 43.5% MgO and 13.1% H_2O, with SiO_2 from 34-44%, MgO from 40-42%, and H_2O from 13-14%. Iron oxides [20] are from 0.6-4%, alumina from 0.3-0.8%, CaO from a trace to 1%, Cr_2O_3 from not reported to 0.3% and other trace elements usually less than 0.1% in the general decreasing order: NiO, Mn_2O_2, TiO_2, Na_2O and K_2O. Chrysotile is the most soluble of the forms of asbestos. The surface of the fibrils is magnesium hydroxide, which is readily dissolved by acid. Strong acids rapidly decompose chrysotile, leaving 40% of the mass with a fragile fibrous morphology [20]. Both in vitro and in vivo studies have shown that chrysotile tends to break down chemically and physically in biological fluids as magnesium is leached away [15]. This characteristic of chrysotile can be used as an index of the airborne concentration of chrysotile in the absence of other magnesium containing minerals, as the amount of magnesium in air samples of asbestos can readily be analyzed by atomic absorption spectroscopy, and chrysotile is approximately 25% magnesium [26].

Physically, the individual fibrils of chrysotile are 30-40 nm in diameter. The fibers that consist of bundles of these fibrils may be as large in diameter as a mm in crudely processed specimens, and fibers may be 5 cm or more in length [27]. Airborne chrysotile fibers are thus bundles of varying numbers of fibrils. At the ends of these bundles, and in fact all along the fiber, smaller bundles of fibrils may project themselves subdividing into still smaller bundles, as shown schematically in Figure 1, and in an electron micrograph [28] of typical fibers collected during air sampling in Figure 2. Aerodynamically, it is not possible to describe chrysotile fibers as easily as, say, glass fibers. The "curliness" of these fibers, as well as the many frayed and projecting fibers greatly increase the probability of capture or removal of these fibers by interception with a surface or with other fibers. Investigators have recognized this problem, and treated chrysotile differently in deposition models. Dement and Harris [29], for example, assume chrysotile to be one-half single curved rods, and one-half crossed rods, following the method of Harris and Timbrell [30].

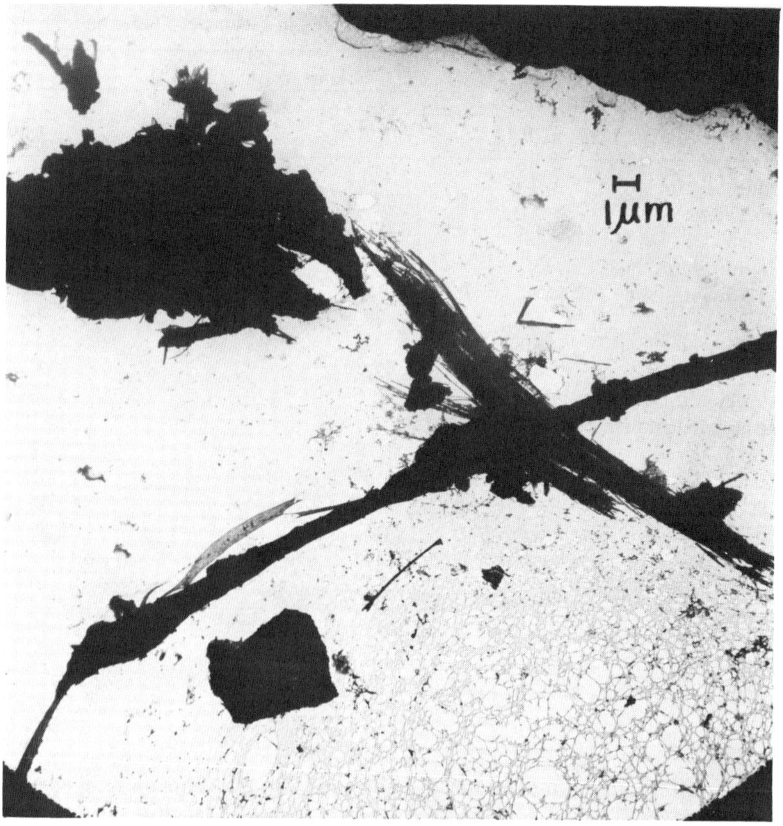

Figure 2. Typical chrysotile fibers from an air sample; electron micrograph.

The difference in size of chrysotile fibers from different operations may be seen in fibers sized in studies conducted by the National Institute for Occupational Safety and Health. Two examples condensed from these data as used by Dement and Harris [29] are presented in Table IV. In both examples most of the fibers are in the finest diameter category, but for the cement pipe finishing, more than half of the fibers are shown in one length-diameter category. In the original table, 53% of the fibers were in the interval centered at 0.08 μm diameter, and 0.35, 0.45, 0.75 and 1.75 length. For long, small diameter fibers, 4.6% of the cement finishing fibers were in the 7.5 μm and 15 μm length category with diameter <0.75 μm, as compared with 12.6% of the textile twisting fibers. It is evident that if length and diameter are important biological factors, then an equivalent concentration of these two fibers could present different hazards.

Table IV. Fiber Size Distribution for Chrysotile, in Fibers per Thousand[a]

Midpoint Interval (μm)	Diameter (midpoint) (μm) 0.15	0.45	0.75	1.50
Twisting Asbestos Yarn Length[b]				
0.15	11	0	0	0
0.45	330	1	0	0
1.75	290	19	2	1
3.25	110	20	5	4
4.75	69	22	5	14
>15	43	27	6	23
Cement Pipe Finishing Length[c]				
0.15	51	0	0	0
0.45	510	1	0	0
1.75	270	10	0	0
3.25	73	6	0	0
4.75	46	11	2	1
>15	9	9	1	3

[a]Totals may not equal 1000 because of rounding.
[b]n = 5817 [29].
[c]n = 4846 [29].

These differing length-diameter distributions also make meaningful specification of concentration by any single number difficult. The most widely used concentration for airborne mineral fibers is number longer than 5 μm/cm^3 as determined by optical microscopy using a 4-mm objective. Regression equations between electron and optical microscopic counts for three operations each in asbestos friction, cement and textile operations showed that a 2-fiber/cm^3 count of fibers longer than 5 μm by optical microscope would correspond to from 1.7 to 5.6 fibers/cm^3 longer than 5 μm by electron microscope [31]. However, five of the nine operations were within 50%, and only two differed by more than a factor of two. It appears that if fibers longer than 5 μm are indeed the most significant criterion, then the optical counting method provides a satisfactory index of concentration. However, from 4.5 to 20.6% of total fibers were counted at the nine operations [31], so the optical index does not describe well the concentration of total fibers.

In a rather different evaluation, asbestos was measured downwind from a mill processing Coalinga chrysotile asbestos in California [32]. The asbestos in ambient air downwind of the plant, believed to be mainly resuspended from ore and tailings piles, ranged in concentration of total fibers up to

1 fiber/cm^3 as determined by electron microscopy. Of these fibers, roughly half were longer than 5 μm. Optical counts were much lower and essentially uncorrelated with the electron microscope counts.

Chrysotile fibers were also collected 1-2 miles from a serpentine rock quarry which contained no commercial asbestos [17]. For these samples, 95% of the fibers were 0.12 μm or less in diameter, and more than 90% were less than 5 μm in length.

The size distribution of chrysotile fibers in human lungs at autopsy is believed to be unrepresentative of the distribution in ambient air because of the physical modification resulting from dissolution of the relatively soluble chrysotile in the biological fluids [15].

Amphiboles

Crocidolite was used in the U.S. most recently as a constituent of some asbestos cement pipe for pressure applications. It has been used for acid-resistant filters, packings and certain types of lagging [34]. It has also been used for respirator filters and, at least in pilot applications, for cigarette filters. It has been used elsewhere as a reinforcing fiber for battery boxes, because of its high acid resistance. Crocidolite is found in Bolivia and Australia as well as in South Africa. Crocidolite has a tensile strength similar to that of chrysotile, but fibers are much "harsher," i.e., stiffer and straighter even in the finest fibers. The difference in "curliness" of crocidolite and thus in ability to penetrate the lung between the two deposits in South Africa has been postulated as a reason for the differences in mesothelioma incidence between the two areas.

Amosite was widely used in the U.S. for thermal insulation, particularly for naval insulation. Like the other amphiboles, the fibers are straighter and stiffer than chrysotile. Data in Table V from Dement and Harris [29] show that fibers are considerably coarser than those encountered in chyrsotile manufacture. In their calculations they assumed amosite to be a straight fiber ordered with the streamlines of airflow.

Although amosite does not occur in the United States in commercial amounts there are exposures to cummingtonite-grunerite fibers as contaminants in other rock [33]. These could be expected to behave as amosite fibers in their chemical, physical and aerodynamic properties.

Anthophyllite has been mined in Georgia and the Carolinas [34] and has been used as a reinforcing fiber in polypropylene [20] as well as in automobile manufacture. Anthophyllite also occurs as a contaminant in talc. Some data on fiber size distribution are available from surveys of a talc mine in upper New York State. Median airborne fiber lengths were 1.5 μm in the mine and 1.4 μm in the mill, with geometric standard deviations of 2.6 and

Table V. Fiber Size Distribution for Amosite, in Fibers per Thousand[a]

Midpoint Interval (μm)	Diameter (midpoint) (μm) 0.15	0.45	0.75	1.50
Pipe Insulation Mfg.– Finishing Length[b]				
0.15	0	0	0	0
0.45	9	4	0	0
1.75	120	92	22	0
3.25	92	61	70	22
4.75	110	120	87	48
15	4	31	39	70
Pipe Insulation Mfg.– Mixing Length[c]				
0.15	0	0	0	0
0.45	35	3	0	0
1.75	120	62	29	3
3.25	94	51	35	21
4.75	83	130	86	46
15	29	59	40	70

[a]Totals may not equal 1000 because of rounding.
[b]n = 229 [29].
[c]n = 373 [29].

2.9 respectively [35]. Median diameters were 0.13 at both locations, with geometric standard deviations of 2.4 and 2.9, respectively. The median aspect ratio of anthophyllite fibers was 9.5 to 1. The distribution is summarized in Table VI from Dement and Harris [29], and typical fibers are shown in Figure 3. It is not known to what extent this size distribution resembles that of airborne fibers from the commercial asbestos.

Table VI. Anthophyllite Fiber Size Distribution[a]: Talc Mining and Milling

Diameter (μm) (midpoint)	Length (interval midpoint) (μm) 0.4	1.3	2.2	3.2	5.0	15
0.09	270	200	61	22	16	4
0.3	8	78	58	82	26	5
0.4	1	14	15	15	16	14
0.8	0	2	6	25	45	14

[a]Fibers per thousand (n = 849) [29].

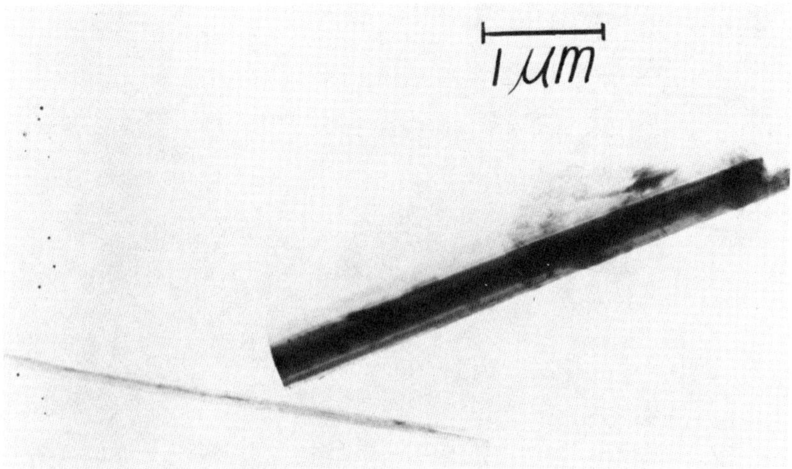

Figure 3. Typical anthophyllite fibers by electron microscope at 20,000X.

Tremolite fibers from the same samples in the talc mine were also identified and sized by electron microscopy [35, 36]. In this talc mine and mill, tremolite fibers had slightly larger median diameters than the anthophyllite, 0.19 μm with geometric standard deviations of 2.3 and 2.4, respectively, and similar median fiber lengths, 1.6 and 1.5, respectively. With lower geometric standard deviations, however, 1.8 and 1.9, respectively, there were fewer long fibers and a median aspect ratio of 7.5. Only 30% of the fibers had an aspect ratio greater than 10 to 1. The fiber size distribution is summarized in Table VII and typical fibers are shown in Figure 4. Tremolite has been processed and repurified by acid treatment for special uses in the filtration field [34].

Actinolite, with tremolite, is classified by Hendry [34] as the least important of the asbestos minerals from an economic standpoint. No data on airborne fiber characteristics were available for this report.

Table VII. Tremolite Fiber Size Distribution[a]: Talc Mining and Milling

Diameter (μm) (midpoint)	Length (interval midpoint) (μm) 0.4	1.3	2.2	3.2	5.0	15
0.09	270	200	74	3	4	0
0.3	17	150	87	25	8	0
0.4	0	21	66	29	8	4
0.8	0	4	33	62	21	4

[a]Fibers per thousand (n = 241) [29].

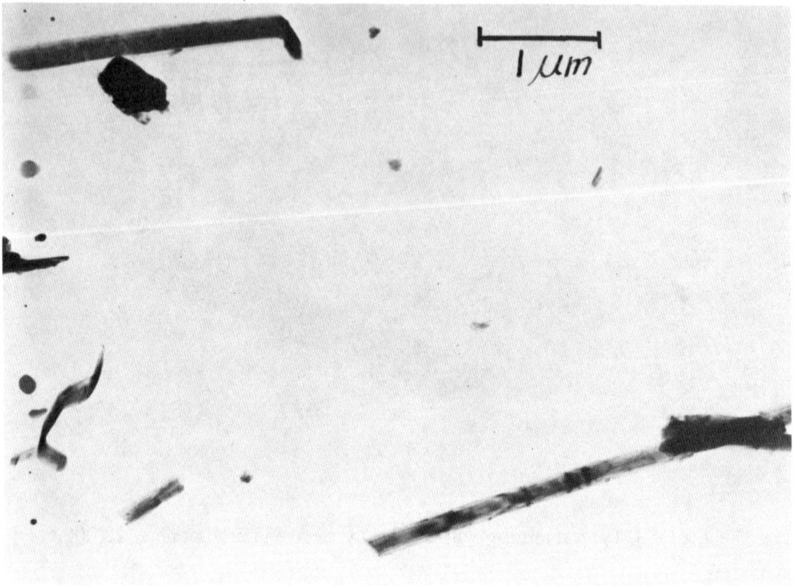

Figure 4. Typical tremolite fibers by electron microscope at 20,000X.

As a group, the amphibole fibers, although much less important than chrysotile economically, are found in significant numbers in human lungs at autopsy. [37]. As indicated by Cralley et al. [3] and by Hendy [34], the less common forms of asbestos, when available in a relatively pure state, have found such uses as inexpensive fillers, reinforcing fibers and thixotropic agents. They can be expected in ambient air from both industrial and commercial uses and from weathering of rock. Their "harsher," i.e., stiff and straight, nature makes them easier to detect in an instrument such as that reported by Lilienfeld and Elterman [38] which rotates fibers by means of an electric field for synchronous detection of the resulting modulation of the light scattered from a continuous wave laser beam by the rotating particles.

Attapulgite

Of the 36 nonasbestos minerals listed by Zumwalde and Dement [22] as examples which may occur in a fibrous state, attapulgite and wollastonite have recenty been studied by the National Institute for Occupational Safety and Health [39,40], and will be discussed here. These minerals both are mined, processed and sold in the United States, and their commercial value is at least partially because of their microfibrous structure.

Attapulgite is the principal mineral component of attapulgus clay. It is a crystalline hydrated magnesium aluminum silicate of the approximate formula: $(MgAl)_2\ [OH\ Si_4O_{10}] \cdot 4H_2O$. It is found in the Soviet Union and India as well as the Georgia-Florida area of the United States and is also known as palygorskite. Both colloidal and sorptive grades are marketed for a wide variety of industrial applications. Airborne fibers were sized by electron microscopy. The median fiber diameter was 0.07 μm, with a geometric standard deviation of 3.14; the largest diameter observed was 0.1 μm. The median fiber length was 0.4 μm with a geometric standard deviation of 3.0; the longest fiber observed was 2.5 μm. The diameters given are for individual fibers which were observed; however there was a strong tendency for the fibers to agglomerate into jagged particulates that ranged in diameter from 0.5 to 5 μm. This may be seen in Figure 5.

It may be noted from the above description that, unlike asbestos, this material would not appear fibrous with either the naked eye or the optical microscope. The designation as fibrous is solely because of the possibility of biological effect when inhaled or ingested.

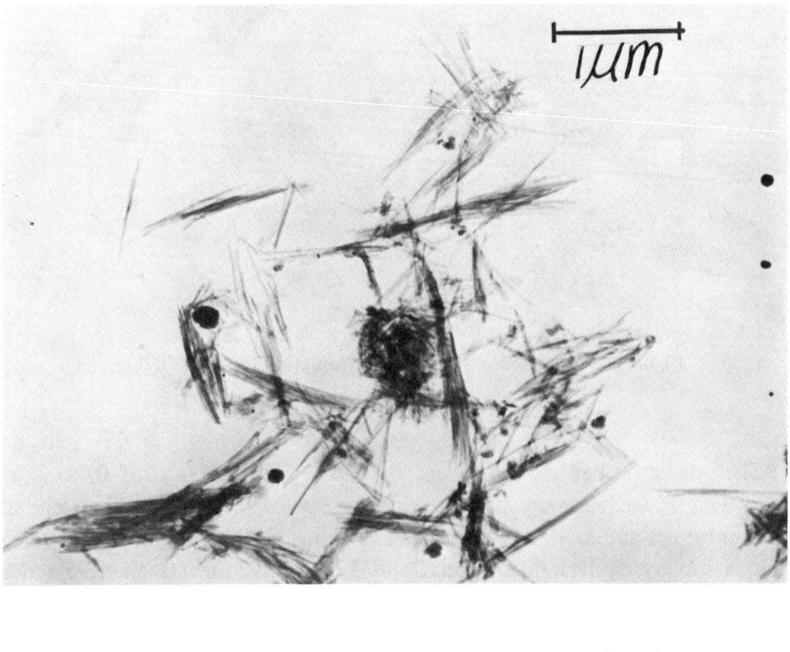

Figure 5. Attapulgite fibers by electron microscope at 20,000X.

Wollastonite

Wollastonite is a naturally occurring calcium silicate, $Ca_3(Si_3O_9)$ mined in Willsboro, NY. It has been used in ceramics, paints, welding fluxes, plastics, cements, wallboards and glass [40]. In the two surveys of the mine and mill reported on by Zumwalde [40], air samples were taken for subsequent fiber counting and characterization. Samples were counted for fibers longer than 5 μm by the NIOSH standard procedure using a 45X phase contrast objective and examined by transmission electron microscopy at 10,000X. Selected area electron diffraction and energy-dispersive X-ray analysis were used for fiber identification. Some typical fibers are shown in Figure 6.

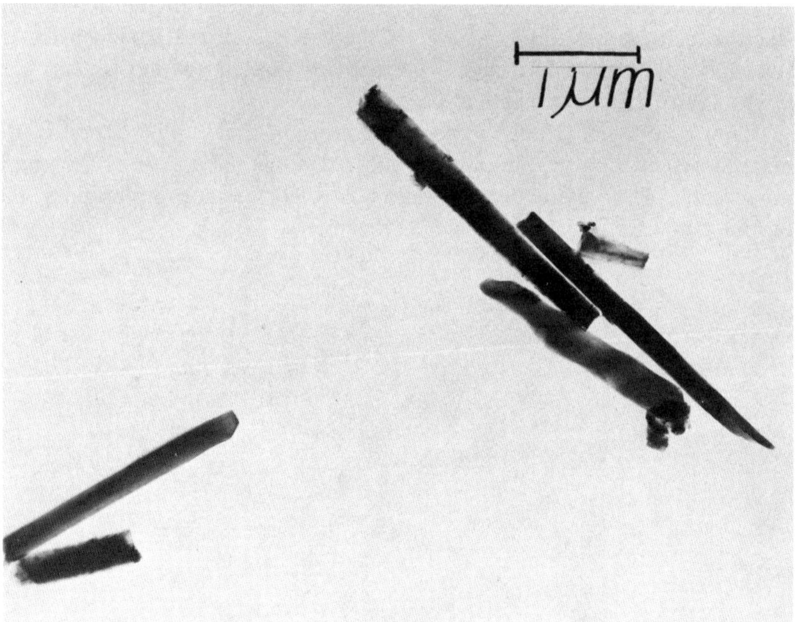

Figure 6. Wollastonite fibers by electron microscope at 20,000X.

The median fiber diameter was 0.22 μm, with a range of 0.1-5.2 μm. Fiber length median was 2.5 μm with a range of 0.3-41 μm. The fiber size distribution is summarized in Table VIII. Some 21% of the fibers were longer than 5 μm by optical microscopy, and 17% were longer than 4.75 μm by electron microscopy. Unlike the attapulgite, there were a number of fibers in these dust clouds which compared in dimension to those longer fibers seen in asbestos dust clouds. If, as the sum of biological, physiological and

Table VIII. Airborne Fiber Size Distribution[a]: Wollastonite Mining and Milling

Diameter (μm (midpoint)	Length (midpoint) (μm)					
	0.15	0.45	1.75	3.25	4.75	15
0.15	0	110	260	52	8	0
0.45	0	0	150	130	32	3
0.75	0	0	2	48	54	9
1.50	0	0	0	4	82	58

[a]Fibers per thousand (n = 1320) [29].

aerodynamic evidence seems to be suggesting, the carcinogenic properties of airborne fibers are more related to their size, shape and aerodynamic behavior than to their chemical or physical properties, then wollastonite should be treated with as much respect as asbestos.

Other Mineral Fibers

As noted by Cralley et al. [3] and amplified by Zumwalde and Dement [22], there are many minerals which may occur in fibrous form. These particles when airborne can, at least in certain sizes, result in the development of ferruginous bodies in the lung [13]. With sufficient exposure and an adequate latent period they may result in an increased probability of lung cancer or cancer of the gastrointestinal tract. The biological and epidemiological evidence is not available to even state with any certainty which of these fibers would create a health hazard. There is certainly no reason to limit the use of materials which are useful in manufacture, commerce and agriculture. It would, however, be foolhardy to treat any airborne fibrous dust as "inert." Wherever significant exposure to such dusts in the working atmosphere or release of the dusts to ambient air is occurring, controls should be instituted to minimize such exposure or release. Measurements of airborne fibers should be made to ascertain the degree of exposure, with help from the National Institute for Occupational Safety and Health or the State Environmental Protection Agency or State Health Department as appropriate. Let us hope our precautions are adequate so that we never can determine with certainty whether these materials are human carcinogens.

REFERENCES

1. Timbrell, V. "The Inhalation of Fibrous Dusts." *Ann. N.Y. Acad. Sci.* 132(1):255-273 (1965).

2. Whitby, K. T., A. B. Ahlgren and R. C. Jordan. "Size Distribution and Concentration of Air-Borne Dust." *ASHAE J. Sec. Heating, Piping and Air Conditioning* (August 1955).
3. Cralley, L. J., R. G. Keenan, J. R. Lynch and W. S. Lainhart. "Source and Identification of Respirable Fibers." *Am. Ind. Hyg. Assn. J.* 29: 129-135 (1968).
4. Stanton, M. F., M. Layard, M. Miller, M. May and E. Kent. "Carcinogenicity of Fibrous Glass: Pleural Response in the Rat in Relation to Fiber Dimension." *J. Nat. Cancer Inst.* 58:587-603 (1977).
5. Pundsack, F. L. "Fibrous Glass–Manufacture, Use and Physical Properties." in *Occupational Exposure to Fibrous Glass*, National Institute for Occupational Safety and Health, HEW Pub. No. (NIOSH) 76-151 U.S. Government Printing Office (1976), p. 11-18.
6. Corn, M., Y. Hammad, D. Whittier and N. Kotsko. "Employee Exposure to Airborne Fiber and Total Particulate Matter in Two Mineral Wool Facilities." *Environ. Res.* 12:59-74 (1976).
7. Fowler, W. P. *Industrial Hygiene Surveys of Occupational Exposure to Mineral Wool.* National Institute for Occupational Safety and Health, Cincinnati, OH (June 1978).
8. Smith, H. V. "History, Processes and Operations in the Manufacturing and Uses of Fibrous Glass," in *Occupational Exposure to Fibrous Glass*, National Institute for Occupational Safety and Health, HEW Pub. No. (NIOSH) 76-151, U.S. Government Printing Office (1976) p. 19-26.
9. Shand, E. B. "Composition and Properties of Fibers," in *Glass Engineering Handbook* (New York: McGraw-Hill Book Company, 1958), p. 375-425.
10. Fowler, D. P., J. L. Balzer and W. C. Cooper. "Exposure of Insulation Workers to Airborne Fibrous Glass," *Am. Ind. Hyg. Assn. J.* 32:86-91 (1971).
11. Dement, J. M. "Environmental Aspects of Fibrous Glass Production and Utilization," in *Occupational Exposure to Fibrous Glass*, National Institute for Occupational Safety and Health, HEW Pub. No. (NIOSH) 76-151, U.S. Government Printing Office (1976), p. 97-109.
12. Balzer, J. L. "Environmental Data: Airborne Concentrations Found in Various Operations" in *Occupational Exposure to Fibrous Glass,* National Institute for Occupational Safety and Health, HEW Publ No. (NIOSH) 76-151, U.S. Government Printing Office (1976), p. 83-89.
13. Gross, P., J. Tuma and R. T. P. deTreville. "Fibrous Dust Particles and Ferruginous Bodies," *Arch. Envir. Health* 21:38-46 (1970).
14. Gross, P., L. J. Cralley, J. M. G. Davies, R. T. P. deTreville and J. Tuma. "A Quantitative Study of Fibrous Dust in the Lungs of City Dwellers," in *Inhaled Particles III*, W. H. Walton, Ed. (Old Woking, England: Unwin Bros. Ltd., 1971), p. 671-679.
15. Langer, A. M., V. Baden, E. C. Hammond and I. J. Selikoff. "Inorganic Fibers, Including Chrysotile, in Lungs at Autopsy," in *Inhaled*

Particles III, W. H. Walton, Ed., (Old Woking, England: Unwin Bros. Ltd., 1971), p. 683-692.

16. Guralnik, D. B., and J. H. Friend, Eds. *Webster's New World Dictionary*, (Cleveland: World Publishing Co., 1962).
17. Campbell, W. J., R. L. Blake, L. L. Brown, E. E. Cather and J. J. Sjoberg. "Selected Silicate Minerals and Their Asbestiform Varieties," U.S. Department of the Interior, Bureau of Mines, Washington, D.C. (1977).
18. Gilson, J. C. "Asbestos," in *Encyclopedia of Occupational Safety and Health* (Geneva, Switzerland: International Labour Organization, 1971), p. 120-124.
19. Hodgson, A. A., "Fibrous Silicates," *Royal Inst. Chem. London Lec., Monog. Rept.*, (1965).
20. Speil, S., and J. P. Leineweber. "Asbestos Minerals in Modern Technology," *Environ. Res.* 2:166-208 (1969).
21. Birks, L. S., M. Fatemi, J. V. Gilfrich and E. T. Johnson. "Quantitative Analysis of Airborne Asbestos by X-Ray Diffraction: Final Report on Feasibility Study" U.S. Environmental Protection Agency, Washington, DC (January 1975).
22. Zumwalde, R. D., and J. M. Dement. "Review and Evaluation of Analytical Methods for Environmental Studies of Fibrous Particulate Exposures," National Institute for Occupational Safety and Health, U.S. Government Printing Office (May 1977).
23. Timbrell, V. "Alignment of Respirable Asbestos Fibers by Magnetic Fields," *Ann. Occup. Hyg.* 18:299-311 (1975).
24. Pooley, F. D. "The Identification of Asbestos Dust with an Electron Microscope Microprobe Analyzer," *Ann. Occup. Hyg.* 18:181-186 (1975).
25. Mueller, P. K., A. E. Alcocer, R. L. Stanley and G. R. Smith. "Asbestos Fiber Atlas," U.S. Environmental Protection Agency, Washington, DC (April 1975).
26. Lynch, J. R., H. E. Ayer and D. L. Johnson. "The Interrelationships of Selected Asbestos Exposure Indices," *Am. Ind. Hyg. Assn. J.* 31: 598-604 (1970).
27. Gaze, R. "The Physical and Molecular Structure of Asbestos," *Ann. N.Y. Acad. Sci.* 132:23-30 (1965).
28. "Determination of Magnitude of Exposure to Asbestos Fibers Associated with a Woman's Coat Containing 8% Asbestos," Unpublished report, National Institute for Occupational Safety and Health, Cincinnati, OH (June 30, 1971).
29. Dement, J. M., and R. L. Harris. "Estimates of Pulmonary and Gastrointestinal Deposition for Occupational Safety and Health (Cincinnati, OH: in press).
30. Harris, R. L., and V. Timbrell. "The Influence of Fiber Shape in Lung Deposition-Mathematical Estimates," in *Inhaled Particles IV*, W. H. Walton, Ed. (Oxford: Pergamon Press, 1977), p. 75-89.

31. Wallingford, K. M., J. M. Dement, R. D. Zumwalde and D. R. Roberts. "Comparison of Phase Contrast and Electron Microscopic Methods for Evaluation of Asbestos Fiber Exposure," paper presented at Annual American Industrial Hygiene Conference, New Orleans, LA, May 24, 1977.
32. John, W., A. Berner, G. Smith and J. J. Wesolowski. "Experimental Determinations of the Number and Size of Asbestos Fibers in Ambient Air," Air and Industrial Hygiene Laboratory (Berkeley, CA: California Dept. of Health, 1976).
33. Gillam, J. D., J. M. Dement, R. A. Lemen, J. K. Wagoner, V. E. Archer and H. P. Blejer. "Mortality Patterns Among Hard Rock Gold Miners Exposed to an Asbestiform Mineral," *Ann. N.Y. Acad. Sci.* 271: 345-352 (1976).
34. Hendry, N. W. "The Geology, Occurrence and Major Uses of Asbestos," *Ann. N.Y. Acad. Sci.* 132:12-21 (1965).
35. Dement, J. M., and R. D. Zumwalde. "Occupational Exposure to Talcs Containing Asbestiform Minerals," paper presented at the Society of Occupational Environment and Health, Washington, DC, Dec. 4-7, 1977.
36. Dement, J. M., and R. D. Zumwalde. "Industrial Hygiene Study of the Gouverneur Talc Company, Number One Mine and Mill," National Institute for Occupational Safety and Health, Cincinnati, Ohio (June 8, 1977), p. 105.
37. Langer, A. M. "Inorganic Fibers, Including Chrysotile, in Lungs at Autopsy: Discussion," in *Inhaled Particles III*, W. H. Walton, Ed. (Old Woking, England: Unwin Brothers, Ltd., 1971), p. 693.
38. Lillienfeld, P., and P. B. Elterman. "Development and Fabrication of a Prototype Fibrous Aerosol Monitor," National Institute for Occupational Safety and Health, Cincinnati, OH, HEW (NIOSH) Pub. No. 78-125 (Springfield, VA: National Technical Information Service, 1977).
39. Zumwalde, R. D. "Industrial Hygiene Study of the Engelhard Minerals and Chemicals Corporation," Unpublished report. National Institute for Occupational Safety and Health, Cincinnati, OH, April 29, 1977, p. 76.
40. Zumwalde, R. D. "Industrial Hygiene Study of the Interpace Corporation," Unpublished report. National Institute for Occupational Safety and Health, Cincinnati, OH, July 29, 1977, p. 79.

CHAPTER 13

FIBER GENERATION AND LENGTH CLASSIFICATION

Kvetoslav R. Spurny

Fraunhofer-Gesellschaft
Institut für Aerobiologie
Schmallenberg, Federal Republic of Germany

INTRODUCTION

Many publications of recent years [1-17] show that the physical properties of respirable fibers (such as fiber shape, fiber length, fiber diameter, fiber firmness) seem to play a more important role in the etiology of diseases developed after inhalation of asbestos or glass fibers (asbestosis, lung cancer, mesothelioma, etc.), than their chemical composition and physicochemical properties (specific surface, porosity, crystalographic structure). It appears that only fibers of minimum fiber length L_F and maximum fiber diameter D_F are responsible for the development of the diseases mentioned. Length and diameter values given are: $L_F > 3\ \mu m$ up to $20\ \mu m$, and $D_F < 2.5\ \mu m$ down to $0.5\ \mu m$. Fibers of length or diameter outside these ranges are considered to be biologically inert–mainly in the case of asbestosis [9]. However, exact data and reliable experimental results supporting the quoted limits have not yet been obtained, because biologists have had no way of conducting experiments with fibers of well-defined lengths and diameters. Exact fiber fractions are not available because of the difficulties in experimentally selecting fibers by length. Some of the problems dealing with the preparation and definition of microfibers for biological experiments have recently been discussed in detail by Friedrichs [17]. A prerequisite for measuring and identifying fibrous particles in ambient air is to obtain further information on the geometrical size of "biologically relevant fibers." In order to determine biological relevance, animal experiments are essential, and may to some extent provide the necessary information. However, a number of technical problems arise in preparing fiber samples for animal experiments: the test samples always cover a range of fiber diameters and lengths. In addition, analytical difficulties have to be resolved when investigating fibers with different types of microscopes (LM, SEM and TEM). Investigations on the dose-effect relationship present different possibilities for interpreting the results. The results obtained must be examined with respect to these limitations.

In practice, two general limits have been assumed for biologically active fiber fractions: $L_F > 5\ \mu m$ and $D_F < 0.5\ \mu m$ (for risk of asbestosis). Since such sharp cut-offs may not be consistent with the laws of nature, Pott [9, 16] has proposed a functional dependency between the biological effects and the parameters L_F, D_F and $\beta = L_F/D_F$ (Figure 1). This model is difficult to prove experimentally, but it seems to be more realistic than any definite cut-off values. In order to get a definition for carcinogenic fibers (e.g., mesothelioma can occur without asbestosis), it seems necessary to move away from the convention which characterizes fibrogenic fibers by threshold values. A convention for suitable fiber size categories should be sought, which is coupled with "carcinogenicity factors."

The greatest problem for the detection of the carcinogenic potency of specific fiber sizes and for the verification of the proposed model (Figure 1) is given by the lack of suitable fiber samples. Existing data for the definition of carcinogenic fibers are insufficient.

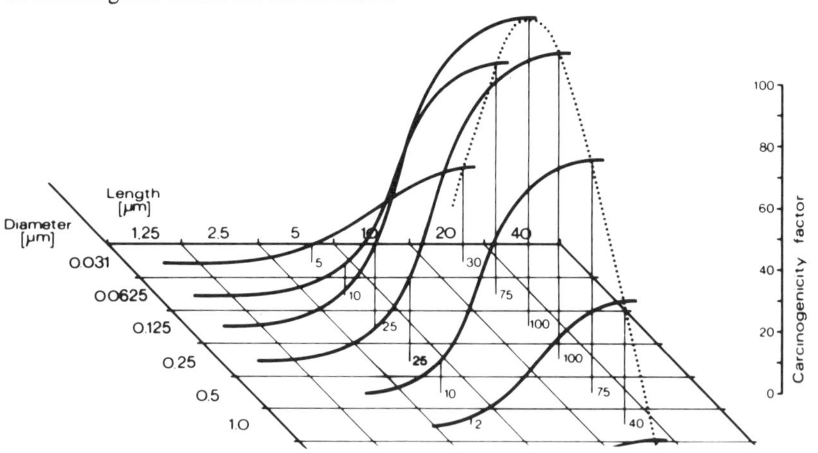

Figure 1. Pott's hypothesis [16] on the carcinogenic potency of a fiber as a function of its physical dimensions with some data on "carcinogenicity factors." This three-dimensional model requires the fiber sizes of a sample to be divided into numerous categories. The size categories include three parameters: length L_F, diameter D_F and aspect ratio $\beta = L_F/D_F$.

Within the constraints of all of these technical, physical and other limitations and problems, we have tried to conduct investigations and develop equipment and procedures for the preparation of fibrous aerosols of narrow size distributions of L_F and D_F as best we can. The constraints were similar when preparing size-selected inorganic fibers for intratracheal and intraperitoneal applications in biological experiments. Moreover, our results cannot be considered to satisfy the wishes of Pott and many other biologists. Nevertheless, we are confident that the fibrous aerosols and the size-selected fibers for biological experiments, that we have described in this chapter, are better size-classified and characterized than the fibrous materials used in biological experiments until now. There is also no doubt, that many more physicochemical investigations in this field need to be conducted in the future.

EXPERIMENTAL PROCEDURES

Preparation of Size-Selected Fibrous Aerosols

To date, there is no satisfactory commercial apparatus available which pulverizes fibrous materials and disperses them into a suspended state. Only a few pulverizers for this purpose have been described in the literature. A very good review has recently been written by Zahradnicek [18]. For example, Timbrell et al. [19] described a mechanical pulverizer in 1968 which incorporates a "moulinex" coffee grinder for pulverizing asbestos fibers. In 1971, Hounam [20] disclosed his "Konimiser," a pneumatic pulverizer which utilizes four compressed airjets [21]. Another fibrous aerosol generator–a modified Timbrell's apparatus–has been constructed and tested by Ortiz et al. [22] and has been used for the pulverization of glass fibers. Its operating principle is based on the controlled feeding of a compact premilled fibrous plug into a rotating multiple-blade assembly. The blades shave off fibers from the continuously advancing plug and an airflow entrains the fibers out of the generator chamber. All of these and many other similar generators were built specially for the preparation of relatively high asbestos dust concentrations for animal experiments. It should also be noted that very dilute liquid suspensions of fine asbestos and glass fibers can be dispersed quite well [23]. Such methods are primarily suitable for the preparation of very low concentrations of fine fibrous aerosols.

As part of our investigations we have constructed a generator for the production of fibrous aerosols (asbestos, glass, etc.). The generator provides low and medium concentrations (i.e., $\leqslant$ 30 mg/m^3) and has a relatively constant output of fiber concentration and fiber size distribution. The technique is based on a vibrating fluidized-bed principle.

A fluidized-bed aerosol generator has been used as early as 1967 to produce sodium chloride aerosols from aqueous solutions [24]. For powdery dust, Guichard has described the "Puldoulit" fluidized-bed aerosol generator [25]. In 1974, Willeke et al. [26] developed an excellent aerosol generator for redispersing coal dust by use of a two-component fluidized bed.

Generator and Measuring Methods.

U.I.C.C. standardized asbestos samples of amosite, anthophyllite, crocidolite and chrysotile [27] as well as glass fibers (MN 100 and MN 104), fine PVC fibers and various powdery dusts have been used for testing our equipment. For two-component fluidized beds, spherical quartz particles (diam = 300 μm) manufactured by the Westdeutsche Quartzschmelze or nickel powder (diam = 100 μm) manufactured by the Sherritt Gordon Mines Company were used as "carrier substances." The exact preparation of these materials for the fluidized-bed experiments has been described elsewhere [42].

The powder to be used for generating the dust cloud is placed into a vibrating cylinder (made of metal or glass) and particle-free dried gas is passed vertically through the powder layer, thus forming a fluidized bed. It appears that the dust or aerosol generation is controlled by the dynamic equilibrium state which develops between disintegration and reagglomeration of the powders or fibers in the surface layer of the vibrating bed. The vibration destroys the cohesion forces between particles or fibers so that they become free and are carried away by the gas stream. This method permits dust clouds of constant concentration to be generated for periods of several hours or days [42].

Figures 2a and b are schematic representations of the generator. The vibration cylinder F in Figure 2a is generally 10 to 100 mm or more in diameter and 50 to 200 mm high. The powder to be dispersed or the powder mixed with the larger beads is inserted into the cylinder onto a Millipore filter (pore diameter of about 0.8 to 1.5 μm) supported by a nylon or metal screen. The bed height ranges from 10 to 100 mm depending on the size of the cylinder. The gas flowrate is typically between 0.5 liters/min to 50 liters/min depending on the diameter of the cylinder. Relatively high flowrates can be used in two-component beds and when an ultrasonic field is applied. The pressure drop across the filter and the powder bed is in the range of 10 to 100 mm H_2O, as indicated by manometer M_2. The air flowrate L is controlled by means of pressure regulator D, manometer M_1, and rotameter R. The air is dried in silica gel column T. Mechanical vibrations in the frequency range of 15 to 300 Hz and amplitude range of 50 to 3000 μm are applied perpendicularly to the axis of cylinder F. The vibrations are

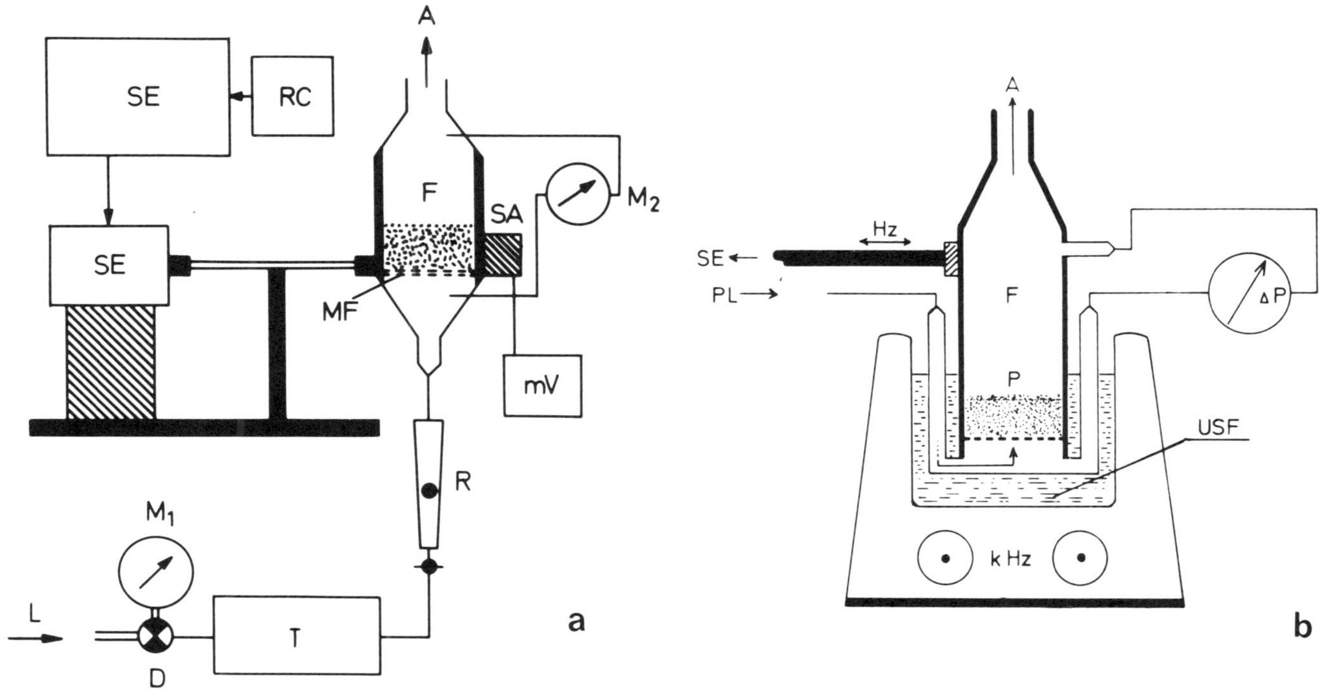

Figure 2. Schematic representation of (a) the fluidized-bed vibrating generator and (b) its ultrasonic modification. Description in text.

produced by an electrodynamic system (RC and SE). The vibration frequency and amplitude are measured by means of a vibration detector system (SA and mV). An ultrasonic field may be applied to this system in addition to the low-frequency mechanical vibrations. As seen in Figure 2a, a cylinder F is then placed into ultrasonic bath USF (frequency - 50 kHz) filled with distilled water. Compressed air PL flows then through bed P which may or may not be mechanically vibrated through vibrator SE. The pressure drop across filter and bed is also measured.

In summary, the generator described can be used as a chemical fluidized-bed generator, as a vibrating-bed aerosol generator (10 to 300 Hz) or as a vibrating-bed ultrasonic generator.

After dry solid powder (dust, fiber, etc.) has been inserted into cylinder F, the frequency and amplitude of mechanical vibration (or of ultrasonic excitation if used) are set and the flowrate is adjusted. The concentration and particle size distribution of the aerosol generated are governed by the physical conditions in the vibrating fluidized bed, especially by the frequency and amplitude of the mechanical vibrations, and thus appear to be a function of the vibration forces acting on the particles or fiber agglomerates. The bed has high porosity ($>$ 90%) and the powder is kept in a "fluidized and disintegrated state" [28]. The binding forces holding agglomerates together can easily be destroyed by the upward gas flow. The cohesion forces between particles or fibers are, to first approximation, a function of particle size or fiber length. Greater vibration forces are therefore needed for the dispersion of large particles and long fibers. It is therefore possible to control the aerosol concentration and particle size quite closely by adjusting the vibration frequency and amplitude. The initial particle- or fiber-size distribution of the powder in the bed is, of course, also very important. The optimum frequency range for the release of fibrous particles (e.g., asbestos fibers) has been found to be about 15 to 50 Hz in the amplitude range of 100 to 1000 μm. Only preliminary applications of ultrasonic fields have been investigated so far. Such investigations will be continued. Thus, any experimental results involving ultrasonic fields have to be considered as preliminary. It appears that ultrasonic energy has, to first approximation, two positive influences: stabilization of the fluidized state and further disintegration of the primary powder and fiber agglomerates. When ultrasonic energy is applied, much higher gas flowrates may be used across the fluidized bed.

The influence of mechanical and ultrasonic vibrations on the physical properties of the aerosols generated was studied first by use of Royco and Climet optical single particle counters. Without applying low vibrational fields to the fiber bed, the generator produces virtually no aerosols. In contrast, isometric particles can be easily dispersed without vibrations after fluidization has been achieved in the fluidized bed.

The characteristics of the dispersed fibrous aerosol effluent were measured relative to the operating conditions, i.e., as functions of vibration frequency f and amplitude y. As mentioned before, light scattering methods were used [29] for rapid indications of aerosol concentration and particle size distribution. An empirical function between the "optical fiber diameter" d_o and $\beta = L_F/D_F$ was estimated, as described in a recent publication [70]. For more accurate measurements of the fiber and mass concentrations, the total aerosol flow was passed through a Nuclepore filter (pore diameter - 0.2 μm). For determining mass concentrations, the Nuclepore filters were weighed on a Cahn electric balance. The fibers on the Nuclepore filter were counted and their lengths and diameters were measured by use of a scanning electron microscope (JEOL 35). The electrostatic charges on the aerosols were estimated by means of a charge spectrometer [30]. The fibers emanating from the generator had a relatively high charge (~ 100e). The number of positively charged fibers was approximately equal to the number of negatively charged fibers. Only a relatively small percentage of the fibers was electrically neutral. Irradiation with a Kr-85 radioisotope (2m Ci) has been found to be effective for charge neutralization of the fibrous aerosols [31]. After evaluation of the fiber samples collected on Nuclepore filters, the aerodynamic fiber diameter was estimated in some cases, e.g., by use of Stöber's formulas for amosite and crocidolite [32].

To date, relatively few experimental and theoretical descriptions are available of the mechanisms for fine powder fluidization and pulverization. The reason for this is the very complicated nature of the interaction of hydrodynamic forces and adhesive forces between particles [33]. The powders are in an aggregated state, so that the hydrodynamic, vibrational and ultrasonic forces, or other kind of forces must overcome the adhesion or cohesion forces between the individual particles and must break down the aggregate state. The conditions are complicated further by the polydispersity of the particulate system. The fluidization conditions in the powder of the fiber bed should be defined and described [34-41]. Further details are discussed elsewhere [42].

Results

The results obtained have been described and discussed in detail in two previous papers [42,43]. These papers demonstrate the function and applicability of such generators. The dependence of aerosol concentration and particle size distribution on the low-vibration frequency and amplitude cannot be predicted theoretically. These dependencies must be determined experimentally and calibration curves must be established for each powdered material such as quartz dust, coal dust, talc, asbestos fiber and glass fiber.

Figure 3 shows an example of the dependence of aerosol concentration on vibration amplitude y. The maximum attainable concentration amounts can be increased, e.g., through application of an ultrasonic bath or by using a larger fluidization cylinder.

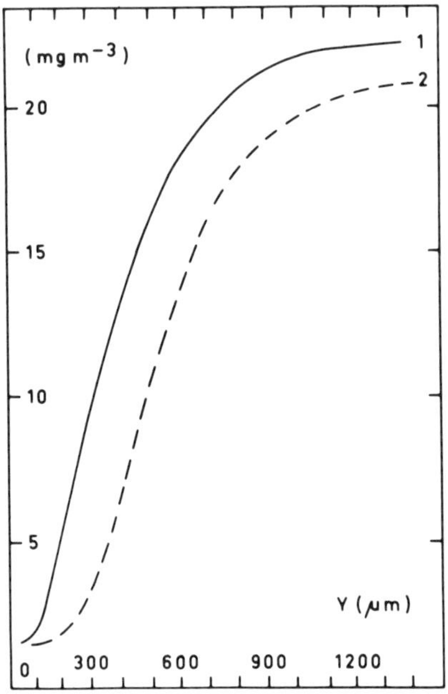

Figure 3. Aerosol concentration (mg/m^3) as a function of vibration amplitude y (frequency f = 30 Hz). Curve 1 is for U.I.C.C. standard chrysotile from Canada; curve 2 is for U.I.C.C. standard chrysotile from Rhodesia.

Figure 4 shows how the fiber size distribution depends on the vibration conditions. As seen in Figure 4a, particle size d_o, measured by light scattering methods, becomes greater as the vibration amplitude y increases. When the vibration amplitudes are low (i.e., $\leqslant$ 500 μm), the asbestos fiber aerosol contains only a small number of aggregates. However, when the vibration amplitude is further increased, the percentage of fiber aggregates also increases (see also Figure 6). Figure 4b clearly demonstrates how the measured fiber size distributions for a specific fluidized-bed condition depend on the methods of measurement.

Some preliminary results on the influence of ultrasonic treatment are shown in Figure 5. Curve 1 in Figure 5a gives the measured size distribution of relatively "coarse" dust from a foundry, dispersed at low-frequency

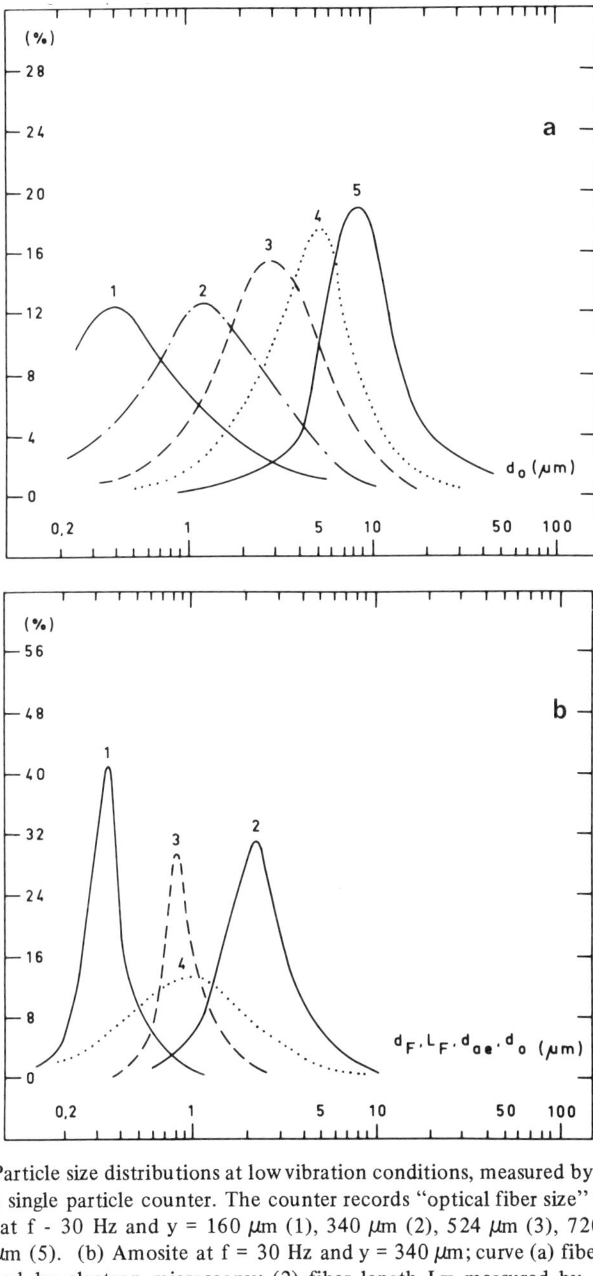

Figure 4. Particle size distributions at low vibration conditions, measured by a Royco 202 optical single particle counter. The counter records "optical fiber size" d_o. (a) Crocidolite at f - 30 Hz and y = 160 μm (1), 340 μm (2), 524 μm (3), 720 μm (4) and 1080 μm (5). (b) Amosite at f = 30 Hz and y = 340 μm; curve (a) fiber diameter d_f measured by electron microscopy; (2) fiber length L_F measured by electron microscopy; (3) aerodynamic particle diameter d_{ae}; (4) optical diameter d_o measured by means of Royco 202 optical particle counter.

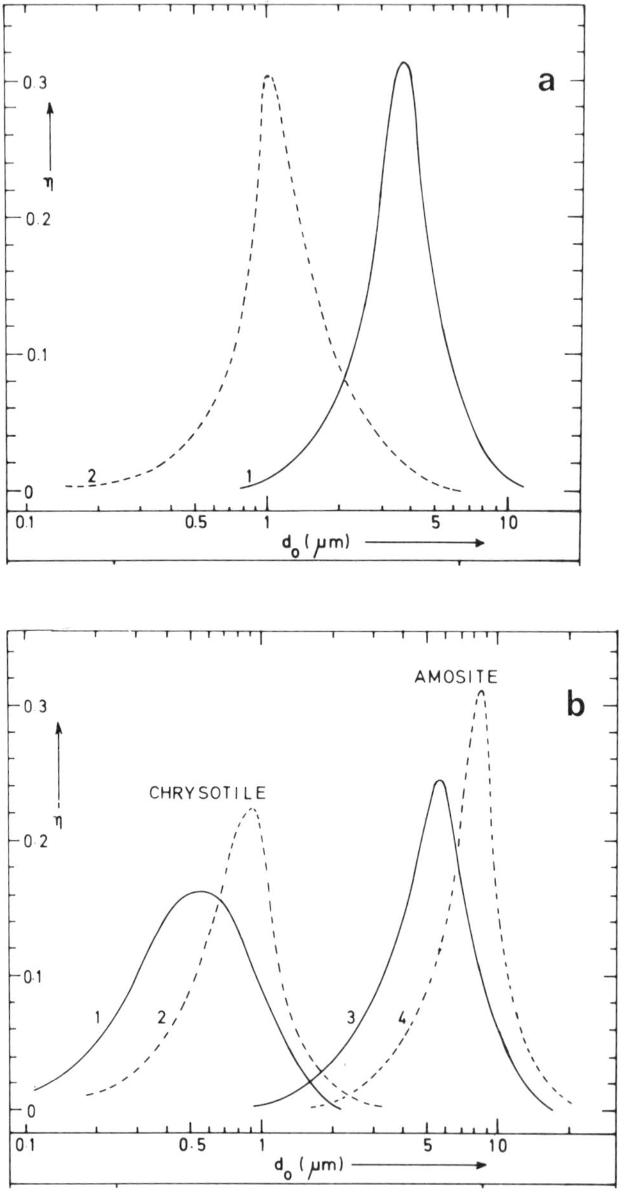

Figure 5. Particle size distribution at low-vibrations without (solid line) and with (dashed line) additional ultrasonic vibrations of 50 kHz. Diameter d_o is the optical fiber diameter recorded by a Royco 202 optical particle counter; (a) powder, (b) chrysotile (1 and 2), and amosite (3 and 4).

vibrations (f = 50 Hz; y = 760 μm). When an ultrasonic vibration of 50 kHz is added, the entire size distribution curve is shifted to a very much smaller particle size range (curve 2). The influence of ultrasonic treatment on the dispersion of fibers is shown in Fibure 5b. U.I.C.C. samples of chrysotile and amosite were used. Curves 1 and 3 are their respective size distributions when dispersed at low-frequency vibrations. The ultrasonic treatment (50 kHz) shifted the dispersion curves towards longer fibers. These experiments are not sufficient to provide a working hypothesis on what is happening in the powder or in the fiber fluidized bed when an ultrasonic field is applied. We think that the most important effect is the disintegration of powder or fiber aggregates. In addition, the "solid bubbles" in the fluidized bed can be disturbed by ultrasonic vibrations. The results of Figure 5 can, for example, be considered to demonstrate the "disintegration effect by ultrasound." In powder agglomerates (Figure 5a), fine isometric particles are released and dispersed, whereas in fiber aggregates (Figure 5b) longer fibers are more readily released. This may explain the apparent discrepancy between Figures 5a and 5b. In any case, further investigations are desirable.

High mass concentrations can be achieved by increasing the vibration amplitude (Figure 3), but a relatively high percentage of fiber aggregates are formed (Figure 6a). Two-component vibrating beds work also very well and may utilize much higher gas flowrates. However, the two-component vibrating bed also has some drawbacks: in our studies, the aerosol concentration did not remain constant for as long a time as with the one-component vibrating bed (the total quantity of fibers in the bed is smaller). Moreover, it was observed that the nickel powder used also contained some very fine particles ($\leqslant$ 0.1 μm) in addition to the larger spherical particles of approximately 100 μm diameter. These small particles cling to the fibers as shown in Figures 6b and 6c.

X-Ray fluorescence analysis has identified these as nickel particles. These fine particles could probably be eliminated from the coarse nickel powder by cleaning in liquids.

Since there are many potential needs for such a generator in long-term inhalation experiments and for its practical use in many physical as well as biological aerosol applications [44, 45], a relatively small portable unit was developed, as seen in Figures 7a, 7b and 7c.

Size-Selective Preparation of Inorganic Fibers for Biological Experiments

Fiber Separation

In characterizing the dynamic behavior of fibers, the basic problem is to relate "aerodynamic" or "optical" diameters to the physical dimensions [46-

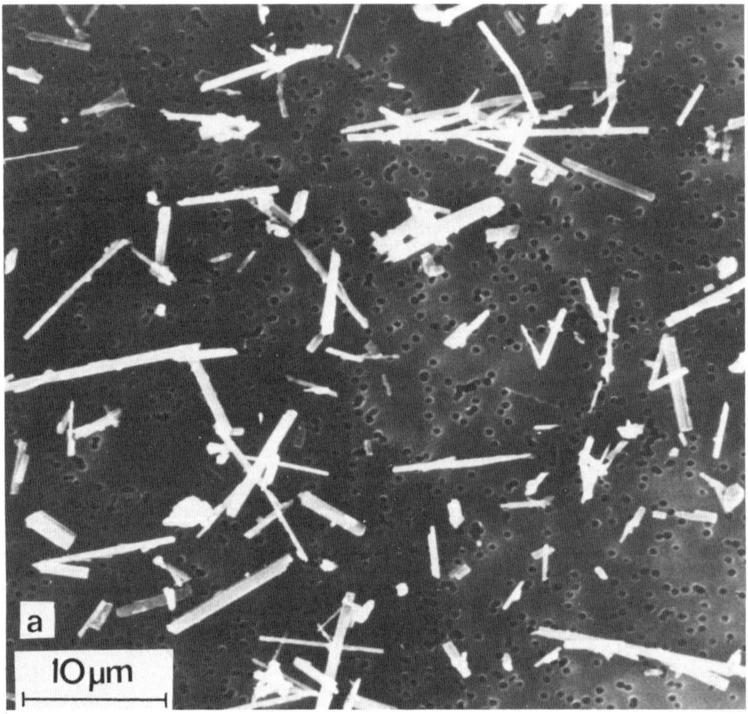

Figure 6. SEM pictures of amosite aerosols dispersed at high-vibration amplitude (a), and by means of a two-component vibrating bed (b, c). "Contaminations" from very fine nickel particles are evident in b and c.

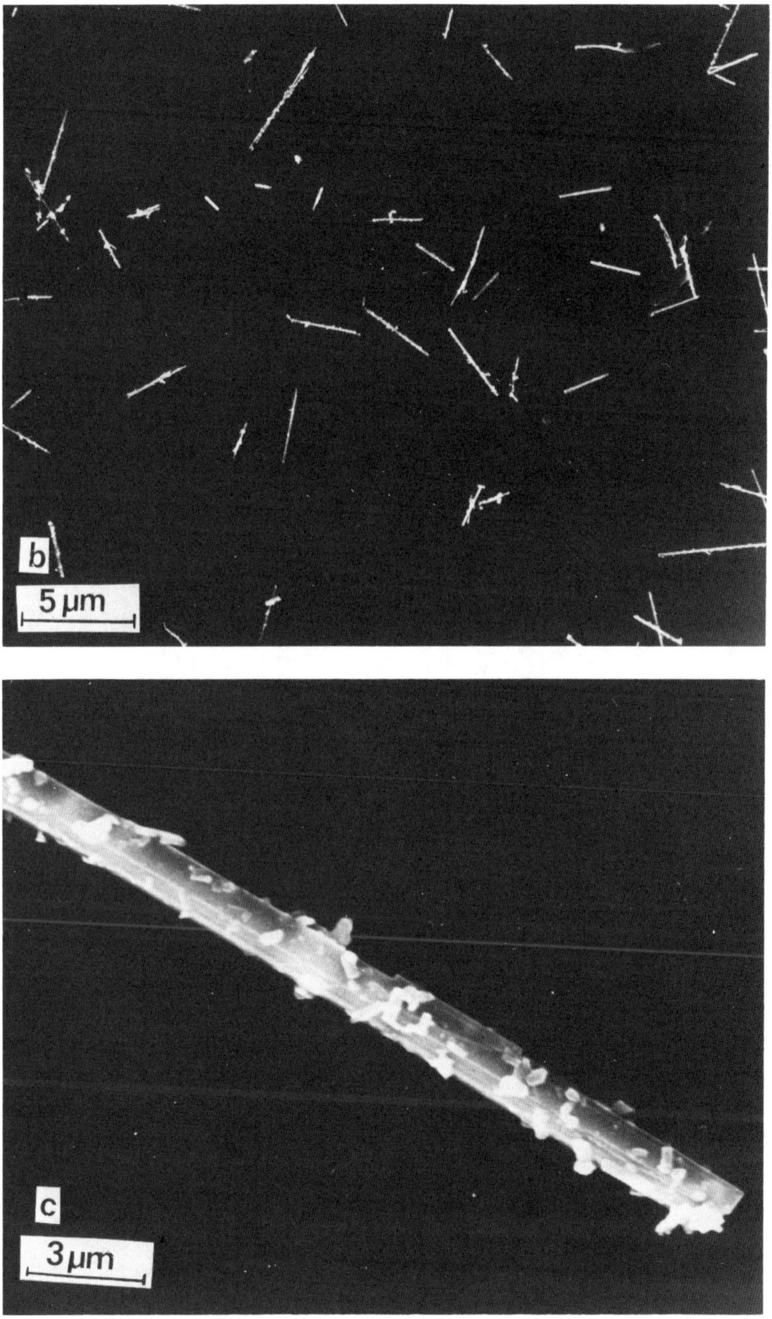
b
5 µm
c
3 µm

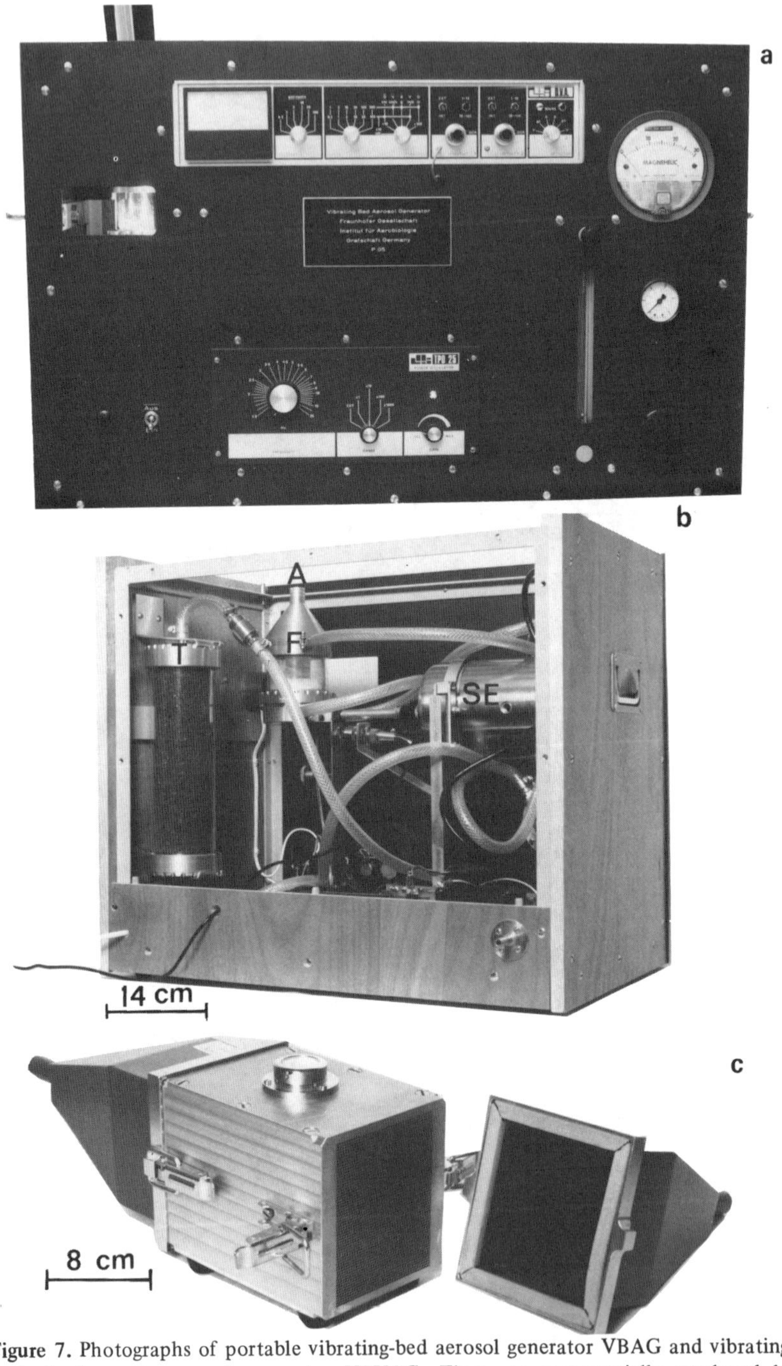

Figure 7. Photographs of portable vibrating-bed aerosol generator VBAG and vibrating-bed ultrasonic aerosol generator VBUAG. They are commercially produced by Wasau Inc., Berlin, Federal Republic of Germany: (a) front view, (b) inside view, (c) sedimentation elutriator. Description in text.

61]. In experimental studies with amosite and crocidolite, correlations were found [32] between aspect ratio $\beta = L_F/D_F$ and aerodynamic diameter D_{ae}. Since chrysotile fibers are not straight, no satisfactory relationship between D_{ae} and particle shape has been found. These experimental results as well as further theoretical considerations [52,57-59] showed that the parameter D_{ae} is almost independent of fiber length L_F and that fiber diameter D_F determines almost uniquely the aerodynamic behavior of fibers. When fibers are subjected to gravitational, electrical, thermophoretic or other external force fields, the behavior of fibrous particles depends strongly on D_F, β and the fiber orientation relative to the external field. The orientation is responsible for the "hydrodynamic instability" of fibers during their motion in force fields (tumbling, orientation changes, descent angle). For example, fibers having the same values of L_F and D_F may differ in their aerodynamic diameter, depending on their orientation in space. The range of aerodynamic diameters for fixed D_F and L_F increases [55] with increasing β.

Since D_{ae} is approximately a function of D_F, the use of sedimentation procedures in gases or liquids can provide fractionation by fiber diameter. For the same reason, sedimentation or impaction in air or liquid is not suitable for fiber length classification because the gravitational settling velocity and the stopping distance are dominated by the fiber diameter. A theory by Gans [46] predicts that fibers settling with their main axis inclined will descend under finite angle α rather than vertically downward. This descent angle α is related to the angle of inclination (measured from the vertical) and to aspect ratio β. Such nonvertical sedimentation of fibers occurs at low Reynolds numbers in the absence of shear flow. Ogden [53] proposed a fiber-length discrimination scheme utilizing the differences in descent angles. If fibers could be oriented at a constant angle of inclination, for example by means of a magnetic field [54], classification by length should be possible. Even if randomly oriented fibers were released from one point, all fibers settling outside a vertical cone with a 12° semi-angle and its apex at the fiber release point would have values of $\beta > 20$, so that fibers settling inside the cone would be depleted of fibers with high β-values. So far, no experimental work utilizing such a procedure has been reported, perhaps because of practical difficulties encountered.

Hardly any data are found in the literature on the behavior of fibrous particles in electric fields. Some preliminary theoretical considerations and laboratory experiments have shown [30, 62, 63] that the electric charge on fibrous particles is proportional to aspect ratio β and, thus, to fiber length. Potential classification schemes utilizing electric fields are described in more detail in a later section. Thermophoretic forces and thermophoretic velocity [64] as well as diffusiophoretic forces and diffusiophoretic velocity [65] of fibrous particles are also very little dependent on L_F. They depend primarily on D_F alone. Therefore, such force fields are not suitable for fiber separation by length.

As shown so far, some physical properties of fibrous particles exist which depend on aspect ratio β and length L_F. Their applicability for fiber separation by length should be fully explored. However, investigators with practical experience in this field are pessimistic about the prospect of producing gram quantities (at an acceptable price) of size-selected microfibers for practical applications.

We have attempted to produce a few grams of asbestos and glass fiber of selected fiber length and diameter for use in biological experiments. First we checked some experimental arrangements already described [5] and then we developed more complex procedures on a semi-empirical basis.

The raw fiber material was prepared by suitable grinding. It was then pre-separated by length in a mechanical process that had to be repeated many times to get satisfactorily separated fiber fractions. Fractionation by fiber diameter was achieved by sedimentation in liquids.

Fiber Size Reduction

U.I.C.C. standard reference asbestos [66,67] samples of amosite, crocidolite and chrysotile were used initially for the preparation of "short" fiber fractions. Long fibrous chrysotile from South Africa and glass fibers from Johns Manville Corporation (types MN 100 and MN 104) were used for the preparation of "short" as well as "longer" fiber fractions. Bidistilled water, ethyl alcohol, methyl alcohol and isobutyl alcohol were used for size separation by settling in liquids.

The biological experiments which will be reported elsewhere required the following fiber size fractions: very fine fibers ($L_F < 3$ μm, $D_F < 0.5$ μm), fine fibers ($L_F < 10$ μm, $D_F < 1.0$ μm), medium long fibers ($L_F = 3$ to 30 μm, $D_F < 2$ μm), and long fibers ($L_F = 5$ to 60 μm, $D_F < 2$ μm). All of these fractions consist of respirable fibers. The initial raw fiber material had to be reduced in size before any size classification could be attempted.

The comminution of the initial fiber sample can be performed in various ways. At first, exploratory experiments were performed with a microtome on amosite and chrysotile fibers. The fibrous material was embedded in polyethylene glycol, a material which is well soluble in water. A futile attempt was made to align the fibers in a magnetic field [54]. The results indicated that the fibers cut in the microtome were shorter than the initial fibers, but they were far from being of uniform length. This polydispersity was apparently due to the random orientation of the embedded fibers and the imperfect cutting of the fibers by the microtome knife. Electron micrographs showed that many of the embedded fibers were not cut, but pulled out of the polyethylene glycol plug. Some authors [23] have reported some success in comminuting fibers in an ultrasonic bath. Our own experience with such a technique, to be reported elsewhere, was not very encouraging.

Dry ball milling was the next size-reduction procedure examined. We used a commercially available vibrating ball mill with agate cylinders. The initial raw material was milled for a period of some hours to several days. The milled samples were studied by electron microscopy and X-ray diffraction analysis. Recently, the physics of milling fibrous material [68], and the variations found in the properties of milled chrysotile asbestos [69] have been thoroughly investigated. The results showed that, under the action of mechanical forces, asbestos fibers change both in diameter and in length, while glass fibers change in length only. It was observed that the number of fibers decreased rapidly during milling. Also, very short-term milling [69] changed many physicochemical, biochemical and molecular biological fiber properties.

Our milling experiments showed a rapid decrease in the number of fibers for all samples of amosite, corcidolite, chrysotile and glass. In addition, the initial crystal structure was degraded. An example of these results is given by Figure 8. The "size-reduced" crocidolite shown in Figure 8a is obviously not a well-defined test material for biological experiments. Therefore, ball milling was abandoned as an acceptable method for the comminution of fibrous material.

Subsequently, a chopping procedure was developed. It provided much better results. In the chopping apparatus, dry fibers were cut by a rapidly rotating set of knives (20,000 to 30,000 rpm). For intervals of some minutes to several hours, the initial fibrous material was homogeneously comminuted, and the portion of nonfibrous isometric fragments produced was rather low ($< 10\%$). Furthermore, there was very little damage done to the initial crystal structure (Figure 8b). Therefore, this method was applied in all our experiments and preparations.

Preselection of Fibers Using the Vibrating-Bed Aerosol Generator and Vibrating-Sieve Procedures

As shown above, our experience with the vibrating-bed aerosol generator showed that the size distribution of the generated asbestos or glass-fiber clouds depends strongly on the frequency and amplitude of vibration. These parameters—as well as the primary fiber size distribution in the fluidized bed—approximately determine the lengths of fibers dispersed into the gas. This method facilitates the production of size-selected particle clouds. The generator could thus be used for preparing fibrous aerosols with a more narrow fiber-length distribution. Under proper operating conditions, the

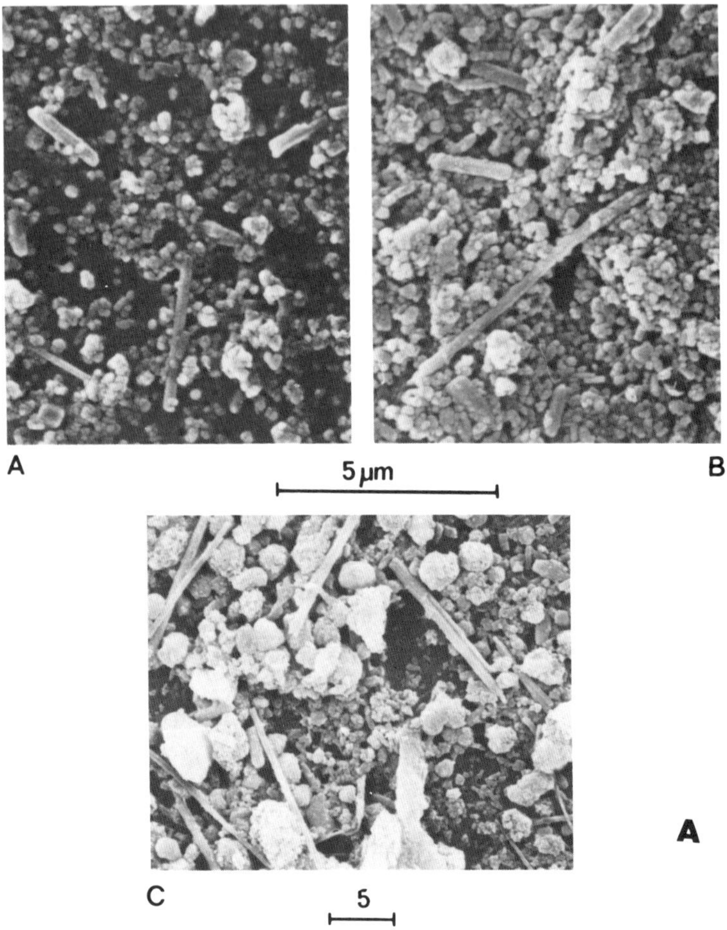

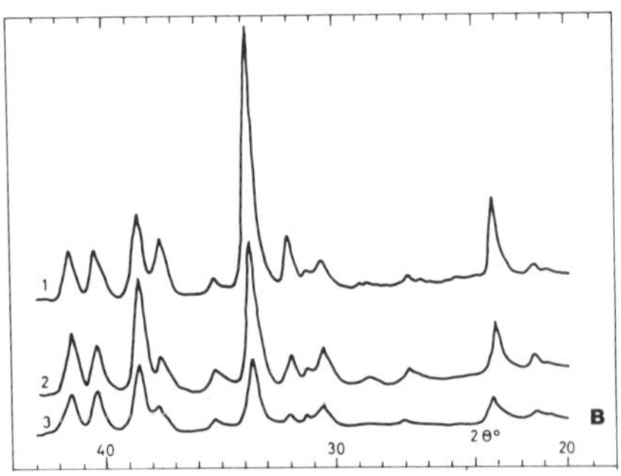

Figure 8. (a) SEM photographs of U.I.C.C. crocidolite asbestos after 120 hr of grinding in a ball mill; (b) X-ray diffraction spectra of U.I.C.C. crocidolite without grinding (1), ground in a "knife mill" (2) and ground in a ball mill (3).

percentage of longer fibers was decreased. The fraction of shorter fibrous particles was then sampled by filtration and used for the next step in the complete procedure: size separation by sedimentation in liquids.

Our experience in filtering fibrous aerosols [70] indicated that direct interception is the most important separation mechanism. In the case of fibers, the efficiency coefficient for this mechanism depends upon fiber length, shape and orientation. When filtration or sieving is used for the selection of fibers by length, the following two important conditions controlling the quality of separation must be met: the filters or sieves must be very homogeneous, and the fibers should be oriented as much as possible perpendicularly to the streamlines of the gas flow.

Experimentally, the first condition was fulfilled by using either metallic or polymer microsieves (mesh diameters = 5, 15, 25 and 50 μm) for sieving in air or Nuclepore filters (pore diameters = 3, 5, 8 and 12 μm) for filtering in liquids. The second condition could not be met exactly. Applications of magnetic and electric fields did not produce the desired orientation to our satisfaction. However, applications of frequency vibrations (100 Hz) apparently distorted the alignment of fibers in the laminar fluid flow which resulted in rather successful fiber length separations by sieving. We built two systems in order to examine the efficiency of this procedure in air and liquid.

The first method was a sieving procedure in air (Figure 9a). We used the commercial microsieving instrument "Analysette" MS with 4 sieves (S_1 = 50 μm, S_2 = 25 μm, S_3 = 15 μm and S_4 = 5 μm). A fifth fraction was collected on a Nuclepore filter (S_5 pore diameter = 0.2 μm). The fibrous material was dispersed by the vibrating-bed aerosol generator SAG. The homogeneous fiber cloud was then drawn into the screening battery. The screens were vibrated at 100 Hz. The air flowrate was 7 liter/min, measured by flowmeter R and maintained by vacuum pump VP. F_1 and F_2 are filter holders. The SEM photographs show size-separated fractions of glass fibers. By utilizing this procedure, five size fractions could be collected. The screens select the fibers approximately by length. However, as the electron micrographs show, some selection by fiber diameter is also evident. By repeating this procedure three times, a relatively narrow separation by length and some selection by diameter could be achieved. The final diameter separation was achieved through repetitive settling of the filter fractions in liquids.

A similar filtration procedure has been developed for liquids by arranging in series Nuclepore filters of different pore sizes (Figure 9b). The arrangement was similar to the previous one. Generally, after fibers had been diameter-classified by gravitational settling, a liquid suspension of fibers R was filtered in filter battery PZ by two or more Nuclepore filters in series. F_1 and F_2 are filters in filter holder FH. The filter holder and the cylinder holding the suspension were vibrated. The vibration frequency and the

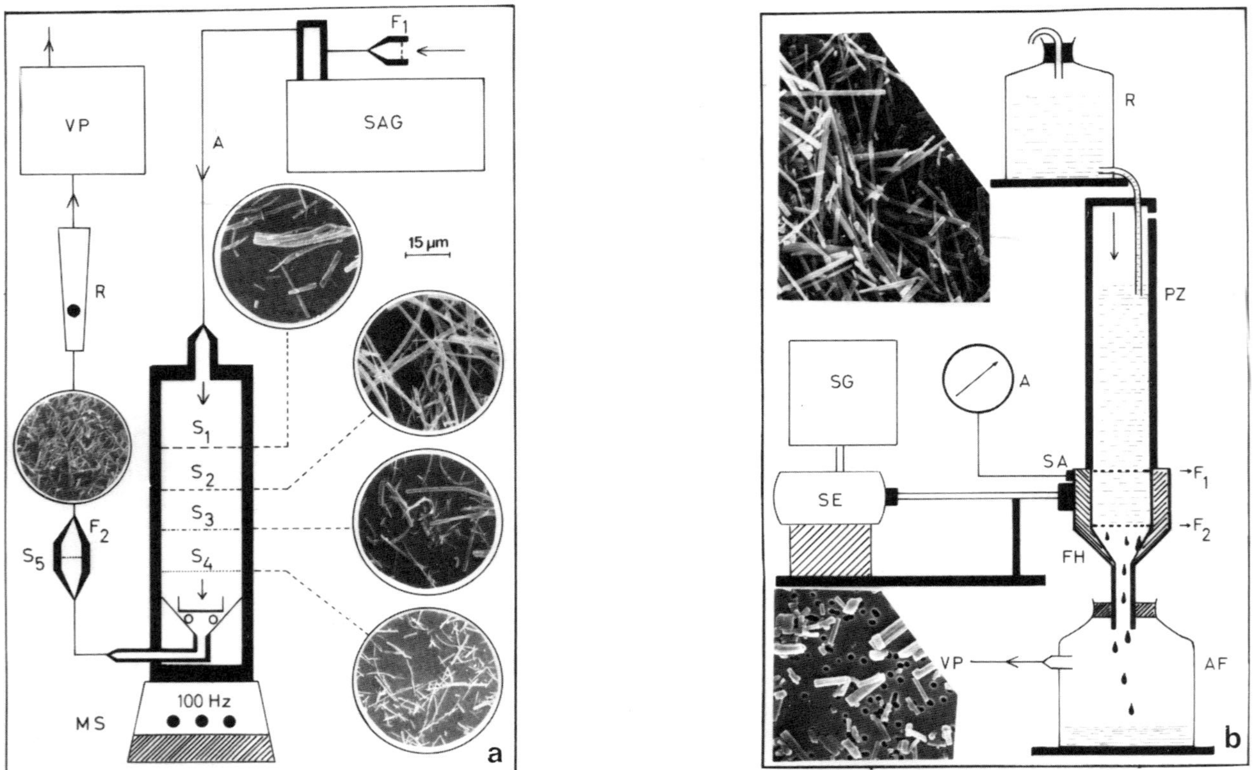

Figure 9. Schematic representation of (a) vibrating-sieve apparatus for particle cloud separation, and (b) vibrating-liquid filtration system. Description in text.

amplitude varied from 25 to 120 Hz and from 0.1 to 1 mm, respectively. The vibrations were generated by the same system described for the aerosol generator. The liquid flowrate was maintained at 0.7 liter/hr by vacuum pump VP. Different Nuclepore filters (pore diameter = 12, 8, 5, 3 and 0.1 μm) were used in series to fractionate fibers by length [79]. The finest fraction penetrated all filters and was found in very low concentrations in filtrate AF. After evaporating the liquid, small amounts of the finest fibers were obtained in dry form. The experiments indicated that both frequency and amplitude of vibration influence the filtration process. Generally, frequencies of about 80 to 120 Hz and amplitudes of less than 0.3 mm were used.

Sedimentation of Fibers in Liquids

For the separation of fibers by diameter, a repetitively applied sedimentation or centrifugation of liquid suspensions seemed to be the most effective method. As already mentioned, the sedimentation velocity of a fibrous particle depends to first approximation on diameter D_F and on the orientation of the fiber relative to the gravitational force field. There is also a slight dependence on length L_F. Figure 10a presents sedimentation distances S calculated for asbestos fibers of length L and diameter D as attained in water after 30 days. For comparison purposes, Figure 10b shows the times t_1 calculated for sedimentation in air. Two orientations are considered: parallel to the direction of settling and perpendicular to the direction of settling. Our experiments have shown that sedimentation of asbestos fibers in highly concentrated (10 to 50 μg/ml) colloidal suspensions took longer than calculated by theory. Furthermore, fibers fractionated by gravitational settling had a more narrow diameter distribution than could be achieved by centrifugation. Therefore, "slow" settling by gravity is preferred, and has been used exclusively in all subsequent experiments.

Water suspensions of fiber samples were found to be most suitable for separation by sedimentation. The behavior of asbestos fibers in a liquid and some of their colloid-chemical as well as surface properties have already been described in the literature [71-78] and could be exploited. For example, in preparing suitable suspensions of chrysotile, the use of wetting agents or organic solvents, such as methyl alcohol, ethyl alcohol and isobutyl alcohol, proved to have many advantages. However, only bidistilled water was used in this study, because of the risk of changing some physicochemical surface properties of the fibrous material which would make it unsuitable for subsequent representative biological experiments. With this restriction, it was important to use very clean water, since the presence of very small amounts of foreign ions induces flocculations of fibers (Figure 11).

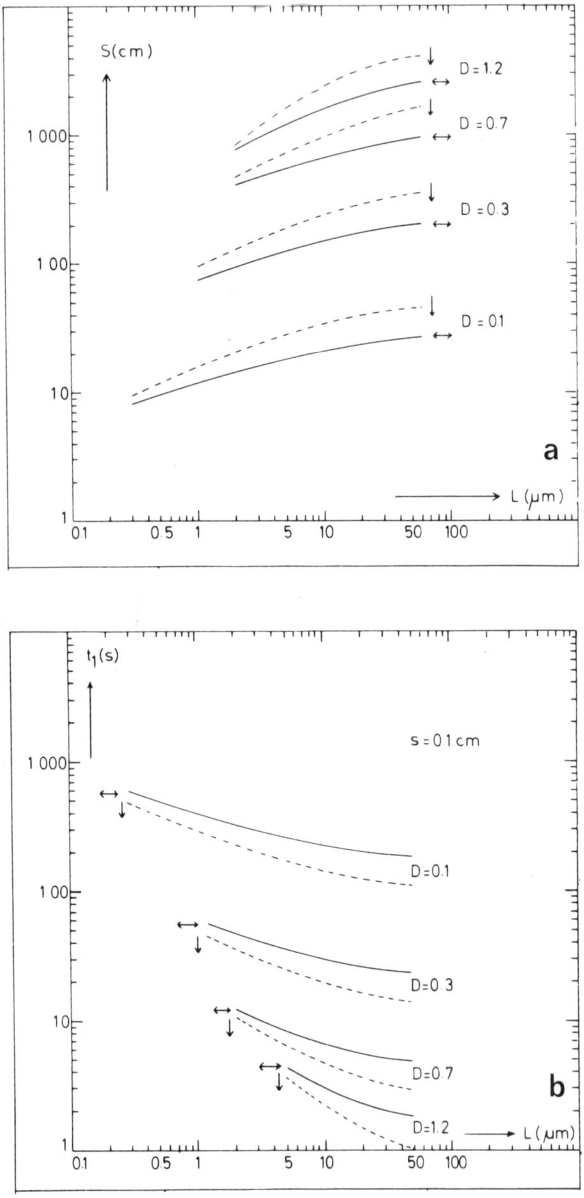

Figure 10 (a) Sedimentation distances S in water plotted as a function of fiber length L for different fiber diameter and fiber orientation. Settling time = 30 days; (b) Sedimentation time t_1 in calm air for sedimentation distance of 1 mm.

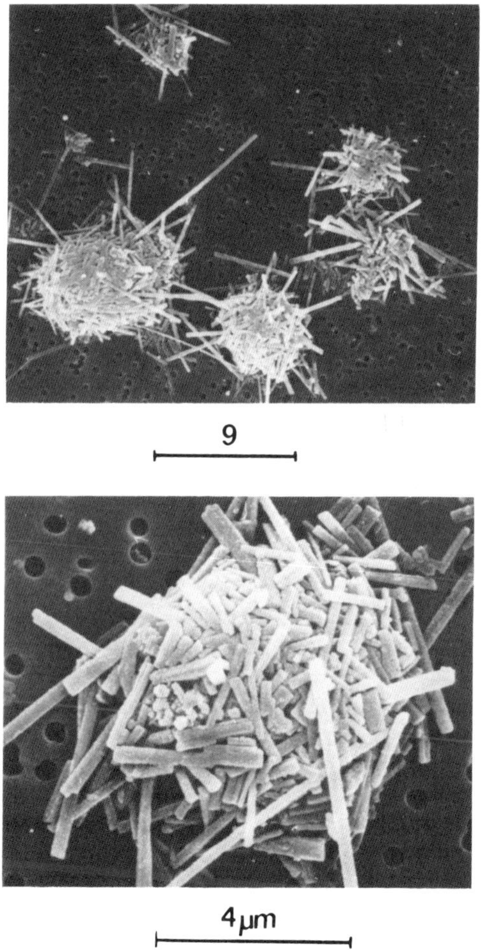

Figure 11. Electron micrographs of flocculated amosite fibers. The water suspension had traces of NH_4Cl.

Preparation of Size-Selected Fiber Samples

Theoretically, the preparation of size-selected fiber samples does not seem to be very complicated: by repetitive screening, the fibrous material is fractionated by length and then, by repetitive settling in liquids, a separation by fiber diameter is achieved, or vice versa. In practice, however, there do exist many difficulties which may disturb the separation process. Therefore, one needs to pay careful attention to all the procedures for the preparation of each narrow size fraction.

Figure 12 shows a schematic diagram of the complex procedure for preparing fine and very fine fractions of asbestos and glass fibers ($L_F < 10$ μm and $L_F < 3$ μm). The initial raw fiber material, e.g., U.I.C.C. asbestos with fiber-size distribution designated as A_o, is reduced in size in the fast-rotating "knife mill" M. The ground material with its size distribution designated by A_1 is then dispersed into a fiber cloud by means of the vibrating-bed aerosol generator SBAG. This fiber cloud is further modified in sedimentation elutriator EL and in the small cyclone Z. It is then drawn through membrane filter F onto which the fibers are collected having size distribution A_2. This fraction is used for the preparation of liquid suspension FL which is given a short ultrasonic treatment US at low frequency. The suspension then settles in cylinder S. Three fractions are collected during sedimentation: a fraction of finely dispersed nonfibrous particle fragments, a fraction of very fine fibers with its size distribution designated by A_3, and a fraction of coarser fibers collected at the bottom of the sedimentation cylinder. The first fraction (see also Figure 15) is discarded. The second fraction is separated from the liquid by ultrafiltration or by evaporation SB and is used for the preparation of very fine fiber samples. This second fraction is homogenized in H and is then evaluated and characterized by SEM, microprobe and X-ray fluorescence analysis with its final size distribution designated by A_4. The fraction of coarse fibers is reprocessed in order to extract the fraction of fine fibers. This sedimentation procedure has been repeated three times in order to get a satisfactorily narrow size distribution. After having standardized the entire procedure, several grams of size-selected fibers have been extracted for special biological experiments (see also Tables I and II).

Fractions of fine and very fine fibers can also be obtained by using the technique of vibrating-liquid filtration. However, the procedures which have been schematically represented in Figure 9 served primarily for the preparation of "medium" and "long" fiber fractions (see also Table III). Both procedures need further improvements to ensure satisfactorily reproducible results so that standard fractions of size-selected fibers can be furnished.

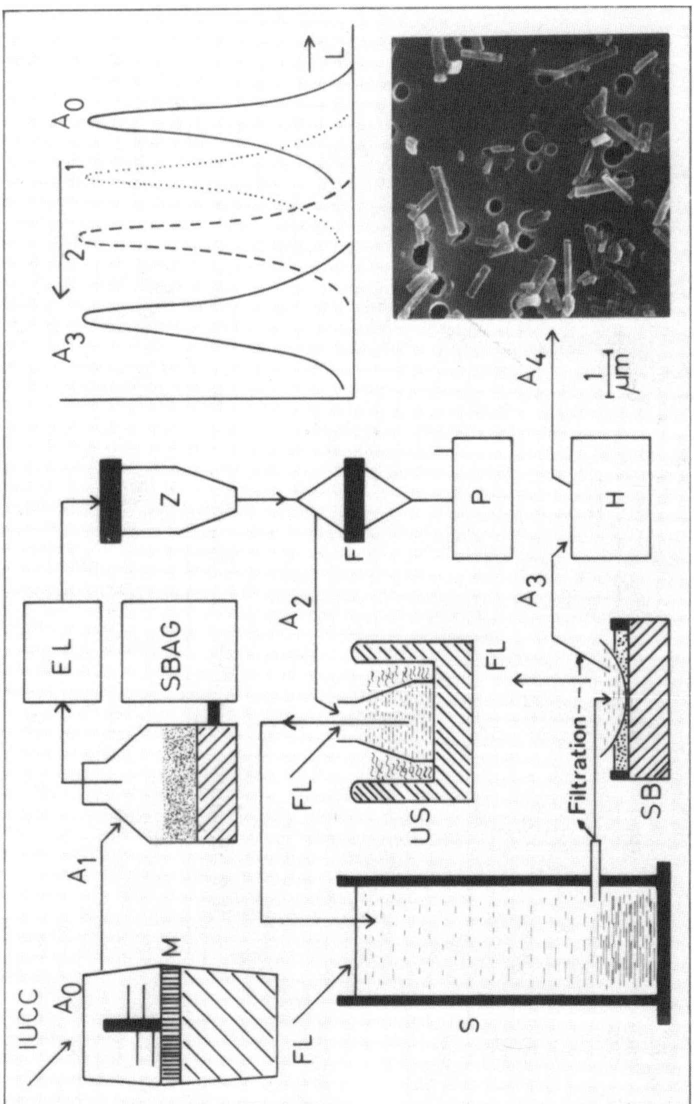

Figure 12. Schematic representation of the procedure for the preparation of fin e and very fine size fractions. Description in tex t.

Table I. Representative Data of Very Fine Fiber Fractions Measured by Electron Microscopy

Fiber	$\overline{L_F}$ (μm)	σ_F (μm)	$\overline{D_F}$ (μm)	σ_D (μm)	$\overline{L_F/D_F}$	$L_F \leqslant$ 1 μm (%)	$L_F \leqslant$ 3 μm (%)	$D_F \leqslant$ 0.1 μm (%)	$D_F \leqslant$ 0.5 μm (%)	L_F/D_F $\leqslant$ 3 μm (%)	L_F/D_F $\leqslant$ 30 μm (%)
Amosite	1.17	0.47	0.23	0.07	5.40	48.3	99.0	2.1	99	6.2	100
Chrysotile	1.11	0.58	0.13	0.04	8.90	58.1	99.0	33.2	100	0.2	100
Crocidolite	1.35	0.83	0.21	0.12	6.75	46.1	97.3	2.1	95	0.5	100
Glass	1.34	0.62	0.19	0.03	7.04	36.2	99.0	3.2	100	0.8	100

$\overline{L_F}$ = mean fiber length.
σ = standard deviation.

Table II. Representative Data of Fine Fiber Fractions Measured by Electron Microscopy

Fiber	$\overline{L_F}$ (μm)	σ_L (μm)	$\overline{D_F}$ (μm)	σ_D (μm)	$\overline{L_F/D_F}$	$L_F \leqslant$ 3 μm (%)	$L_F \leqslant$ 10 μm (%)	$D_F \leqslant$ 0.5 μm (%)	$D_F \leqslant$ 1.0 μm (%)	L_F/D_F $\leqslant$ 5 μm (%)	L_F/D_F $\leqslant$ 50 μm (%)
Amosite	2.52	1.44	0.47	0.17	5.29	72.1	100	66.4	98.1	57.2	100
Chrysotile	1.42	0.83	0.16	0.04	9.18	95.1	100	100.0	–	17.3	99
Crocidolite	1.39	0.89	0.23	0.06	6.98	93.4	100	99.0	100.0	46.8	100
Glass	3.17	2.78	0.43	0.27	7.20	68.9	96	78.1	96.0	32.2	100

Table III. Representative Data of Medium- and Long-Fiber Fractions Measured by Electron Microscopy

Fibers	L_F (µm)	D_F (µm)	$L_F \leqslant 5$ µm (%)	L_F = 1 to 30 µm (%)	L_F = 1 to 60 µm (%)	$D_F \leqslant 2$ µm (%)
Glass Medium	18.7	0.52	1.3	97.8	100	98
Glass Long	23.4	0.54	0.8	69.8	99	98
Chrysotile Medium	17.4	0.14	4.6	98.4	100	100
Chrysotile Long	18.1	0.14	2.3	72.3	100	100

Length-Selected Separation of Fibers in Electric Fields

We consider this paper as a limited review as well, and will therefore also briefly mention some other interesting results achieved at our Institute. An improved sampling device schematically shown in Figure 13 has recently [80] been described. It deposits airborne fibers and isometric particles onto different parts of a membrane filter. The aerosol flows through an electrical aerosol charger [81] at a rate of about 4 liter/min. The flow rate of particle-free air PA is 22 liter/min. Both, the aerosol charger and the spectrometer have rotational symmetry. In the charger, ions from a corona discharge migrate through a grid, thus charging the aerosol particles unipolarly. In the spectrometer, fibers and isometric particles are separated from each other by their different mobility in the applied electric field. Figure 13b shows an example of such separation or fractionation. Furthermore, a separation and fractionation by fiber length is also evident and is in agreement with theory [62,63].

A sedimentation elutriator with an applied electric field also has potential for future development [82]. In such a device the fibers would be unipolarly charged and would then be drawn through a chamber which acts simultaneously as a horizontal elutriator and as a counteractive electrostatic collector. Equilibrium between gravitational and electric forces should exist only for a specific fiber length fraction. Depending upon the flowrate, this fraction would penetrate the separator.

Preliminary experiments have shown that an electrophoretic separation method similar to the one shown in Figure 13a, but in a liquid, could also be an interesting prospect. In any case, all of the last methods mentioned

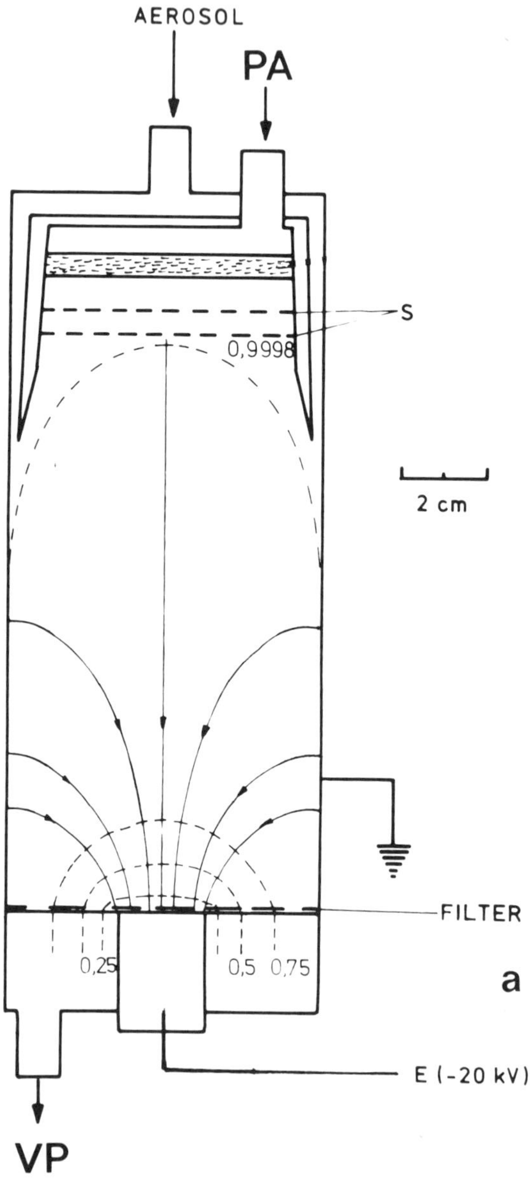

Figure 13. (a) Schematic representation of a special mobility spectrometer for fibrous aerosols [80]. Equipotential curves with their dimensionless values are shown by dashed lines. Electric field lines are shown by solid lines. S = metal sieve, E = elec-

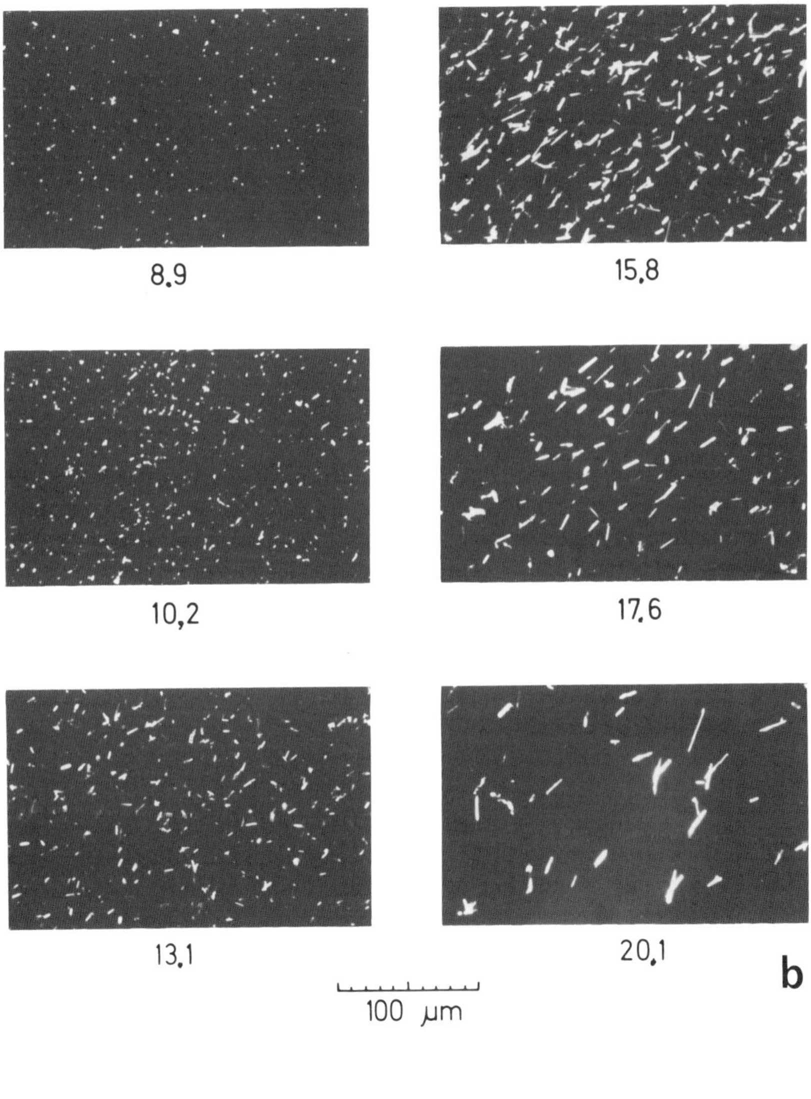

trode and VP = vacuum pump; (b) particles and asbestos fibers separated onto a Millipore filter in the mobility spectrometer. The numbers refer to the distances of the probes from the filter's circumference.

need further intensive investigations. The accuracy and sensitivity of their length fractionation is not predictable at this time.

Sample Evaluation and Characterization

As mentioned before, electron microscopy, microprobe analysis, X-ray fluorescence spectroscopy and X-ray diffraction were used for the evaluation and characterization of fiber preparations. Collected samples for all these methods were taken from Nuclepore filters [79] with pore diameters of 0.1 to 0.4 μm. Electron micrographs with magnifications from 3000 to 30,000 were used for determining the size distribution curves. Some 500 to 800 fibers were measured by length and diameter for each fiber-size evaluation. The other instruments were used for the chemical and crystalographic analysis of the samples.

Results

The procedure described for the production of fine and very fine fiber fractions is cumbersome and time-consuming, but was otherwise successful. Sedimentation of fibers in liquids primarily resulted in separation by fiber diameter, but a weak dependency upon fiber length was observed as well, primarily for amphibole asbestos and glass. A positive correlation between D_F and L_F could be confirmed for all fiber preparations except chrysotile (Figure 14). This is also demonstrated very well in Figure 15b which shows four fractions of amosite settled in water. The thicker fibers were also longer. Thus, by virtue of the initial knife-mill treatment, the separation by diameter through sedimentation in liquids was accompanied by a slight separation by length.

The expected biological activity of fibers with $L_F < 3$ μm should be low, and the biological activity of fibers with $L_F < 5$ μm should be rather moderate. To test this assumption, several grams of these fiber fractions were prepared according to the procedure demonstrated in Figure 12. Very fine fiber fractions were prepared from amosite, crocidolite, chrysotile and glass. In all of these samples, the number content of fibers with $L_F < 3$ μm was about 98% or more, and the fractions with $D_F < 0.5$ μm ranged between 95% and 100% (Table I and Figure 16). The fine fiber fractions were prepared from the same materials (Table II). In these samples, the number content of fibers with $L_F < 5$ μm were between 85% (glass) and 99% (crocidolite) and the fractions with $D_F < 1$ μm ranged between 96% (glass) and 100% (chrysotile). Small amounts of fiber fractions with $L_F < 5$ μm (99 ± 1%) and $D_F < 0.8$ μm (99 ± 1%) were also prepared by the same procedure.

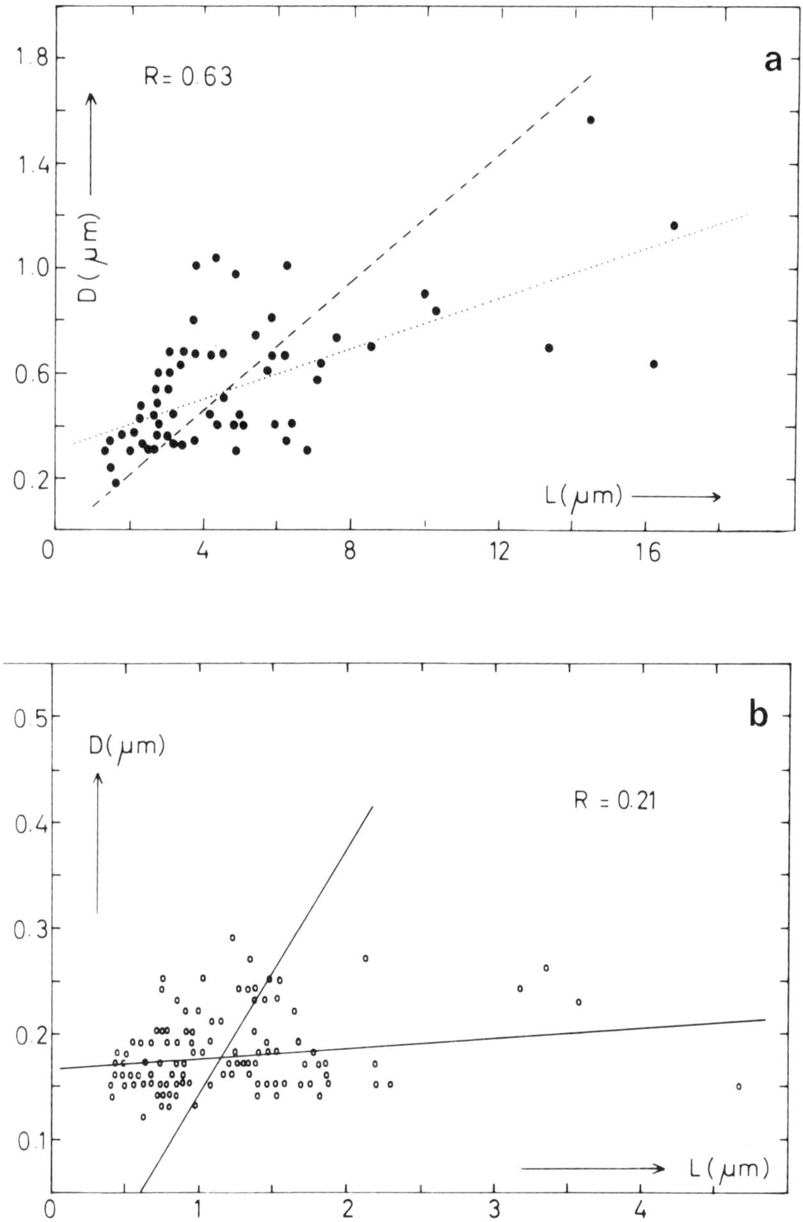

Figure 14. Experimental correlation between fiber diameter D and fiber length L for (a) U.I.C.C. amosite and (b) U.I.C.C. chrysotile. R = correlation coefficient.

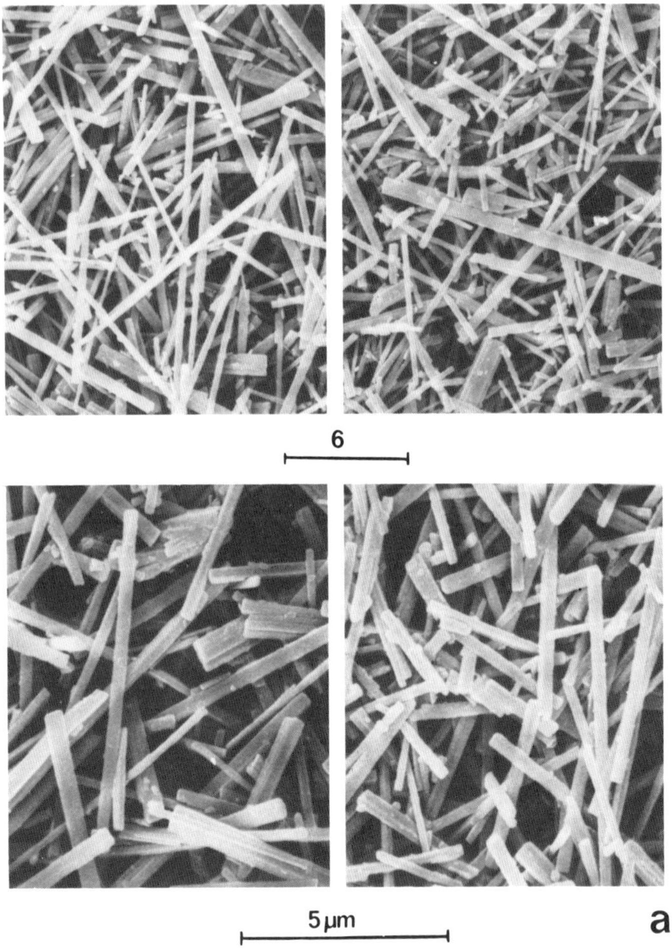

Figure 15. Separation of U.I.C.C. amosite fibers by settling in water: (a) initial material;

(b) 4 fractions of ground amosite samples onto Nuclepore filters.

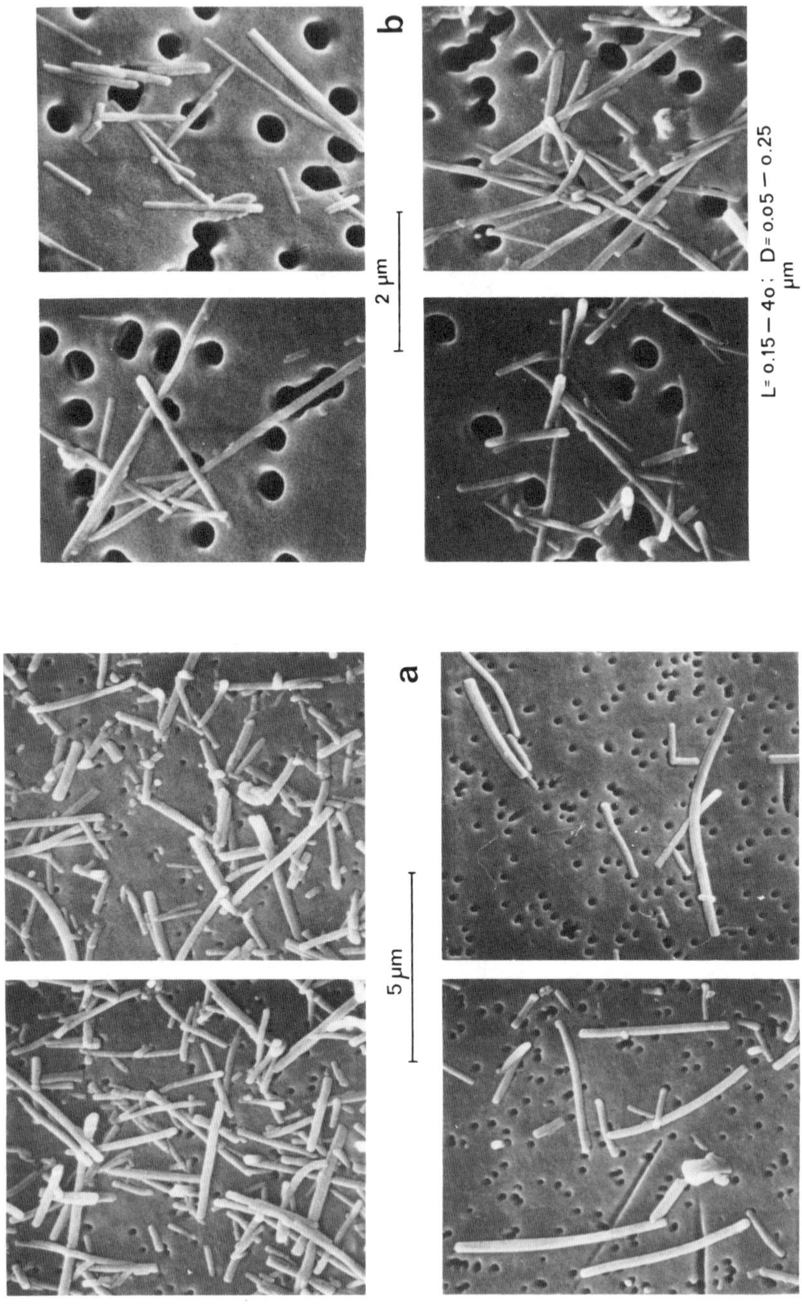
b
2 µm
L= o.15 — 4o ; D= o.o5 — o.25 µm
a
5 µm

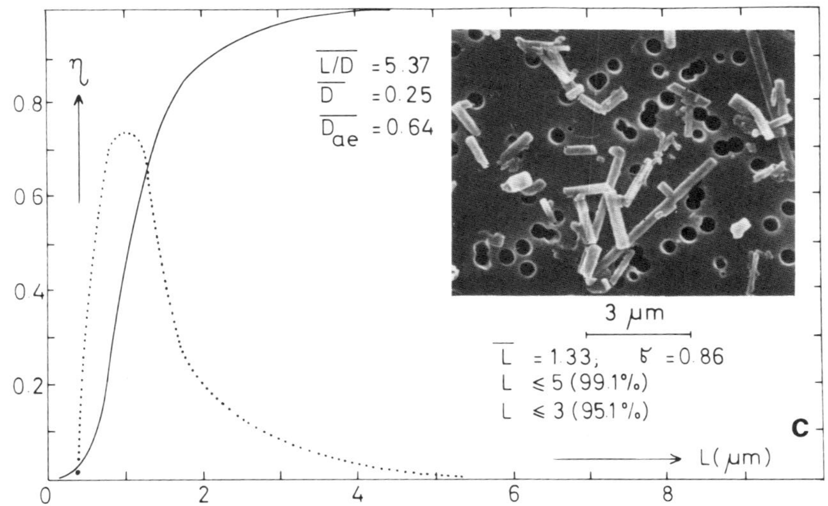

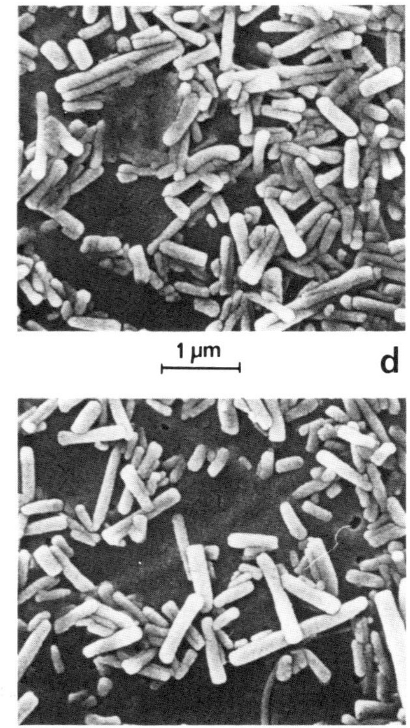

Figure 16. SEM pictures of very fine fractions of different fibers: (a) MN 100 glass fiber; (b) Canadian chrysotile fiber; (c) amosite fiber and dispersion curves; (d) superfine MN 100 glass fiber.

The methods of vibrational screening in air or water were used primarily for the preparation of "medium" and "long" fiber fractions. Both methods (Figure 9) were successful for this purpose. Sieving in air was advantageous, because it facilitated the preparation of more fractions and larger amounts of each fraction in a shorter time. However, the quality of separation was better using the method of vibrational filtration in liquids (Figure 17). Both methods appear to be promising, but need further investigation and improvement for reliable standardization. The efforts in preparing "long" fibers by size separation are rather recent and will, therefore, continue (Table III). Physicochemical examinations and analyses of the fiber fractions showed that the surface properties of the samples remained practically unchanged during the separation process.

CONCLUSIONS

A vibrating-bed aerosol generation has been described which permits pulverization of powdery as well as fibrous material. A portable generator suitable for physicochemical as well as biological investigations has also been developed. It can produce stable and reproducible concentrations, with constant and reproducible particle size distributions. The mean particle size—e.g., fiber length—of this aerosol is dependent on the frequency and amplitude of the vibrational field. This dependence varies and depends on the type of powder or fibrous material. Some preliminary experiments have also shown a positive influence of an ultrasonic field on the dispersion process.

Size-selective procedures have been developed for the preparation of fine and very fine fibers of chrysotile, crocidolite, amosite and glass. U.I.C.C. standard asbestos samples as well as MN 100 and MN 104 glass-fiber samples were ground in a special mill. This material was then prescreened in the vibrating-bed aerosol generator, suspended in liquids by ultrasonic treatment and size-separated by slow sedimentation. Subsequently, different fiber-size fractions were prepared by means of high-efficiency filtration or slow evaporation. Size-distribution curves for fiber diameters and fiber lengths of each fraction were plotted after electron microscopic evaluation. For the preparation of "longer" size-selected fibers, new promising methods were developed utilizing vibrational screening in air and vibrational filtration in liquids. A length-selected separation of fibers in electric fields has also been mentioned and discussed. The latter work is, however, still in the preliminary investigation stage.

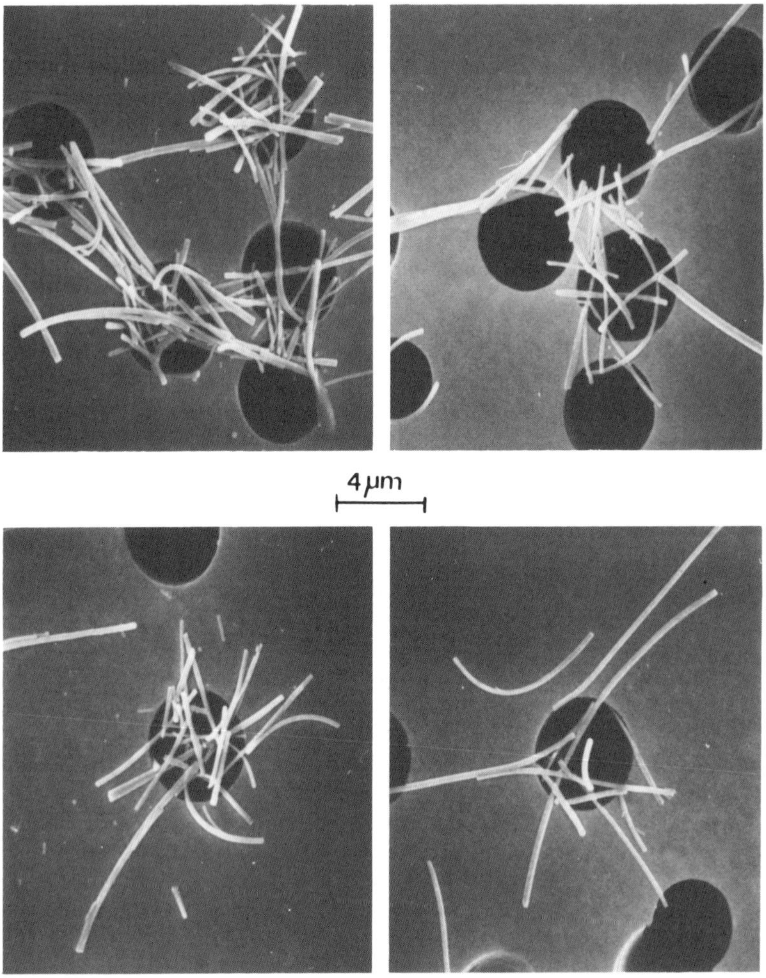

Figure 17. Electron micrographs of "sieved" glass fibers on a Nuclepore filter (pore diameter = 5 μm). The sieving procedure of Figure 9(b) was used.

ACKNOWLEDGMENTS

These studies were supported by a contract from the German Environmental Protection Agency in Berlin (Umweltbundesamt).

REFERENCES

1. Klostorkötter, W., and K. Robock. "Experimentelle Untersuchungen zum Wirkungsmechanismus von Asbest unter besonderer Berücksichtigung der Faserlänge," *Jahr. Gesell. Arbeitsmed.* 111 (1969).
2. Beck, E. G., P. F. Holt and N. Manojlovic. "Comparison of Effects on Macrophage Cultures of Glass Fibre, Glass Powder and Chrysotile Asbestos," *Brit. J. Ind. Med.* 29:280 (1972).
3. Stanton, M. F., and C. Wrench. "Mechanisms of Mesothelioma Induction with Asbestos and Fibrous Glass," *J. Nat. Cancer Inst.* 48:797 (1972).
4. Wagner, J. C., G. Berry and V. Timbrell. "Mesotheliomata in Rats after Inoculation with Asbestos and other Materials," *Brit. J. Cancer* 28:173 (1973).
5. Pelnar, P. V., Ed. "Fibers for Biological Experiments," *IOEH Conference*, Montreal, October 29-30 (1973).
6. Robock, K. "Die Wirkung mechanischer, thermischer und chemischer Behandlungen von Siliciumdioxid- und Asbeststäuben auf Zytotoxizität und Elektronenstruktur," *Beitr. Silikose-Forsch. (Pneumokon.)* 26:112 (1974).
7. Beck, E. G. "Biologische Wirkung von faserförmigen Stäuben," *Arbeitsmed. Sozialmed. und Präventivmed.* 10:178 (1975).
8. Pott, F., K.-H. Friedrichs and F. Huth. "Ergebnisse aus Tierversuchen zur kanzerogenen Wirkung faserförmigen Stäube und ihre Bedeutung im Hinblick auf die Tumorenentstehung beim Menschen," *Zbl. Bakt. Hyg. I. Abt. Orig.* B162:467 (1976).
9. Pott, F. "Krebserzeugende faserige Feinstäube," *Arbeitsmed. Sozialmed. Praventivmed.* 12:172 (1977).
10. Pooley, F. D. "An Examination of the Fibrous Mineral Content of Asbestos Lung Tissue from the Canadian Chrysotile Mining Industry," *Environ. Res.* 12:281 (1976).
11. Ulmer, W. T., Ed. "Asbest und Asbestose," Colloquium, Berlin, November 21-22, 1975, *Hefte Unfallheilkunde* 39H126:511 (1976).
12. Finklea, S. F., Ed. "Occupational Exposure to Fibrous Glass," *Public Health Science Publ., Proceedings of the Symposium*, Maryland, June 26-27, 1974. (Washington DC, 1976).
13. Dobbertin, S. "Gesundheitsgefahren durch Luftverunreinigungen mit Asbest," *Staub-Reinhalf.-Luft* 37:65 (1977).
14. Finklea, S. F., Ed. "Devised Recommended Asbestos Standards," Pub. 77-16a, U.S. Dept. of Health, Education and Welfare, Washington, D.C. (1976).
15. Wright, G. W., and M. Kuschner. "The Influence of Varying Lengths of Glass and Asbestos Fibers on Fissure Response in Guinea Pigs," in *Inhaled Particles IV*, W. H. Walton, Ed. (Oxford: Pergamon Press, 1977), p. 455.

16. Pott, E. "Some Aspects on the Dosimetry of the Carcinogenic Potency of Asbestos and Other Fibrous Dusts," *Staub-Reinhalt.-Luft* 38:486 (1978).
17. Friedrichs, K.-H. "Probleme bei der Anwendung von Fasern im Tierversuch und Möglichkeiten zur Interpretation der Ergebnisse," *Staub-Reinhalt.-Luft* 38:490 (1978).
18. Zahradnicek, A. "Methoden zur Aerosolherstellung aus vorgegebenen Feststoffhaufwerken," *Staub-Reinhalt.-Luft* 35:226-231 (1975).
19. Timbrell, V., A. W. Hyett and J. W. Skidmore. "A Simple Dispenser for Generating Dust Clouds from Standard Reference Samples of Asbestos," *Ann. Occup. Hyg.* 11:273 (1968).
20. Hounam, R. F. "The Konimiser–A Dispenser for the Continuous Generation of Dust Clouds from Milligram Quantities of Asbestos," *Ann. Occup. Hyg.* 14:329 (1972).
21. Craig, D. K., A. P. Welner and W. G. Morrow. "The Generation and Characterization of a Respirable Aerosol of Chrysotile for Chronic Inhalation Studies," *Amer. Ind. Hyg. Assoc. J.* 33:283 (1972).
22. Ortiz, L. W., H. E. Black and J. R. Coulter. "A Modified Fibrous Aerosol Generator," *Ann. Occup. Hyg.* 20:25 (1977).
23. Samudra, A. V., F. C. Bock, C. F. Harwood and J. W. Stockham. "Evaluation and Optimizing EM-Methods for Characterizing Airborne Asbestos," EPA-Report 600/2-78-038, Research Triangle Park, NC (1978).
24. Spurny, K. R., and J. P. Lodge. "A Contribution to the Mechanism of Genesis of the Condensation Nuclei from Sea Water," *J. Rech. Atmos.* 3:117 (1967).
25. Guichard, J. C. "Use of Fluidized-Bed Cleaning in the Production of Aerosols," *Dechema-Monographien* 59:247 (1968).
26. Willeke, K., C. S. K. Lo and K. T. Whitby. "Dispersion Characteristics of a Fluidized Bed," *J. Aerosol Sci.* 5:449 (1974).
27. "Data Sheets of Physical and Chemical Properties of U.I.C.C. Standard Reference Asbestos Samples," Report of the National Research Institute of Occupational Disease, Johannesburg, pp. 1-30 (1972).
28. De Jong, J. A. H., and S. F. Nomden. "Homogeneous Gas-Solid Fluidization," *Powder Technol.* 9:91 (1974).
29. Van de Hulst, V. C. *Light Scattering by Small Particles* (London: Schapman and Hall, 1957), p. 297.
30. Megaw, W. J., and A. C. Wells. "High Resolution Charge and Mobility Spectrometer for Radioactive Submicrometer Aerosols," *J. Sci. Instrum.* 2:1013 (1969).
31. Liu, B. Y. H., and D. Y. H. Pui. "Electrical Neutralization of Aerosols," *J. Aerosol Sci.* 5:465 (1974).
32. Stöber, W., H. Flachsbart and D. Hochrainer. "The Aerodynamic Diameter of Latex Aggregates and Asbestos Fibers," *Staub-Reinhalt.-Luft* 30:227 (1970).

33. Fuchs, N. A. *Mechanics of Aerosols* (Moscow: Academy of Science, 1955), p. 325.
34. Soo, S. L. "Formation of Bubbles in a Fluidized Bed," *Powder Technol.* 10:211 (1974).
35. Davies, C. N. *Air Filtration* (London: Pergamon Press, 1973), p. 35.
36. Derryagin, B. V. and A. D. Zimon. "Adhesion of Powder Particles to Plane Surfaces," *Kolloid Zh. (USSR)* 23:544 (1961).
37. Zimon, A. D., and G. A. Serebryakov. "Adhesion of Cylindrical Particles to Rough Surfaces," *Kolloid Zh. (USSR)* 35:26 (1973).
38. Gucker, F. T., and G. S. Doyle. "The Amplitude of Vibration of Aerosol Droplets in Sonic Field," *J. Phys. Chem.* 60:989 (1965).
39. Harwood, C. F. "Powder Segregation due to Vibration," 2nd Int. Conference on the Compaction and Consolidation of Particulate Matter, Brighton, England, September 2-4 (1975).
40. Larsen, R. I. "The Adhesion and Removal of Particles Attached to Air Filter Surfaces," *Amer. Ind. Hyg. Assoc. J.* 19:265 (1958).
41. Nguyen, X., and S. M. Beckmans. "Sonically Enhanced Depth Filtration of Aerosols," *J. Aerosol Sci.* 5:133 (1974).
42. Spurny, K. R., C. Boose and D. Hochrainer. "On the Pulverization of Asbestos Fibers in a Fluidized-Bed Aerosol Generator," *Staub-Reinhalt.-Luft* 35:440 (1975).
43. Spurny, K. R., C. Boose, D. Hochrainer F.-J. Mönig. "A Note on the Dispersing of Fibrous Powders," *Ann. Occup. Hyg.* 19:85 (1976).
44. Spurny, K. R., C. Boose, D. Hochrainer and F.-J. Mönig. "Ein Schwingbett-Aerosolgenerator für faserige und pulverförmige Stäube," *Zbl. Bakt. Hyg. I. Abt. Orig.* B161:362 (1976).
45. Spurny, K. R., G. Weiss, H. Opiela. "Zur Herstellung von Feinstaubproben für biologische Untersuchungen," *Staub-Reinhalt.-Luft* 38:417 (1978).
46. Gans, R. "Wie fallen Stäbe und Scheiben in einer reibenden Flüssigkeit?" *Sitzungsberichte math. phys. Klasse Akad. Wiss.* 41:191 (1911).
47. Ostwald, W. "Zur allgemeinen Theorie der raschen Koagulation," *Kolloidchemische Beihefte* 27:223 (1928).
48. Harris, R. L. "A Model for Deposition of Microscopic Fibers in the Human Respiratory System," Ph.D. Thesis, School of Public Health, University of North Carolina, Chapel Hill (1972).
49. Bogowski, P., V. Timbrell, J. C. Gilson and J. C. Wagman, Eds. *Biological Effects of Asbestos* (Lyon: IARC, 1973).
50. Peddieson, J. "Impulsive Axial Motion of an Infinite Cylinder in a Suspension," *Lett. Appl. Eng. Sci.* 2:11 (1974).
51. Bragg, G. M., L. van Zuiden and C. E. Hermance. "The Free Fall of Cylinders at Intermediate Reynold's Numbers," *Atm. Environ.* 8:755 (1974).
52. Horvath, H. "Das Fallverhalten nicht-kugelförmiger Teilchen," *Staub-Reinhalt.-Luft* 34:251 (1974).

53. Ogden, T. L., and W. H. Walton. "The Descent Angle of Inclined Cylinders: a Possible Means of Separating Filters by Length," *Ann. Occup. Hyg.* 18:157 (1975).
54. Timbrell, V. "Alignment of Respirable Asbestos Fibers by Magnetic Fields," *Ann. Occup. Hyg.* 18:299 (1975).
55. Gallily, I., and A. Cohen. "On the Stochastic Nature of the Motion of Nonspherical Aerosol Particles," *J. Colloid Interface Sci.* 56:443 (1976).
56. Harris, R. L., and A. A. Fraser. "A Model for Deposition of Fibers in the Human Respiratory System," *Am. Ind. Hyg. Assoc. J.* 37:73 (176).
57. Walkenhorst, W. "Modellversuche zur Bestimmung des dynamischen Formfaktors nicht isometrischer Teilchen," *Staub-Reinhalt.-Luft* 36:149 (1976).
58. Walkenhorst, W., and W. Coenen. "Zusammenhänge zwischen den äusseren Abmessungen von Aggregaten und ihrem aerodynamischen Durchmesser," *Staub-Reinhalt.-Luft* 37:106 (1977).
59. Walkenhorst, W. "Aerodynamisches Verhalten von Fasern und Probenahme von Faserstaub," *Staub-Reinhalt.-Luft* 37:323 (1977).
60. Wouda, T. W. M., K. Rietema and S. P. P. Ottengraf. "Continuous Sedimentation Theory," *Chem. Eng. Sci.* 32:351 (1977).
61. Harris, R. L., and V. Timbrell. "The Influence of Fiber Shape in Lung Deposition: Mathematical Estimates," in *Inhaled Particles IV*, W. H. Walton, Ed. (Oxford: Pergamon Press, 1977), p. 75.
62. Zebel, G., D. Hochrainer and C. Boose. "A Sampling Method with Separated Deposition of Airborne Fibers and Other Particles," *J. Aerosol Sci.* 8:205 (1977).
63. Laframboise, J. G., and Chang, J.-S. "Theory of Charge Deposition on Charged Aerosol Particles of Arbitrary Shape," *J. Aerosol Sci.* 8:331 (1977).
64. Yalamov, Yu. I., and A. M. Afanasev. "Thermophoresis of a Cylindrical Aerosol Particle," *J. Tech. Phys. (USSR)* 47:1998 (1977).
65. Yalamov, Yu. I., and A. M. Afanasev. "Diffusiophoretic Force on a Cylindrical Aerosol Particle," *J. Tech. Phys. (USSR)* 47:2001 (1977).
66. Timbrell, V., and R. E. G. Rendall. "Preparation of the U.I.C.C. Standard Reference Samples of Asbestos," *Powder Technol.* 5:279 (1972).
67. Rendall, R. E. G. "Data Sheets of Physical and Chemical Properties of U.I.C.C. Standard Reference Asbestos Samples," Report of the National Research Institute of Occupational Diseases, Johannesburg (1972).
68. Assucao, J., and M. Corn. "The Effects of Milling on Diameters and Lengths of Fibrous Glass and Chrysotile Asbestos Fibers," *Amer. Ind. Hyg. Assoc. J.* 36:811 (1975).
69. Langer, A. M., M. S. Wolff, A. N. Rohl and I. S. Selikoff. "Variation of Properties of Chrysotile Asbestos Subjected to Milling," *J. Toxicol. Environ. Health* 4:173 (1978).

70. Gentry, J. W., and K. R. Spurny. "Comparisons of Measurement Techniques for Penetration of Amosite Fibers through Nuclepore Filters," *J. Colloid Interface Sci.* 65:174-180 (1978).
71. Parfitt, G. D., and E. T. Wharton. "The Dispersion of Rutile Powder in Aqueous Surfactant Solutions," *J. Colloid Interface Sci.* 38:341 (1972).
72. Johan, Z., J. Goni, C. Sarcia, G. Bonnaud and J. Brignon. "Influence de certains acides organiques sur la stabilité du réseau du chrysotile," *Adv. Organic Geochem.* 41:883 (1973).
73. Chowdhury, S., and J. A. Kitschener. "The Zeta-Potentials of Natural and Synthetic Chrysotiles," *Int. J. Min. Processing* 2:277 (1975).
74. Parfitt, G. D. "The Stability of Colloidal Dispersions in Non-Aqueous Media," *J. Colloid Interface Sci.* 54:4 (1974).
75. Light, W. G., and E. T. Wei. "Surface Charge and Asbestos Toxicity," *Nature* 265:537 (1977).
76. Parfitt, G. D. "The Dispersion of Powders in Liquids," *Powder Technol.* 17:157 (1977).
77. Aichele, W., E. Echmollmeyer and H. Herlinger. "Elektrokinetische Untersuchungen an Fasern," **Makromol. Chem.** 178:2011 (1977).
78. Lin, I. S. "Wet Dielectric Separation," *Powder Technol.* 17:95 (1977).
79. Spurny, K. R. "Nuclepore Siebfilter-Membranen: Zehn Jahre Anwendung für Staub- und Aerosolmessungen," *Staub-Reinhalt.-Luft* 37:328 (1977).
80. Hochrainer, D., G. Zebel and V. Prodi. "Ein Gerät zur Trennung von Fasern und isometrischen Partikeln bei der Probenahme," *Staub-Reinhalt.-Luft* 38:425 (1978).
81. Liu, B. Y. H., and D. Y. H. Pui. "On the Performance of the Electrical Analyzer," *J. Aerosol Sci.* 6:249 (1975).
82. Holländer, W. Private Communication.

CHAPTER 14

SULFUR COMPOUNDS IN AMBIENT ENVIRONMENTS AND THEIR SIMULATION IN THE LABORATORY

D. P. Y. Chang
Department of Civil Engineering
University of California
Davis, California 95616

INTRODUCTION

The process of utilizing available information on sulfur-containing compounds in the atmosphere to develop methods for preparing laboratory-generated aerosols for exposure studies presents a number of philosophical as well as practical problems. These problems stem chiefly from the incomplete chemical characterization of ambient sulfur aerosols and the labile and reactive nature of sulfur compounds. The characterization of ambient aerosols is inherently incomplete because of the varied origin and history of each sample obtained. Further complicating matters are the possibility of artifact formation during the collection and analysis of complex aerosol mixtures and the difficulty of identifying the chemical species *originally* present in the air. Sulfur compounds, in particular, may undergo rapid changes (relative to the time scale of generation and delivery of the compound to a subject) involving physical state, acid-base reaction, oxidation-reduction reaction and complexation. Therefore, additional care and effort are required to deliver the desired physical form and chemical species to the test subject. In fact, it appears that on occasions in the past, research groups, including our own, have incorrectly or incompletely identified the chemical species delivered to the test subject. The purpose of this article is to summarize and cite selected references describing the characteristics of ambient sulfur compounds, present methods which have been used to generate sulfur-containing aerosols, discuss characteristics of the test aerosol which should be determined and relate examples of pitfalls which should be avoided along the way.

PHYSICAL AND CHEMICAL CHARACTERISTICS OF AMBIENT SULFUR COMPOUNDS

Much of the knowledge concerning ambient sulfur compounds is summarized in "Sulfur in the Atmosphere" proceedings of the International Symposium held in Dubrovnik, Yugoslavia, in 1977 [1]. The following synopsis draws heavily from that reference, especially the works by R. J. Charlson et al. [2] and K. T. Whitby [3]. The interested reader is directed to those works.

Naturally occurring oxidation states of sulfur which have been identified in the atmosphere and which occur in significant quantites are S(-II), S(O), S(IV) and S(VI)(2). In addition it has been hypothesized that S(V) in the form of dithionate ion, $S_2O_6{}^{2-}$, may be present in ambient air samples, but may have been mistakenly identified as $SO_4{}^{2-}$ or missed entirely in the past [2,4]. The more reduced states of sulfur, S(-II) and S(O), are thought to be of less potential health consequence because of their relatively low ambient concentration and rapid oxidation in the atmosphere. Health effects research efforts, therefore, have focused primarily on the S(IV) and S(VI) oxidation states, particularly $SO_{2\,(g)}$, $SO_{2\,(g)}+$ particles at various relative humidities, metallic sulfites, metallic sulfates, sulfuric acid and its neutralization products with ammonia.

Sulfur (VI)

Particulate sulfate is believed to be formed by a number of different mechanisms which affect size and chemical distribution. Sulfate in the coarse particle mode ($> 2\ \mu m$) is typically associated with sea salt or suspended products of weathering, hence with the cations Na^+, Ca^{2+} and Mg^{2+}. The bulk of the sulfur-containing aerosol is found in the fine particles ($< 2\ \mu m$). Homogeneous oxidation of SO_2 tends to produce sulfuric acid in an ultrafine or nuclei mode (~ 0.01-$0.1\ \mu m$) as does homogeneous condensation of SO_3 from power plant exhaust or catalytic converter-equipped automobiles [3]. Smog chamber studies have shown that aerosols other than sulfuric acid, such as nitrosyl sulfuric acid, $NOHSO_4$, are also formed [5]. Direct evidence for the existence of these species in ambient air are currently lacking. Whitby [3] reported the geometric mean size by volume (DGV) or volume median diameter (VMD) of nuclei-mode sulfate to range from 0.015 μm to 0.04 μm with a geometric standard deviation (σ_g) of 1.6. The total volume concentration of nuclei-mode particles has been observed to be as large as 9 $\mu m^3/cm^3$, though only a fraction of the mass was identified as sulfate (~ 2 $\mu g/m^3$). Coagulation of these nuclei-mode particles with, or direct condensation upon particles in the accumulation mode (~ 0.1 to 2 μm) leads to sulfate in combination with cations other than H^+. In internal mixtures of this type it is not always feasible to assign a coordination partner to the sulfate ion. There is evidence that sulfate will tend to coordinate with the alkaline earth cations present, e.g., Ca^{2+} in fly ash [6], as would be anticipated from solubility considerations. Whitby [3] reported the geometric mean size by volume for the accumulation mode ranges from 0.15 to 0.5 μm with a geometric standard deviation ranging from 1.6 to 2.2. Sulfate exhibited an average aerodynamic geometric mean diameter (DGMA) or mass median aerodynamic diameter (MMAD) of 0.48 ± 0.1 μm and a geometric standard deviation of 2.0 ± 0.29. The mass concentration was observed to range from 0.04 $\mu g/m^3$ up to 80 $\mu g/m^3$ under polluted conditions. Strongly acidic sulfate particles can further react to form organosulfates, but there is a lack of data concerning the presence or identity of these compounds in ambient air samples. If such compounds form, it is anticipated that they would be found among the fine particles.

Although the mass concentration of nuclei-mode aerosol was found to be about a factor of ten lower than the accumulation mode and can be shown to have a short half-life in polluted air on theoretical grounds, it nevertheless exists. Furthermore, it can constantly be replenished by direct emission sources and homogeneous oxidation reactions, thereby providing a possible means for chronic human exposure. It has been suggested that these nuclei-mode sulfate aerosols represent a greater inhalation hazard because of their greater acidity and deposition efficiency in the lower respiratory tract [7]. This has not yet been demonstrated, the chief difficulties being the rapid rate of coagulation as the mass concentration approaches levels above about 100 $\mu g/m^3$, and the possibility of neutralization by volatile bases below that level.

Acidic sulfate-containing particles can readily absorb ammonia. Molar ratios of NH_4^+ to SO_4^{2-} ranging from 1 to 2 are frequently observed. The problem of neutralization reactions in exposure chambers was recognized in the early work of Pattle [8,9], but has been neglected until recently. It is interesting to note that the lower level at which effects of sulfuric acid begin to vanish (~ 1 $\mu mol/m^3$ to 10 $\mu mol/m^3$) more or less coincides with the level of ammonia determined in exposure chambers (1 $\mu mol/m^3$ up to 50 $\mu mol/m^3$) [10-12]. This may be purely coincidental; nevertheless, care must be exercised to prevent neutralization reactions. Stevens [13] has observed that the degree of neutralization of ambient air samples by ammonia may be reduced by placing an ammonia "denuder" in front of the filter sample to prevent neutralization on the collection medium. One can infer then that the atmosphere should not be considered well-mixed, but comprised of regions of incompletely neutralized acidic aerosol and regions of neutralizing gases such as ammonia. This appears to be the case in various designs of exposure chambers as well [10-13]. Precautions such as an ammonia denuder are necessary if the equivalents and degree of neutralization of the acid inhaled by the test subjects are to be determined. Furthermore, care must be exercised in the determination of the acid concentration. Some method such as the Gran titration for strong acid [14] or a separate volatile base determination such as a gas-purge technique [12] should be employed. Colorimetric methods for acid concentration must be used cautiously as the pH shift observed in partially neutralized acid solutions may be small (e.g., equimolar solutions of H_2SO_4 and NH_4HSO_4 in the range 10^{-3} to 10^{-5} *M* exhibit only a 0.3 pH unit difference).

The hygroscopic nature of most sulfate particles affects their size and physical state in humid atmospheres. Deposition site and efficiency may be altered by the initial particle size, solution composition and humidity. Usually this does not pose a problem which cannot be taken into account during the generation of sulfates. However, the physical state of the particles may not be that expected at equilibrium for a given humidity because of hysteresis exhibited in the transition from solution drops to solid particles. The situation can be further complicated by hydrolysis reactions of the associated cation, e.g., Fe(III), which result in particles which are difficult to dry out. It is in fact questionable whether anyone has produced a pure ferric sulfate $Fe_2(SO_4)_3 \cdot x\,H_2O$ aerosol by nebulization! The rapid evaporation which normally occurs in nebulizer systems can also lead to microcrystalline or amorphous particles with different physicochemical properties, e.g., solubility or hydration state, than those which have aged in the ambient air. Careful micrographic studies are suggested as a check for gross differences.

Sulfur (IV)

Sulfur dioxide is the predominant form of S(IV) in ambient air. It can exist in the gas phase, is readily absorbed in solution, and can be adsorbed on solids. In solution an equilibrium exists among $SO_{2(aq)}$, HSO_3^-, SO_3^{2-}, $HS_2O_5^-$ and $S_2O_5^{2-}$ and $SO_{2(g)}$ over the solution phase. Figure 1 illustrates the distribution of species as a function of solution pH as might occur in a large aqueous droplet in equilibrium with 1.5 ppm of $SO_{2(g)}$ (range of SO_2 observed during the London episode). It should be noted that the species H_2SO_3 does not exist to any appreciable extent, but that which is commonly referred to as "H_2SO_3" is a complex mixture of dissolved $SO_{2(aq)}$, $HS_2O_5^-$, and HSO_3^- at low-to-moderate pH [15,16]. If one imagines that the droplet containing dissolved $SO_{2\,(g)}$ is evaporating, the total dissolved sulfur species $S(IV)_T$ would be shifted upward. The partial pressure of $SO_{2\,(g)}$ would necessarily need to be increased if equilibrium were to be maintained. If $P_{SO_{2\,(g)}}$ did not increase, the $S(IV)_T$ would diminish. This would result from the loss of $SO_{2\,(g)}$ from the droplet. The solution pH would tend to rise if nonvolatile cations, e.g., Na^+, are present in the droplet resulting in a lesser loss of SO_2. Experience has demonstrated that aerosolized solution droplets will move toward $SO_{2\,(g)}$ equilibrium rapidly relative to the residence time in a typical exposure chamber (order of a few minutes). Therefore it is practically impossible to conduct exposures with acidic droplets having as their principal components dissolved $SO_{2\,(aq)}$, $HS_2O_5^-$ or HSO_3^- ions without the presence of considerable concentrations of SO_2 in the gas phase. Furthermore, at high pH values SO_3^{2-} undergoes rapid air oxidation in solution drops, resulting in the production of various amounts of SO_4^{2-}[17]. The utility of nebulizers to produce S(IV)-containing aerosol is limited without extraordinary precautions for the above reasons.

An exception to these statements may exist in the case of metal complexes of SO_3^{2-} whose stability constant may be sufficiently large to prevent air oxidation and $SO_{2\,(g)}$ evolution as the solution droplet evaporates[2].

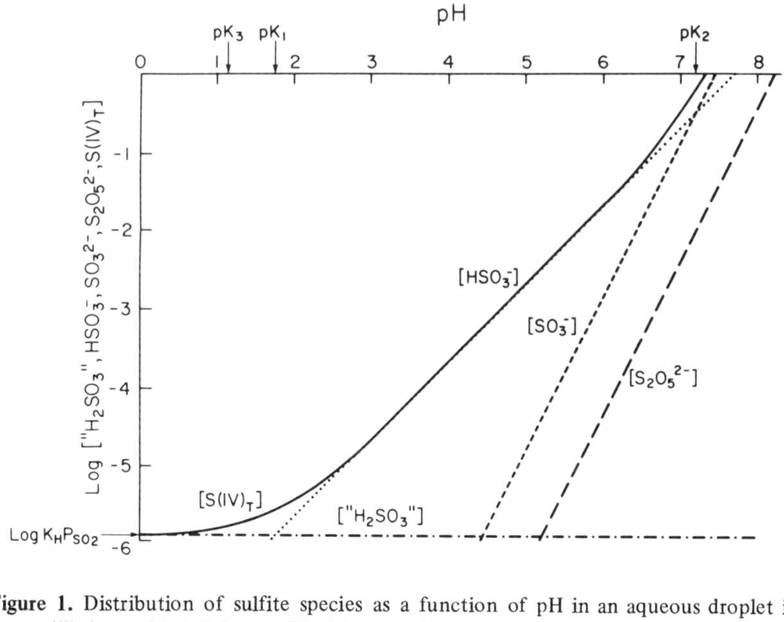

Figure 1. Distribution of sulfite species as a function of pH in an aqueous droplet in equilibrium with ∿1.5 ppm SO_2(g). Henry's law constant, K_H and acidity constants K_1, K_2, K_3 adapted from Reference 2. ["H_2SO_3"], —. —. —; [HSO_3^-], ...; [SO_3^{2-}], ---------; [$S_2O_5^{2-}$], — — — —; [$S(IV)_T$], ——————.

An example would be the Hg^{2+} -$SO_3{}^{2-}$ complex formed in the absorption step of the West-Gaeke analysis. Additional metal ions claimed to possess this ability are Fe^{3+} and Cu^{2+} [18, 19]. There is currently some question as to the stability of the latter complexes to internal oxidation-reduction reactions. Much of the previous literature suggests that Fe^{3+} and Cu^{2+} will promote oxidation of $SO_3{}^{2-}$ under the proper conditions, though a strong sulfite complex of Cu^+ is formed as well[4]. It has been hypothesized that the species reported to be present in environmental samples[19] is in reality dithionate, $S_2O_6{}^{2-}$, whose formal oxidation state is S(V)[4]! Further research is required before the issue can be resolved.

Sulfur dioxide can be adsorbed onto particles. The nature of the particles and environmental conditions such as humidity and temperature affect the rate of transformation to higher oxidation states, primarily sulfate. Carbonaceous particles (soot and graphitic carbon) and fly ash have received the most attention, though artificially produced metal-containing aerosols have also been utilized to study catalysis of SO_2 to $SO_4{}^{2-}$. The reader is referred to references 1 and 20 as starting points to the vast literature on the subject.

Unlike sulfate, the mass of "sulfite" per unit mass of particle was reported to be distributed more uniformly as a function of particle size in samples obtained from copper smelter plumes[19]. The authors also believed that inorganic sulfite and possibly an organic adduct of SO_2 was present in the majority of ambient air samples obtained from sites located in New York City. Furthermore, that the majority of these species were contained in the respirable size range. The exact identity of these compounds is not known. Until such time as they have been identified, exposures to simple compounds or compounds formed under simulated conditions would appear to be appropriate.

The difficulties encountered in the analytical determination of S(IV) are indicative of the need for careful chemical characterization of the species delivered to subjects in exposure studies. Acid-base, complexation, redox reactions and phase transformations proceed rapidly in the sulfur system. They are subject to changes which, from a practical perspective, are difficult to control to a sufficient degree of precision, e.g., volatile base production, temperature and humidity, to assure reproducibility of the aerosol's chemical speciation. It is therefore imperative that careful characterization studies of the aerosol *delivered to the test subject* be conducted if exposures are to be reproduced. For sulfur compounds these should include:

1. oxidation state of the sulfur atom and other cations such as Fe(II) or Fe(III);
2. acidity of the aerosol produced;
3. physical state, i.e., liquid or solid;
4. presence of $SO_{2(g)}$ in exposure chamber atmosphere;
5. size distribution of particles.

GENERATION OF SULFUR-CONTAINING AEROSOLS

The following discussion relates principally to factors which should be considered in selecting an aerosol generation system. No attempt is made to exhaustively review generators employed in biological studies. References to some aerosol generation methods employed in exposure studies involving sulfur-compounds are presented in Table I. An excellent review of aerosol generation methods is already available [47]. A brief description of a relatively new generation system for sulfuric acid particles is presented. Aerosol generators for reduced sulfur are currently under development and have not been described in the literature. It is questionable whether those which have been previously employed [42-44] actually produced the particles for which they were designed! An effort to briefly describe approaches to S(IV) generation is given, however.

S(VI)

The high solubility of sulfate salts combined with the relatively stable oxidation state of S(VI) have made generation of sulfate aerosols by mechanical dispersion from solution attractive. The major factors in selecting from among the array of liquid aerosol generators available are the desired size distribution, mass concentration and duration of exposure. Monodisperse aerosol generators such as the vibrating-orifice or spinning-disk generators tend to have low mass output rates so that high mass concentrations can only be achieved in mask or small chamber exposures. Polydisperse aerosol generators such as the Collison, Dautrebande, Devilbiss, Ultrasonic and Babbington can achieve higher mass concentrations, but have fixed size distribution characteristics associated with them. Fortunately, ambient sulfate compounds in the accumulation mode exhibit geometric standard deviations about the same range as these generators, $\sigma_g \sim 1.6\text{-}2.0$, and can therefore be suitably simulated. For extended exposures, suitable modification of the liquid feed system, such as that described by Liu [48], must be incorporated to reduce solution concentration effects. Aerosols generated from mechanical disintegration must also be checked for electrical charge equilibration.

Two problems are encountered with the use of mechanical dispersion generators in the production of nuclei-mode sulfuric acid aerosol. First, the mass concentration from a single generator is limited and a large quantity of solvent (H_2O) must be absorbed to accomplish the necessary size reduction from the original droplet. Second, any solvent impurity will be concentrated as well. Such impurities are difficult to avoid, as even glass has limited solubility in water.

Table I. References to Some Aerosol Generation Methods for Sulfur Compounds

Compound Generated	Method	Reference	Comments
H_2SO_4	SO_3 vapor from fuming sulfuric acid reacting with H_2O.	21,22,23	Long-term stability exhibits some variation because of changing vapor pressure of SO_3. Early investigations produced accumulation-mode particles, nuclei mode possible.
H_2SO_4	Rappaport-Weinstock generator	24	Accumulation mode particles were produced. Nuclei mode possible.
	La Mer generator (heated concentrated H_2SO_4 vapor).	25	
H_2SO_4	Stabilized liquid SO_3 as source of SO_3 vapor.	26,27	Acceptable long-term stability as SO_3 vapor pressure constant. Both nuclei and accumulation mode aerosol were produced.
H_2SO_4	Nebulized dilute solution of H_2SO_4 followed be heating and controlled condensation and coagulation.	28	Nuclei mode can be produced. Adjustment of size independent of residence time of chamber.
H_2SO_4	Same as above, but uncontrolled coagulation.	29	High mass concentration limited size to accumulation mode.
H_2SO_4	Nebulization of dilute solutions of H_2SO_4 and H_2O evaporation.	21,30	Custom glass nebulizer; accumulation mode.
		31	Babbington glass nebulizer; accumulation mode.
		32	Custom glass nebulizer; accumulation mode.
		33	10 N H_2SO_4 with unspecified nebulizer; accumulation mode.
		34	Devilbiss #40

Soluble sulfates	Nebulization	35	Dautrebande D_{30}.
		36	Spherical maxi cool nebulizers–Collison type.
		37	Environmental Research Corporation–Collison type.
		34	Monagham #670 ultrasonic.
		38	Heat treated $(NH_4)_2SO_4$.
Soluble sulfates	Vibrating-orifice	39	Monodisperse vibrating orifice of Berglund-Liu design. Low mass output.
Ammonium bisulfate	Heating of solid $(NH_4)_2SO_4$ drives off NH_3 and produced $NH_4HSO_{4(s)}$ by condensation.	38	Well-characterized with nuclei mode capability.
Sulfate	Dark-phase chemical rxn product of O_3/trans-2-butene/SO_2.	40	Probably produces nuclei-mode H_2SO_4 and possibly organo sulfonates.
H_2SO_4 + carbon particles	Nebulization of dilute H_2SO_4 solution mixed with carbon aerosol, heated and recondensed.	41	Carbon particles served as carriers of H_2SO_4.
Bisulfite, sulfite	Nebulization of H_2O/polyethylene mixture.	42,43	Dautrebande D_{18}. Probably did not produce much bisulfite. SO_2 would have evolved.
Ferric sulfite complex	Nebulization	44	Possible reduction of Fe^{3+} to Fe^{2+} and oxidation of $SO_3{}^{2-}$ to $SO_4{}^{2-}$ makes product composition uncertain.
SO_2 + fly ash	Mix SO_2 + fly ash at high relative humidity.	45	Developmental stage.
SO_2 + iron oxide	Passed SO_2 over catalyst and mixed with nebulized iron oxide.	46	Characterization stage.

Condensation of vapor from the gas phase is best suited to the production of nuclei-mode aerosols. There are, however, difficulties associated with precise temperature and humidity control in the rate of nucleation of sulfuric acid aerosols [25, 49]. Stable reproducible generation of sulfuric acid aerosols relies upon the rapid nucleation of sulfuric acid upon itself to form such a high initial number concentration that coagulation kinetics dominates the size distribution of the resulting particles [26-28]. One condensation generator system employing liquid sulfur trioxide, $SO_{3(l)}$, as the source of sulfuric acid will be discussed briefly [26, 27]. A second system [28] will be discussed at length in a subsequent chapter.

The liquid $SO_{3(l)}$ system is conceptually simple. Temperature control of the $SO_{3(l)}$ provides a constant vapor pressure of $SO_{3(g)}$, unlike fuming sulfuric acid aerosol generators which have been used previously. Sulfur trioxide vapor is conveyed by a metered stream of dry nitrogen gas to the exposure chamber where it is mixed with the chamber air supply (Figure 2). Water vapor in the chamber air supply causes rapid nucleation and subsequent coagulation of H_2SO_4 aerosol. Relative humidity and particle size determine the composition of the droplet [25]. Turbulence in the duct provides rapid mixing and the swirling action, imparted by the tangential entry, even distribution within the exposure chamber. The initial aerosol number concentration is sufficiently large such that coagulation kinetics dominates the size distribution and number concentration, leading to an approximately log-normally distributed aerosol with geometric standard deviations of about 1.6-2.1. In such a system the mass concentration, residence time in the chamber, temperature and relative humidity basically fix the size distribution produced.

The advantages of the liquid SO_3 system are numerous:

1. It has essentially unlimited mass output by simply adjusting the nitrogen carrier gas flowrate (concentrations as high as 700 mg/m^3 in a volumetric flowrate of about 2 m^3/min have been attained).
2. The mass of SO_3 evaporated can be determined by weight loss, and a direct measure of average mass concentration can be made when chamber air flowrate is known.
3. The SO_3 in N_2 gas stream is easily conveyed and can be metered to several chambers to provide essentially identical conditions in each chamber.
4. The mass concentration has been held within close limits ($\lessdot$30%) for periods up to one-half year, operating 23 hours a day, simply by controlling $SO_{3(l)}$ temperature and nitrogen flowrate.
5. Nuclei-mode aerosols can be simulated when mass concentrations are lowered to about 100 μg/m^3, count median diameter $\sim$0.04-0.05 μm, $\sigma_g \sim$1.5-1.6.

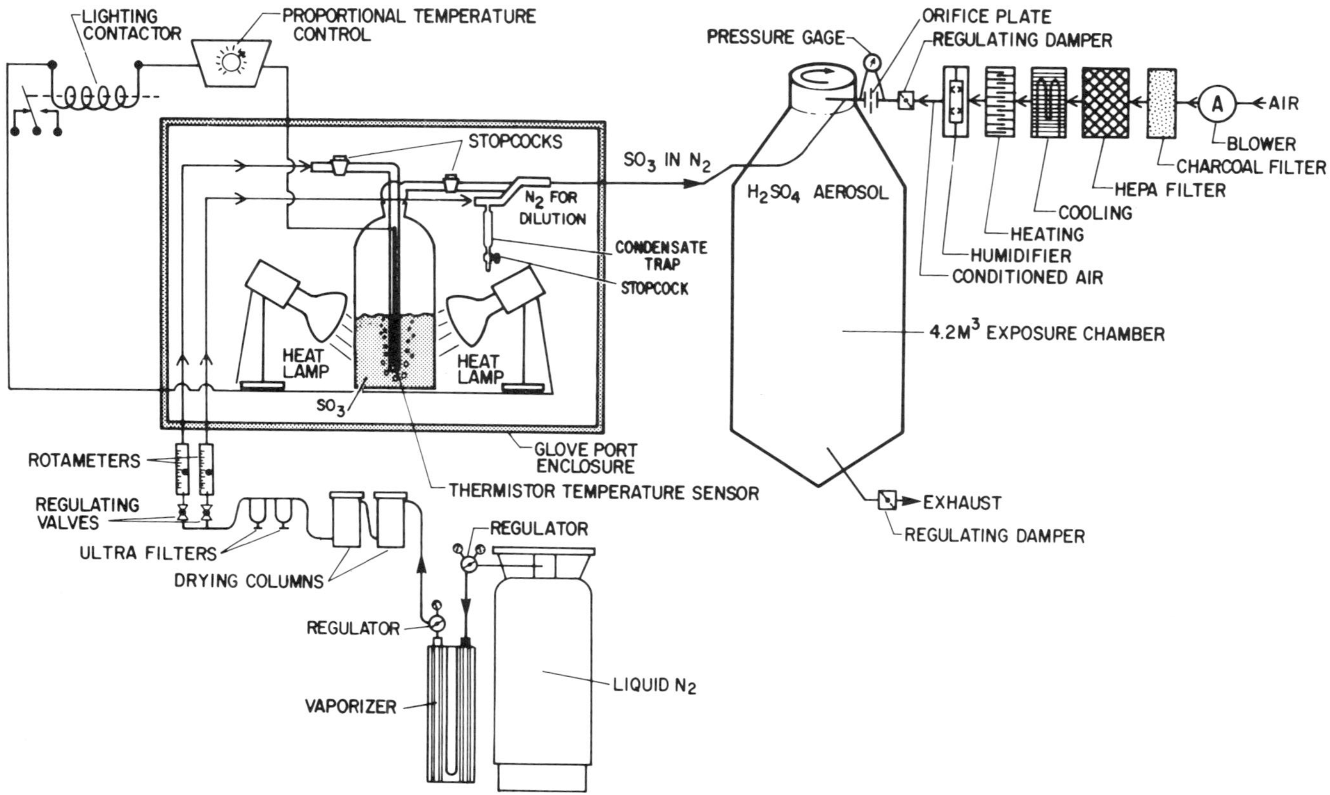

Figure 2. Schematic diagram of stabilized liquid sulfur trioxide aerosol generating system and exposure chamber. Not shown are the distribution manifold for multiple-chamber exposures and heating tape on joint where $SO_{3(g)}$ meets dilution N_2.

6. The SO_3 vapor stream can be combined with particle streams to achieve condensation of H_2SO_4 on carrier particles.
7. The system is inherently simple in design and can be constructed from readily available materials.

Disadvantages of the liquid SO_3 system are:

1. Liquid SO_3 is a potential fire and safety hazard; precautions such as a suitable containment vessel for the generation system are required.
2. Liquid SO_3 is a difficult material to work with initially because it tends to crystallize at room temperature, especially when exposed to even traces of moisture, hence transfer operations require special techniques.
3. The Teflon delivery lines need to be conditioned by operating the system for about one day during low concentration exposures.

The design of the above generator has not been optimized. Certain improvements which could be made include:

1. Replacement of infrared lamps for temperature control by temperature-regulated forced air convection.
2. Sheathing Teflon lines with water-impermeable material or using all glass lines to minimize H_2SO_4 condensation in delivery lines. This is particularly a problem during low concentration exposures; the SO_3 is absorbed by the condensed H_2SO_4 causing fluctuations in the generator output as the line's temperature changes. Alternately, the lines could be heated to ensure vaporization of H_2SO_4 and prevent its buildup; this would also shorten the conditioning time.
3. A precoagulation chamber could be installed in low concentration studies to maintain an initial nuclei formation rate sufficiently high to ensure coagulation dominated size distribution or to grow the aerosol into a desired size range.

PROPOSED METHODS FOR GENERATION OF SULFUR IN LOWER OXIDATION STATES

The earlier discussion regarding the chemistry of sulfur in lower oxidation states indicated that simple nebulization of S(IV) solutions might not produce the desired chemical speciation. Many of the difficulties are encountered while S(IV) is in the solution phase. If one is careful to exclude oxidizing agents by working in an inert atmosphere prior to evaporation to the solid state, it is possible to produce metal sulfite particles. The metal must also be one which will not undergo a rapid internal redox reaction, e.g., Na^+ or Fe^{2+}. It is convenient then to have a drying column (either heat or diffusion drying) following the generator to ensure dryness before mixing with chamber air. Chamber relative humidity must be maintained below the deliquescence point of the sulfite salt to assure delivery of sulfite to the respiratory tract. Attempts to nebulize bisulfite solutions to produce bisulfite salts lead to sulfite particles, as $SO_{2\,(g)}$ is evolved as the particle evaporates.

A potential method for circumventing the difficulties of oxidation and $SO_{2\,(g)}$ evolution is simply to avoid the solution phase altogether by aerosolizing dry sulfite compounds. Pure solids, including bisulfites or dithionates, could be brought to the desired size range by a milling operation. A dust feed system could then be employed to disperse the particles. Provision of an inert atmosphere for either of the above operations would prevent contact with moisture and oxidation. Material compatibility would be a major concern because of the potential for corrosion.

Another alternative to this approach would be to simulate the formation processes which lead to sulfur of lower oxidation states in the atmosphere. An adsorption/absorption system onto carrier particles such as fly ash, transition metal compounds or graphitic carbon might serve as appropriate models. The major difficulty would be chemical characterization of the exact compound formed and the quantity of the material delivered. At least two systems of this variety are being developed in various laboratories across the country [46, 47]. There are certain limitations to the above approaches with regard to the rate of chemical reaction. Atmospheric transformation of $SO_{2\,(g)}$ to a metastable state in power plant plumes appear to have rates on the order of a few percent per hour [1], though initial rates of reaction may be higher near the stack exit. In an exposure facility an appropriate means for increasing the rate of transformation, without altering the products of reaction, would be needed.

CONCLUSION

Simulation of atmospheric sulfur compounds in exposure systems, even those involving single pure compounds, is not as straightforward as one might wish. Appropriate methods are available for simulation of size distribution; however, crystal structure and solubility may not be adequately reflected. Increased information concerning speciation of ambient sulfur compounds is badly needed to guide biological research. Sulfur compounds of lower oxidation state than S(VI) add a new dimension of complexity to generation and characterization of test aerosols by introducing the possibility of oxidation-reduction reactions and gas-solution solid-phase transformations. Experience indicates that these are relatively rapid when the solution state is present and must be properly taken into account. Aerosolization of solid compounds appears to be attractive for the above reasons.

In the past a somewhat naive approach to exposures has been taken. This does not appear to be acceptable in the present or future.

ACKNOWLEDGMENTS

This review was made possible by Environmental Protection Agency Contract No. 68-02-2496 and by Electric Power Research Institute Contract No. RP-1112-2. The author wishes to express his sincere appreciation to the contributions of his colleagues at the California Primate Research Center and Radiobiology Laboratory: Mr. Brian K. Tarkington, Dr. Otto Raabe and especially Dr. Purnendu Dasgupta, whose knowledge of sulfur chemistry was extremely enlightening. Ms. Mary Baker, Donna Consiglio and Dottie Morrissey were of greatly valued assistance in the preparation of this manuscript.

REFERENCES

1. Husar, R. B., Lodge, J. P., Jr., and Moore, D. J., Eds. "Sulfur in the Atmosphere," *Atmos. Environ.* 12(1-3):1-796 (1978).
2. Charlson, R. J., Covert, D. S., Larson, T. V., and Waggoner, A. P. "Chemical Properties of Tropospheric Sulfur Aerosols," *Atmos. Environ.* 12:39-53 (1978).
3. Whitby, K. T. "The Physical Characteristics of Sulfur Aerosols," *Atmos. Environ.* 12:135-159 (1978).
4. Dasgupta, P. K., Mitchell, P. A., and West, P. W. "Sulfur Compounds in the Atmosphere: Study of Metal Sulfite Systems," *Atmos. Environ.* In Press (1979).
5. Eggleton, A. E. J., and Cox, R. A. "Homogeneous Oxidation of Sulfur Compounds in the Atmosphere," *Atmos. Environ.* 12:227-230 (1978).
6. Fisher, G. L., Chang, D. P. Y., and Brummer, M. "Fly Ash Collected from Electrostatic Precipitators: Microcrystalline Structures and the Mystery of the Spheres," *Science* 192:553-555 (1976).
7. Wilson, W. E., et al. "General Motors Sulfate Dispersion Experiment: Summary of EPA Measurements," *J. Air Poll. Control Assoc.* 27: 46-51 (1977).
8. Pattle, R. E., Burgess, F., and Cullumbine, H. "The Effects of a Cold Environment and of Ammonia on the Toxicity of Sulphuric Acid Mist to Guinea Pigs," *J. Path Bact.* 72:219-232 (1956).
9. Sim, V. M., and Pattle, R. E. "Effect of Possible Smog Irritants on Human Subjects," *J. Am. Med. Assoc.* 165-1908 (1957).
10. Stevens, R. K., Dzubay, T. G., Russworm, G., and Rickel, D. "Sampling and Analysis of Atmospheric Sulfates and Related Species," *Atmos. Environ.* 12:55-68 (1978).
11. Bailey, R. M., Bell, K. A., Karlsen, U. D., and Kleinman, M. T. "Strong Acid Measurements of Sulfate Exposure Aerosols," Presented at the American Industrial Hygiene Conference, Paper #305 (Los Angeles, CA: May 7-12, 1978).
12. Chang, D. P. Y., Tarkington, B. K., and Rodgers, P. A. "Neutralization of Sulfuric Acid Aerosol by Volatile Bases in Exposure Chambers," To be presented at the American Industrial Hygiene Conference (Chicago, IL: May 27-June 1, 1979).

13. Barrow, C. S., and Dodd, D. E. "Ammonia Production in Inhalation Chambers and Its Relevance to Chlorine Inhalation Studies," *Toxicol. Appl. Pharmacol.* In press (1979).
14. Askne, C., and Brosset, C. "Determination of Strong Acid in Precipitation, Lake Water, and Airborne Matter," Letters to the Editors. *Atmos. Environ.* 6:695-696 (1972).
15. Schmidt, M. In: *Sulfur in Organic and Inorganic Chemistry, Vol. 2,* A. Senning, Ed. (New York: Marcel Dekker, Inc., 1972), p. 71.
16. Falk, M., and Giguere, P. A. "On the Nature of Sulphurous Acid," *Can. J. Chem.* 36:1121-1125 (1958).
17. Larson, T. V., Horike, N. R., and Harrison, H. "Oxidation of Sulfur Dioxide by Oxygen and Ozone in Aqueous Solution: A Kinetic Study with Significance to Atmospheric Rate Processes," *Atmos. Environ.* 12:1597-1611 (1978).
18. Hansen, J. D., Whiting, L., Eatough, D. J., Jensen, T. E., and Izatt, R. M. "Determination of Sulfur (IV) and Sulfate Aerosols by Thermometric Methods," *Anal. Chem.* 48:634-638 (1976).
19. Eatough, D. J., Major, T., Ryder, J., Hill, M., Mangelson, N. F., Eatough, n. L., Hansen, L. D., Meisenheimer, R. G., and Fischer, J. W. "The Formation and Stability of Sulfite Species in Aerosols," *Atmos. Environ.* 12:263-271 (1978).
20. Judeikus, H. S., Stewart, T. B., and Wren, A. G. "Laboratory Studies of Heterogeneous Reactions of SO_2," *Atmos. Environ.* 12:1633-1641 (1978).
21. Alarie, Y., Busey, W. M., Krumm, A. A., and Ulrich, C. E. "Long-Term Continuous Exposure to Sulfuric Acid Mist in Cynomologous Monkeys and Guinea Pigs," *Arch. Environ. Health* 27:16-24 (1973).
22. Hinners, R. G., Burkart, J. K., and Punte, C. L. "Animal Inhalation Exposure Chambers," *Arch. Environ. Health* 16:194-206 (1968).
23. Loscutoff, S. M., Burton, F. G., and Killand, B. W. "Neutralization of Inhaled Sulfuric Acid Aerosols by Ammonia in the Lung," *Fed. Proc.* 37(3):867 (1978).
24. Amdur, M. O., Dubriel, M., and Creasia, D. A. "Respiratory Response of Guinea Pigs to Low Levels of Sulfuric Acid," *Environ. Res.* 15: 418-423 (1978).
25. La Mer, V. K., Inn, E. C. Y., and Wilson, I. B. "The Methods of Forming, Detecting and Measuring the Size and Concentration of Liquid Aerosols in the Size Range of 0.01 to 0.25 Microns Diameter," *J. Coll. Sci.* 5:471 (1950).
26. Chang, D. P. Y., and Tarkington, B. K. "Experience With a High Output Sulfuric Acid Aerosol Generator," *Am. Ind. Hyg. Assoc. J.* 38: 493-497 (1977).
27. Chang, D. P. Y., Tarkington, B. K., and Duvall, T. R. "Generation of Nuclei-Mode Sulfuric Acid Aerosols for Health Effects Studies," *J. Air Poll. Control Assoc.* 28:1137-1139 (1978).

28. Liu, B. Y. H., and Levi, J. "Generation of Sulfuric Acid Aerosols for Health Effects Studies," U.S. EPA, Report No. EPA-600/2-78-104 (Research Triangle Park, N.C.: Environmental Sciences Research Laboratory, 1978).
29. Juhos, L. T., Evans, M. J., Mussenden-Harvey, R., Furiosi, N. J., Lapple, C. E., and Freeman, G. "Limited Exposure of Rats to H_2SO_4 With and Without O_3," *J. Environ. Sci. Health* C13(1):33-47 (1978).
30. Cavender, F. L., Steinhagen, W. H., Ulrich, C. E., Busey, W. M., Cockrell, B. Y., Haseman, J. K., Hogan, M. D., and Drew, R. T. "Effects in Rats and Guinea Pigs of Short-Term Exposures to Sulfuric Acid Mist, Ozone and Their Combination," *J. Toxicol. Environ. Health* 3:521-533 (1977).
31. Avol, E., Bailey, R., and Bell, K. A. "Sulfate Aerosol Generation and Characterization for Controlled Human Exposures," Presented at the American Industrial Hygiene Conference, Paper #311 (Los Angeles, CA: May 7-12, 1978).
32. Crocker, T. T., Phalen, R. F., and Wilson, A. F. "Absence of Additive Effects on Rat Lung from Ozone in the Presence of Sulfuric Acid Mist," Presented at the American Industrial Hygiene Conference, Paper No. 255 (Los Angeles, CA: May 7-12, 1978).
33. Fairchild, G. A., Stultz, S., and Coffin, D. L. "Sulfuric Acid Effect on the Deposition of Radioactive Aerosol in the Respiratory Tract of Guinea Pigs," *Am. Ind. Hyg. Asso. J.* 36:584-594 (1975).
34. Schlesinger, R. B., Lippmann, M. L., and Albert, R. E. "Effects of Short-Term Exposures to Sulfuric Acid and Ammonium Sulfate Aerosols upon Bronchial Airway Function in the Donkey," *Am. Ind. Hyg. Assoc. J.* 39:275-286 (1978).
35. Amdur, M. O., Bayles, J., Ugro, V., and Underhill, D. W. "Comparative Irritant Potency of Sulfate Salts," *Environ. Res.* 16:1-8 (1978).
36. Crider, W. Personal Communication, U. S. EPA Health Effects Research Laboratory (Research Triangle Park, N. C.: October, 1978).
37. Kenoyer, J. L. "Effect of Sulfate Aerosol and Ozone on the Clearance of Radioactive Tracer Particles in Rats," Presented at the Eleventh Aerosol Technology Meeting (University of California, Davis: September 5-8, 1978).
38. Tu, K. W., and Kanapilly, G. M. "Generation and Characterization of Submicron Ammonium Sulfate and Ammonium Hydrogen Sulfate Aerosols," *Atmos. Environ.* 12:1623-1629 (1978).
39. Tarkington, B. K., Chang, D. P. Y., and Raabe, O. G. "The Effect of Humidity on the Size Distribution of Sulfate and Sulfuric Acid Aerosols," Presented at the American Industrial Hygiene Conference, Paper No. 310 (Los Angeles, CA: May 7-12, 1978).
40. Kirtz, M. Personal Communication, U.S. EPA, Health Effects Research Laboratory (Research Triangle Park, N.C.: November, 1978).
41. Fenters, J. D., Bradof, J., Aranyi, C., Ketels, K., Ehrlich, R., and Gardner, D. "Health Effects of Long-Term Inhalation of Sulfuric Acid Mist-Carbon Particle Mixtures," submitted to *Environ. Res.* (1978).

42. Alarie, Y., Wakisaka, I., and Oka, S. "Sensory Irritation by Sulfite Aerosols," *Environ. Physiol. Biochem.* 3:182-184 (1973).
43. Wakisaka, I. "Sensory Irritation by Sulfite Aerosols on the Upper Respiratory Tract in Mice," *Japan. J. Hyg.* 31:347-352 (1976).
44. Hilton, C. M., Christiansen, J. J., Eatough, D. J., and Hansen, L. D. "Fe(III)-S(IV) Aerosol Generation and Characterization," *Atmos. Environ.* In press (1979).
45. Raabe, O. G. Personal Communication. Radiobiology Laboratory (University of California, Davis, CA: 1979).
46. Lippmann, M. L. Personal Communication. Institute of Environmental Medicine, New York University Medical Center (New York, N.Y.: 1979).
47. Raabe, O. G. In: *Fine Particles,* B. Y. H. Liu, Ed. (New York: Academic Press, Inc., 1979), p. 60.
48. Liu, B. Y. H., and Lee, K. W. "An Aerosol Generator of High Stability," *Am. Ind. Hyg. Assoc. J.* 36:861-865 (1975).
49. Coutarel, L., Matijevic, E., Kerker, M., and Huang, G. M. "Aerosol Studies by Light Scattering. VI: Preparation and Growth of Sulfuric Acid Aerosols," *J. Coll. Sci.* 24:38 (1967).

CHAPTER 15

GENERATION OF SUBMICRON SULFURIC ACID AEROSOL BY VAPORIZATION AND CONDENSATION

B. Y. H. Liu and J. Levi*

Particle Technology Laboratory
Mechanical Engineering Department
University of Minnesota
Minneapolis, Minnesota 55455

INTRODUCTION

Under the sponsorship of the Environmental Protection Agency, the Particle Technology Laboratory, University of Minnesota undertook a study on methods for generating sulfuric acid aerosols. The goal of the study was to develop a sulfuric acid aerosol generator for use in health effects studies. The desired particle size range was from 0.02 to 0.2 μm, and the concentration from about 0.1 to several mg/m^3.

The impetus for the present research came from the finding that sulfuric acid particles are emitted by catalyst-equipped automobiles. The particles are formed apparently by the catalytic conversion of SO_2 in the catalytic converter. Studies conducted by EPA and others concerning this "sulfuric acid problem" [1] have shown that the particles produced are very small in size, typically in the 0.02 and 0.2 μm diameter range [2, 3]. Since small aerosol particles can penetrate deeply into the human respiratory system, the potential health effect of these fine sulfuric acid aerosols is of some concern. The purpose of the present study is to develop a generator for producing such fine sulfuric acid aerosols for use in laboratory health effect studies on small test animals.

*Present address: Whirlpool Corp., P.O. Box 43380, St. Paul, Minnesota 55164.

METHODS FOR GENERATING SULFURIC ACID AEROSOLS

The techniques for generating sulfuric acid aerosols have been reviewed recently by Chang and his coworkers [4, 5]. Therefore, only a brief description of the principal approaches taken is given here.

The most common methods used for generating sulfuric acid aerosols are mechanical atomization and vapor condensation. In the case of vapor condensation, the desired vapor can be generated either by heating or by bubbling an inert carrier gas through fuming sulfuric acid. When very high sulfuric acid concentration ($\sim$ 100 mg/m^3) or large particle size ($>$ 0.1 μm diameter) is desired, the use of fuming sulfuric acid is indicated, since large amounts of sulfuric acid vapor can be introduced into the carrier gas stream relatively easily. However, when the desired particle size is small, as in the present case, or when moderate concentration levels are required, mechanical atomization may be indicated.

The method chosen for the present study involves mechanically atomizing a dilute solution of H_2SO_4 followed by vaporization of the solution droplets and recondensing the vapor in a turbulent jet. The advantage of this approach is that relatively dilute H_2SO_4 solutions are used, which are easy to handle in the laboratory. Also, the pneumatic atomizer is inherently stable, and the resulting aerosol generator can also be expected to give stable operating conditions.

DESIGN CRITERIA AND THEORETICAL CONSIDERATIONS

The specific generator described in this paper was designed for a small-animal exposure chamber with an internal volume of 330 liters. The generator was designed to supply test aerosols at a rate of 330 liter/min in order to provide a complete air change per minute in the chamber. The basic design parameters were chosen as follows:

Exposure chamber volume: 0.33 m^3
Chamber air flow: 330 liter/min
Aerosol diameter range: 0.02 to 0.2 μm
Sulfuric acid aerosol concentration: 0.1 to several mg/m^3

Although the aerosol generated did not need to be monodisperse, both the size distribution and aerosol concentration should be stable. In addition, it was desired that the generator be capable of unattended operations and be reasonably maintenance- and trouble-free.

In generating fine sulfuric acid aerosols of high concentration, coagulation limits the maximum particle concentration that can be obtained. For a monodisperse aerosol undergoing Brownian coagulation, the number concentration, N, at time t is related to the number concentration, N_o, at t = 0 and the coagulation constant, K, as follows:

$$N = \frac{1}{Kt + \frac{1}{N_o}} \tag{1}$$

which reduces to

$$N = \frac{1}{Kt} \tag{2}$$

for $N_o \gg 1/Kt$. Thus, for high enough initial concentrations, the aerosol number concentration, N, at time t becomes independent of the initial concentration, being a function of the coagulation time only.

The coagulation coefficient for monodisperse aerosols is a function of particle size. The following table, taken from Fuchs [6], gives the coagulation coefficient for various particle sizes:

D_p, μm	0.02	0.04	0.1	0.2
K, cm^3/sec	12×10^{-10}	11×10^{-10}	7.2×10^{-10}	5.2×10^{-10}

Over the particle size range from 0.02 to 0.2 μm, the coagulation coefficient is seen to not vary greatly. For the present purpose of making an estimate of the limiting particle concentration that can be obtained, $K = 10^{-9}$ cm^3/sec will be used as an approximate average for the coagulation coefficient in the abovementioned size range.

Using Equation 2 and the coagulation coefficient of $K = 10^{-9}$ cm^3/sec, we obtain for the coagulation-limited number concentrations at various coagulation times:

t, sec	1	10	100
N, cm^{-3}	10^9	10^8	10^7

The corresponding mass concentration of the aerosol is related to the number concentration by:

$$M = \frac{1}{6} \pi D_p^3 \rho N \tag{3}$$

where ρ is the particle density.

In Figure 1, the mass concentration of a coagulation-limited aerosol based upon the above analysis is shown as a function of particle size for various coagulation times. The particles are sulfuric acid droplets in equilibrium with ambient air at 50% relative humidity. At this humidity, the equilibrium

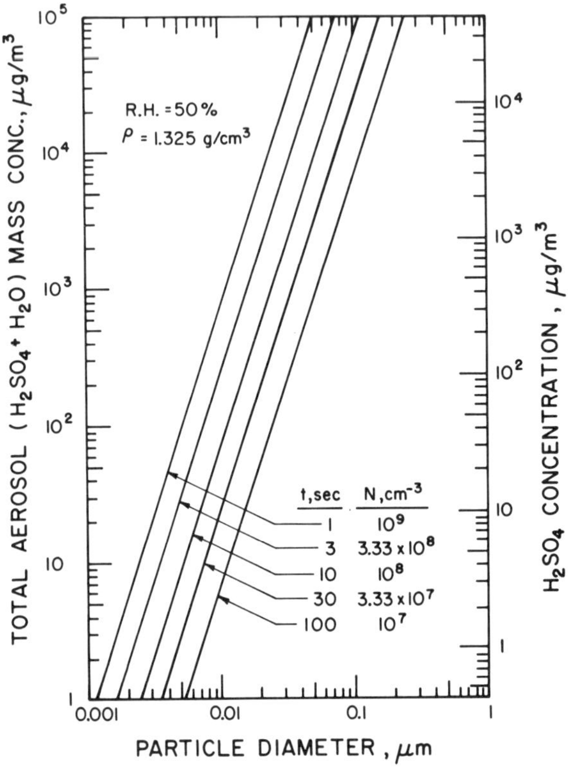

Figure 1. Coagulation limited aerosol concentration for H_2SO_4 particles at 50% relative humidity

sulfuric acid concentration in the solution is 42.5% by mass, and the solution density is $\rho = 1.325$ g/cm^3. Similar graphs for other humidity conditions can easily be constructed.

The above analysis shows that coagulation imposes a limit on the maximum aerosol concentration that can be obtained. Consider, for instance, the hypothetical case of an animal exposure experiment in which sulfuric acid particles are transported from where they are formed to where they are inhaled by the test animals in 30 seconds. Then, according to Figure 1, the maximum sulfuric acid concentration will be 79 μg/m^3 if the particle diameter is 0.02 μm. If the ambient relative humidity is 50%, the maximum aerosol mass concentration (sulfuric acid plus associated liquid water) will be 185 μg/m^3, also from Figure 1. Any attempt to increase the aerosol concentration to a level above these limits would only result in more coagulation and the production of larger particles. This is an important factor that must be taken into account both in the design of exposure facilities and in the design of human or animal exposure experiments.

PROPERTIES OF SULFURIC ACID SOLUTION DROPLETS

Sulfuric acid droplets are hygroscopic particles whose size will change in response to changes in relative humidity. Figure 2 summarizes the important sulfuric acid properties useful for working with sulfuric acid aerosols.

In Figure 2, the pertinent sulfuric acid solution properties are shown plotted as a function of x, the solution concentration expressed in g H_2SO_4/g solution. Included in the plot are the equilibrium relative humidity over a flat solution surface, the solution density, ρ, the boiling point, b.p., and the surface tension, σ. The data were taken from the Chemical Engineers' Handbook [7] and from Reference 8, as indicated in the figure. In addition, several calculated parameters are shown. These include the solution normality, the product, $x\rho$, the particle growth factor, D_p/D_{po}, and the Kelvin effect parameter, D_{po} *ln* p/p_∞, which are defined below. To illustrate the use of the various plotted quantities, the following example is given.

Consider a 1-μm diameter sulfuric acid solution droplet in equilibrium with relative humidity (r.h.) at 50%. Using the curve denoted by r.h. in Figure 1, we find that the solution concentration, x, is 0.425 g of H_2SO_4 per g of solution, or a concentration of 42.5% by mass. At this concentration, the solution has a density, ρ, of 1.32 g/cm^3, a boiling point, b.p., of 115°C, a surface tension, σ, of 76 dyne/cm, a normality of 11 *N*, and a mass concentration, $x\rho$, of 0.55 g H_2SO_4/cm^3 of solution. The particle size growth factor, D_p/D_{po}, is 1.48, and the Kelvin effect parameter, D_{po} *ln* (p/p_∞), is 11.3×10^{-4} μm. The last two parameters are explained in more detail below.

The particle size growth factor, D_p/D_{po}, shown in Figure 2, is the ratio of the actual diameter of the particle, D_p, to the diameter of the particle, D_{po}, if the water associated with the droplet is completely removed. In the example given above, the 1-μm diameter sulfuric acid solution droplet will become a droplet of pure sulfuric acid of 1/1.48 = 0.68 μm if water is removed. Similarly, if the ambient humidity is now increased to 90%, then according to Figure 2, the equilibrium solution concentration is 17% by mass and the particle size growth factor is 2.12. Consequently, the 1-μm diameter particle at 50% r.h. will now become a particle of (2.12/1.48)(1) = 1.43 μm diameter at 90% r.h. Thus, the particle size growth factor can be used to predict changes in particle size when ambient relative humidity around the particles is changed.

In the above discussion of particle size change, the Kelvin effect has been

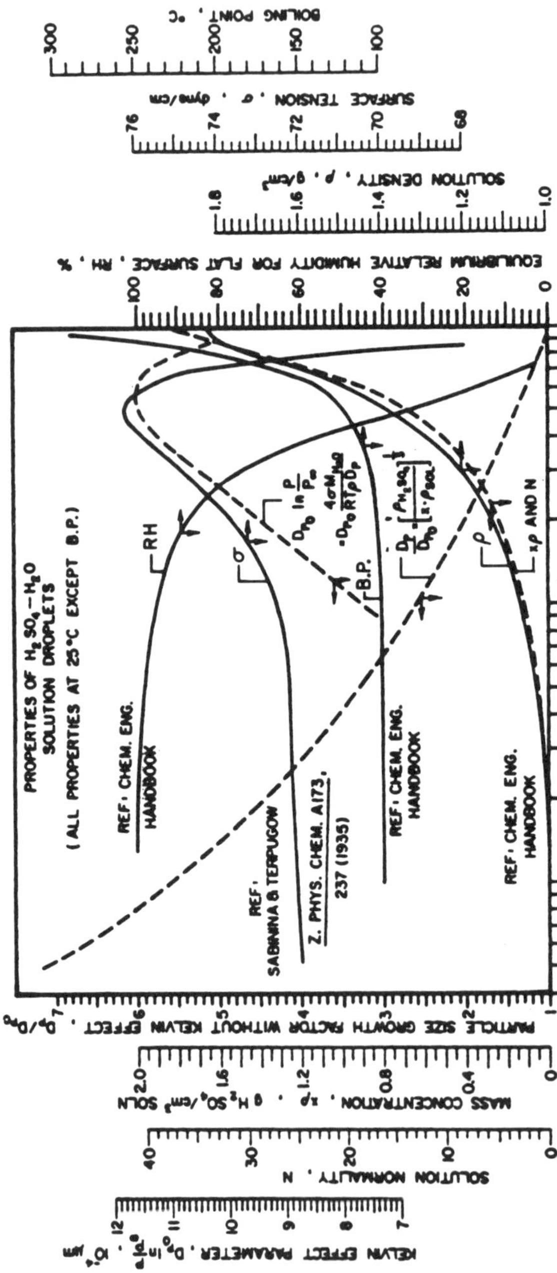

Figure 2. Properties of sulfuric acid droplets.

ignored. The Kelvin effect relates the increase in equilibrium vapor pressure, p, over a curved surface to that over a flat surface, p_∞,

$$ln(p/p_\infty) = \frac{2\,\sigma\, M}{RT\,\rho\, a} \tag{4}$$

where σ is the surface tension of the solution
M is the molecular weight of water
R is the gas constant for water
T is the absolute temperature
ρ is the solution density and
a is the particle radius.

The quantity plotted in Figure 2 is the ratio $D_{po}\ ln(p/p_\infty)$ which, according to Equation 4 is:

$$D_{po}\ ln(p/p_\infty) = \frac{4\sigma M}{RT\rho}\left(\frac{D_{po}}{D_p}\right) \tag{5}$$

where D_{po} has the same meaning as that discussed above. It should be noted that the quantity on the right-hand side of Equation 5 involves only solution and vapor properties, which are functions of x, the solution concentration. Consequently, the Kelvin effect parameter is also a function of x.

To illustrate the use of the Kelvin effect parameter, let us consider the example discussed earlier, viz., a 1-μm diameter sulfuric acid solution droplet at 50% r.h. Since D_{po} = 0.68 μm, and the Kelvin effect parameter is 11.3 x 10^{-4} μm, we have $ln(p/p_\infty) = 11.3 \times 10^{-4}/0.68 = 16.62 \times 10^{-4}$. Consequently, $p/p_\infty = 1.00166$. For this case, the Kelvin effect is negligible; the increase in water vapor pressure is only 0.166% above that for a flat surface.

Let us now consider a 0.01 μm diameter sulfuric acid particle at 50% humidity. Since D_{po} = 0.01/1.48 = 0.0068 μm in this case, we have $ln(p/p_\infty)$ = $11.3 \times 10^{-4}/0.0068 = 0.166$, which gives $p/p_\infty = 1.18$. Consequently, the Kelvin effect will cause the equilibrium vapor pressure over the curved droplet surface to be 18% higher than that over a corresponding flat surface. Thus, the effect is significant and generally should not be ignored.

SYSTEM DESCRIPTION

The system chosen for the present study is comprised of an atomizer, a tube furnace, and a mixing nozzle connected as shown in Figure 3. A dilute solution of H_2SO_4 is placed in the atomizer reservoir and atomized to form an H_2SO_4 spray. The spray is then vaporized in the tube furnace. The vapor is then injected into the mixing nozzle where it is mixed with particle-free air at ambient temperature to form a high concentration of small H_2SO_4

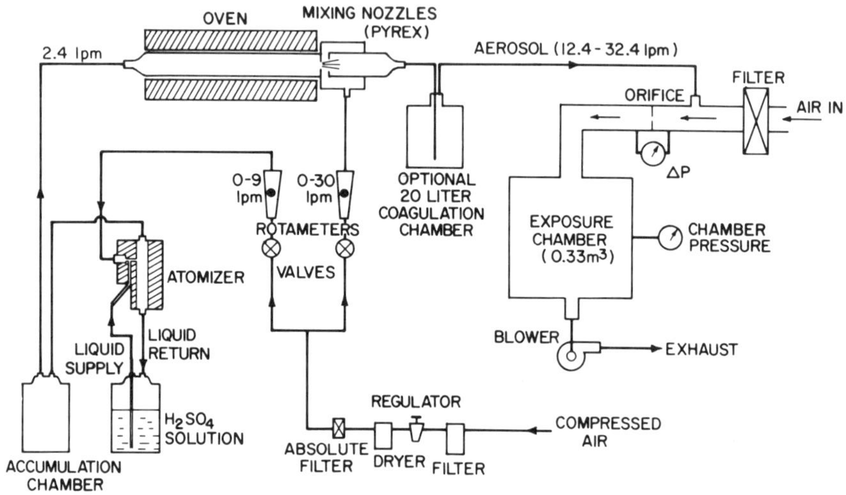

Figure 3. Schematic diagram of sulfuric acid aerosol generating system.

particles. The resulting aerosol is then introduced either directly into the exposure chamber, or first passed through an optional coagulation chamber to further increase the particle size.

The atomizer chosen for the present study is similar in design to the one described earlier by Liu and Lee [9], the only difference being that in the earlier design, a syringe pump is used to supply the liquid to the atomizer, whereas in the present case, a suction feed system is used and the liquid is recirculated in the atomizer. The atomizer produces aerosols at a rate of 2.4 liter/min when operated with compressed air at a pressure of 241 kPa (35 psig).

The vaporization of the H_2SO_4 droplets takes place in a Pyrex glass tube placed in a laboratory tube furnace (Lindberg Model 55035, Solar Basic Industries, 304 Hart St., Watertown, WI 53094). The furnace is heated to approximately 204°C. At this temperature, the gas is heated to approximately 104°C at the exit, the residence time of the gas in the furnace being approximately 3 seconds.

The hot gas from the furnace is injected into the mixing nozzle through a small 1.59-mm diameter orifice. The gas issues from the furnace in the form of a turbulent jet, which then mixes with the primary dilution air supplied through the rotameter at a rate of 10 to 30 liter/min. The mixing of the vapor-laden gas with the cold ambient air produces the high supersaturation needed for the formation of small H_2SO_4 particles. The resulting aerosol is then introduced into the exposure chamber by means of a 1.27-cm i.d., 60-cm long flexible tubing. The total volume of the tubing and that of the mixing nozzle is 125 cm^3. This results in a residence time of 0.25 sec for a primary dilution air flow of 30 liter/min.

The exposure chamber used in the present study is approximately cubic in shape, with an internal volume of 0.33 m^3. An exhaust blower at the chamber outlet provides the needed air flow of 330 liter/min. This air flow is metered by a 2.54-cm diameter orifice placed at the chamber inlet. The aerosol from the generator is introduced into the chamber just upstream of the orifice. The rapid mixing of the aerosol with the main chamber flow quickly dilutes the aerosol and causes the aerosol size distribution to be "frozen" at that point. Since the total chamber airflow of 330 liter/min is much higher than the aerosol flow of 12.4 to 32.4 liter/min, the humidity of the aerosol in the chamber is approximately the same as the humidity of the air entering the chamber. In the experiments described below, the main chamber air was taken from the air-conditioned laboratory room, and the relative humidity was between 42 and 48%, with a mean of 45%.

PERFORMANCE OF THE AEROSOL GENERATOR

Several preliminary experiments were performed in order to establish the basic operating parameters for the system.

The temperature of the furnace necessary for vaporizing the H_2SO_4 droplets from the atomizer was established by turning on the power supply to the furnace and monitoring the aerosol in the exposure chamber with an electrical aerosol analyzer (EAA) (Model 3030, TSI, Inc., P.O. Box 43394, St. Paul, MN 55164). A 0.1 N H_2SO_4 solution was used in the atomizer and the collector rod voltage in the EAA was set to zero so that the total EAA electrometer current was monitored.

Figure 4 shows that as the furnace temperature increased, there was initially no change in the total EAA current. This indicated that the size distribution of the aerosol was constant during this initial warmup period. However, after 33 minutes, and as the exit gas temperature had reached 59°C (130°F), the total EAA current began to rise sharply. This rise in EAA current is attributed to the vaporization of the H_2SO_4 droplets in the tube furnace and the subsequent recondensation of the vapor to form a high

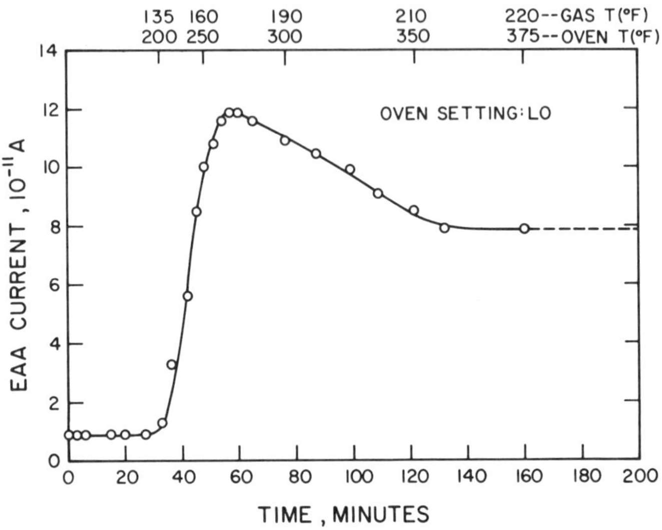

Figure 4. Variation of total EAA current during furnace warm up (furnace temperature setting "lo").

concentration of small H_2SO_4 particles. The experiment was repeated for three furnace power settings, and the results are quite similar, as Figures 4, 5, and 6 show.

On the basis of the above experiments, it was determined that a minimum temperature of 59°C (130°F) was needed to vaporize the H_2SO_4 solution droplets. In the subsequent experiments, it was decided to operate the furnace with a fixed power setting–"lo" on the specific furnace used. This produced a steady state temperature of 204°C (400°F) on the furnace and an exit gas temperature of 104°C (220°F), which was above the minimum temperature needed for the vaporization of the H_2SO_4 droplets.

Following the preliminary experiments described above, the following conditions were chosen as the standard operating conditions for the aerosol generation system. The system was then evaluated under these standard conditions.

Atomizer pressure: 241 kPa (35 psig)
Furnace temperature: Power control set to "lo" with steady-state furnace temperature of 204°C and exit gas temperature of 104°C
Chamber air flow: 350 liter/min

Mass Concentration

Four different methods were used to measure the aerosol mass concentration in the exposure chamber. In the gravimetric and chemical analyses,

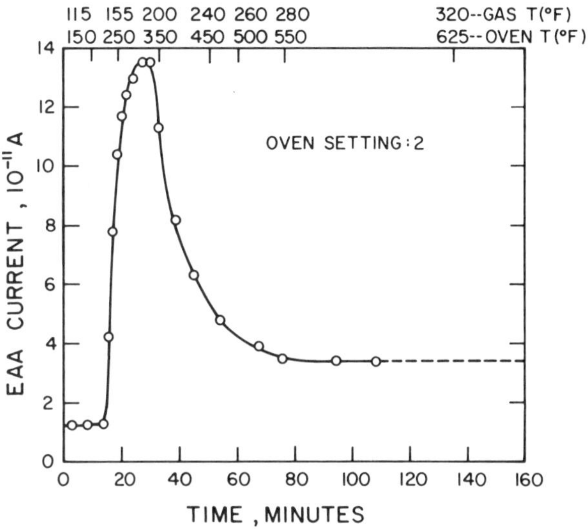

Figure 5. Variation of total EAA current during furnace warm up (furnace temperature setting "2").

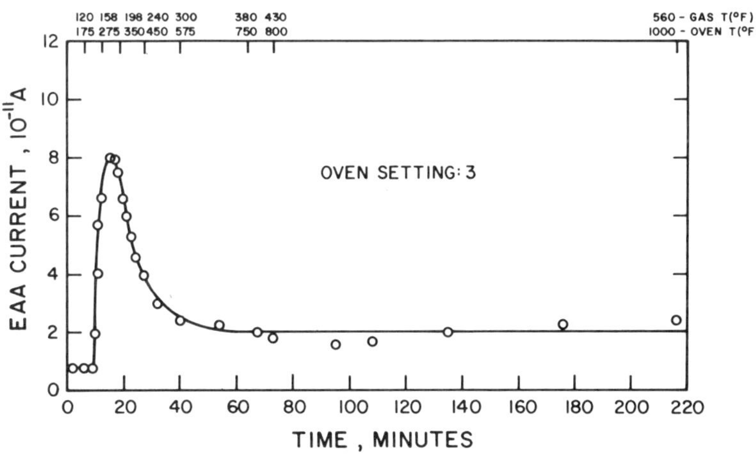

Figure 6. Variation of total EAA current during furnace warm up (furnace temperature setting "3").

an aerosol sample of at least 1 m^3 volume was obtained by drawing the aerosol through a 1.0 μm pore Fluoropore filter (Millipore Corp., Bedford, MA 01730). This was achieved by timing a flow of 14 liter/min through a critical orifice for a minimum of 71 minutes. The filter was weighed before and after the experiment, and the mass concentration of the aerosol was deduced. An electronic microbalance (Model AD-1, Perkin Elmer Corp., Main Ave., Norwalk, CT 06856) was used for this purpose. The filters were subsequently analyzed by wet chemical methods at EPA to give the $SO_4{}^2$ concentration. Previous experiments [10] have shown that the efficiency of the Fluoropore filter is in excess of 99.9%.

A quartz-crystal mass monitor (QCMM) (Model 3500 Piezobalance, TSI Inc., P.O. Box 43394, St. Paul, MN 55164) was used simultaneously during the experiments. The instrument measures the aerosol mass concentration by sensing the change in vibrational frequency of a quartz crystal as particles are deposited onto its surface by a small electrostatic precipitator. The sampling period of the instrument is 24 seconds for high mass concentrations or 120 seconds for low concentrations. Several readings were taken for each run and averaged to give a representative concentration.

The fourth method used was the electrical aerosol analyzer, or EAA. This instrument measures the size distribution of the aerosol in the 0.0032- to 1.0-μm diameter range by charging the particles and measuring the resulting particle mobilities. The operating principle of the instrument has been described by Liu and Pui [11] and Liu et al. [12], who also presented the basic calibration data for the instrument. However, in order to convert the instrument readings into a size distribution, a "data reduction scheme" must be used. In the present case, the "simplex minimization" technique described by Liu and Kapadia [13] was used. From the measured size distribution, the aerosol volume was calculated. The mass concentration was then calculated by multiplying the aerosol volume concentration by 1.35 g/cm^3, which is the density of sulfuric acid droplets at 45% relative humidity.

The EAA measurement was made by drawing all the required 50 liter/min flow from the exposure chamber. Except for the case of the 0.1 N solution, all EAA samplings required a dilution. A dilution ratio of 130 to 1 was used in these experiments.

The experiments were performed with 0.1 N and 1.0 N H_2SO_4 solutions. Table I shows a comparison of the measured aerosol mass by filter weighing and by $SO_4{}^2$ analysis. The agreement between the two methods is seen to be satisfactory—the mean difference is about 13%, and the maximum difference is 29%.

The discrepancy between the two methods of measurement for samples A and C may have been due in part to the partial neutralization of the

Table I. Comparison of Aerosol Mass Measurement by Weighing and by SO_4^{2-} Analysis

Sample	H_2SO_4 Conc.	Primary Dilution Air (liter/min)	Sample Volume (m^3)	Aerosol Mass (mg)	SO_4^{2-} Mass by Aerosol[a] Mass M_1 (mg)	SO_4^{2-} Mass Measured, M_2 (mg)	Ratio M_1/M_2
A	0.1 *N*	10	1.0	0.24	0.106	0.121	0.88
B	0.1 *N*	20	1.24	0.34	0.150	0.148	1.01
C	0.1 *N*	30	1.06	0.25	0.110	0.154	0.71
D	1.0 *N*	10	1.35	4.41	1.945	1.66	1.17
E	1.0 *N*	20	1.48	4.87	2.148	1.97	1.09
F	1.0 *N*	30	1.44	4.92	2.170	1.98	1.10

[a]The values given are obtained by multiplying the measured aerosol mass by the factor [(96/98)(0.45)] = 0.441, where 96 and 98 are, respectively, the molecular weights of SO_4^{2-} and H_2SO_4, and 0.45 is the equilibrium mass concentration of H_2SO_4 in the solution droplet at a relative humidity of 45%.

collected H_2SO_4 by ammonia in the laboratory air. Neutralization of H_2SO_4 on the filter would not affect the measured SO_4^{2-} mass by chemical analysis. However, if the collected H_2SO_4 particles had been partially neutralized by ammonium at the time of weighing of the filter, the particle mass would be reduced. The calculated SO_4^{2-} mass would then also be lower, since it was calculated assuming no neutralization had taken place. This may have been a factor contributing to the discrepancy between measurement methods for samples A and C, since the total collected mass for these samples was quite small (about 0.1 of that for samples D, E, and F).

In Table II, the aerosol mass concentrations measured by the several techniques described above are summarized and compared. Again, there is good agreement among the four independent measurement techniques. In view of the differences in the operating principles involved, the agreement may indeed be considered as excellent.

The experimental results obtained above are summarized and plotted in Figure 7, which shows the aerosol concentration in the exposure chamber as a function of the normality of the H_2SO_4 solution in the atomizer. Also shown in Figure 7 are the results obtained by the EAA for several intermediate solution normalities. These additional EAA measurements are described in more detail in the next section.

It is interesting to note in Figure 7 that the aerosol mass concentration is proportional to the solution normality, and that over the normality range of 0.1 *N* to 1.0 *N*, the H_2SO_4 mass concentration in the chamber varied from 0.13 to 1.3 mg/m^3.

Size Distribution

The aerosol size distribution was measured by the EAA for various sulfuric acid concentrations and for primary dilution airflow of 10, 20 and 30 liter/min. The data were reduced by means of the "Simplex Minimization" technique. The results are summarized in Table III, and a few selected size distribution curves are plotted and compared in Figures 8 and 9.

Although the aerosol volume concentration is not greatly affected by primary dilution airflow, the particle size is seen to decrease with increasing dilution airflow. This is to be expected, since the aerosol volume concentration is determined primarily by the atomizer output, which was unchanged in these experiments. The smaller particle size obtained at the higher primary dilution airflows is due to the more rapid mixing of the vapor with the dilution air. With more rapid mixing, a greater number of small particles are produced, resulting in a decrease in the mean particle size. In all cases, the σ_g of the aerosol was not greatly affected, the σ_g being in the range from 1.5 to 2.0 with an average of about 1.7.

Experience with the aerosol generator shows that the generator has good operational stability when the primary dilution air was kept above 20 liter/min. The result in Table III shows that the measured size distribution of the aerosol was quite stable when primary dilution airflows of 20 and 30 liter/min were used, whereas at a primary dilution airflow of 10 liter/min, less stable results were obtained. Consequently, 30 liter/min was then chosen as the standard operating flowrate for the primary dilution air. In Figure 10, the variation of the geometrical mean volume diameter of the aerosol with solution normality is shown for a primary dilution airflow of 30 liter/min.

COAGULATION CHAMBER

The result of the above-described studies shows that the aerosol generator is capable of generating sulfuric acid aerosols in the 0.04 to 0.087 μm diameter range when sulfuric acid solution in the 0.1 N to 1.0 N range is used. Although larger particles can be produced by means of more concentrated solutions, it was decided to investigate the use of a coagulation chamber for increasing the size of particles produced.

The experiment was performed with 1.0 N solution in the atomizer. The size distribution of the aerosol was measured with and without the optional 20 liter coagulation chamber shown in Figure 3. It was found that the geometrical mean volume diameter increased from 0.087 μm when no coagulation chamber was used, to 0.130 μm when a 20 liter chamber was inserted into the system. It was found that the measured aerosol volume concentration

Table II. Comparison of Aerosol Mass Concentration Measurement by Several Different Techniques[a]

H_2SO_4 Concentration	Aerosol Mass Concentration (H_2SO_4 + H_2O) (mg/m^3)					
Primary Dilution Air	0.1 *N*			1.0 *N*		
(liter/min)	10	20	30	10	20	30
Filter Weighing	0.24	0.27	0.24	3.26	3.29	3.42
SO_4^{2} Analysis[b]	0.27	0.27	0.33	2.79	3.02	3.12
QCMM	0.16	0.25	0.24	2.75	3.00	3.00
EAA[c]	0.25	0.30	0.27	2.73	3.00	2.65

[a]Measured at 45% relative humidity.

[b]Obtained by multiplying the measured SO_4^{2-} concentration by (1/0.45)(98/96) = 2.27, where 0.45 is the mass concentration of H_2SO_4 in the solution droplet at 45% r.h., 98 is the molecular weight of H_2SO_4, and 96 is that of SO_4^{2-}.

[c] Obtained by multiplying the measurement aerosol volume concentration by 1.35 g/cm^3, the density of the sulfuric acid droplet at 45% r.h.

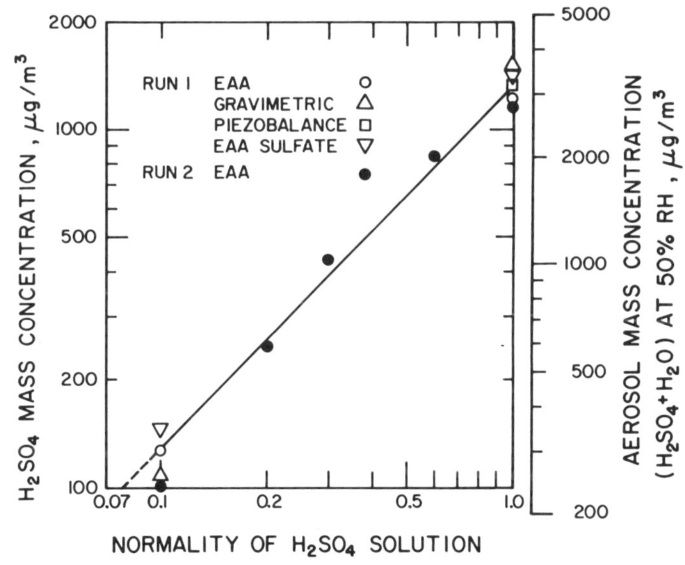

Figure 7. Sulfuric acid aerosol concentration in exposure chamber as a function of normality of H_2SO_4 solution in atomizer.

Table III. Summary of Aerosol Size Distribution Parameters for Various Generator Operating Conditions

Solution Normality	0.1 N			0.2 N			0.3 N			0.6 N			1.0 N		
Primary Dilution (liter/min)	10	20	30	10	20	30	10	20	30	10	20	30	10	20	30
VMD[a] (1)[b]	0.14	0.056	0.041				0.077	0.058	0.056				0.088	0.082	0.082
VMD (2)[b]	0.082		0.038	0.077	0.051	0.049	0.083	0.058	0.072	0.093	0.080	0.081	0.16	0.11	0.089
σ_g (1)	1.9	1.7	1.5				1.7	1.6	1.7				1.5	1.5	1.7
σ_g (2)	1.8		1.6	1.7	1.6	1.7	1.7	1.6	1.8	1.6	1.7	1.8	2.0	1.8	1.7
V_t[c] (1)	189	226	204				823	680	622				2060	2264	2000
V_t (2)	215		167	424	425	408	787	640	718	1530	1430	1400	1830	2077	1910

[a]VMD = geometrical volume mean diameter, μm.
[b](1) and (2) refer to the first and second runs, respectively.
[c]V_t, total volume in $\mu m^3/m^3$.

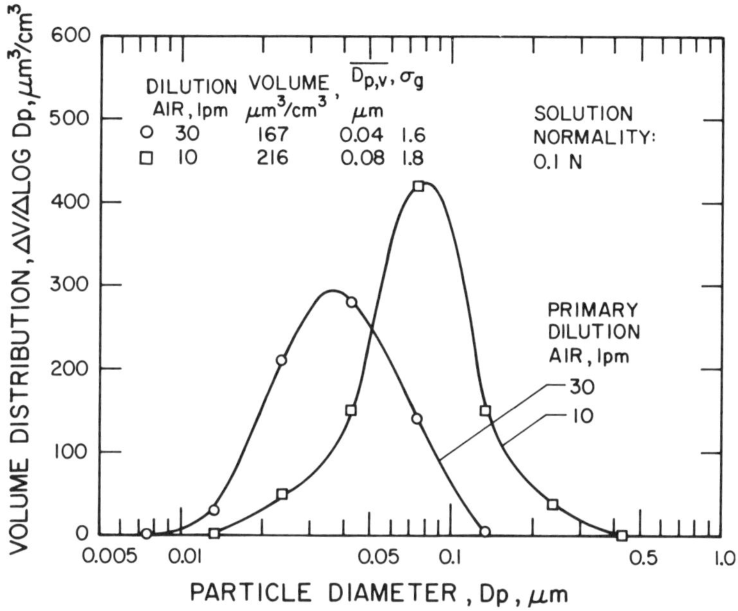

Figure 8. Aerosol size distribution for 0.1 N H_2SO_4 solution.

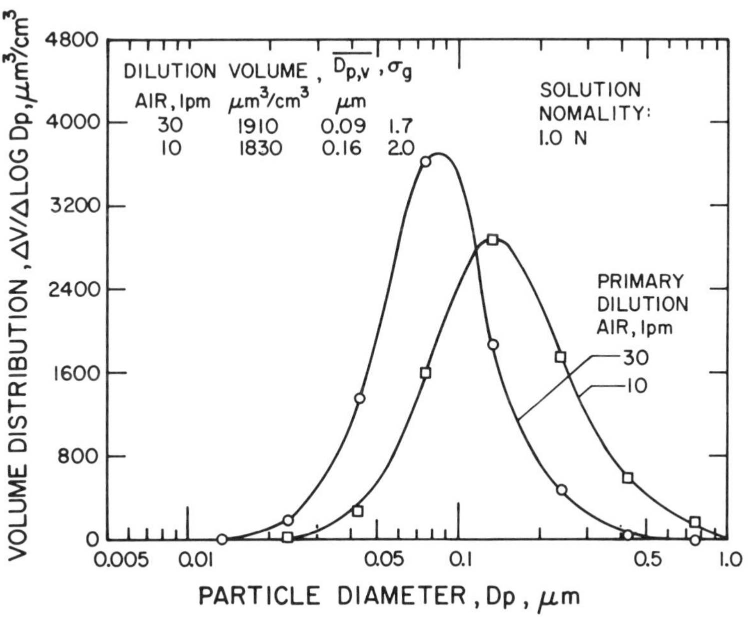

Figure 9. Aerosol size distribution for 1.0 N H_2SO_4 solution.

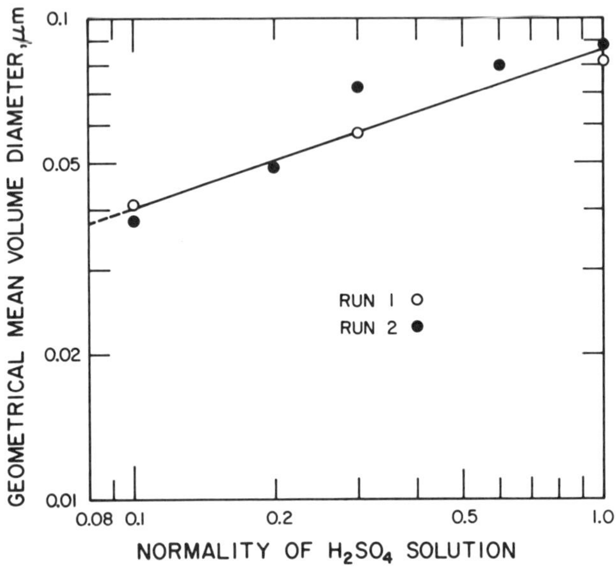

Figure 10. Mean volume diameter of aerosol as a function of solution normality for primary dilution airflow of 30 liter/min.

remained essentially unchanged, indicating that there was negligible aerosol loss in the coagulation chamber.

To determine the aerosol particle diameter for other coagulation chamber volumes, we note that for a coagulation-limited aerosol, the number concentration N is inversely proportional to the coagulation time t according to Equation 2. Since $t \propto V'$, we have

$$N \propto (1/V') \tag{6}$$

where V' is the total effective coagulation volume in the system. In addition, the aerosol mass concentration is given by

$$m = (\pi D_p{}^3/6)(N)(\rho) \tag{7}$$

If there is no loss of aerosol mass during coagulation, we have, according to Equation 7, $D_p \propto N^{-1/3}$, and with Equation 6,

$$D_p \propto (V')^{1/3} \tag{8}$$

The total effective coagulation chamber volume in the system is the sum of the coagulation chamber volume V_o in the absence of a coagulation chamber, and the volume V of the coagulation chamber, or

$$V' = V + V_o \tag{9}$$

By means of Equations 8 and 9, we have

$$D_p \propto (V + V_o)^{1/3} \tag{10}$$

Using the experimental result,

$$D_p = 0.087 \ \mu m \text{ for } V = 0 \tag{11}$$
$$D_p = 0.13 \ \mu m \text{ for } V = 20 \text{ liters}$$

and treating V_o in Equation 10 as an empirically adjustable parameter, we have

$$D_p = \left(\frac{V + 8.55}{8.55}\right)^{1/3} (0.087) \ (\mu m) \tag{12}$$

which can be used as an interpolation or extrapolation formula to estimate the aerosol particle size for various coagulation chamber volumes. The result shows that with a 30-liter coagulation chamber, particles as large as 0.144 μm can be obtained using a 1.0 *N* solution in the atomizer.

SUMMARY AND CONCLUSIONS

The sulfuric acid aerosol generator described in this paper has been developed to supply sulfuric acid aerosols to an animal exposure chamber at a rate of 330 liter/min. The particle size range of the generator is from 0.04 to 0.15 μm, and the concentration range from 0.13 to 1.3 mg/m^3. The size range distribution of the aerosol is approximately log-normal with a mean geometrical standard deviation of 1.7.

The generator has been found to have good operational stability. It should be useful for health effect and other laboratory studies.

ACKNOWLEDGMENTS

This research was supported by grant #R801301 from the U.S. Environmental Protection Agency. Further details concerning this work may be found in EPA-Report No. 600/2-78-104 (June 1978).

REFERENCES

1. Maugh, T. H., II. "Sulfuric Acid from Cars: A Problem that Never Materialized," *Science* 198:280-284 (1977).

2. Whitby, K. T., D. B. Kittelson, B. K. Cantrell, N. J. Barsic and D. F. Dolan. 'Aerosol Size Distributions and Concentrations Measured during the General Motors Proving Grounds Sulfate Study," presented before the Division of Environmental Chemistry, American Chemical Society, San Francisco, CA (1976).
3. Wilson, W. E., L. L. Spiller, T. G. Ellestad, P. J. Lamothe, T. G. Dzubay, R. K. Stevens, E. S. Macias, R. A. Fletcher, J. D. Husar, R. B. Husar, K. T. Whitby, D. B. Kittelson and B. K. Cantrell. "General Motors Sulfate Dispersion Experiment: Summary of EPA Measurements," *J. Air Poll. Control Assoc.* 27:46-51 (1977).
4. Chang, D. P. Y., and B. K. Tarkington. "Experience with a High Output Sulfuric Acid Aerosol Generator," *Am. Ind. Hyg. Assoc. J.* 38:493-497 (1977).
5. Chang, D. P. Y., B. K. Tarkington and T. R. Duvall. "Generation of Nuclei-Mode Sulfuric Acid Aerosols for Health Effects Studies," *J. Air Poll. Control Assoc.* 28:1137-1139 (1978).
6. Fuchs, N. A. *The Mechanics of Aerosols* (New York: The MacMillan Company, 1964).
7. Perry, R. H., and C. H. Chilton, Eds. *Chemical Engineers' Handbook*, 5th ed. (New York: McGraw-Hill Book Company, 1973).
8. Sabinina, L., and L. Terpugow. "Die Oberflachenspannung des Systems Schwefelsaure-Wasser," *Z. Phys. Chem.* A173:237-241 (1935).
9. Liu, B. Y. H., and K. W. Lee. "An Aerosol Generator of High Stability," *Am. Ind. Hyg. Assoc. J.* 36:861-865 (1975).
10. Liu, B. Y. H., and G. A. Kuhlmey. "Efficiency of Air Sampling Filter Media." In: *X-Ray Fluorescence Analysis of Environmental Samples,* T. G. Dzubay, Ed. (Ann Arbor, MI: Ann Arbor Science Publishers, Inc., 1977), pp. 107-119.
11. Liu, B. Y. H., and D. Y. H. Pui. "On the Performance of the Electrical Aerosol Analyzer," *J. Aerosol Sci.* 6:249-264 (1975).
12. Liu, B. Y. H., D. Y. H. Pui and A. Kapadia. "Electrical Aerosol Analyzer: History, Principle and Data Reduction," presented at the Aerosol Measurement Workshop, University of Florida, Gainesville, FL (1976). to be published in the proceedings of the workshop.
13. Liu, B. Y. H., and A. Kapadia. "A Computer Data Reduction Program for the Electrical Aerosol Analyzer." Publication Number 325, Particle Technology Laboratory, Department of Mechanical Engineering, University of Minnesota, Minneapolis, MN 55455 (1977).

CHAPTER 16

DIESEL EXHAUST AEROSOLS

David B. Kittelson and Daniel F. Dolan

Particle Technology Laboratory
Mechanical Engineering Department
University of Minnesota
Minneapolis, Minnesota 55455

INTRODUCTION

The need for improved passenger car fuel economy has led to a search for more efficient power plants. Diesel engines, which have an approximately 30% fuel economy advantage over comparable spark ignition engines, are thus seeing wider use in passenger cars. In addition to consuming less fuel, diesel engines have lower gaseous hydrocarbon and carbon monoxide emissions than equivalent spark ignition engines. On the other hand, their emissions of oxides of nitrogen, odorants and particles are higher.

This chapter is an examination of the nature of diesel exhaust aerosols (particles suspended in a gas). Particle size, structure and chemical composition will be discussed as well as the influence of coagulation, condensation, adsorption and chemical reactions on the nature of diesel exhaust particles. These processes are fast and may lead to improtant differences between the physical and chemical character of these particles as they exist in the atmosphere and as they exist in a sampling system or exposure facility.

SOME CHARACTERISTICS OF DIESEL EXHAUST PARTICLES

Diesel engine exhaust particles consist primarily of roughly spherical carbonaceous nuclei which have grown by coagulation into chain aggregates.

A transmission electron microscope photograph of a representative particle is shown in Figure 1. Typically, 80% by mass have aerodynamic diameters smaller than 1.0 μm. Mass mean diameters range from about 0.05 to 0.30 μm, depending upon engine design and operating conditions [1-3]. These particles are primarily elemental carbon, but also may contain significant quantities of adsorbed or condensed hydrocarbons and hydrocarbon derivatives, sulfur compounds and other materials. Solvent extractable organics comprise about 5-50% of the particle mass [4-6] and may contain mutagenic agents [7,8].

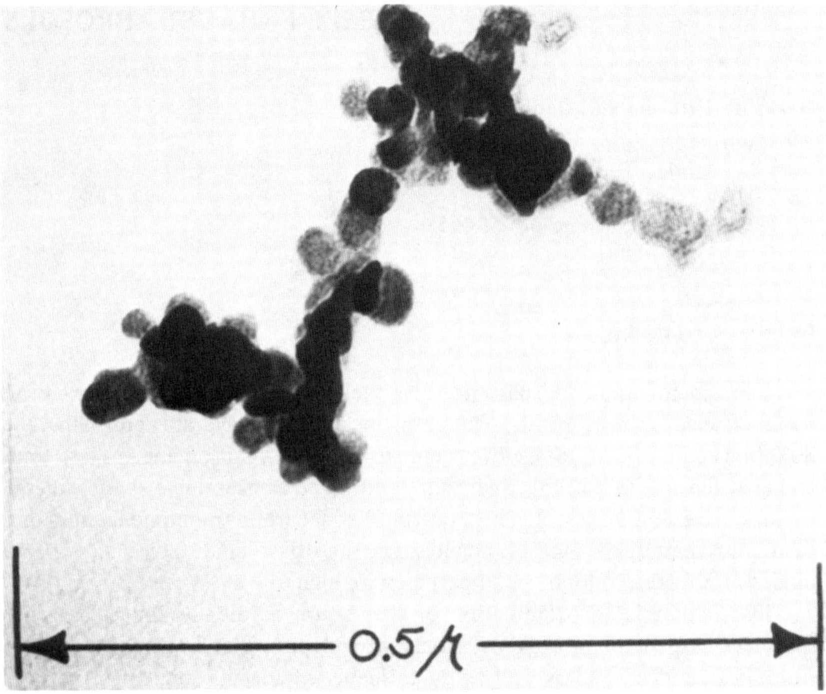

Figure 1. Typical large diesel exhaust particle.

There are a number of potential environmental problems associated with diesel exhaust particles. These include the following:

> Light Absorption–diesel aerosols are strongly absorbing, and high concentrations of these particles in the atmosphere could lead to severe visibility reductions. Diesel particles extinguish light primarily by absorption, in contrast to urban smog aerosols, which extinguish light primarily by scattering. Consequently, visibility reductions produced by these particles might be associated with significant local atmospheric heating.

- Inhalation Hazard–particles in this size range penetrate deep into the pulmonary region (functional area of the lungs), and a significant fraction may be deposited there [10]. Thus, they may represent a direct health hazard, especially if mutagenic or carcinogenic agents are adsorbed on their surfaces.
- The particles have large specific surface areas and may act as sites for heterogeneous conversion of atmospheric pollutants into other more hazardous or objectional pollutants [11].

All of these phenomena will depend upon the physical and chemical properties of the particles *as they exist in the atmosphere*. These properties will depend not only upon processes taking place in the engine and exhaust system, but also in the exhaust plume and the atmosphere itself.

DIESEL EXHAUST AEROSOL DYNAMICS

It is important to examine the relationship between the roadway dilution process and the laboratory dilution process. Figure 2 compares the roadway

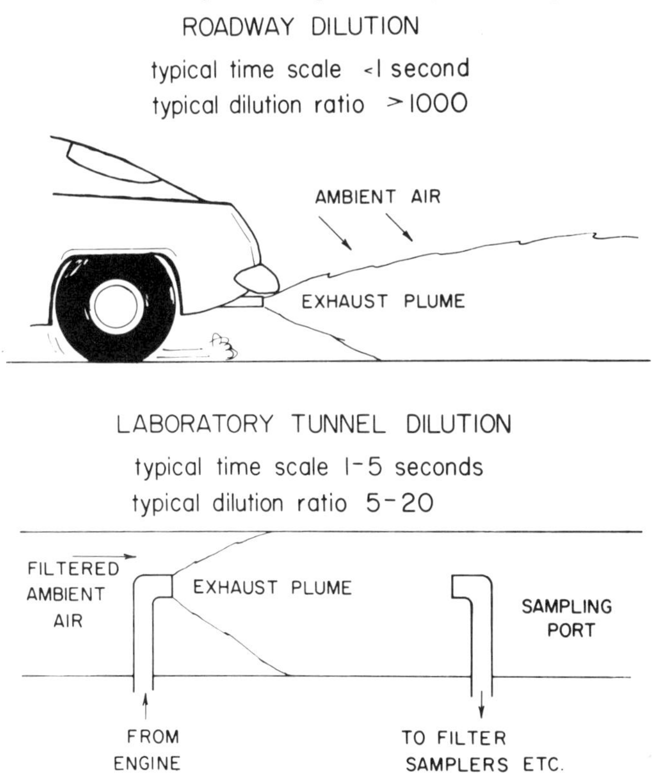

Figure 2. Comparison of roadway and laboratory dilution processes.

dilution process with a typical laboratory dilution tunnel measurement system. Measurements made in the exhaust plume of diesel cars on roadway by Dolan and Kittelson [2] show that under typical conditions, dilution ratios of several thousand to one are reached in less than 1 second. Dilution tunnels, on the other hand, typically produce dilution ratios between 5 and 20 and have transit times of 1-5 seconds. These diluted streams may then be passed through impactors and/or filters for sampling and analysis or to exposure chambers for animal studies. The total residence time of the particles in such systems may be many minutes. For example, in the animal exposure facility described by Hinners et al. [12] the average residence time in the tunnel/mixing chamber/exposure chamber system is about 5 minutes. Diesel aerosols are highly dynamic nonlinear systems. Processes taking place in the relatively concentrated streams produced by laboratory dilution systems may lead to particle size distributions and chemical compositions different from those produced under roadway dilution conditions. These processes include coagulation, condensation/adsorption, evaporation/desorption and chemical reactions.

Coagulation

Coagulation, which results from interparticle collections, has a strong influence on the diesel exhaust particle size distribution [13]. It leads to particles, like the one shown in Figure 1, which are agglomerates of small (~ 0.02 μm diameter) nuclei. The governing equation for the change in the aerosol size distribution with time as a result of coagulation can be given in the following form for the continuous case:

$$\delta n(v,t)/\delta t = 1/2 \int_0^v B(\bar{v},v-\bar{v})n(v-\bar{v},t)n(\bar{v},t)d\bar{v} - \int_0^\infty B(\bar{v},v)n(\bar{v},t)n(v,t)d\bar{v} \qquad (1)$$

where $n(v,t)dv$ represents the number concentration of particles in the size range v to $v+dv$ at time t. $B(\bar{v},v-\bar{v})$ represents the collision frequency factor for particles of size $\bar{v}$ and $v-\bar{v}$.

This equation can be given for the discrete case by:

$$\Delta n_k/\Delta t = \sum_{i+j=k} K_{c(i,j)} n_i n_j - 2\sum_{j=1}^{\infty} K_{c(j,k)} n_j n_k$$

where $K_{c(i,j)} = 2\pi(r_i + r_j)(D_i + D_j)\beta$ = coagulation coefficient defined by Fuchs [14].

To examine the influence of coagulation, this equation has been solved for a diluted diesel exhaust aerosol representative of what might be expected from a typical dilution tunnel system. The exhaust aerosol parameters used closely approximate the aerosol produced by a light-duty diesel engine running under moderate load. Each mole of exhaust is assumed to be mixed with nine moles of clean, particle-free air. The resulting diluted exhaust aerosol is used as a starting point for the coagulation calculations. Table I lists the main properties of the initial diluted exhaust aerosol, as well as the properties of the aerosol after coagulating for times ranging from 5 to 600 sec. Figures 3, 4 and 5 are plots showing the number, surface and volume weighted size distributions corresponding to each of the times listed in Table I.

Table I. Parameters Describing the Evolution of a Diluted Diesel Exhaust Aerosol Undergoing Brownian Coagulation[a]

t (sec)	t/τ	N (cm^{-3})	D_{pN} (μm)	S ($\mu m^2/cm^3$)	S/m (m^2/gm)	D_{pS} (μm)	D_{pV} (μm)	σ_{gV}
0	0	9.96×10^7	0.040	8.81×10^5	74.7	0.071	0.092	1.67
5	0.68	5.91×10^7	0.052	7.75×10^5	65.7	0.080	0.099	1.58
30	4.11	2.27×10^7	0.078	5.99×10^5	50.8	0.105	0.120	1.44
60	8.22	1.41×10^7	0.095	5.22×10^5	44.2	0.120	0.134	1.40
300	41.1	4.01×10^6	0.152	3.58×10^5	30 3	0.170	0.188	1.37
600	82.2	2.39×10^6	0.184	3.11×10^5	26.3	0.222	0.241	1.33

[a]Total volume concentration is 11800 $\mu m^3/cm^3$ and the mass concentration is 11.8 mg/m^3 (assuming a particle density of 1.0 gm/cm^3).

t = time
τ = initial coagulation time constant = 7.3 sec
N = number concentration
D_{pN} = geometric number mean diameter
S = aerosol surface concentration
S/m = surface area per unit particle mass
D_{pS} = geometric surface mean diameter
D_{pV} = geometric volume mean diameter
σ_{gV} = geometric standard deviation

The shortest aging time shown is 5 sec. This corresponds to what might be expected if the aerosol sample passed directly from the tunnel to measuring instruments. The number, surface and volume size distribution all shift to the right as a result of coagulation. The number distribution is the most strongly influenced and shows a 41% decrease in total number concentration and an increase in the number mean diameter from 0.040 to 0.052 μm. The surface area is also decreased, but by only 12%. Volume (or mass) is, of course, conserved, but the volume mean diameter increases from 0.092 to 0.099 μm.

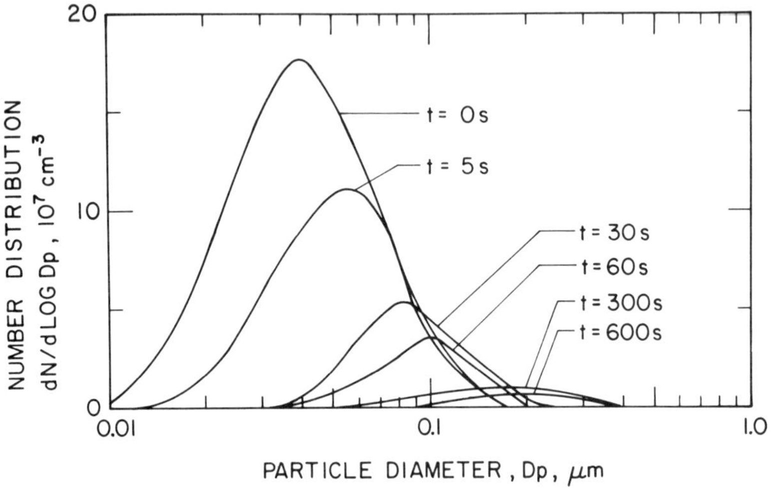

Figure 3. Calculated particle number distribution functions for a diluted diesel exhaust aerosol undergoing Brownian coagulation after various aging times (t). The total number of particles is proportional to the area under the curve.

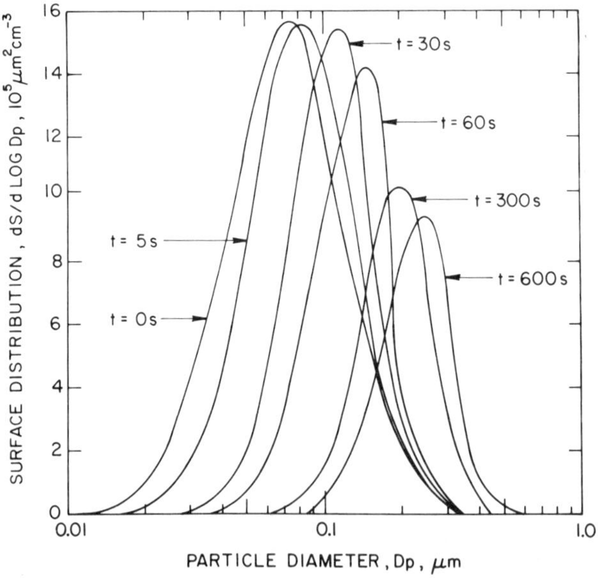

Figure 4. Calculated particle surface distribution functions for a diluted diesel exhaust aerosol undergoing Brownian coagulation after various aging times (t). The total surface of particles is proportional to the area under the curve.

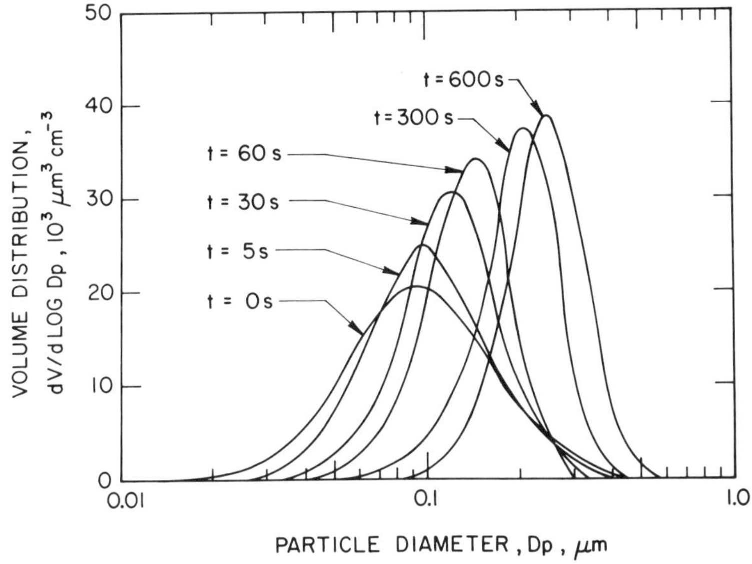

Figure 5. Calculated particle volume distribution functions for a diluted diesel exhaust aerosol undergoing Browning coagulation after various aging times (t). The total volume of particles is proportional to the area under the curve.

The results obtained for five minutes aging time are probably fairly representative of what might happen in an animal exposure facility. Under these conditions, all three weightings of the size distribution have been profoundly changed by coagulation. The number concentration has decreased by 96%, and the geometric number mean diameter has increased from 0.04 to 0.152 μm. The surface area has decreased by 59%, and the geometric surface mean diameter has increased from 0.069 to 0.170 μm. Although the total volume remains constant, the geometric volume mean diameter has increased from 0.092 to 0.188 μm.

Thus, these results suggest that coagulation may very significantly change aerosol number and surface concentrations, and number, surface and volume size distributions under the dilution conditions and residence times characteristic of exposure chambers.

On the other hand, coagulation should be much less important under roadway dilution conditions. Coagulation is a second order process and strongly influenced by dilution. Equation 1 can be rewritten in terms of N, the total number concentration. Then,

$$\frac{dN}{dt} = -k\,N^2 \tag{2}$$

where k is the size distribution-dependent average coagulation coefficient. If k is assumed to be constant, integrating Equation 2 gives

$$\frac{N}{N_o} = \frac{1}{1 + k\,N_o t} = \frac{1}{1 + t/\tau}$$

and

$$\tau = \frac{1}{k\,N_o}$$

where N_o is the initial number concentration and τ is the coagulation half-life. Although k depends upon the size distribution, it does not change greatly over the range of conditions expected for diesel aerosols. In the numerical experiment described above, it changed from an initial value of 1.4×10^{-9} cm^3/sec to a final value of 0.6×10^{-9} cm^3/sec. This inital value of k gives a half-life of $\tau = 7.3 \times 10^8/N_o$ (sec). Thus, for an exhaust number concentration of $10^9/cm^3$ and a dilution ratio of 10, $\tau_{1/2} = 7.3$ sec. Under roadway conditions of 1000:1 dilution ratio, $\tau_{1/2} = 730$ sec. The residence time of diluted exhaust aerosols over roadways is probably not much longer than a few minutes, e.g., a 2 m/sec crosswind would clear a 30 m wide roadway in 15 sec. Thus, except under exceptional conditions, such as those found in tunnels or parking lots, coagulation should not significantly alter the character of roadway aerosols. Of course, these aerosols ultimately become part of the atmospheric aerosol, in which coagulation will continue at a much reduced rate.

Condensation and Adsorption

Condensation/adsorption and evaporation/desorption are considerably more difficult to quantify because neither the condensing species nor its concentration is known. However, as indicated above, many investigators have reported significant organic extractable fractions in diesel exhaust particles collected on filters. This material apparently condenses or adsorbs on the surfaces of primary combustion particles sometime during the exhaust, dilution or sample collection processes. The amount of material adsorbed or condensed upon the particles should depend upon the particle surface area available for adsorption and the saturation ratio, i.e., the ratio of the partial pressure to the vapor pressure, of the adsorbing or condensing species.

Figure 6 is a sketch of a generalized BET [15] type II adsorption isotherm. Such isotherms are observed experimentally for many systems. Three

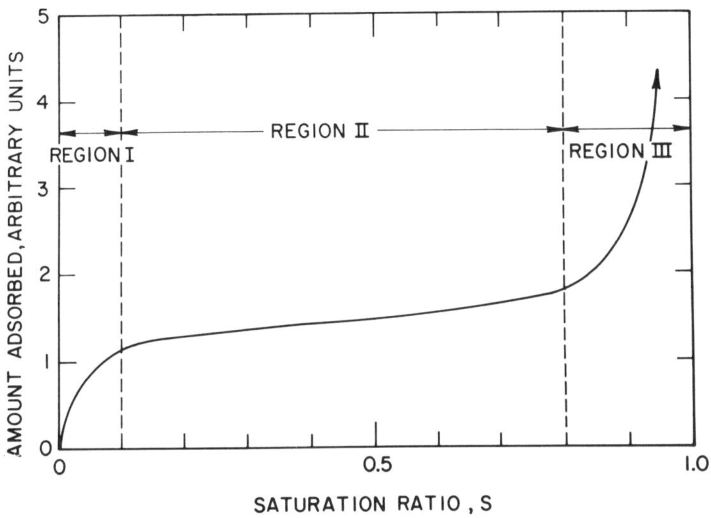

Figure 6. Generalized BET Type II isotherm. This type of isotherm is fairly representative of the adsorption of hydrocarbons on carbon. Here S is the ratio of the partial pressure of the component being adsorbed to its vapor pressure.

regions may be identified. In region I, the available surface adsorption sites are being filled. The amount adsorbed increases as saturation ratio S increases. In region II, these sites are essentially filled, i.e., a monolayer has formed, and the amount adsorbed only increases slowly with increasing S. In region III, multilayer adsorption becomes important and the amount adsorbed again increases with increasing S until, as S approaches one, adsorption gradually gives way to condensation.

It is apparent that the saturation ratio may affect the extractable organic present on the particles. In order to determine how the saturation ratio is influenced by the dilution process, a simple model has been constructed. The following assumptions were made:

1. The exhaust aerosol consists of particles, noncondensibles and a single component which may condense or adsorb onto the particles. The concentration of this species in the undiluted exhaust is assumed to be 1.0 ppm.
2. The exhaust is adiabatically diluted with clean dry air.
3. The specific heats of the dilution air and exhaust are constant and equal.
4. The condensible material obeys Trouton's rule, i.e., $\Delta H_{vap}/RT_B = 10.5$ where ΔH_{vap} is the molar heat of vaporization, R is the gas constant, and T_B is the boiling temperature.

5. The variation of vapor pressure with temperature is given by the Clasius-Clapeyron equation, with ΔH_{vap} assumed constant.
6. The dilution air temperature is 300°K.

These assumptions have been used to calculate the variation of the saturation ratio S of the condensible species with dilution ratio for various exhaust temperatures and condensible boiling points.

Figure 7 shows variation of S with dilution ratio for exhaust temperatures of 480°K and 600°K and condensible boiling points of 720°K and 810°K. The highest saturation ratios occur at dilution ratios in the range of about 5 to 50. Although these calculations are based on a simple one-component system, the general trends should be the same in a multicomponent system. Thus, the strongest driving force for condensation or adsorption occurs in the moderate dilution ratio range, roughly the same range produced by a typical dilution tunnel system.

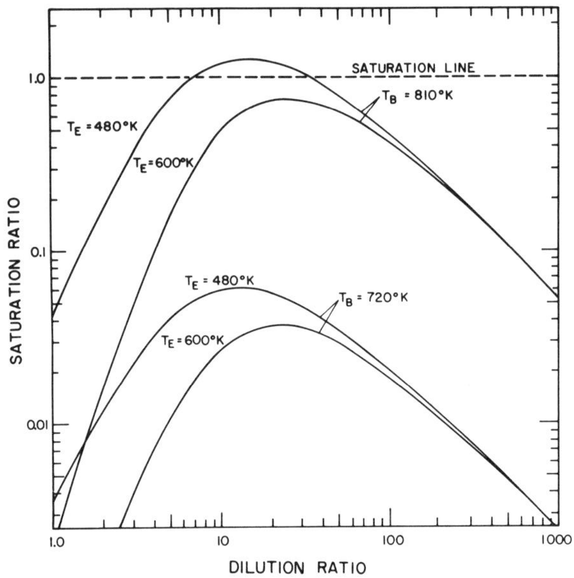

Figure 7. Calculated saturation ratios for condensible hydrocarbons in an exhaust stream undergoing adiabatic dilution with clean ambient air. The saturation ratio is the ratio of the partial pressure of the condensible to its vapor pressure. T_E is the exhaust temperature and T_B is the boiling point temperature of the condensible species. The exhaust mole fraction of condensible is assumed to be 1.0 ppm. Dilution air is assumed to be at 300°K.

The model described above is based strictly on equilibrium considerations. However, it is also important to examine the rate at which condensation or adsorption might be expected to occur. The rate of mass transfer by diffusion to submicron particles may be calculated using expressions given by Fuchs [14]. These expressions may be rearranged to give:

$$(1/m)\ dm/dt = \frac{12\ D(\rho_\infty - \rho_s)}{D_p^2\ \rho_p\ (1 + \ell K_n)}$$

$$\ell = \frac{1.33 + 0.71\ K_n^{-1}}{1 + K_n^{-1}}$$

where m = particle mass
t = time
D = diffusion coefficient
ρ_∞ = particle density
D_p = particle diameter
K_n = Knudsen number.

Diesel exhaust particles often contain roughly 10% solvent extractable mass. The time required to increase the particle mass by 10% has been calculated using the expressions above. The conditions which produced the maximum saturation ratio shown in Figure 3 were used in the calculation, along with a molecular weight of 400, a diffusion coefficient of 0.03 cm^2/sec, and a particle diameter of 0.1 μm. This gives a mass transfer time of 0.27 sec.

Under roadway conditions, dilution ratios greater than 1000 are achieved in less than 1.0 sec; thus, the time available at dilution ratios between about 5 and 50, where the highest saturation ratios are obtained, may be long enough to lead to some condensation or adsorption. This may be followed by reevaporation or desorption at higher dilution ratios, and thus, lower saturation ratios. Dilution tunnels operating in the dilution ratio range of 5 to 20 with residence time of 1-5 sec make long times available at high saturation ratios and thus maximize the possibility of condensation or adsorption. In addition, particle samples are usually obtained using filters which are exposed to a stream of aerosol from the tunnel for many minutes. This should further encourage condensation or adsorption. The same considerations should apply to animal exposure chambers where residence times of several minutes are common.

This simple model suggests that the extractable fraction on diesel exhaust particles collected from dilution tunnels should depend upon the dilution ratio. At this time, the data available on the influence of dilution ratio on the organic extractable fraction is limited and somewhat contradictory. Frisch et al [16] show a slight dependence of extractables on dilution ratio, while Black and High [17] and Bradow et al. [18] show none. The model presented here, at least in a qualitative sense, encompasses all of these results. Thus, experiments performed under conditions corresponding to region II of the BET isotherm, shown in Figure 5, would be expected to show little influence of dilution ratio, whereas those performed under conditions corresponding to regions I, III or to condensation ($S > 1$) should exhibit a dilution ratio dependence. Thus, the question of the influence of dilution ratio on the extractable fraction has not yet been resolved, and results of studies done at moderate dilution ratios should be interpreted with care.

Other Considerations

In addition to influencing coagulation and gas-to-particle mass transfer processes, the dilution ratio may influence chemical reactions taking place on the particles themselves. Under typical dilution tunnel conditions, concentrations of NO_X, SO_X, and hydrocarbons are 10-100 times higher than under roadway conditions, the temperature is 10-15°C above ambient and the relative humidity is nearly 100%. Carbon particles are known to be catalytic [11], and reactions might take place on their surfaces under dilution tunnel and exposure chamber conditions which do not occur under roadway dilution conditions. For example, Pitts [19] has suggested that PAH might be converted to much more biologically active nitrated forms under dilution tunnel-like conditions.

Thus, diesel exhaust aerosols are highly dynamic and subject to rapid physical and chemical changes. Moderately diluted exhaust aerosols produced by dilution tunnels may be quantitatively and qualitatively different from aerosols diluted under roadway conditions.

Unfortunately, to obtain results in reasonably short times, animal exposure facilities are often operated at high aerosol concentrations, i.e., low dilution ratios. In view of the changes expected to take place in an aerosol as it ages under these conditions, great care should be taken in interpreting the results of such experiments.

RECENT STUDIES OF DIESEL AEROSOLS AT THE UNIVERSITY OF MINNESOTA

In this section, some techniques for sampling and physical analysis of diesel exhaust aerosols which have been developed at the University of Minnesota will be discussed. This discussion will take the form of an examination of the influence of fuel on the particle emissions from a small diesel engine.

Apparatus and Procedure

The engine employed in this study was a single cylinder 30-in.^{3} (492-cm^3) displacement Onan Model 3MDJA marine diesel. It was coupled to a 3-kW alternator and resistance bank.

The exhaust dilution and sampling system is illustrated in Figure 8. The engine and measurement system have been discussed previously [20]. The sampling-measurement system is a modified version of the one used by Verrant and Kittelson [21], by Dolan [13], by Dolan and Kittelson [22], and by Kittelson et al. [23]. The major components of the system are a Krypton-85 neutralizer, an electrical aerosol analyzer (EAA), a filter mass sampling system, electron microscope samplers and a dilution system.

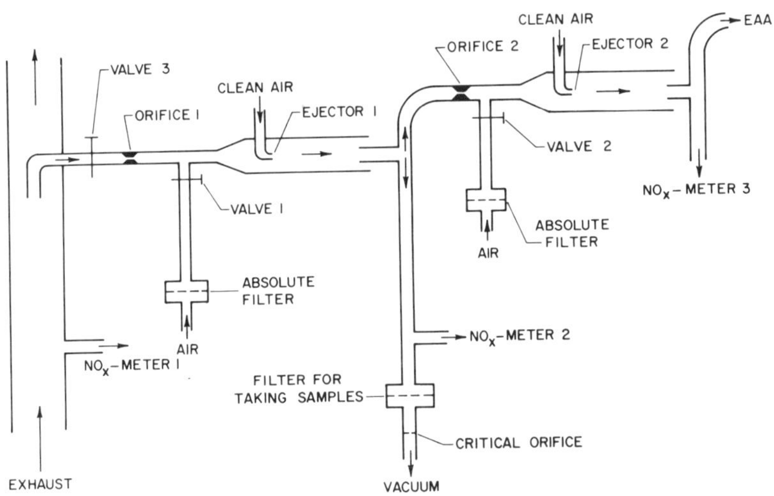

Figure 8. Dilution and sampling system.

Size distributions in the 0.01-1.0 μm diameter range were determined using the EAA (TSI, Inc. Model 3030). The size distributions are determined by charging the incoming aerosol and measuring the electrical mobility of the charged particles. The principle of operation and calibration of the EAA have been discussed in detail elsewhere [13,21,24]. Diesel exhaust particles are highly charged [25], and this may bias the charging process in the EAA and, consequently, the measured size distributions. To prevent this, a Krypton-85 neutralizer is used in conjunction with the EAA. Its function is to reduce the natural charge on the diesel exhaust particles to Boltzmann equilibrium charge distribution [26]. This ensures proper EAA operation and, in addition, minimizes the loss of charged particles to the walls of the sampling system.

The filters used were 37 mm diameter Fluoropore filters (Catalog #FHLP 03700). The pore size for these filters is listed as 0.5 μm. They have sufficiently high sampling efficiency for the particle size range of interest, 0.02-1.0 μm (> 99%) [27]. Millipore ambient aerosol monitor (Model M000 037A0) filter holders were used. A critical orifice was used to give a constant flow of 0.14 liter/sec through the filters.

A TSI, Inc. Model 3100 electrical aerosol sampler and an American Instrument Co. Model 5-3900 thermal precipitator were utilized in taking electron microscope samples.

The use of a rapid dilution system was necessary in order to minimize coagulation and condensation in the sampled exhaust. Additionally, because of an upper limit on the concentration of particles the EAA can measure, dilution was necessary to utilize this instrument. The dilution system consisted of two stages. In the first stage, the raw exhaust was diluted with approximately 20 parts clean air. It was from this stage that the filter and microscope samples were taken. The second stage diluted the already diluted exhaust sample another 40 to 100 parts clean air.

Each stage of the dilution system consists of three main parts: an orifice (1 mm diameter), an air ejector and a mixing chamber. The ejector serves three functions: to provide dilution air, to pump the exhaust sample through the system and to mix the aerosol with the dilution air. Compressed air was supplied to the ejectors from the building compressed air supply. Before entering the dilution system, it was filtered, dried with silica gel and filtered again with an absolute filter.

Calculations have been made to estimate particle losses within the system. Diffusion, impaction and turbulent deposition were considered. For particles between 0.01 and 1 μm in diameter, the calculated losses were less than 12%.

Dilution ratios were determined by comparing the NO_X concentration at each stage of the dilution system with that in the undiluted exhaust. Thus, the first stage dilution ratio was

$$\frac{NO_X \text{ Exhaust}}{NO_X \text{ 1st Stage}}$$

Second stage dilution ratio calculations required that background NO_X be taken into account, thus:

$$\text{Dilution Ratio, 2nd Stage} = \frac{NO_X \text{ Exhaust}}{NO_X \text{ 2nd Stage} - NO_X \text{ Background}}$$

Background NO_X was determined by measuring second stage NO_X concentrations when the engine exhaust flow was cut off through the dilution system.

Sampling diesel particulate emissions without introducing some falsifications is a virtually impossible task. Particle sizes range from a few molecular diameters to tens of microns in diameter, therefore making it impossible for any one instrument to give the overall particle size distribution. Particles larger than about 1 μm diameter require isokinetic sampling techniques, and the effects of the sampling system on the particles are difficult to deal with. Despite these problems, it is believed that the system used here provided reasonably representative samples.

The fuels used in this study included 40 and 50 cetane number secondary reference fuels (Phillips check fuels 40 and 50), and No. 1 and No. 2 diesel fuel. Net fuel flow to the engine was measured by volume using a burette and stopwatch. The physical and chemical properties of the fuels examined are summarized in Tables II and III.

Table II. Properties of Fuels Tested

Parameter	Fuel			
	1D	2D	Check 40	Check 50
Specific Gravity @ 15.6°C (60°F)	0.823	0.847	0.793	0.807
% Sulfur	0.058	0.28	0.011	0.016
Hydrogen/Carbon Atomic Ratio	1.89	1.81	2.02	1.93
% Paraffin Content	75	66	87	82
% Aromatic Content	23	30	11	16
% Olefin Content	2	4	2	2

Table III. Fuel Distillation Curves (°C)

% Vaporized	Fuel			
	1D	2D	Check 40	Check 50
Initial Boiling Point	176	172	183	182
10	196	212	196	197
20	203	227	199	203
30	211	237	199	203
40	218	248	210	216
50	226	258	217	222
60	233	268	224	228
70	243	278	229	234
80	254	291	246	236
90	271	303	261	252
End Point	302	336	293	273

Five load settings from idle to nearly full load were utilized. Engine speed was maintained at 1800 rpm by its governor. Loads were selected in random order to minimize systematic errors. All runs were made under steady state conditions.

Results

In order to describe fully the effects of these fuels on exhaust particle emissions, the results are best presented in several forms. Many characteristics of the exhaust particle emissions have been measured. For our purposes, the most important of these are number, mass and volume concentrations. Number and volume concentrations were obtained using the EAA, while mass concentrations were obtained using gravimetric filter analysis.

Figure 9 is a plot of the total particle mass concentration measured in the exhaust against brake mean effective pressure (BMEP) for the four undoped fuels used in these experiments. It can be seen that the results are very similar for fuels 1D, 2D, and reference 50, but that the mass emissions from reference 40 are significantly lower. This same tendency is seen in the emission index which is plotted in Figure 10. The emission index is here defined as the grams of particulate matter emitted per kilogram of fuel consumed, or grams per megajoule energy input in the fuel.

The exhaust particle size distributions provide additional insights into the nature of these aerosols. Figure 11 shows the volume size distribution

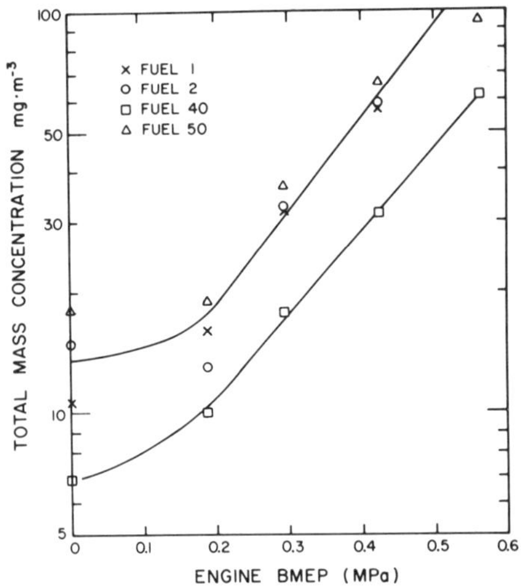

Figure 9. Exhaust aerosol particle mass concentration plotted against brake mean effective pressure (load). The mass concentrations were determined by filter measurements. Engine speed: 1800 rpm.

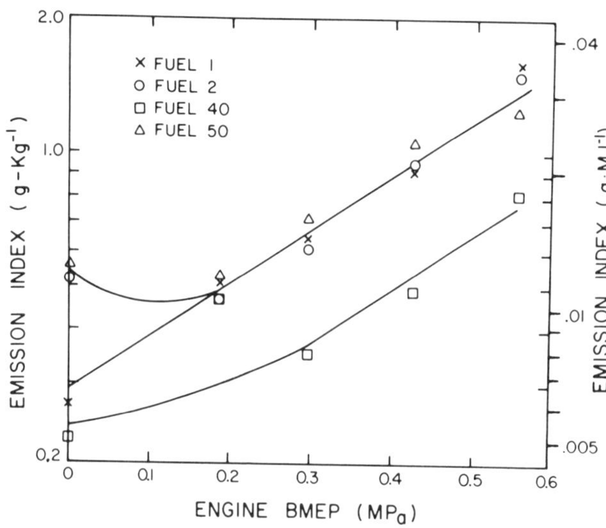

Figure 10. Particle mass emission index plotted against brake mean effective pressure (load). Engine speed: 1800 rpm.

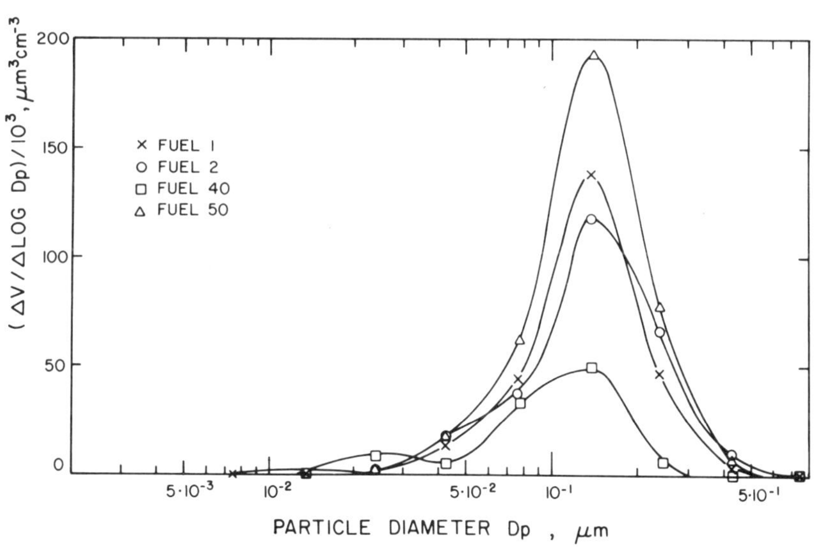

Figure 11. Measured particle volume distribution ($\mu m^3/cm^3$) for the four fuels tested. Engine speed: 1800 rpm; 3/4 Load (BMEP - 0.43 MPa).

obtained at an engine load of 0.43 MPa (3/4 load). This type of size distribution was obtained at all loads investigated. Here $\Delta V/\Delta \log D_p$ is plotted against $\log D_p$. The ordinate ΔV is the particle volume ($\mu m^3/cm^3$) contained in the logarithmic diameter increment $\Delta \log D_p$ and the abscissa is the geometric mean diameter corresponding to each diameter increment. The area under any part of the distribution curve is proportional to the aerosol volume contained in that diameter interval [28]. It is apparent that the volume distribution obtained while using 40 cetane number fuel is bimodal. Although this is not so apparent for the other three fuels, they also produce exhaust particle size distributions which are bimodal. Size distributions can be presented in terms of number, surface area or volume (as given here). By considering all three forms of the size distribution, the bimodality becomes apparent as shown by Dolan et al. [28]. There is a small mode in the range of 0.01 to 0.03 μm diameter, usually called the "nuclei mode," and there is a mode in the range of 0.1 to 0.2 μm diameter, usually called the "accumulation mode." The nature of these modes and the theoretical basis for their existence have been discussed in several previous publications [13,20-23, 28,29].

The observed size distributions fit rather well to bimodal log-normal distribution functions which are conveniently described by five parameters: the fraction of volume contained in the nuclei mode, volume weighted mean diameter of each mode, and the geometric standard deviation of each mode.

Figure 12 is a plot of the volume fraction in the nuclei mode against BMEP for the various fuels. Note that the results of No. 1 and No. 2 diesel fuels

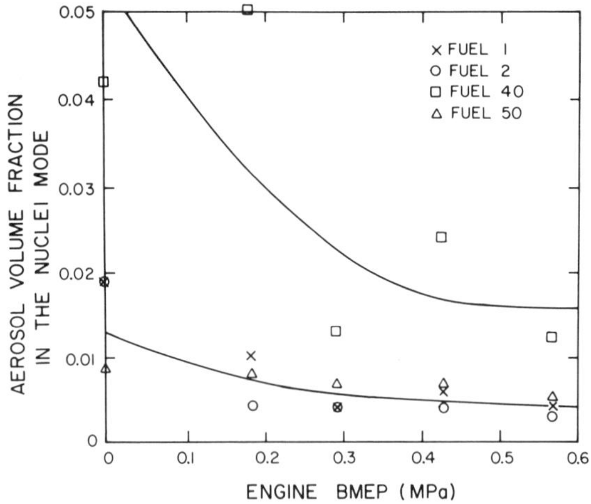

Figure 12. Aerosol volume fraction in the nuclei mode plotted against brake mean effective pressure (load). Engine speed: 1800 rpm.

and 50 cetane number reference fuel are quite similar, with about 1% of the aerosol volume appearing in the first mode. The 40 cetane number reference fuel exhibits quite different behavior. The volume appearing in the nuclei mode decreases with load from about 4% at idle to just over 1% at 0.6 MPa BMEP. In terms of the model described by Dolan [13], the larger volume fraction found in the nuclei for the 40 cetane fuel is consistent with the smaller mass and volume concentrations in the exhaust with this fuel. His model assumes that primary particles formed by combustion are in the nuclei mode and grow into the accumulation mode by coagulation. Coagulation is a second order process and should be faster for the higher particle concentrations observed with the other three fuels and at high loads with the 40 cetane fuel.

Figure 13 consists of plots of volume geometric mean diameter of both the nuclei and accumulation modes against load for the different fuels tested. In general, the modal diameters increase with load, i.e., with increasing particle concentrations. These trends probably result from particle coagulation.

Another interesting characteristic of the exhaust aerosols is the particle mass-to-volume ratio. The mass is obtained from filter samples and the volume, by integration of the volume size distribution given by the EAA.

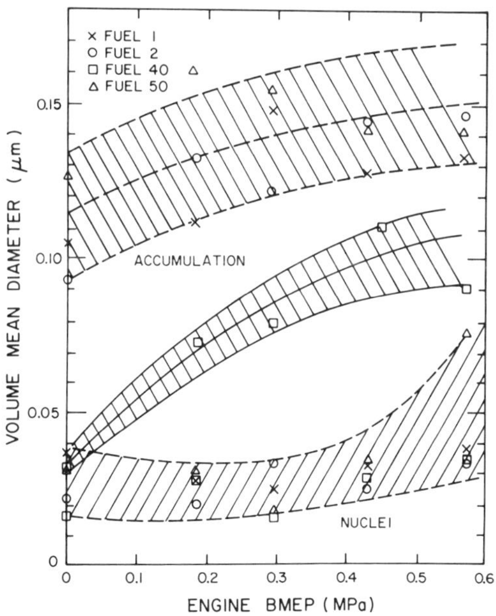

Figure 13. Volume mean diameters of the nuclei and accumulation modes plotted against engine brake mean effective pressure (load). Engine speed: 1800 rpm.

This mass-to-volume ratio is plotted in Figure 14 for all fuels and loads. These curves should not be taken to represent particle density, but rather a comparison between two independent measurement techniques. A comparison of Figures 13 and 14 reveals that the m/v ratio generally decreases with increasing volume mean diameter. Transmission electron microscope photographs, of which Figure 1 is an example, show that the smallest diesel exhaust particles (~ 0.03 μm diameter) are roughly spherical clusters of a few primary particles. As particle size increases, more primary particles are involved, and the agglomerate becomes quite irregular and porous. These observations are consistent with the experimentally observed decrease of particle m/v ratio with increasing particle size.

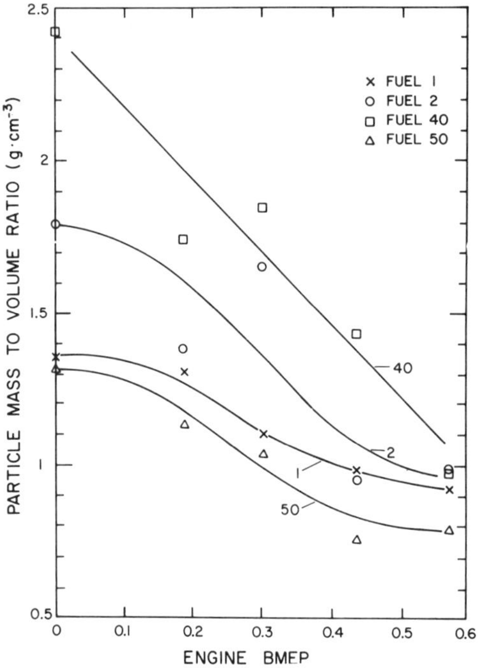

Figure 14. Ratio of filter mass concentrations to electrical aerosol analyzer volume concentrations plotted against brake mean effective pressure (load). Engine speed: 1800 rpm.

Discussion

The results described above show that for the engine and operating conditions examined here, the properties of the fuel can strongly influence both the size and concentration of the particles emitted. For this set of fuels, the most important property appears to be the fuel cetane number. Reducing the cetane number decreases particle mass and volume concentrations in the exhaust. A possible explanation for this decrease is that the longer ignition delay periods associated with reduced cetane number fuels allow more complete mixing of the fuel and air prior to ignition, thus minimizing the amount of fuel burnt under the rich conditions that lead to particle formation.

CONCLUSIONS

The main conclusions of this chapter are given below:

1. Diesel exhaust particles are primarily submicron agglomerates of carbonaceous nuclei. From 5 to 50% of the mass of these agglomerates is solvent extractable. The extracts are mainly hydrocarbons, but also contain oxygen, nitrogen and sulfur.
2. Diesel exhaust aerosols are much more dynamic and susceptible to artifact formation than are gaseous pollutants. Thus, differences between processes which occur during and after dilution in the laboratory and processes which occur over roadways must be carefully considered. Calculations of aerosol growth by coagulation show that markedly larger particles should be formed under typical dilution tunnel/exposure chamber conditions than under roadway dilution conditions. In addition, adsorption/condensation processes and chemical reactions may lead to chemical differences between dilution tunnel and roadway aerosols.
3. The exhaust particle size distributions produced by a small diesel engine were found to depend upon fuel and operation conditions. Such size distribution measurements are important not only for assessing the environmental impact of diesel particles, but also aid in understanding how they are formed by the engine.

ACKNOWLEDGMENTS

This work was supported by a grant from General Motors Research Laboratories. Additional equipment support was provided by U.S. Environmental Protection Agency Grant No. R803851 and Onan Corporation.

REFERENCES

1. Kittelson, D. B., D. F. Dolan, R. B. Diver and E. Aufderheide. "Diesel Exhaust Particle Size Distributions–Fuel and Additive Effects," Society of Automotive Engineers, Technical Paper No. 780787 (1978).
2. Dolan, D. F., and D. B. Kittelson. "Study of Roadway Diesel Aerosols," Society of Automotive Engineers, Technical Paper No. 790492 (1979).

3. Groblicki, P. J., and C. A. Begeman. "Particle Size Variation in Diesel Car Exhaust," Society of Automotive Engineers, Technical Paper No. 790421 (1979).
4. Hare, C. T., K. J. Springer and R. L. Bradow. "Fuel and Additive Effects on Diesel Particulate Development and Demonstration of Methodology," Society of Automotive Engineers, Technical Paper No. 760130 (1976).
5. Laresgoiti, A., A. C. Loos and G. S. Springer. "Particulate and Smoke Emission from a Light-Duty Diesel Engine," *Environ. Sci. Technol.* 11(10):973-978 (1977).
6. Khatri, N. J., J. H. Johnson and D. G. Leddy. "The Characterization of the Hydrocarbon and Sulfate Fraction of Diesel Particulate Matter," Society of Automotive Engineers, Technical Paper No. 780111 (1978).
7. Huisingh, J., et al. "Application of Bioassay to the Characterization of Diesel Particulate Emissions: Part I. Characterization of Heavy-Duty Diesel Particle Emission," presented at Symposium on Application of Short-Term Bioassays in the Fractionation and Analysis of Complex Environmental Mixtures, Williamsburg, VA (1978).
8. Huisingh, J. et al. "Application of Bioassay to the Characterization of Diesel Particulate Emissions: Part II. Application of a Mutagenicity Bioassay to Monitoring Light-Duty Diesel Particle Emissions," presented at Symposium on Application of Short-Term Bioassays in the Fractionation and Analysis of Complex Environmental Mixtures, Williamsburg, VA (1978).
9. Bergstrom, R. W. "The Effect of Carbon Particles on Atmospheric Radiation," Paper No. 38, presented at the Conference on Carbonaceous Particles in the Atmosphere, Berkeley, CA (1978).
10. Lippmann, M. "Size Selective Sampling for Inhalation Hazard Evaluations," in *Fine Particles: Aerosol Generation, Measurement, Sampling, and Analysis*, B. Y. H. Liu, Ed. (New York: Academic Press, Inc., 1976), pp. 287-310.
11. Chang, S.-G., R. Brodzinsky. R. Toossi, S. S. Markowitz and T. Novakov. "Catalytic Oxidation of SO_2 on Carbon in Aqueous Suspensions," Paper No. 19, presented at the Conference on Carbonaceous Particles in the Atmosphere, Berkeley, CA (1978).
12. Hinners, R. G., J. K. Burkart, M. Malanchuk and W. D. Wagner. "Animal Exposure Facility for Diesel Exhaust Studies," Chapter 26, this volume.
13. Dolan, D. F. "Experimental Theoretical Investigation of Diesel Exhaust Particulate Matter," Ph.D. Thesis, University of Minnesota (1977).
14. Fuchs, N. A. *The Mechanics of Aerosols* (New York: Pergamon Press, 1964).
15. Clark, A. *The Theory of Adsorption and Catalysis* (New York: Academic Press, Inc., 1970).

16. Frisch, L. E., J. H. Johnson and D. G. Leddy. "Effect of Fuels and Dilution Ratio on Diesel Particulate Emissions," Society of Automotive Engineers, Technical Paper No. 790417 (1979).
17. Black, F., and L. High. "Methodology for Determination of Diesel Hydrocarbon Emissions–Particulate and Gas Phase," Society of Automotive Engineers, Technical Paper No. 790422 (1979).
18. Bradow, R., F. Black, P. Gabele and J. Ma. "Sampling Diesel Particles," Society of Automotive Engineers, Technical Paper No. 790423 (1979).
19. Pitts, J. N. et al. "Atmospheric Reactions of Polycyclic Aromatic Hydrocarbons: Facile Formation of Mutagenic Nitro Derivatives," *Science* 202:215 (1978).
20. Kittelson, D. B., and D. F. Dolan. "Dynamics of Sampling and Measurement of Diesel Engine Exhaust Aerosols," Paper No. 33, presented at the Conference on Carbonaceous Particles in the Atmosphere, Berkeley, CA (1978).
21. Verrant, J. A., and D. B. Kittelson. "Sampling and Physical Characterization of Diesel Exhaust Aerosols," Society of Automotive Engineers, Technical Paper No. 770720 (1977).
22. Kittelson, D. B., D. F. Dolan and J. A. Verrant. "Investigation of a Diesel Exhaust Aerosol," Society of Automotive Engineers, Technical Paper No. 780109 (1978).
23. Dolan, D. F., and D. B. Kittelson. "Diesel Exhaust Aerosol Particle Size Distributions: Comparison of Theory and Experiment," Society of Automotive Engineers, Technical Paper No. 780110 (1978).
24. Liu, B. Y. H., D. Y. H. Pui and A. Kapadia. "Electrical Aerosol Analyzer: Principle and Data Reduction," presented at Aerosol Measurement Workshop, University of Florida, Gainesville, FL (1976).
25. Verrant, J. A. "Development of a Dilution System for Measuring Diesel Exhaust Particulate Matter," M.S. Thesis, University of Minnesota (1976).
26. Whitby, K. T., and B. Y. H. Liu. "Polystyrene Aerosols–Electrical Charge and Residue Size Distribution," *Atmos. Environ.* 2:106-116 (1968).
27. Liu, B. Y. H., and K. W. Lee. "Efficiency of Membrane and Nuclepore Filters for Submicrometer Aerosols," *Environ. Sci. Technol.* 10:345-350 (1976).
28. Dolan, D. F., D. B. Kittelson and K. T. Whitby. "Measurement of Diesel Exhaust Particle Size Distributions," ASME Paper No. 75-WA/APC-5 (1975).
29. Whitby, K. T. "On the Multimodal Nature of Atmospheric Aerosol Size Distributions," presented at the VIII International Conference on Nucleation, Leningrad, USSR (1973).

CHAPTER 17

METAL-FUME GENERATION AND ITS APPLICATION TO INHALATION EXPERIMENTS

Katsunori Homma, Kiyoyuki Kawai and Kosuke Nozaki

National Institute of Industrial Health
Ministry of Labour
Kawasaki, Japan

INTRODUCTION

The health effects caused by exposure to various kinds of metal fumes in industrial and ambient air environments have recently become of increasing concern in Japan. Experimental studies on this subject have therefore been undertaken. It has been widely recognized that the design of the metal fume generator and the inhalation facility are of great importance for the performance of inhalation toxicology experiments with metal fumes.

Several kinds of generators for experimental inhalation studies with metal fumes have been described in the literature [1-6]. In one of the most commonly used generation methods, metal aerosols are vaporized and subsequently condensed. Frequently this has the disadvantage of yielding polydisperse and/or aggregated particles, depending on the design of the heating column. To overcome this difficulty, Homma [7] designed a new type of generator using a high-frequency induction furnace. The principal mechanism of this metal-fume generation method is the condensation of metal vapors without foreign nuclei.

This report deals with the recent studies and developments in the authors' laboratory and includes a description of the evaporation-condensation type metal-fume generator and of the inhalation chambers used. The principle of application with this and other types of generators for short- and long-term inhalation experiments is also described.

PRINCIPLE AND CONSTRUCTION OF THE GENERATOR

The principal components of the evaporation-condensation type generator are (1) a high-frequency oscillator, (2) a metal-fume-generating furnace, and (3) a carrier-gas flow system, as schematically illustrated in Figure 1. Details of the induction-furnace construction and a cross-sectional view are shown in Figure 2.

A temperature gradient is established in the center portion of the long cylindrical furnace through high-frequency inductive heating. Any metal placed into the graphite crucible at the bottom of the furnace is thus heated to the melting point and is subsequently evaporated. Before the metal vapor reaches the cooling and condensation zone at the top of the furnace, the vapor is carried through a zone of further heating, in which the temperature is higher than at the molten-metal surface. Because of this additional heating, metal fumes with a narrow size distribution have been generated successfully. This method has been applied satisfactorily to metals having low melting points and large temperature differences between melting and boiling points.

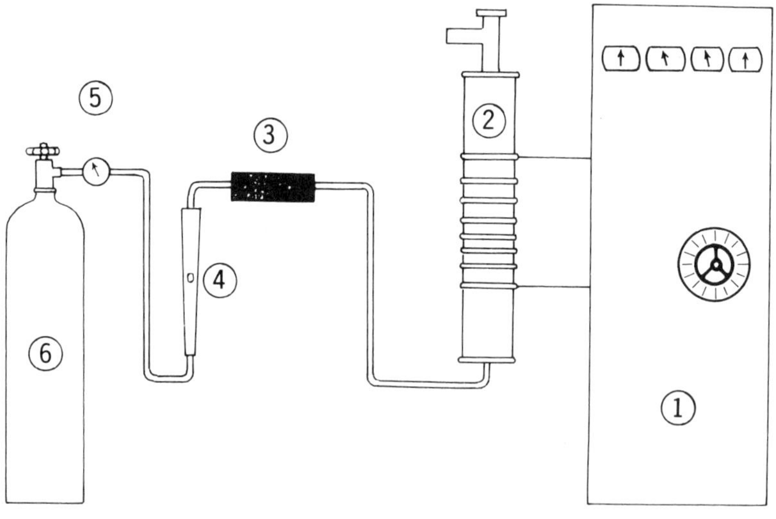

Figure 1. Schematic diagram of the metal-fume-generating apparatus: 1. high-frequency oscillator, 2. metal-fume-generating furnace, 3. gas drier (silica gel), 4. rotameter, 5. pressure gauge and regulator, 6. nitrogen gas.

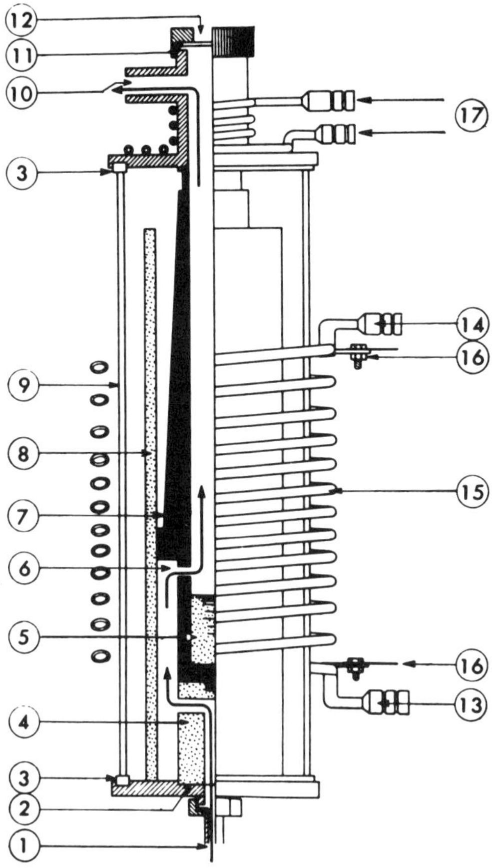

Figure 2. Design details of the fume-generating furnace: 1. carrier-gas inlet, 2. high-efficiency glass-fiber filter, 3. silicone rubber packing, 4. ceramic base, 5. melting metallic material, 6. opening on graphite crucible for carrier-gas inlet, 7. graphite crucible, 8. ceramic tube, 9. quartz tube, 10. outlet of generating fumes, 11. silica glass plate, 12. inserting hole for thermocouple, 13. inlet of cooling water for heating coil, 14. outlet of cooling water, 15. heating coil, 16. terminal of high-frequency output, 17. cooling pipe for top flange.

The graphite crucible has an inner diameter of 18 mm, a wall thickness of 5 mm and a depth of 230 mm. It is placed inside a quartz tube. Placed between the graphite crucible and the quartz tube is a ceramic tube, to prevent heat loss from the graphite crucible and to protect the quartz tube from thermal damage. The top and bottom of the quartz tube are tied to flanges through packings of silicone rubber. The top flange is cooled with running water through the cooling pipe.

There are two holes at the top of the furnace. One is for the outflow of the generated fumes, and the other is for inserting a thermocouple or an observation window for optical pyrometry. The latter hole is kept closed during fume generation.

The furnace is heated by loading the graphite crucible with a high frequency wave, which is induced by putting a corresponding current of high frequency through the heating coil around the quartz tube. A furnace with a frequency of 400 kHz is able to provide a maximum temperature of the metal in the crucible of up to 1700°C when the routine output power of 3.0 kW is applied.

The temperature in the crucible is proportional to the voltage applied to the heating coil, and a temperature gradient is formed along the whole length of the furnace, as seen in Figure 3. The temperature measurements in

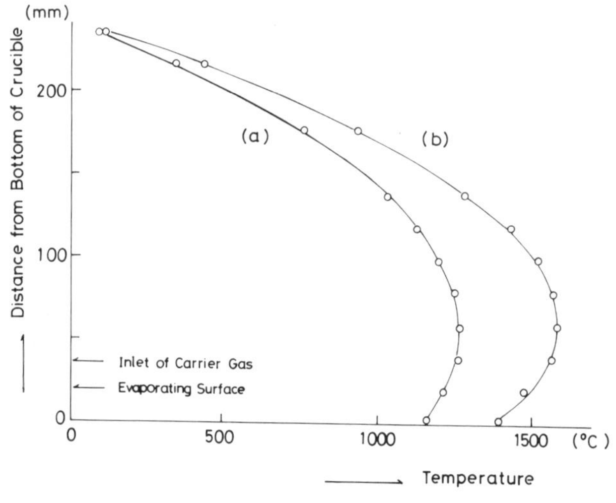

Figure 3. Temperature gradient in graphite crucible at routine output voltages. (a) 4.2 kV, (b) 5.3 kV.

the center duct were made with a thermocouple (Pt 60-Rh 40, Pt 80-Rh 20; Johnson Matthey Metals Ltd., 100 High St., Southgate, London N14 6ET, England). The temperature of the evaporating surface of the molten metal was measured with an optical pyrometer during fume generation.

Our usual choice of carrier gas is notrogen, which is passed through a drier and a high-efficiency air filter (Gelman Type A) before entering the furnace.

PERFORMANCE CHARACTERISTICS OF THE APPARATUS FOR LEAD-FUME GENERATION

The fume-generating characteristics of this apparatus will be described for lead as a representative example. A precise amount, 60 g of 99.99% pure metallic lead (Johnson Matthey Chemicals Ltd., Orchard Rd., Roystone, Hertfordshire, SG8 5HE, England), was loaded into the graphite crucible in order to fill up to a predetermined level of the evaporating surface (approximately 20 mm from the bottom of the crucible). The temperature at the surface of the molten lead reached its equilibrium about 20 min after establishing the output voltage of the high-frequency oscillator.

The mass median aerodynamic particle diameter (MMAD) and the size distribution of the generated lead fumes were examined for various temperatures and for various flow rates of the carrier gas. The lead fumes were collected by use of a thermal precipitator (Shibata Chemical Apparatus Mfg. Co., Ltd., 1-25 Ikenohata 3, Taitoh-ku, Tokyo, Japan), operated at a sampling flowrate of 0.5 liter/min. The fumes were collected onto specimen grids covered with collodion film.

The count median particle diameter (CMD) and the particle-size distribution were measured electronmicroscopically. Then, the mass median diameter (MMD) was calculated using the Hatch-Choate equation [8]. The corresponding aerodynamic resistance diameter was calculated as described by Raabe [9]. The density, ρ, of particles was assumed to be the bulk density [10] of lead of 11.34 g/cm^3.

The relationship between the MMAD of the generated lead fumes and the temperatures of the evaporating surface are shown in Figure 4. Particle-size distribution curves, plotted on log-normal probability paper, are shown in Figure 5. As seen from those figures, the most favorable size distributions with geometric standard deviation (σ_g) values under 1.3 were obtained when the flowrates of the carrier gas were kept between 0.5 and 1 liter/min. Variations in MMAD were achieved by changing the furnace temperature.

The amount of generated lead fume, i.e., the generation rate, was dependent on the evaporation temperature as well as the flowrate, as shown in Figure 6. When the flowrate was chosen in the range of 0.5 to 1 liter/min, the concentration of generated fume showed a linear correlation with the MMAD from 0.3 to 3 μm (Figure 7).

The flowrate of the carrier gas played a significant role in the operation of this generator. When the flowrate was below 0.5 liter/min, the generated fumes could not be conveyed to the outlet of the generator because the particles have too high a gravitational settling velocity at the temperatures used.

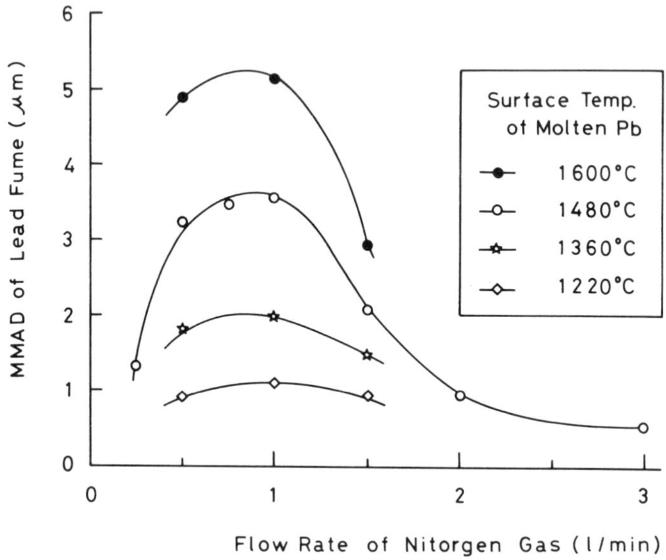

Figure 4. Relationship between MMAD of lead fume and carrier-gas flowrate for different generating temperatures.

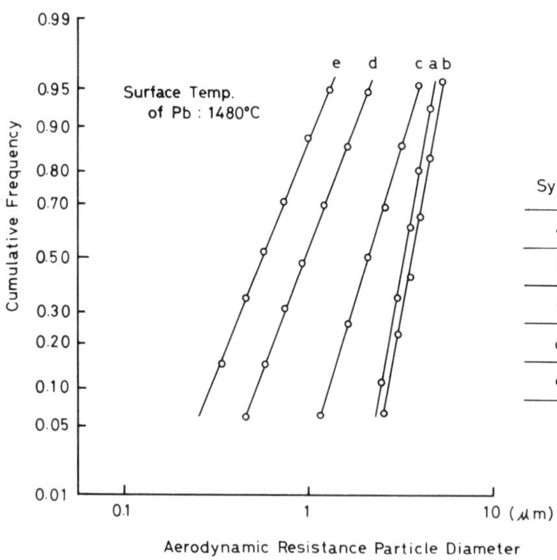

Symbol	Flowrate (l/min)	MMAD (μm)	σ_g
a	0.5	3.25	1.24
b	1.0	3.57	1.26
c	1.5	2.07	1.45
d	2.0	0.95	1.60
e	3.0	0.55	1.69

Figure 5. Size distribution curves of lead fumes for different flowrates of nitrogen carrier gas.

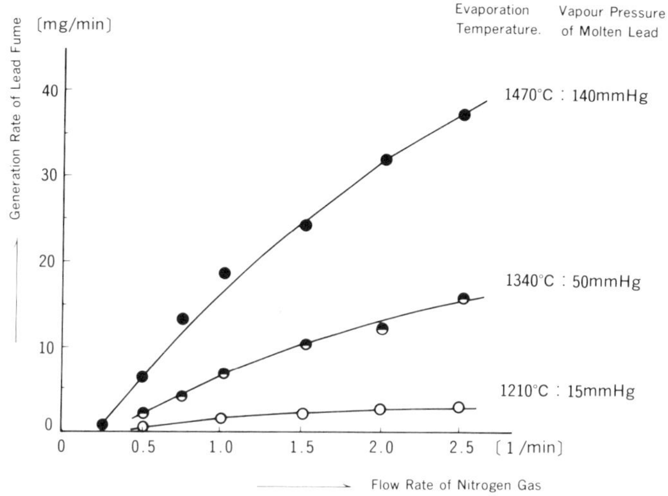

Figure 6. Relationship between generation rate of lead fume and flowrate of nitrogen carrier gas for different temperatures.

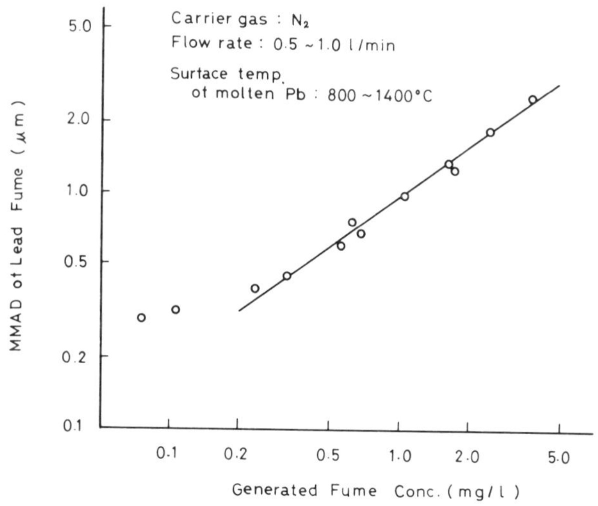

Figure 7. Relationship between MMAD of lead fume and generated fume concentration.

On the other hand, an increase in flowrate of the carrier gas resulted in an increase in the amount of lead fumes generated at every evaporating temperature above about 1200°C. When the flowrate exceeded 2 liter/min, the particle size decreased, since the flowing nuclei and the lead fumes did not stay in the growing zone of the crucible long enough. The values of the geometric standard deviation (σ_g) are also correspondingly increased (Figure 5).

Electron micrographs of various lead fumes for different flowrates of the carrier gas at the same evaporating temperature are shown in Figure 8. The size distribution of the particles in the upper photograph, labeled 1, appears to be nearly monodisperse.

Since the volume of the carrier gas expands to six times its original volume when brought to the maximum temperature of 1600°C in the heating zone, the transport velocity of the carrier gas, together with the lead vapor and the generating condensation nuclei, may be increased nearly six times. For flowrates exceeding 2 liter/min the residence time of the flow through the condensation zone is assumed to be less than 0.3 sec, which is too short to give a favorable size distribution. Turbulent flow in the crucible may play a role in not achieving homogeneous condensation.

From the results shown, it is reasonable to conclude that the flowrate of the carrier gas should be chosen to be between 0.5 and 1 liter/min to obtain a narrow size distribution when generating lead fumes. It was also shown that for this optimum range of flowrates, the relationship between the MMAD and the concentration of lead fumes is linear, irrespective of the rather wide range of temperatures of the molten lead surface (Figure 7).

FUME GENERATION FROM OTHER METALS OR METAL HALIDES

Fumes were also generated successfully from other metals or metal halides having sufficiently high vapor pressures to produce fumes under about 1500°C when the generator was operated at the temperature corresponding to their melting point.

Metals with rather low melting points and large temperature differences between their melting and boiling points, e.g., bismuth and indium, also resulted in satisfactorily spherical and monodisperse aerosol particles, under almost the same generating conditions as lead. In the case of tellurium, it was necessary to add electrically-sputtered gold as condensation nuclei to obtain sufficiently monodisperse aerosols.

From some metals, e.g., zinc, cadmium and antimony, the results obtained have not been satisfactory so far in that they have yielded crystalline and polydisperse aerosols.

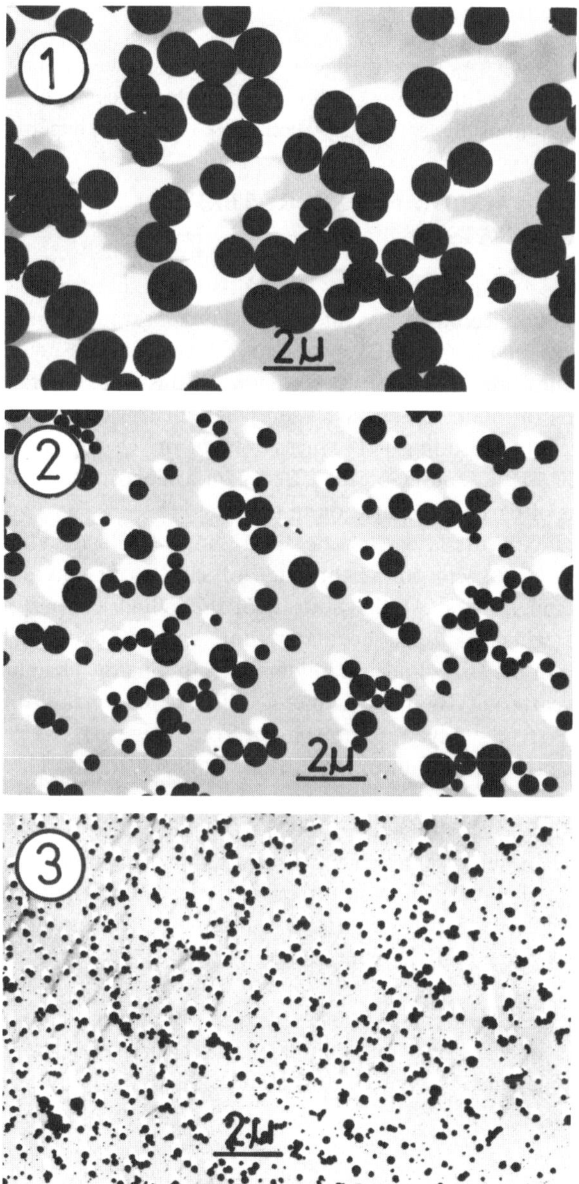

Figure 8. Electron micrographs of lead fumes generated at different flowrates of the carrier gas at a constant temperature of 1480°C. 1. 0.5 liter/min, 2. 1.5 liter/min 3. 3.0 liter/min.

Furthermore, this generator is able to generate fumes of several metal halides such as NaCl, KCl, PbI_2 and $CdCl_2$ without changing their original chemical character. As seen in Figure 9, the aerosols were spherically shaped or were of crystalline form. Deliquescent substances such as $SbCl_3$ and thermally decomposable substances such as VCl_3 were an exception.

INHALATION FACILITY FOR SHORT-TERM EXPOSURE TO RATS

When this generator was used for short-term exposure of two to three hours, it was connected to a cylindrical head-exposure-type inhalation chamber, as shown in Figure 10. Appropriate diluting devices could be inserted instead of the aerosol control valve shown. In this way, several studies with various kinds of metal fumes have been performed in our laboratory: examination of acute poisoning [11], measurement of the deposition fraction in the lung [12, 13], studies of the effects on alveolar macrophage [14] and observations of particle localization in the lung [15].

The head-exposure-type chamber is a simple vertical cylinder of polyvinyl chloride, 36 cm in height and 40 cm in diameter, with conical shaped in- and outlets. At the bottom of the cylindrical part is a fine wire mesh, onto which rats can be placed unrestrained. Usually, however, six rats are held in the six cylindrical holders surrounding the chamber.

A light-scattering Digital Dustmeter (Shibata Chemical Apparatus Mfg. Co., Ltd., 1-25, Ikenohata 3, Taitoh-ku, Tokyo, Japan), has generally been used for continuous monitoring of fume concentrations. The weight concentrations have been determined by sampling onto glass fiber filters (Millipore AP 20055) with a flowrate of 20 liter/min.

When the outlet of this generator was connected directly to the top of the exposure chamber, the concentrations of lead fumes reached an equilibrium state in less than 10 minutes. For instance, the concentrations of lead fumes of 0.45-μm MMAD remained constant at 18 ± 1.4 mg/m^3 during 90 minutes of exposure.

This head-exposure method resulted in more uniform deposited fractions in the lungs of individual rats compared to the individual variations found when the same chamber was used as a common whole-body-exposure chamber (Table I).

INHALATION FACILITY FOR LONG-TERM EXPOSURE

In applications of the above and other types of fume generators to long-term inhalation studies we have used a combination of chambers for concentration regulation and whole-body exposure. The principal components of

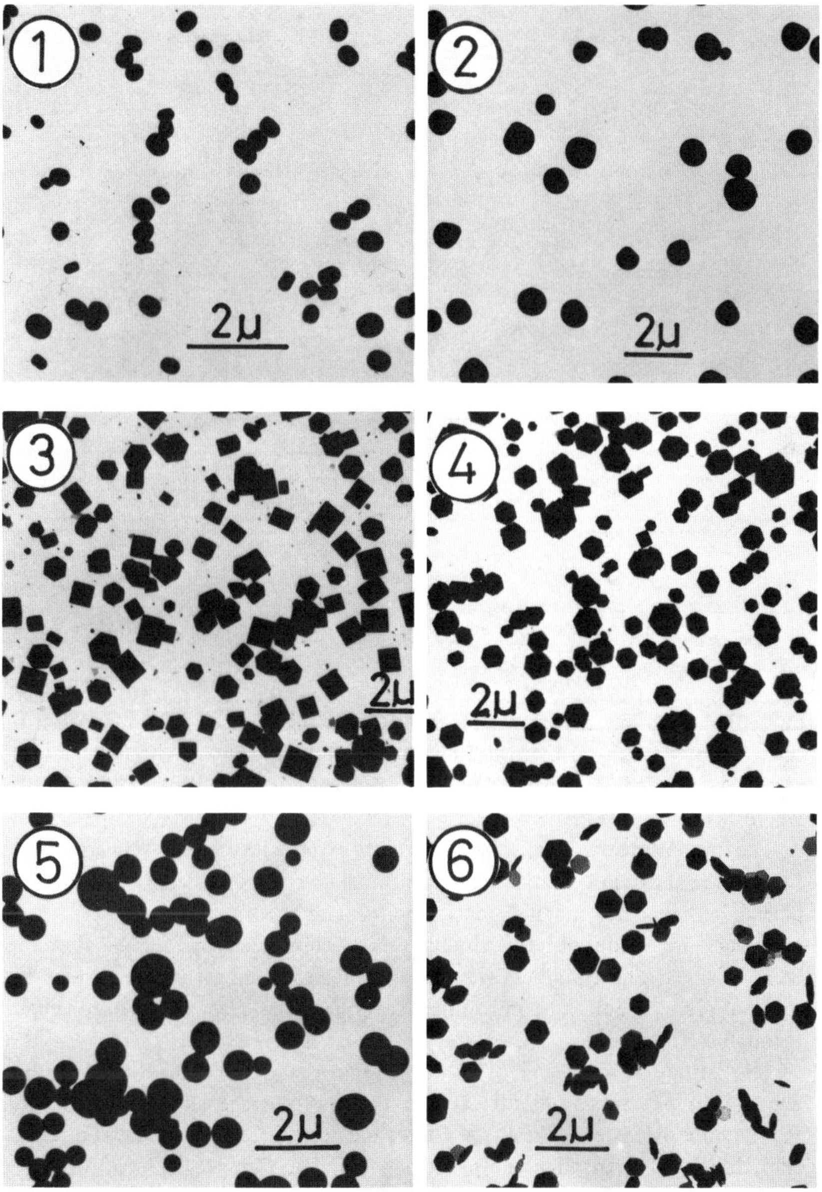

Figure 9. Electron micrographs of several kinds of metal fumes and metal halide fumes: 1. indium, 2. tellurium (with sputtered gold nuclei), 3. cadmium, 4. zinc, 5. potassium chloride, 6. lead iodide.

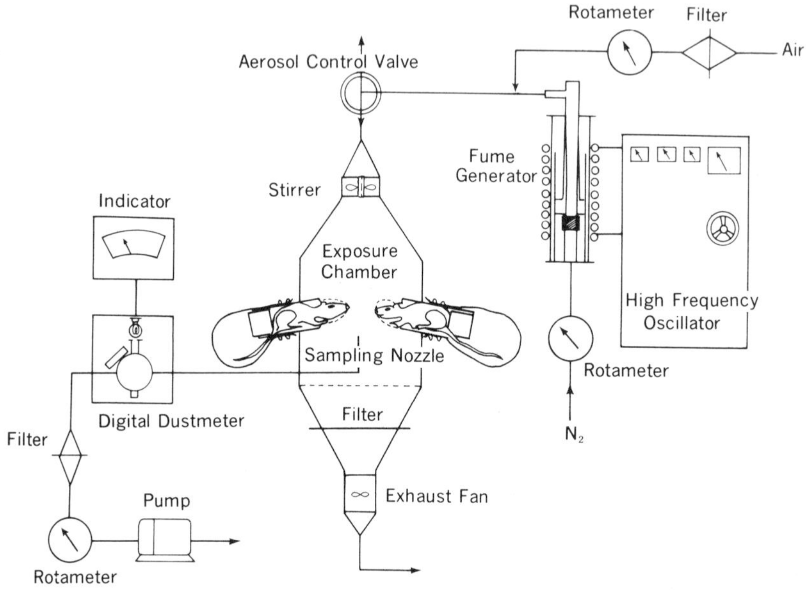

Figure 10. Block diagram of metallic-fume inhalation apparatus.

our inhalation system are: (1) metal fume generator; (2) concentration control apparatus; (3) exposure chambers; and (4) exhaust or suction apparatus. A detailed description of our inhalation system has appeared elsewhere [16].

The system for concentration regulation is schematically illustrated in Figure 11. Two large chambers, a storage and a regulation chamber, are placed between the metal fume generator and the exposure chamber. Both chambers are made of polyethylene sheet supported by a cubical framework of steel. The chambers were kept under slight negative pressure by the suction system, and air-conditioned room air was allowed to enter freely through a HEPA filter.

The outlet of any kind of fume generator, including the one described above, can be connected directly to the first storage chamber (K). After this chamber has been filled, the fumes are transferred to the second regulation chamber (L) through the processor (N). The fume concentration in the regulation chamber is monitored continuously by use of a light-scattering-type dustmeter. The aerosol concentrations in the regulation chamber are continuously compared in the controller with two previously selected values, which serve as the upper and the lower limits of allowable fluctuations.

Table I. Lead-Fume Deposition in the Rat Lung–Comparison Between Whole-Body-Exposure and Head-Exposure Inhalation

Particle Size[a] (μm)	Lead Fume Concentration (mg/m^3)	Exposure Period (min)	Inhalation form	Deposited Lead Fume in Lung[b] (μg)	Deposition Fraction in Lung[c] (%)	Coefficient of Variation (%)
0.45	18	90	Whole-body-exposure	40.9 ± 8.5	25.2	20.8
0.45	18	90	Head-exposure	50.3 ± 7.2	32.7	13.6
2.0	50	90	Whole-body-exposure	53.6 ± 18.0	8.9	33.3
2.0	50	90	Head-exposure	82.6 ± 23.0	13.8	27.6

[a]MMAD.
[b]Mean ± standard deviation.
[c]Deposition fraction of lead fume was calculated by assuming 100 ml/min of breathing air volume in the rats of about 280 g body weight used in these experiments.

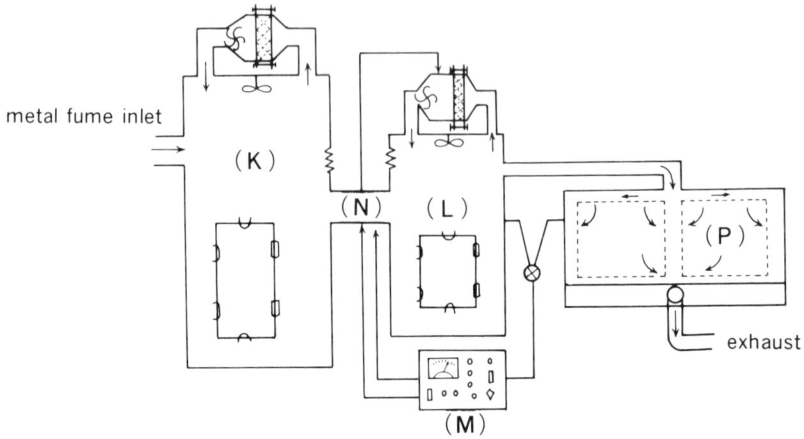

Figure 11. Aerosol concentration control system. K: fume storage chamber, L: regulation chamber, M: controller, N: processor, P: exposure chamber.

When the measured concentration deviates from the narrow fluctuation range, the feedback processor between the storage and regulation chambers adjusts the concentration in the regulation chamber to the appropriate level. The working range of this regulator is 0.05-40 mg/m^3. It is also effective to some extent below 0.05 mg/m^3.

Figure 12, obtained [17] as part of an experimental inhalation study with cadmium fumes at low concentration (0.05 mg/m^3), illustrates the regulation capacity of the above system. In this intermittent daily exposure experiment for three months, cadmium fumes were supplied from an automated electric-furnace generator, which yielded particles of 0.089 μm MMAD with σ_g = 1.78. Observed average concentrations in the exposure chambers were 56.8 ± 5.4 μg/m^3 (mean ± SD) and 27.3 ± 6.1 μg/m^3 corresponding to desired levels of 50 and 25 μg/m^3 respectively. The fluctuation range did not exceed ±10% during exposure as partly shown in Figure 12. As another example for the performance of this type of regulation system, 0.47 ± 0.026 mg/m^3 were maintained versus a desired concentration level of 0.5 mg/m^3 throughout 171 weeks of life-long continuous exposure to soot containing heavy-oil combustion products [18]. The regulator kept the chamber concentration steady, although the soot generator output fluctuated widely.

The above examples were selected to show the concentration regulation facility of the authors' system, especially at rather low concentrations of particulate matter, including fumes. The space necessary for the storage and regulation chambers can be estimated before the start of an experiment by calculating the necessary volumetric flowrate which is determined by the

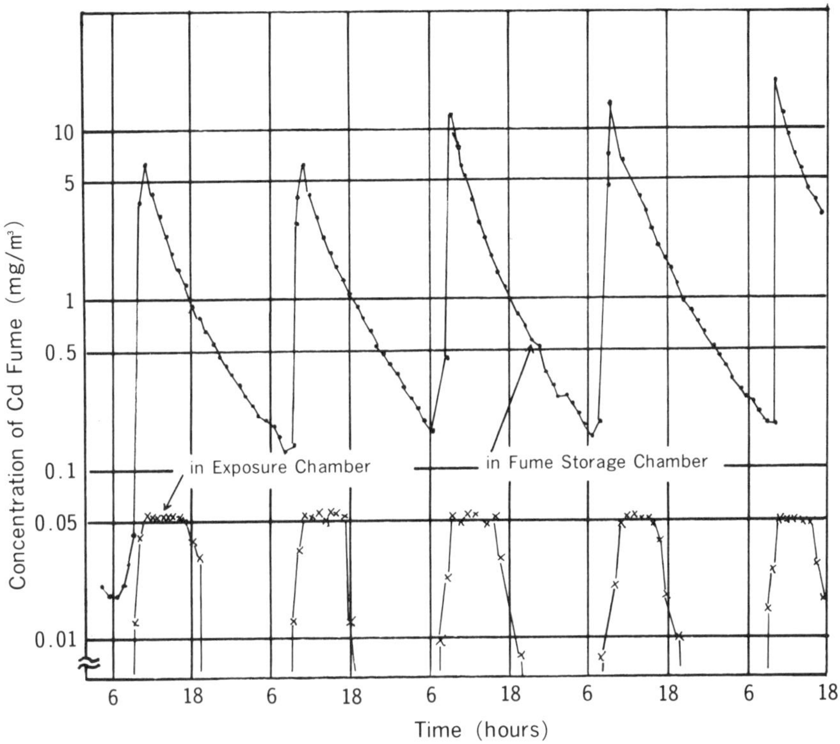

Figure 12. Regulating characteristics of the regulation chamber. Intermittent daily (6 hrs) exposure to cadmium fumes. Top: Fume storage chamber, Bottom: Animal exposure chamber.

number and space of exposure chambers and by the air exchange rate. Reconstruction of the polyethylene-sheet covered chamber is easy and inexpensive.

SUMMARY AND CONCLUSION

This chapter has presented recent studies and developments in the authors' laboratory on metal-fume aerosol generation and its application to animal exposure experiments.

The design and performance characteristics of a newly developed metal-fume generator which uses a high-frequency induction furnace for heating have been described. This generator performs well in yielding monodisperse metal fume aerosols of several kinds of metals and metal halides having rather low melting points and rather large differences between their melting and boiling points.

Applications of this and other types of fume generators to short- and long-term exposure experiments with animals have been described briefly. A concentration-regulation system for long-term inhalation studies has also been described with a brief presentation of its regulating capacity during actual long-term inhalation experiments.

REFERENCES

1. Espenscheid, W. F., E. Matijevic and M. Kerker. "Aerosol Studies by Light Scattering, III. Preparation and Particle Size Analysis of Sodium Chloride Aerosols of Narrow Size Distribution," *J. Phys. Chem.* 68: 2831-2842 (1964).
2. Sutugin, A. G., and N. A. Fuchs. "Formation of Condensation Aerosols at High Vapor Supersaturation," *J. Colloid Interface Sci.* 27(2): 216-228 (1968).
3. Sano, I., Y. Hobo, Y. Fujitani and A. Kozima. "An Experimental Study on the Formation of Lead Aerosols," *J. Res. Assoc. Powder Tech.* (in Japanese) 7:23-28 (1970).
4. Kanapilly, G. M., O. G. Raabe and G. J. Newton. "A New Method for the Generation of Aerosols of Insoluble Particles," *J. Aerosol Sci.* 1:313-323 (1970).
5. Friedman, J. M., and S. W. Horstman. "A Simple Method for the Generation of Metallic Oxide Aerosols," *Am. Ind. Hyg. Assoc. J.* 35:825-831 (1974).
6. Sokolowski, M., A. Sokolowska, A. Michalski and B. Gokieli. "The In-Flame-Reaction Method for Al_2O_3 Aerosol Formation," *J. Aerosol Sci.* 8:291-230 (1977).
7. Homma, K. "Experimental Study for Preparing Metal Fumes," *Ind. Health* 4:129-137 (1966).
8. Hatch, T., and S. P. Choate. "Statistical Description of the Size Properties of Non-uniform Particulate Substance," *J. Franklin Inst.* 207: 369-387 (1929).
9. Raabe, O. G. "Aerosol Aerodynamic Size Conventions for Inertial Sampler Calibration," *J. Air Poll. Control Assoc.* 26:856-860 (1976).
10. Japan Chemical Society, Ed., *Handbook of Chemistry: Fundamental Part I* (Tokyo: Maruzen Co., Ltd., 1970), p. 44.
11. Yoshikawa, H., and K. Homma. "Toxocity of Inhaled Metallic Cadmium Fumes in Rats," *Jap. J. Ind. Health* (in Japanese) 16(3):212-215 (1974).
12. Homma, K. "Lung Retention of Metallic Fumes, II. Lung Retention Rate of Monodisperse Spherical Metallic Fumes in the Lungs of Rats," *Jap. J. Hyg.* (in Japanese) 30(6):632-638 (1976).

13. Homma, K. "Lung Retention of Metallic Fumes, III. Lung Retention Rate of Polydisperse Irregular Fumes in the Lungs of Rats," *Jap. J. Hyg.* (in Japanese) 30(6):639-644 (1976).
14. Koshi, K., K. Homma and H. Sakabe. "Response of Alveolar Macrophage to Metallic Fume," *Ind. Health* 13:37-49 (1975).
15. Kyono, H., K. Homma, T. Nagatani, T. Watanabe and K. Kawai. "Localization of Inhaled Lead Particles in the Rat Lung," *Ind. Health* 12:49-72 (1974).
16. Nozaki, K. "Inhalation Apparatus for Long-term Exposure with Aerosol Concentration Regulation Facility," *J. Ind. Hyg. Jap.* (in Japanese) 12:35-40 (1971).
17. Yoshikawa, H., K. Kawai, K. Nozaki and M. Ohsawa. "Report on the Health Effect of Cadmium," Japan Environment Agency (Tokyo, Japan 1976).
18. Nozaki, K. "Experimental Studies of Long-term Aerosol Inhalation," *Jap. J. Ind. Health* (in Japanese) 21:349-360 (1979).

CHAPTER 18

CONSUMER PRODUCT AEROSOLS: GENERATION, CHARACTERIZATION AND EXPOSURE PROBLEMS

Brian V. Mokler, Edward G. Damon,
Thomas R. Henderson and Robert K. Jones

Inhalation Toxicology Research Institute
Lovelace Biomedical and Environmental Research Institute
Albuquerque, New Mexico 87115

INTRODUCTION

The widespread use of aerosolized products represents a potential inhalation exposure hazard for consumers. The diversity of products available leads to a possibility of exposure for nearly all segments of the population. Although the rate of growth of aerosolized product use has recently slackened in the U.S., it is still strong in many other countries. The use of aerosolized products is accompanied by the generation of small particles that, rather than serving the intended application of the product, may reach the consumer's breathing zone and be inhaled. Many of the constituents of these particles are relatively nonvolatile materials that are also insoluble under physiological conditions. If these materials are deposited in the nonciliated portion of the lung, they may remain there for a long time. Whether long-term retention of these materials represents a threat to the health of the individual exposed during the normal use of pressurized products is not presently known. The experience of such users suggests there is no short-term, immediate hazard, but the results of 30 or more years exposure are not known, nor can they be estimated on the basis of presently available published information.

The purpose of this paper is to show how a general approach to inhalation toxicology studies was used in a study of aerosolized consumer products. The approach provides a means for including factors related to aerosol size, deposition and clearance characteristics, exposure conditions and patterns, and species differences in studies of the effects of inhaled materials. The experimental results, because these factors have been explicitly considered, should be applicable to the assessment of long-term health consequences of repeated, normal use of aerosolized products.

GENERAL APPROACH

The application of results from experimental toxociology studies to human health and safety problems is beset by a host of difficulties and uncertainties. The approach outlined in Figure 1 emphasizes points that must be explicitly addressed to minimize the uncertainties. Central features of the approach are development of data to permit estimation of the concentration of material retained in the human lung after protracted exposure and use of

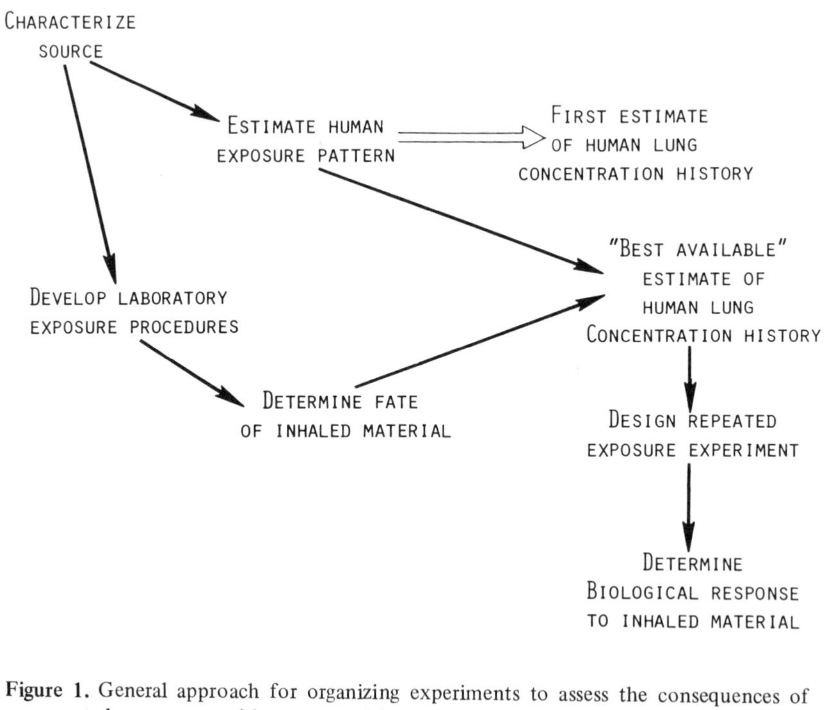

Figure 1. General approach for organizing experiments to assess the consequences of repeated exposure to airborne materials.

this concentration as the basis for designing repeated exposure studies in experimental animals. The required information is developed from the results of two lines of investigation that begin with the characteristics of a product or material of interest. The physical, chemical and biological data are integrated through application of a numerical simulation model. One output from the simulations is a "best estimate" of the concentration of retained material that will develop in the human lung as exposure is continued. Another output from the model can be an exposure regimen for experimental animals such that the pattern of accumulation in their lungs will mimic that estimated for humans.

The application of this general approach in studies with aerosolized consumer products will be described. The various steps, through the repeated exposure design phase, will be illustrated with results for a single product type, antiperspirants.

SOURCE CHARACTERISTICS

The first information necessary in describing human exposure is the nature of the particles that might be deposited in the respiratory tract. This includes data on the aerosol size distribution, particle concentration, and particle composition.

When we initiated these studies in 1974, almost no data were available describing these parameters for either product users or others in the vicinity of product use. Characteristics have typically been obtained for the size distribution of the entire product spray [1-4], although some data are available describing the aerosol in the user's breathing zone [5] or the entire respirable mass output of a pressurized package [6]. The breathing zone data, unfortunately, were obtained under use conditions for only one product and reported in terms of particle number concentration, all particles between 0.3 and 5 μm diameter being treated as a single class. The available respirable mass measurements are also not satisfactory for estimating human exposure conditions because the entire output has been collected rather than only that reaching the user's breathing zone. The total spray measurements are clearly not applicable in describing the aerosol to which users might be exposed. Our first step, therefore, was to develop a laboratory method capable of providing information on the characteristics of particles that might be inhaled by the user of an aerosolized product.

The breathing zone sampling approach of occupational exposure monitoring was adapted to provide experimental conditions that were both physically similar to those under which people might be exposed and reproducible. The arrangements of product, sampler and intended spray target were designed to simulate the "worst reasonable" conditions under which a product might be used. We have not attempted to simulate accidental or deliberate misuse of the products studied. The primary design requirements, in addition to geometrical similarity, were that ventilation of the test area be controlled and reproducible and that the sampling system be convenient to use. These considerations led to the use of a 1-m^3 chamber with different arrangements appropriate to the product type being studied. The arrangement for characterizing antiperspirants is shown in Figure 2. A detailed description of the experimental approach and procedures has been published [7].

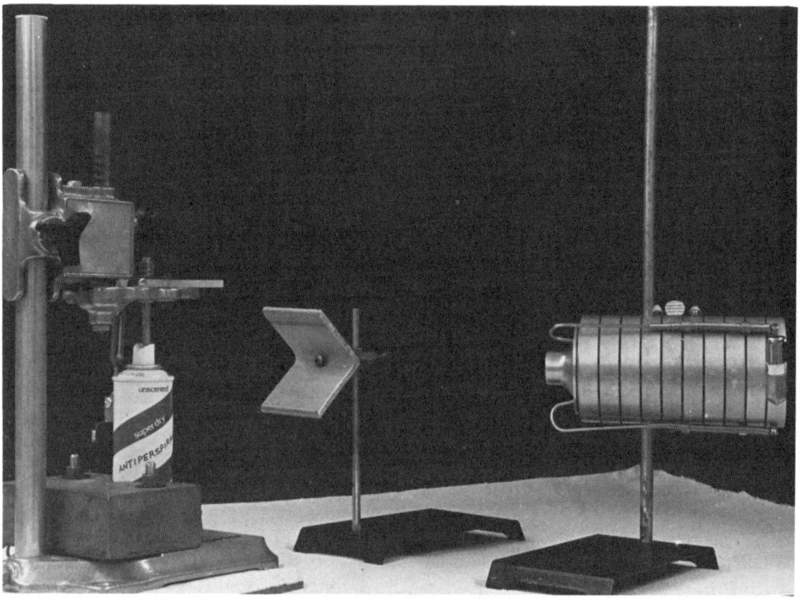

Figure 2. Experimental arrangement for simulated breathing zone characterization of antiperspirants. The distance from the can nozzle to the target is 15 cm. The distance from the nozzle to the impactor inlet (simulated breathing zone) is 33 cm.

The simulated breathing zone approach has been used to characterize a wide variety of pressurized consumer products. Results obtained under standardized conditions are summarized in Table I. Each product type was represented by a single, major brand. The aerosols did not generally appear to be well described by the log-normal distribution function used to fit the cascade impactor data because of an enhancement of the fine particle fraction. It appeared more useful, in estimating human exposure conditions, to deal with the concentration of particles in a few size intervals spanning the

Table I. Simulated Breathing Zone Aerosol Characteristics

Product Type	Typical Mass Median Aerodynamic Diameter (μm)	Concentration of Particles (mg/m^3)		
		<1 μm	1-3 μm	3-6 μm
Air freshener	6	1	4	9
Antiperspirant	6	7	42	74
Dusting aid	7	2	9	26
Fabric protector	3	1	3	2
Furniture wax	3	2	6	7
Hair spray	6	2	5	9
Paint	8	4	15	42
Wood panel wax	8	<1	1	3

respirable range. As seen in Table I, the potential exposure environment varies widely between different product types.

One of the difficulties in describing human exposure to these aerosols is the range of conditions, techniques and other variables that can affect the size and concentration of particles reaching the breathing zone. The experimental chamber allowed these factors and the magnitude of their influence to be studied in a well-controlled manner [7, 8].

The product types listed in Table I contain a variety of ingredients. Depending on the product, pressurized packages may, in addition to the propellant and volatile materials, include inorganic compounds, pigments, synthetic resins, waxes, surfactants, high boiling point hydrocarbons and a broad spectrum of organic compounds.

Considering antiperspirants specifically, results from trials with a major brand under standard conditions are shown in Figure 3 together with the log-normal distribution line for the average value of the fitted parameters. The deviation from log-normality is typical of that observed with most of the products characterized. The size distribution characteristics are essentially independent of the prior consumption of the can contents (Figure 4) and are only slightly affected by the antiperspirant brand (Figure 5). Experiments in a small (8-m^3), well-ventilated bathroom showed that the size distribution characteristics were not altered by the change of surroundings. The concentration, although it dropped from 246 mg/m^3 in the chamber to 52 mg/m^3 in the bathroom, was not reduced to the extent that might be expected on the basis of the volume and ventilation changes. A smaller range of concentration values was observed for the different brands measured in the chamber. Results from the 1-m^3 chamber are thus representative of the particle size distribution to which a consumer is exposed but the concentrations exceed those likely during normal use.

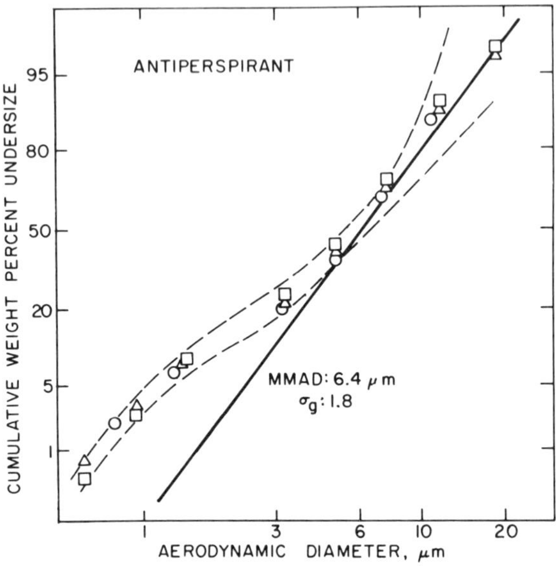

Figure 3. Cumulative size distribution data summary for an aerosolized antiperspirant. Data from only three of the six trials are shown for clarity; all data lie between the two dashed lines. The solid line is the log-normal distribution calculated from the average values of the fitted parameters.

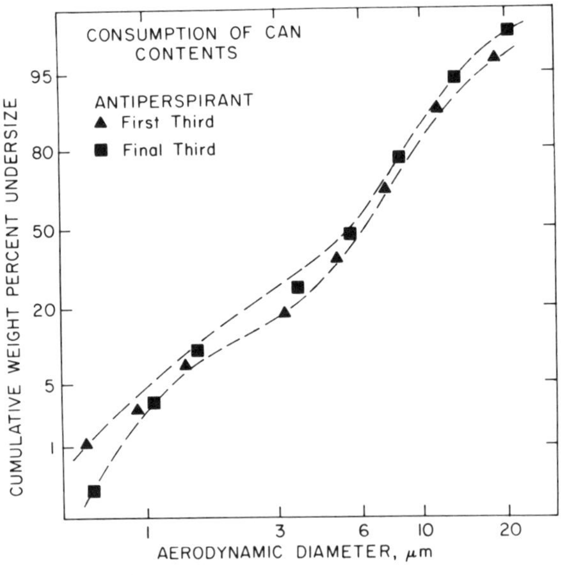

Figure 4. Effect of the consumption of antiperspirant can contents on simulated breathing zone size distribution. The data from the middle third also fall within the dashed lines.

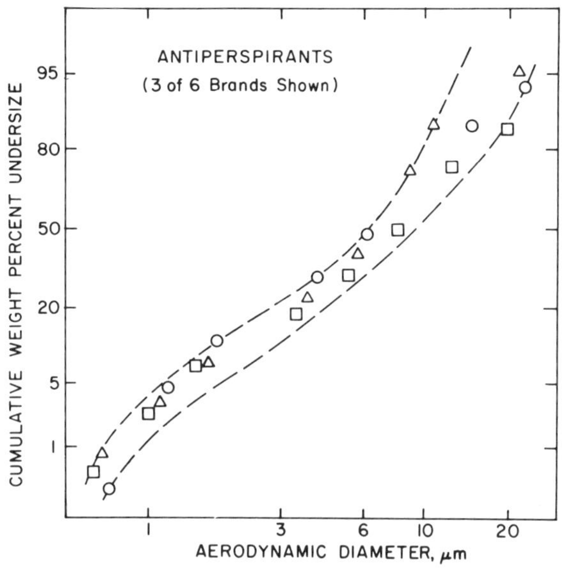

Figure 5. Simulated breathing zone size characteristics of six brands of antiperspirant. All data lie within the dashed lines but only three sets are shown for clarity.

Antiperspirants are generally formualted with an aluminum salt as the active ingredient. Aluminum chlorhydrate (ACH), a partially hydrated form of the composition $Al_2(OH)_5Cl{\cdot}nH_2O$ (n = approximately 4), is currently the most widely used compound. The ACH is suspended as small solid particles in the liquefied propellant in a pressurized package. The other principal nonvolatile ingredient in antiperspirants is an emollient, typically a fatty acid ester such as isopropyl myristate (IPM). Because many of the original droplets in the discharged spray are quite small, ACH particles may be excluded from them. Thus, although the product and the major portion of the spray contained approximately equal proportions of ACH and IPM, particles smaller than 5 μm were approximately 10-25% ACH [9].

HUMAN EXPOSURE PATTERN

The second type of information necessary in describing human exposure is the frequency and duration of exposure. The amount of material deposited in the lungs during each exposure is governed by the aerosol characteristics and the duration of exposure. The nature of the deposited material and the frequency of exposure are important factors in determining the amount of material retained after protracted exposure. We have been unable to find data describing either frequency of aerosolized product use or the duration of exposure after each use. These factors probably cannot be studied reliably in the laboratory. One of the major difficulties is the variability around the "average use" pattern. Neglecting those who never use it, the normal user of a given product type might be exposed as briefly as a few minutes per week or as long as several hours per day every day. Consideration of the various types of users suggests that a 16-fold range of exposure severity should include most normal users [7]. Because of the breadth of this range and the difficulty of obtaining more exact information, we believe it best to assume a use pattern unlikely to be exceeded by any but a small fraction of the population of product users.

Antiperspirants are typically used on a daily basis and thus represent a relatively straightforward situation for estimating the human exposure pattern. We assumed in our calculations an exposure of 1 min/day repeated on a daily basis throughout the entire adult life of the exposed individual. The exposure aerosol was assumed to be of 4 μm aerodynamic diameter at a concentration of 100 mg/m^3. Because ACH is partially excluded from the smaller spray particles, the aerosol was assumed to be 25% ACH by weight.

FIRST ESTIMATE OF HUMAN LUNG CONCENTRATION

Repeated exposures, if continued long enough, will lead to a stable concentration of material retained in the lungs. This concentration is determined by the exposure conditions, the exposure pattern, and the rates of clearance of deposited particles. These factors can be combined and the concentration estimated through application of a numerical simulation model. It is generally assumed in such models that deposition and mechanical clearance to the gastrointestinal tract are independent of the exposure conditions or the material used. The approach, in its simplest form, is most applicable to exposures involving insoluble, nonhygroscopic materials under nonirritating conditions. Although estimates of lung concentration should be based on biological data for the material of interest, a useful first approximation can be obtained without additional empirical data on nonmechanical clearance rates or estimates based on in vitro dissolution studies.

The numerical simulation model used in our studies of aerosolized products is shown schematically in Figure 6. The model is concerned almost exclusively with the respiratory system because that was the region of primary interest to us and the anticipated source of analytical data for comparison to model estimates. Exposure and the addition of material to the different compartments were treated as a discrete event in the calculations. Clearance processes were treated as first order rate processes. The choice of compartments and their arrangement vary between models depending on the purpose of the model and the input data available [10-12]. The compartments

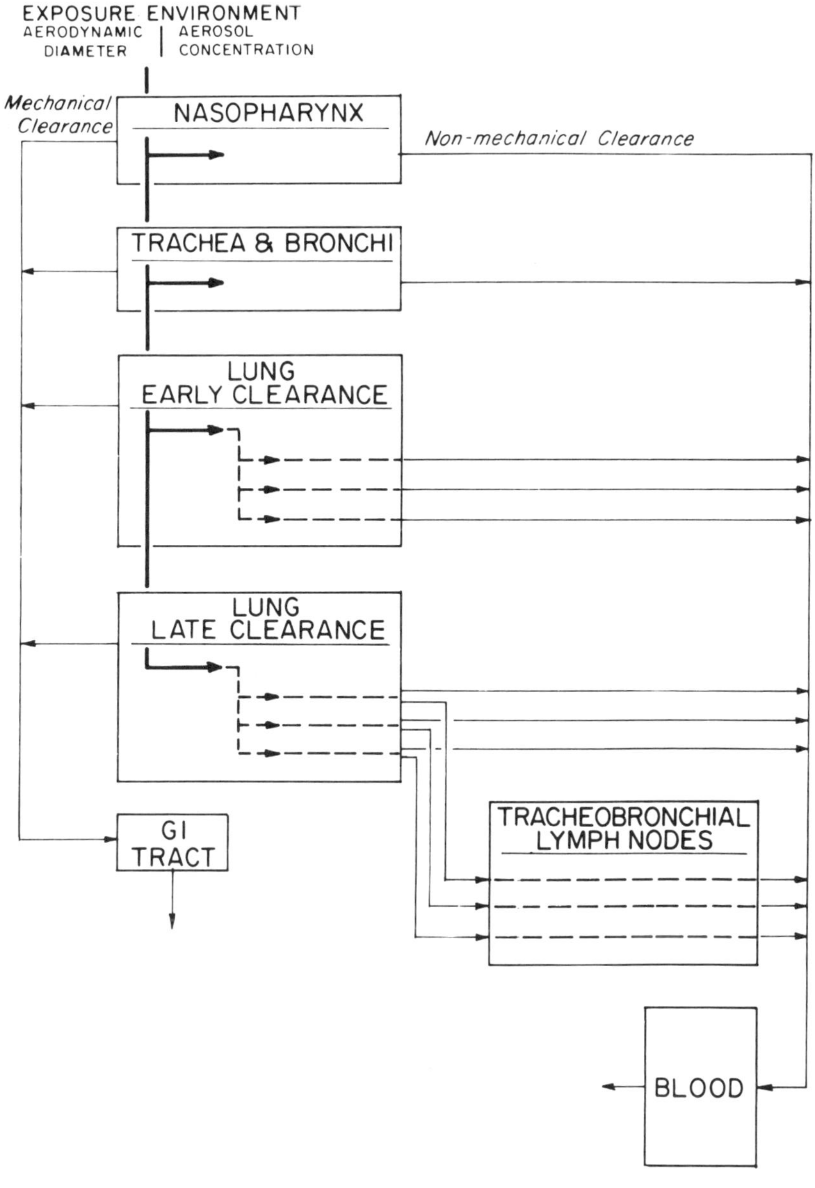

Figure 6. Structure of the numerical simulation model used for estimating lung concentrations, obtaining nonmechanical clearance rate parameters, and scheduling repeated exposures. The heavy lines indicate aerosol deposition processes, the light ones clearance processes.

and clearance descriptions used here were chosen to simplify the calculations needed to bring the simulation results into agreement with tissue analysis data from both single and repeated exposure studies.

Aerosol deposition and mechanical clearance are dependent on the species being considered. Many experimental studies of the amount of material deposited in the respiratory system, where it is deposited and how rapidly the deposited material leaves the lung have been reported. These studies have described deposition for small rodents [13], larger mammals [14, 15] or have summarized experimental results and theoretical calculations to describe deposition in humans [10]. These results have typically provided average values describing a given species. There is, however, a considerable amount of animal-to-animal variation in each of the important parameters so that exact comparison between observation and calculation is difficult.

Although deposition and mechanical clearance have been treated as species-dependent factors, nomechanical clearance appears to proceed primarily by dissolution, a physiochemical process that should be essentially independent of species. Although dissolution can be studied either in vivo or in vitro, it appears best to combine these approaches in obtaining rate constants. Dissolution in vivo is a complex process and its effect on lung clearance kinetics is difficult to quantitate [16]. A combination of limited in vitro data with in vivo observations should provide the most reliable estimate for simulation calculations and extrapolation between species.

If the aluminum in ACH is totally insoluble in the lung, there will be no clearance by the nonmechanical route. The human exposure conditions outlined above yield an estimated daily deposition of ACH equivalent to 0.039 μg Al/g lung. The estimated lung concentration of Al due to the assumed antiperspirant exposure rises to a steady state value of 10.6 μg Al/g as shown in Figure 7.

LABORATORY PROCEDURES DEVELOPMENT

Before laboratory studies with experimental animals are initiated, interrelated decisions concerning species, exposure method, aerosol material, aerosol generation method and method of tissue analysis must be made.

The first step should be selection of the species to be used in the experiments. Considerations involved in species selection are generally beyond the scope of this discussion. It is important, however, that information be available on the respiratory, deposition and clearance parameters of the selected species so that modeling can proceed without the need to conduct additional experiments to obtain this basic information. Respiratory paramcters must also be considered, together with the number of animals to be included in an exposure group, when selecting the exposure and aerosol generation methods.

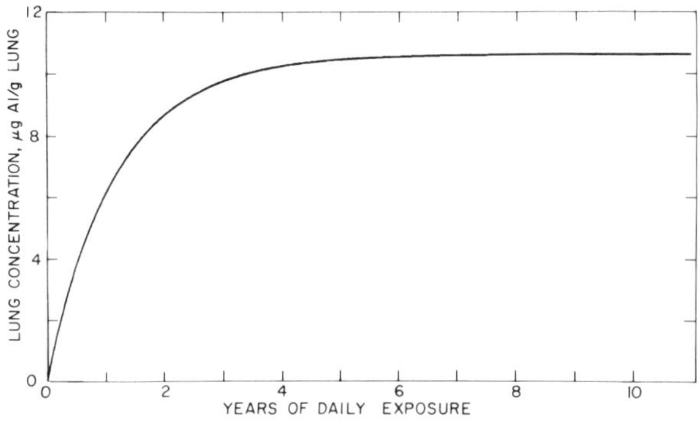

Figure 7. First estimate of the history of the human lung concentration of Al due to repeated exposure to antiperspirants. The Al has been assumed to have essentially no nonmechanical clearance.

The second step, selection of the exposure method, is a choice between whole-body and nose-only systems. Either is practical if small rodents are being exposed. Factors other than size, however, also enter into the decision. For example, what pattern of human exposure is to be simulated in the experiments. Protracted, low-level human exposure effects cannot be studied with short animal exposures at high concentrations. Whole-body exposures are the only reasonable choice in this case. If short exposures are acceptable, both methods can be considered. Whole-body methods are generally more practical than nose-only if large numbers of rodents (> 100) are required. As the experimental animals become larger, however, or the toxicity of the exposure material increases, nose-only methods become more attractive. The nose-only approach enjoys several advantages with rodents. First, less space and equipment are required for the same number of animals. Second, the amount of aerosol deposited on the pelt is greatly reduced compared to that with whole-body exposures. Rodents clean their fur by licking, thus providing a second route of exposure if the exposure material is absorbed from the GI tract. Finally, exposures are of relatively short duration (order of 30 minutes) and the required period of stable aerosol generation is much shorter than for whole-body exposures.

All our studies with aerosolized consumer product ingredients were conducted using Syrian hamsters and nose-only exposures.

Selection of the exposure material and the aerosol generation method are intermeshed. If the exposure must be to a product as formulated, there is essentially no choice concerning either the exposure material or the method of aerosol generation. The characteristics of many formulations are poorly suited to nearly all laboratory methods of generating respirable aerosols.

In such cases it may be necessary to use pressurized packages and accept the loss of the majority of the dispensed material in particles too large for inhalation and the inclusion of vaporized propellant in the exposure atomsphere. Detailed toxicological studies of product ingredients, on the other hand, are probably best carried out using aerosols generated by conventional laboratory methods. The method selected must be suitable to provide a particle size and concentration that will deliver, within a reasonable exposure time, the desired amount of material to the previously selected animals. It is also essential, when exposures are repeated over a long period such as the adult lifespan of the experimental animals, that the generation method reliably deliver aerosols with the same size and concentration characteristics.

It appears that lifespan repeated exposure studies using formulated products have not been reported in the literature. Published studies of the effects of inhaled aluminum salts have typically used single aluminum compounds. Drew and coworkers [17] used a glycol complex of aluminum-chloride-hydroxide that is not widely used in aerosolized antiperspirants. Steinhagen et al. [18] used ACH as supplied by the manufacturer in a size suitable for aerosolized antiperspirants. The exposure aerosol was generated with a Wright dust feeder. As noted earlier, respirable antiperspirant particles reaching the breathing zone are rarely pure ACH. In fact, many of these particles may be pure emollient. It is presently not clear if the emollient influences the biological response to the inhaled particles, although it is known that IPM is rapidly metabolized and excreted [19].

In developing the exposure aerosol generation system, we wished to produce either plain ACH particles or ACH particles coated with IPM. The plain ACH particles could be generated by nebulizing a solution and drying the droplets at moderate temperatures. The generation system for the coated particles posed several problems. First, we needed an aerosol in which the characteristics of the ACH particles were essentially the same as those of the uncoated particles. Second, the fraction of particles that were pure IPM needed to be minimized. The final approach was generation of plain ACH particles as outlined above followed by the addition of IPM vapor which condensed on the ACH particles as the mixture cooled. The generation system, shown schematically in Figure 8, has been described and details of its operation will be published [20].

Tissue analysis methods are necessary to determine the concentration of exposure material retained following one or several exposures. This information serves both as a means of relating biological observations to the dose received and as verification of the performance of the exposure system and simulation calculations. The analytical procedure must be extremely sensitive and capable of distinguishing the exposure material from a biological background that is frequently both high and variable.

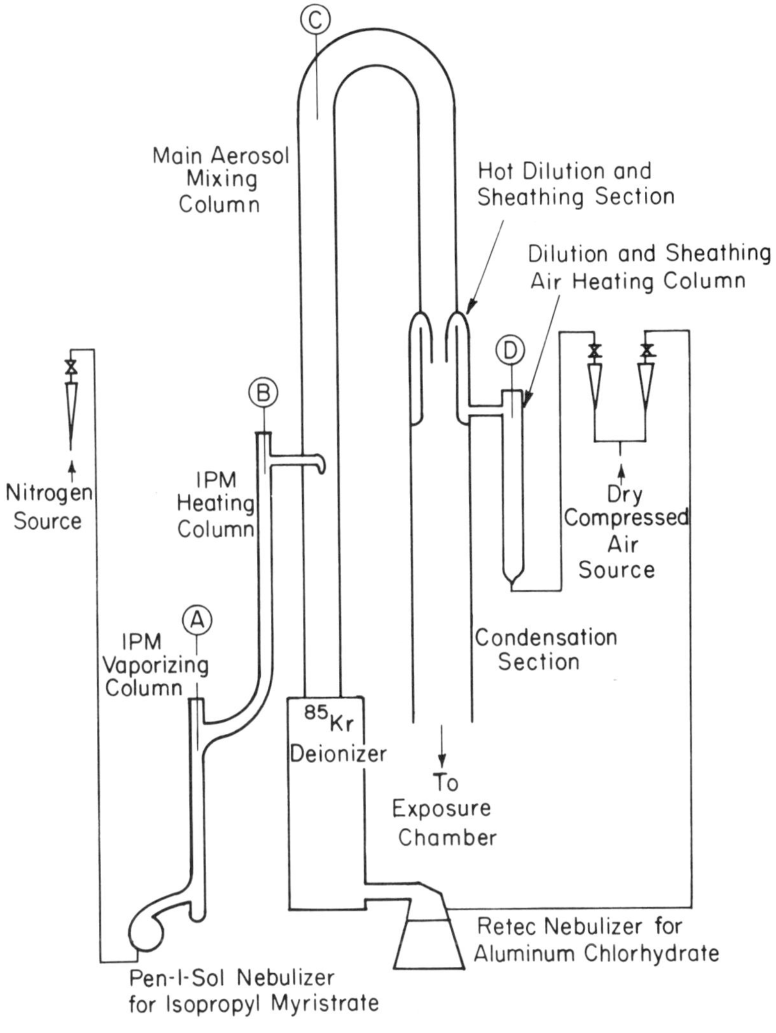

Figure 8. Diagram of the generation system for producing exposure aerosols of ACH coated with IPM.

The procedure to determine aluminum concentration due to ACH involved selective extraction in 0.1 *M* NaOH to eliminate naturally occurring aluminosilicates followed by flameless atomic absorption spectrometry. The lower detection limit of the method was approximately 0.5 μg Al per gram of fresh tissue; background levels in unexposed hamsters were 2-3 times this value.

FATE OF INHALED MATERIAL

The biological response to many inhaled particles is related to the amount deposited, how long the particles remain in the lung, and the biological pathways followed by material leaving the lung. Although deposition and mechanical clearance can be estimated from aerosol characteristics and the species exposed, nonmechanical clearance must be treated on a case-by-case basis. Some information can be derived from in vitro studies, but the necessary rate constants are best obtained by exposing animals and analyzing tissue samples (primarily lung) from animals sacrificed at various times after exposure. The exposure and analysis approach is most applicable to unreactive materials because there is no provision in the model, as described, for transformation of the exposure material to some other chemical form. Metabolic processes could be included in the simulation model, but the added studies to substantiate the reactions and products assumed could be very costly.

Several nose-only exposures were required to increase the lung concentration to a level that would leave sufficient material for accurate analysis in hamsters sacrificed as long as 2 months after the last exposure. Exposures were conducted on five successive days. The exposures were each of 35 minutes duration and the aerosols had an MMAD of approximately 2 μm and a concentration of 0.23 mg/l. Analytical results, shown as the range of observations for 4-7 animals at each sacrifice, are shown in Figure 9 together with the lung concentration values estimated with the simulation model. The nonmechanical clearance parameters used in the calculations were based on this and several other exposure experiments with ACH.

"BEST AVAILABLE" ESTIMATE FOR HUMANS

The first estimate of the human lung concentration was calculated primarily as an aid in establishing the magnitude of concentrations likely to be of interest. The assumptions in that calculation were such that, for the same exposure pattern, a higher concentration would be unlikely. The nonmechanical clearance parameters obtained in determining the fate of the inhaled material can be used to obtain a second estimate. Under the assumptions included in the simulation model, these parameters and the calculated result provide the best estimate available for humans so long as the material is relatively unreactive and there are no species specific metabolic transformations.

Using the antiperspirant exposure pattern and conditions described earlier for humans and, in place of the assumption of total insolubility, the nonmechanical clearance rate parameters derived from the experiment with

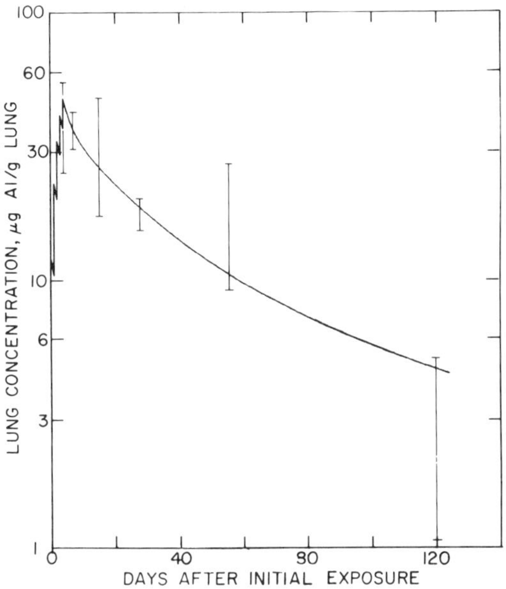

Figure 9. Lung concentrations of Al in hamsters exposed five times to an aerosol of ACH coated with IPM. The bars show the range of analytical results for four to seven hamsters at each sacrifice and the line gives the concentration estimated with the numerical simulation model.

hamsters exposed to ACH coated with IPM, the "best available" estimate of the human lung concentration of Al due to repeated exposure to antiperspirants was calculated. The results were similar in form to those shown in Figure 7 but the steady state value was 1.43 μg Al/g lung tissue rather than the 10.6 μg/g calculated earlier. Based on this substantial change in the estimated concentration, it appears dissolution is an important factor in the clearance of ACH from the human lung.

DESIGN OF REPEATED EXPOSURE EXPERIMENT

Repeated exposure experiment design poses both biological and physical problems. The biological problems are more difficult to deal with as they involve issues of species differences and extrapolation between species. Two aspects of these problems, species differences in respiratory, deposition and clearance rate parameters and length of lifespan, must be considered in selecting the exposure regimen for long-term experiments. The physical problems relate primarily to the reliability of exposure aerosol generation and provision of adequate exposure facilities.

The respiratory rate of rodents, considered in terms of liters/min/kg body weight, is approximately threefold higher than for humans but the pulmonary deposition efficiency is generally lower. Rodents also generally have more efficient and protracted mechanical clearance processes than humans. These differences influence the lung concentration of inhaled material both immediately after exposure and at later times after clearance processes have removed some of the deposited material.

Commonly used laboratory animals have lifespans much shorter than humans. The most widely used rodents typically have lifespans of 2 to 3 years as compared to approximately 70 years for humans. Similarly, rodents are considered sexually mature at approximately 90 days of age whereas this value is approximately 14 years for humans. The best method of allowing for these differences has not been established, especially when considering effects that may occur in humans several years after they have reached a stable lung concentration through repeated daily exposures. Use of exaggerated concentrations is one method, although it is recognized that the type and magnitude of biological response to many materials is dependent on the concentration in the target tissue.

The interaction of respiratory system and lifespan differences can be illustrated by returning to the estimate of human lung concentration of aluminum due to antiperspirant exposure. Assume a hamster is exposed to the same aerosol as the human, also for 1 min/day. However, to conform to common experimental practice, assume the hamster is exposed 5 days/week rather than seven as for the human. To facilitate plotting on the same scale, time will be phrased in dimensionless form as the fraction of the median lifespan of the species that has elapsed since exposures were initiated. The results of these calculations are shown in Figure 10. The pattern of the lung concentration history of the hamster is quite different from that of the human. In fact, even when the animals are quite old the lung concentration has not reached a stable value. This difference in pattern is independent of the exposure level.

Simulation model calculations were used to provide closer agreement between the temporal pattern of the Al concentration in repeatedly exposed hamsters and that estimated for humans. This was achieved by using exposure aerosols with characteristics chosen to maximize pulmonary deposition and allowing the exposure schedule to vary from a rigid, once-a-day pattern. The goal was to begin exposing hamsters at 12 to 14 weeks of age, rapidly bring their lung concentration to a preselected value, and then maintain them at that value for the remainder of their adult lives. The value selected was approximately 40 times the "best available" estimate of thc steady state value in humans. The estimated concentration in the hamster lungs was allowed to vary from 80% (before each exposure) to 120% (after exposure)

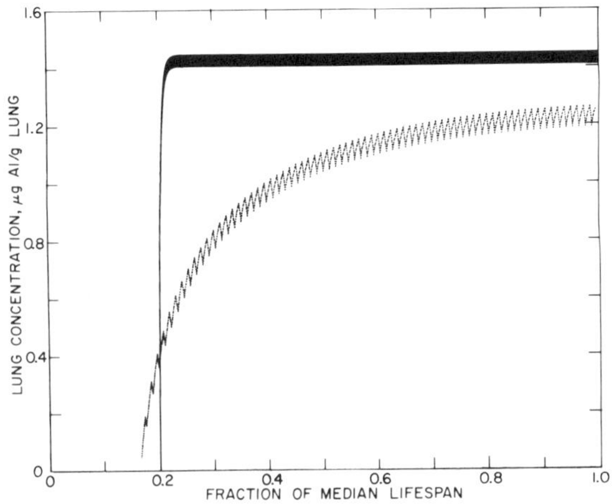

Figure 10. Calculated lung concentrations in man (solid line) and hamster (dotted saw tooth line), exposed to the same antiperspirant aerosol. It has been assumed that the man is exposed seven and the hamster five days each week of their adult life. The median lifespan of hamsters was taken to be 600 days; the value for man was 70 years.

of the target value. The simulation program allowed for the characteristics of past exposures and scheduled the next exposure, including the anticipated exposure conditions, and the length of exposure required.

RESULTS OF THE REPEATED EXPOSURES

An initial group of 72 hamsters was exposed repeatedly to ACH coated with IPM according to the conditions outlined above. Reliable operation of the aerosol generation system was essential in achieving the desired pattern of lung concentration. Exposures were typically of 35 minutes duration with aerosols of 2 μm MMAD and a concentration of 0.25 mg/l. Ninety-four exposures were carried out over a period of slightly greater than 21 months. Aerosol characteristics were determined for each exposure. The MMAD's derived from fitting the impactor data to a log-normal function had a mean value of 2.2 μm with a standard deviation of 0.3 μm; the mean aerosol concentration was 0.27 mg/l with a standard deviation of 0.04 mg/l. The overall IPM content of the aerosol was approximately 10%.

It is essential to verify that a repeated exposure regimen has maintained the lung concentration at the desired level. Analytical results from this series of exposures are summarized in Figure 11. Animals were sacrificed 3, 6 and 9 months after the beginning of the exposures and again when only 4 of the original 72 still survived. The observed lung concentrations of Al in each of these animals are shown together with the concentrations estimated before and after each exposure by the simulation model. The agreement is generally good, although there is a suggestion of some divergence at 180 days. The source of this difference is not known.

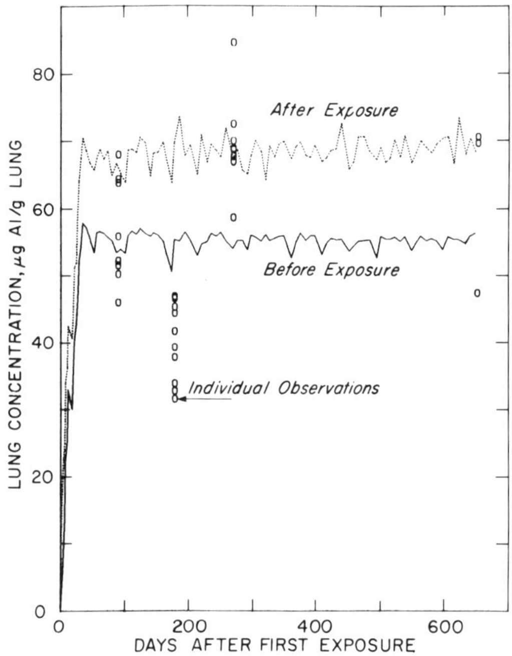

Figure 11. Lung concentrations of Al in hamsters repeatedly exposed to ACH coated with IPM. Analytical results from individual animals are shown as circles. The solid line is the estimated lung concentration before each exposure, the dotted line that immediately after exposure.

SUMMARY

Inhalation toxociology studies of aerosolized consumer products and assessment of the potential hazard associated with their repeated use pose several problems. Recognition of these problems and minimization of their influence through detailed aerosol characterization measurements and application of numerical simulation models should enhance the applicability of experimental results to the setting of human exposure limits. The general approach outlined in this paper can be applied to other human exposure situations.

ACKNOWLEDGMENTS

The work described here was carried out under U.S. Department of Energy Contract EY-76-C-04-1013 through interagency agreements with the Food and Drug Administration and the Consumer Product Safety Commission. The research benefitted from numerous specific, short-term contributions by many members of the Inhalation Toxicology Research Institute staff and, for the majority of the four and one-half years required, the expert technical assistance of R. K. White and B. A. Wong.

REFERENCES

1. Sciarra, J. J., P. McGinley and L. Izzo. "Determination of Particle Size Distribution of Selected Aerosol Cosmetics. I. Hair Sprays," *J. Soc. Cosmet. Chem.* 20:385-394 (1969).
2. Hathaway, D. "Particle Size Measurement of an Aerosol Deodorant Using Laser Holographic Microscopy," *Aerosol Age* 18(9):28-33 (1973).
3. Vos, K., and D. B. Thomson. "Particle Size Measurement of Eight Commercial Pressurized Products," *Powder Technol.* 10:103-109 (1974).
4. Pengilly, R. W., and J. A. Keiner. "The Influence of Some Formulation Variables and Valve/Actuator Designs on the Particle Size Distributions of Aerosol Sprays," *J. Soc. Cosmet. Chem.* 28:641-650 (1977).
5. Berres, C. R., K. D. Vos, D. B. Thomson and E. D. Baretta. "In-home Measurement of Background Particles and Particulates and Propellents Produced by an Air Freshener," *Am. Ind. Hyg. Assoc. J.* 37:305-310 (1976).
6. Wells, A. B., and D. J. Alexander. "Methods for Determining the Respirable Fraction of Aerosols from Commercial Pressurized Toiletries Products," *Aerosol Age* 21(11):20-25 (1976).
7. Mokler, B. V., B. A. Wong and M. J. Snow. "Respirable Particulates Generated by Pressurized Consumer Products. I. Experimental Method and General Characteristics," *Am. Ind. Hyg. Assoc. J.* 40:330-338 (1979).
8. Mokler, B. V., B. A. Wong and M. J. Snow. "Respirable Particulates Generated by Pressurized Consumer Products. II. Influence of Experimental Conditions," *Am. Ind. Hyg. Assoc. J.* 40:339-347 (1979).
9. Mokler, B. V., E. G. Damon, T. R. Henderson, R. L. Carpenter, S. A. Benjamin, A. H. Rebar and R. K. Jones. "Inhalation Toxicology Studies of Aerosolized Products," LF-66 (Springfield, VA: National Technical Information Service, 1979).
10. Task Group on Lung Dynamics. "Deposition and Retention Models for Internal Dosimetry of the Human Respiratory Tract," *Health Phys.* 12:173-207 (1966).

11. Thomas, R. G. "An Interspecies Model for Retention of Inhaled Particles," in *Assessment of Airborne Particles*, T. T. Mercer, P. Morrow and W. Stöber, Eds. (Springfield, IL: Charles C. Thomas, 1972), pp. 405-420.
12. Brain, J. D., and P. A. Valberg. "Models of Lung Retention Based on ICRP Task Group Report," *Arch. Environ. Health* 28:1-11 (1974).
13. Raabe, O. G., H.-C. Yeh, G. J. Newton, R. F. Phalen and D. J. Velasquez. "Desposition of Inhaled Monodisperse Aerosols in Small Rodents," in *Inhaled Particles IV*, W. H. Walton, Ed. (Oxford: Peragmon Press, 1977), pp. 3-21.
14. Kajland, A., M.-L. Edfors, L. Friberg and B. Holma. "Radioactive Monodisperse Test Aerosols and Lung Clearance Studies," *Health Phys.* 10:941-945 (1964).
15. Albert, R. E., M. Lippmann, H. T. Peterson, J. Berger, K. Sanborn and D. Bohning. "Bronchial Deposition and Clearance of Aerosols," *Arch. Intern. Med.* 131:115-127 (1973).
16. Moss, O. R., and G. M. Kanapilly. "Dissolution of Inhaled Aerosols," Chapter 4, this volume.
17. Drew, R. T., N. G. Bhola, J. R. Bend and G. E. R. Hook. "Inhalation Studies with a Glycol Complex of Aluminum-Chloride-Hydroxide," *Arch. Environ. Health* 28:321-326 (1974).
18. Steinhagen, W. H., F. L. Cavender and B. Y. Cockrell. "Six Month Inhalation Exposure of Rats and Guinea Pigs to Aluminum Chlorhydrate," *J. Environ. Path. Toxicol.* 1(3):267-277 (1978).
19. Finkelstein, P., and R. J. Wulf. "The Uptake, Distribution, and Excretion of a Commercial Aerosol Antiperspirant by the Monkey," *J. Soc. Cosmet. Chem.* 25:645-654 (1974).
20. Carpenter, R. L., and B. V. Mokler. "Generation of Aerosols of Two Immiscible Components," *Am. Ind. Hyg. Assoc. J.* (in preparation).

CHAPTER 19

RADIOACTIVE LABELING OF AEROSOLS: GENERATION METHODS AND CHARACTERISTICS

George J. Newton, George M. Kanapilly and Bruce B. Boecker

Inhalation Toxicology Research Institute
Lovelace Biomedical and Environmental Research Institute
Albuquerque, New Mexico 87115

Otto G. Raabe

University of California at Davis
Davis, California 95616

INTRODUCTION

General

This chapter is a review of radioactive labeling of aerosols primarily intended for the investigator interested in laboratory aerosol studies using radioactive labeling. Although the emphasis is on fine particles or respirable aerosols, many of the techniques could be adapted for generation of larger particles. Most of the generation techniques discussed herein were developed at the Lovelace Inhalation Toxicology Research Institute (ITRI). Some particularly unique methods developed in other laboratories for special applications are also included. Radioactively labeled aerosols are useful for basic and applied studies in: inhalation toxicity studies using laboratory animals, aerosol science, industrial hygiene, health physics and atmospheric sciences.

Fine particle aerosols are significant environmental pollutants and frequently are inhalation hazards. If aerosols are inhaled, significant amounts can be deposited in the lung. Depending upon their physicochemical properties, these deposited particles can be transported at different rates to other body organs. Subsequent biological effects may be observed in these organs as well as in the lung. Assessment of potential inhalation hazards from these environmental pollutants requires information on the deposition, retention, translocation and excretion of these aerosols after inhalation exposure. This difficult scientific problem frequently can best be addressed using radioactively labeled aerosols. The currently most reliable data on deposition, tissue distribution, retention and clearance of inhaled particles have been obtained using radioactive aerosols. In basic and applied aerosol studies, radioactive aerosols provide improved accuracy and speed of measurement. Radiolabeling can provide a detection sensitivity in parts per billion or smaller concentrations for a large number of elements. Industrial hygiene-oriented use has employed radioactive aerosols for filter testing to extend the sensitivity.

Radioactive aerosols are also a separate class of environmental pollutants. Some radioactive aerosols are formed naturally but most are the result of industrial activities ranging from underground mining to nuclear fuel cycles in nuclear-powered central electric generating stations.

One of the simplest techniques for generating radiolabeled aerosols is the nebulization of solutions containing dissolved salts such as stable CsCl with radioactive labels such as ^{137}Cs, ^{90}Sr, ^{91}Y, ^{144}Ce and ^{95}Zr. The resultant labeled aerosols are the residue particles after solvent evaporation. A useful method for generation of "insoluble" fine particles is the use of fused montmorillonite clay containing various beta-gamma-emitting fission products. Even soluble cations such as ^{137}Cs have been made into insoluble forms by incorporation into fused montmorillonite clay [1]. Heat treatment of other aerosols is a general technique that has been used to change physicochemical properties. An example is ^{95}Zr-labeled zirconium oxalate which, when heat treated at different temperatures, yielded ^{95}Zr-labeled zirconium carbonate or oxide aerosol particles [2].

An extremely useful technique is gamma labeling (^{51}Cr and ^{169}Yb) of actinide oxides to provide for estimates of initial lung burden and early retention data for inhalation toxicity studies in laboratory animals. Combinations of methods can be used to produce monodisperse "insoluble" radioactive aerosols for laboratory animal inhalation studies and other applications [3]. Methods for radioactive labeling of standard laboratory aerosols such as monodisperse polystyrene latex (PSL) and polyvinyl toluene (PVT) are also reviewed.

Aerosol Characteristics

Parameters used to characterize aerosols described in this chapter are primarily measurements of particle aerodynamic behavior or basic physicochemical properties such as specific activity, density, chemical composition, dissolution rate, Brownian diffusion, surface area and shape. Aerodynamic properties determined by inertial sampling devices and radioactive counting using aerosol centrifuges [3] or cascade impactors [4] are expressed as activity median aerodynamic diameter (AMAD) with an associated geometric standard deviation (σ_g). Density [5] and dissolution rates [6-8] were determined from aerosol centrifuge samples. Specific activity (SA) was determined from radioactive counting data and is expressed as curies per gram (Ci/g). Half-lives of the labels range from a few hours to many years.

METHODS

Nebulization of Solutions and Suspensions

A simple method of generating radioactive labeled aerosols is nebulization of solutions or suspensions. Particle size of aerosols produced by nebulization is a function of concentration and droplet size distribution and is related by:

$$D_p^3 \rho = D_d^3 c \quad (1)$$

where: D_p = the physical or real diameter of the spherical particle
ρ = density (g/cm^3)
D_d = the droplet diameter
c = the solution concentration (g/ml)

Nebulizers produce a distribution of droplet sizes and the resultant aerosol after solvent evaporation will have a geometric standard deviation (σ_g) nearly equal to the σ_g of the original droplet distribution.

Suspensions of insoluble materials will also behave similarly if the micelle size is small enough. If the micelle size is large, the lower size range of the droplet distribution may not contain particles and the residue particle size distribution will be correspondingly truncated. In this case, the particle size cannot be predicted using Equation 1.

Various systems of solvents, solutes and suspensions can be used to produce a wide variety of aerosols. At ITRI, relatively soluble aerosols of CsCl labeled with fission product radionuclides such as ^{137}Cs-^{137}Ba, ^{144}Ce-^{144}Pr, ^{90}Sr-^{90}Y, ^{91}Y, ^{90}Y and ^{140}Ba-^{140}La have been used in both dose-response and radiation dose pattern studies following inhalation by laboratory animals. Cesium chloride was chosen as a common vector for the soluble fission product radionuclides instead of NaCl because CsCl containing droplets are easier to dry than are NaCl containing droplets.

Insoluble aerosols have also been produced at the ITRI by nebulization of solutions or suspensions with subsequent heat treatment. Among insoluble aerosols produced are:

1. $^{144}CeO_2$ from $^{144}CeCl_3$ [9]
2. ^{60}CoO and $^{60}Co_3O_4$ from $^{60}Co(No_3)_2$ [10]
3. $^{95}ZrO_2$ from ^{95}Zr-oxalate

Kanapilly et al [2] produced various forms of ^{95}Zr oxides, carbonates, and oxycarbonates from Zr-oxalate treated at different temperatures shown in Figure 1.

Alteration of the nebulized aerosol by heat treatment produces amorphous forms (Figure 2, $^{95}ZrO_2$) and in some cases, crystalline forms (Figure 3, ^{198}Au-gold). Densities of aerosols from nebulization techniques usually are about three-fourths of the theoretical crystalline density [1,2]. If heat treatment approaches the melting point, densities are near the theoretical crystalline density of the material.

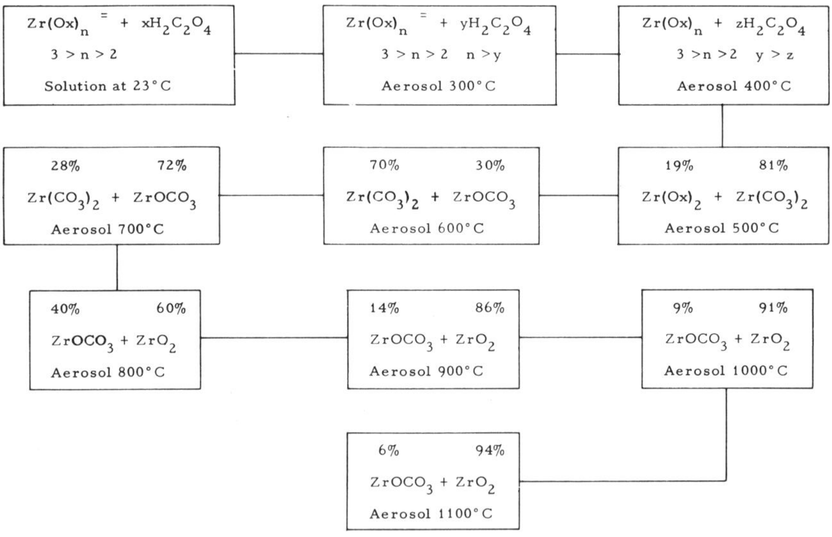

Figure 1. Block diagram showing production of different aerosols formed by decomposition of zirconium oxalate particles subjected to increasing temperatures. Data obtained by quantitating ^{14}C collected on particles formed through thermolysis of ^{14}C-Zr-oxalate and IR spectroscopy.

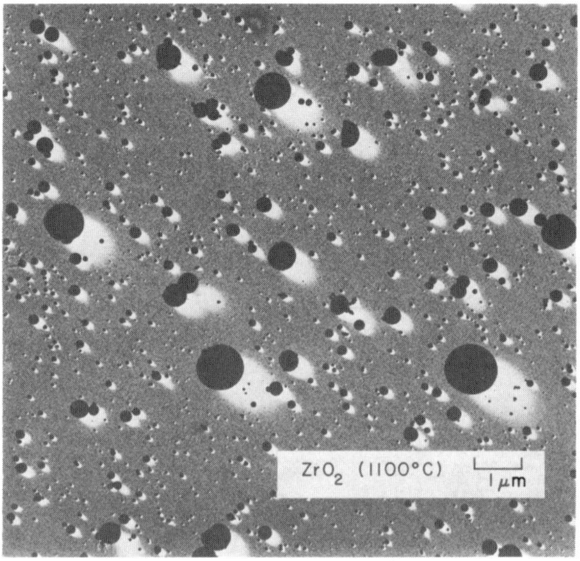

Figure 2. Transmission electron photomicrograph of very insoluble ^{95}Zr-ZrO_2 particles formed by heat treatment (1100°C) of nebulized ^{95}Zr-Zr-oxalate.

Radioactive Labels for Standard Particles

Monodisperse suspensions of polystyrene latex (PSL) and polyvinyl toluene (PVT) are available (Dow Diagnostics, The Dow Chemical Company, Indianapolis, IN 96206) in size ranges of 0.088 μm to about 3.8 μm. These particles have been widely used in laboratory studies and several methods of labeling with radionuclides have been developed [11-14]. A technique described by Hinrichs et al [13] adapted from Szende and Udvarheli [14] is a method for labeling PSL particles with ^{51}Cr; a modified flowchart of the procedure is illustrated in Figure 4. Labeling was by emulsion polymerization of chromium acetylacetonate dissolved in styrene monomer and used Dow PSL microspheres. Two steps were required: (1) radiation excitation with an 11,000-Ci ^{137}Cs source to an accumulated dose of $\cong 1.3 \times 10^7$ rads, and (2) radical polymerization initiated with potassium persulfate. Repeated wash-centrifugation cycles removed unattached ^{51}Cr. In vitro tests indicated a leach rate of less than 0.19% per day. Starting with 1.099 μm diameter PSL microspheres, Hinrichs et al [13], using this procedure, produced 1.4-μm count median diameter (CMD) ^{51}Cr-labeled particles with a $\sigma_g < 1.2$.

Various techniques [15,16] have been developed to label particles with ^{99m}Tc for deposition and early clearance studies in laboratory animals and

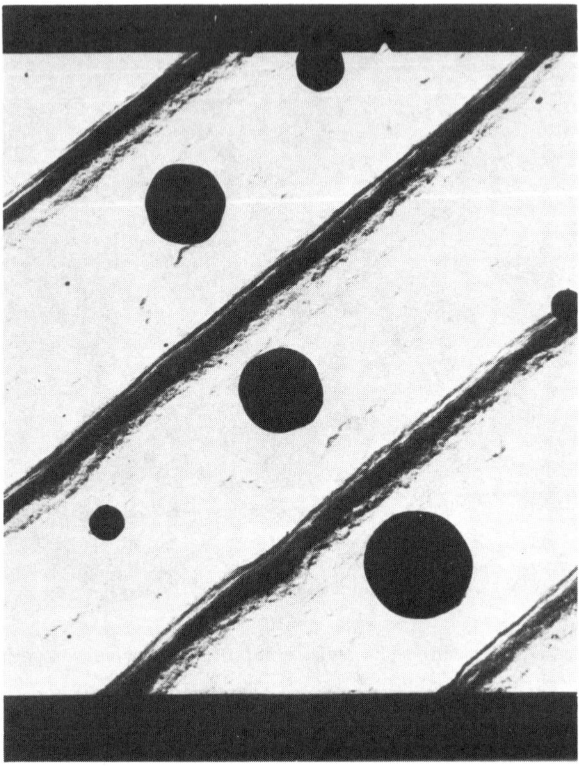

Figure 3. Transmission electron photomicrograph of ^{198}Au-gold particles showing crystalline structure. Parallel lines are from a diffraction grating replica used for accurate sizing and are 11,340 lines/cm.

man. Technetium-99m is suited for these studies due to its 6-hr half-life and 140 keV gamma emissions. Few et al [17] described a useful procedure in which monodisperse polystyrene particles labeled with ^{99m}Tc were generated with a spinning disc generator. A dry form of tetraphenylarsonium pertechnetate was dissolved in a mixture (4:1 v/v) of xylene and methyl isobutyl ketone containing the polystyrene. Particles were produced in sizes of 2.1-8.6 μm diameter and indicated a leach rate in saline of 4% for the first day and < 1% per day thereafter.

Camner et al [18] generated 4-μm monodisperse particles of Teflon labeled with ^{51}Cr and coated with silver, beryllium or carbon for use in inhalation studies in rabbits. In an earlier work, Camner et al [19] generated 7-μm monodisperse fluorocarbon resin particles tagged with ^{18}F. Both types of particles were generated using the spinning disc.

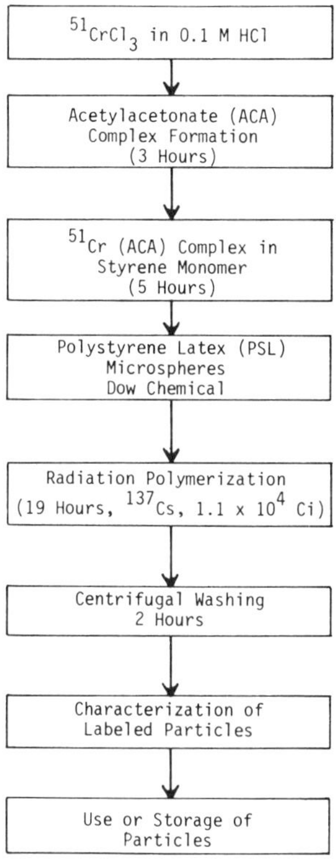

Figure 4. Flowchart of procedure used to label monodisperse polystyrene latex particles with ^{51}Cr. Original size was 1.099 μm diameter and final monodisperse ($\sigma_g < 1.2$) particles were 1.4 μm diameter. Flow chart modified from Hinrichs et al. [13].

Radiolabeled Aluminosilicate Particles

Adapting methods of Ginnell and Simon [20], a variety of aerosol types have been prepared with radioactive tags for use in laboratory animal inhalation studies and in other basic aerosol science applications. The method uses cation exchange properties [20-27] of montmorillonite clay where the structure consists of alternating layers of silicate tetrahedra and aluminate octahedra. The basic formula for montmorillonite clay is $Si_8Al_4O_{20}(OH)_4 \cdot nH_2O$. The clay structure acquires a net negative charge when ions such as Mg(II) and Fe(II) replace Al(III) or Al(III) replaces Si(IV) in the clay network. This acquired negative charge is accommodated by cations which move freely between layers. Natural clay contains calcium or magnesium in the exchange sites. Soluble cations in solution can replace natural ions in the exchange sites and the reactions are usually reversible. In time ion exchange equilibrium is attained. A simple model for these reactions can be represented for cation exchange by the formula

$$\overline{mA}^{+n} + nB^{+m} \rightleftharpoons mA^{+n} + \overline{nB}^{+m} \tag{2}$$

where A and B represent exchangeable cations and quantities with bars refer to the exchange sites inside the clay. Two factors which increase the selectivity for exchange uptake of a given cation are higher valence state and smaller solvated volume. The selectivity sequence of montmorillonite clay for various cations, based on thermodynamic data of several researchers [24-27], is

$$Li^{+} < Na^{+} < K^{+} < Cs^{+} < Mg^{++} < Ca^{++} < Sr^{++} < Y^{+++} < Ce^{+++} \tag{3}$$

The exchange capacity of montmorillonite clay from Bayard, New Mexico, has been reported to range from 1.17 to 1.36 milliequivalents (meq) per gram at room temperature. The average exchange capacity for the clay is approximately 1.1 meq per gram of clay when the cation of interest is present at a concentration of about 2 meq per gram of clay [28]. Exchange can be maximized by prior packing of the exchange sites with ions of low exchange rank such as Na^{+} for the Ca, Mg or other divalent or polyvalent cations normally present in the exchange sites.

Clay Preparation and Heat Treatment

Figure 5 is a composite of a flow chart showing procedures developed at ITRI for preparation of fused aluminosilicate particles labeled with a variety of radioactive cations and electron micrographs of raw and heat-treated particles. Montmorillonite clay was:

1. ground and suspended in water;
2. the small particle suspension decanted;
3. treated with excess 30% hydrogen peroxide to remove organic matter;
4. exchange sites packed with sodium;
5. excess sodium removed by running water dialysis;
6. radionuclide cation exchanged into clay replacing sodium;
7. labeled clay suspension vacuum filtered, washed and resuspended in water; and
8. aerosolized and heat treated (1100°C).

Heat treatment is necessary to firmly fix the radionuclide in the clay particle. Studies [20-27] have shown that when montmorillonite clay is heated to 400°C, water is irreversibly lost from the clay lattice structure and the particle becomes relatively insoluble. Clay particles assume spherical shapes

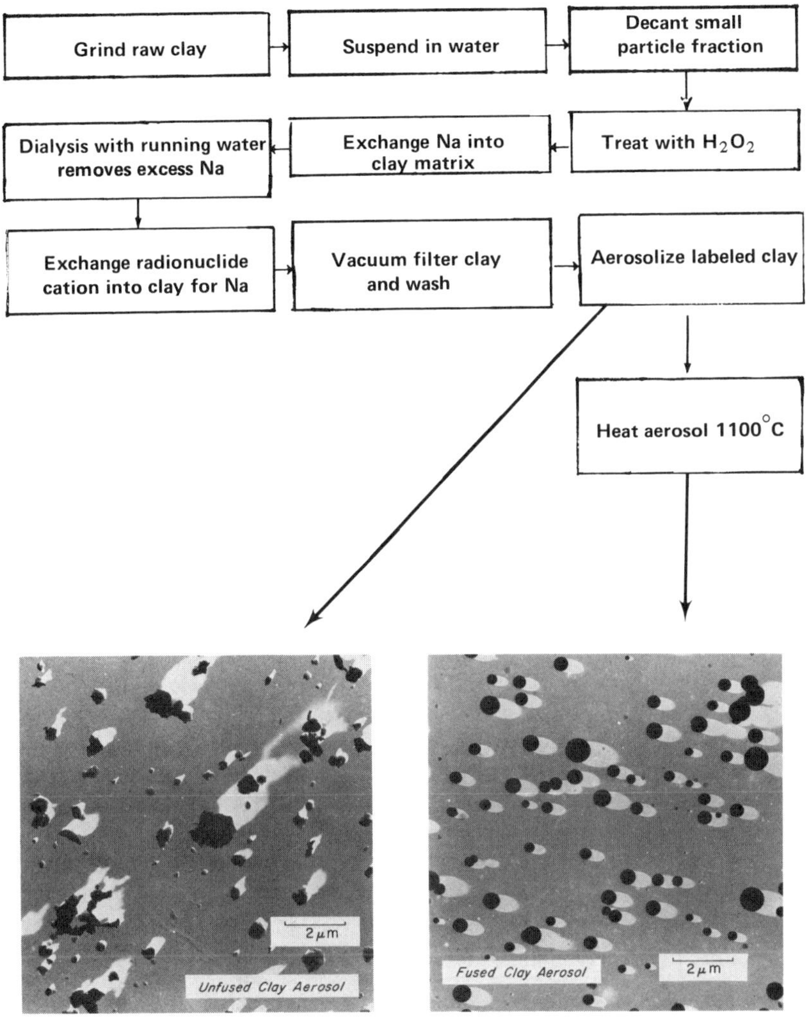

Figure 5. Composite illustrating flowchart for production of radioactively labeled fused aluminosilicate particles and transmission electron photomicrographs of irregular clay particles before heat treatment and final spherical particles.

and appear "fused" at temperatures above 1000°C. The relatively insoluble nature of these particles in Beagle dog lungs was shown by Boecker et al. [29] where ^{137}Cs in fused aluminosilicate particles was retained with an effective half-life of about 400 days. The rate constant of dissolution [30] of these fused aluminosilicate aerosols measured in vitro was about 3×10^{-8} g/cm^2/day .

Table I lists specific activities of tagged clay obtained and resultant aerosol concentrations achieved in animal inhalation studies.

Examples and Properties of Some Tagged Clay Aerosols

Yttrium-labeled fused aluminosilicate aerosols have been particularly useful at the ITRI. This is due to the physical characteristics such as half-life, radioactive emissions and specific activities that are available for the various isotopes of Y. Table I lists three isotopes of Y that have been used as tags for aerosols prepared for laboratory animal inhalation studies. The extremely high specific activity (1.7 x 10^3 Ci/g clay) attained using an ^{90}Y label was possible only after development of a procedure for separation[31] of the pair, ^{90}Sr-^{90}Y. Figure 6 is a flow chart for the multicurie (150 Ci) separation of ^{90}Y from ^{90}Sr. The nominal cost of ^{90}Sr-^{90}Y, the short half-life of ^{90}Y (64 hr) and the high specific activities attainable might be a useful "tag" for other applications such as in atmospheric sciences. Yttrium-90 tagged clay has been prepared in 100-Ci batches with a ^{90}Sr contamination of less than 0.01 ppm.

The decay of ^{88}Y is accompanied by emission of hard gamma photons with energies of 0.9, 1.8 and 2.8 MeV. These characteristics of the decay process make ^{88}Y a useful tag for ^{91}Y, an important fission product radionuclide (beta emitter) that has been employed extensively in inhalation studies at the ITRI. In addition, the physical characteristics of an ^{88}Y-tagged clay aerosol make it useful for applications requiring a hard gamma emitter. Yttrium-88 is produced in a cyclotron bombardment reaction

$$^{88}\text{Sr (p, n)}\ ^{88}\text{Y} \qquad (4)$$

To obtain high specific activity material suitable for exchange into clay, the separation procedure outlined in Figure 6 was used to separate ^{88}Y from the target material, ^{88}Sr. A modified procedure was also used to separate ^{140}La from the mixture ^{140}Ba-^{140}La.

Gallium-67, a radionuclide widely used in diagnostic nuclear medicine, is produced in a bombardment reaction, ^{68}Zn (p, 2n) ^{67}Ga. The resultant mixture containing ^{67}Ga and massive quantities of target ^{68}Zn is unsuitable for exchange into clay. Newton et al. [32] developed a procedure (Figure 7) for separation of 100-mCi batches of ^{67}Ga from ^{68}Zn which enabled preparation of high specific activity ^{67}Ga labeled fused aluminosilicate aerosols. Gallium-67 decays with a 78.1-hr half-life by electron capture to stable ^{67}Zn, has no particulate emissions (excepting Auger electrons) and is accompanied by emission of three major gamma photon groups at 92, 182 and 300 keV with 69, 24 and 22% yield, respectively. These gamma photons have an average energy of 150.5 keV and could deposit a maximum dose to a 1000-g

Table I. Specific Activity and Emission Characteristics of Tagged Clay Used in Animal Inhalation Exposures

Radionuclide	Specific Activity (S.A.) Ci/g Clay	Aerosol[a] Concentration (μCi/liter)	T-1/2 (days)	Major Emissions: β (max) (keV)	Major Emissions: γe (keV)	Major Emissions: γ-yield %
^{137}Cs-^{137}Ba	2.0	0.025	10,957	510-656	662	85
^{144}Ce-^{144}Pr	6.0	0.150	285	2990	134	11
^{88}Y	1.0	0.013	108	B^+ 780	1840	100
^{90}Y	1.7×10^3	4.250[b]	2.7	2280	None	-
^{91}Y	6.0	0.150	58	1540	1210	very low (0.3)
^{147}Pm	14.8	0.185	950	224	None	-
^{169}Yb	8.2	0.103	31.3	None	110,131, 177,198, 308	18,11 22,35 10
^{90}Sr-^{90}Y	1.2	0.015	10,519	2280	None	-

[a] Lovelace Nebulizer used. 10 mg clay/ml in generator and 20 l/min diluting air → AMAD ∿ 1.8 μm, σ_g ∿ 1.7.
[b] 1.0 mg clay/ml in generator → AMAD ∿ 0.9 μm, σ_g ∿ 1.7.

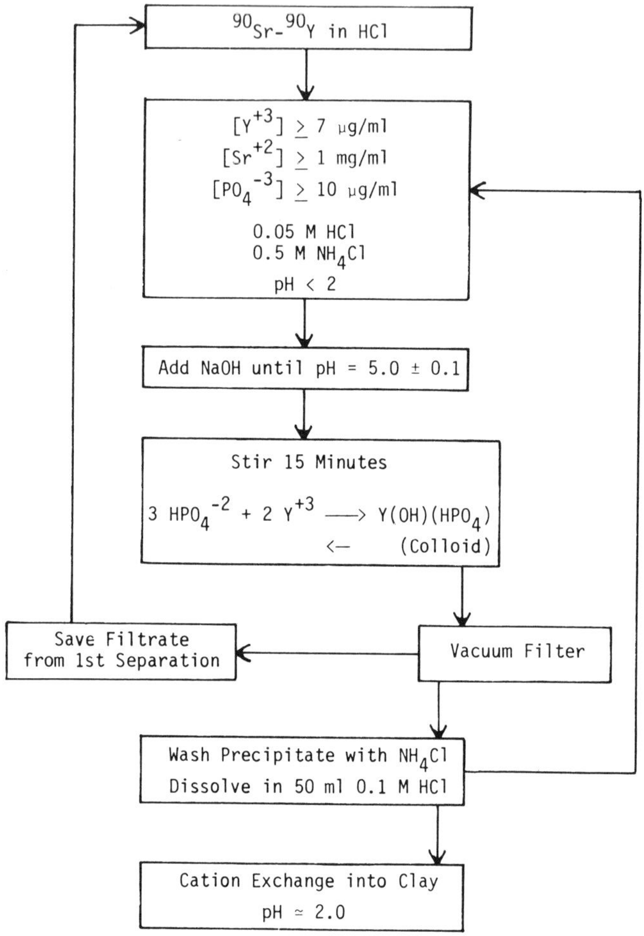

Figure 6. Flow chart of ^{90}Sr-^{90}Y separation procedure used to obtain multi-Curie quantities of pure, high specific activity ^{90}Y.

human lung of 36 mrad/μCi. This low dose commitment and the fact that ^{67}Ga-labeled materials are already widely used in humans makes the use of ^{67}Ga-labeled fused aluminosilicate particles a good candidate for human deposition and early clearance studies. Calculated activity levels [32] obtainable in both human and Beagle dog inhalation exposures using monodisperse ^{67}Ga-labeled fused aluminosilicate particles are listed in Table II.

Copper-64 is another radionuclide that has been exchanged into clay and has potential for uses where its unique characteristics are desirable. Copper-64

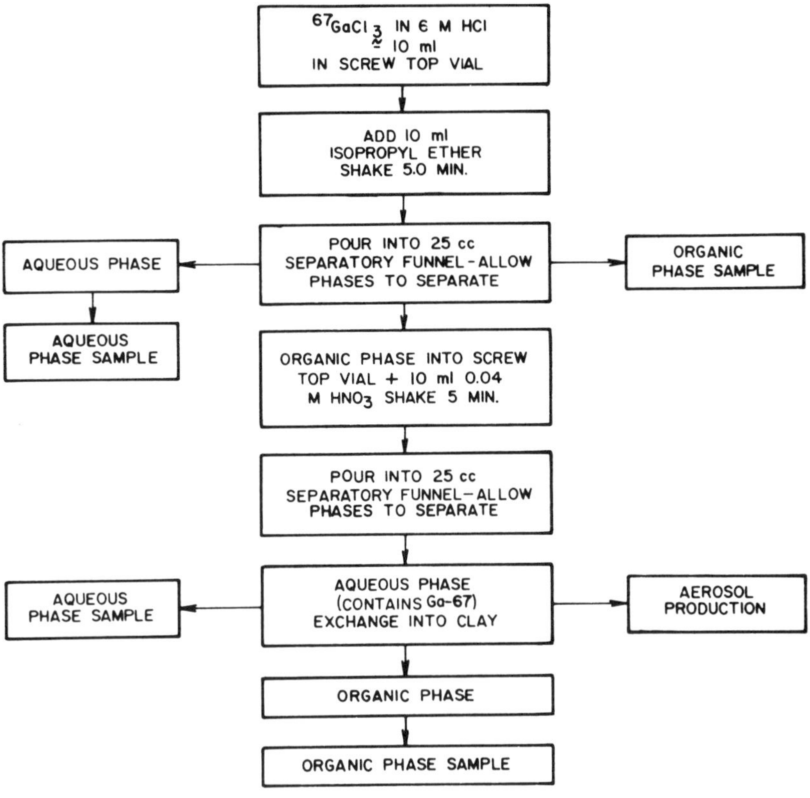

Figure 7. Flowchart of ^{67}Ga-^{68}Zn solvent extraction based separation procedure used to obtain pure, high specific activity ^{67}Ga.

Table II. ^{67}Ga-Labeled Monodisperse Aluminosilicate Spheres for Inhalation Studies with Dogs and Humans[a]

Generator Concentration (μCi/ml at 3.5×10^8 Particles/ml)	Aerodynamic[b] Resistance Diameter D_{ar} (μm)	Geometric (real) Diameter (μm)	Expected Pulmonary Deposition in a Dog (μCi/min)	Expected Pulmonary Deposition in a Human (μCi/min)
1400	3.00	1.88	2.50	5.30
380	2.00	1.22	0.67	1.40
200	1.66	1.00	0.37	0.79
20	0.80	0.44	0.03	0.07
6	0.60	0.30	0.01	0.02

[a]Assumes a Lovelace Nebulizer, 20 liter/min diluting air and 25% pulmonary deposition.
[b]Aerodynamic resistance diameter $D_{ar} = D\sqrt{\rho C(D)/\rho^*}$ as defined by Raabe [54].

decays by both beta and positron emission to stable nuclides, has a 12.9-hr half-life and has been primarily used at ITRI as a short-lived, polydisperse, insoluble test aerosol. Neutron activation of stable ^{63}Cu produces ^{64}Cu which cannot be separated from the target material using chemical procedures; therefore, high specific activity levels are not possible.

Radon-220 (Thoron) from decay of natural ^{232}Th has also been used to tag clay aerosols. A short-lived alpha-emitting aerosol in respirable size ranges was needed to calibrate a monitoring device. Agopsowicz and Newton [33] aerosolized a suspension of montmorillonite clay into a holding chamber which contained ^{232}Th (either ~100 Welsbach gas mantles or 100 g of natural ThO_2). Thoron daughters became attached to the aerosolized clay particles and this system, schematically illustrated in Figure 8, resulted in an alpha-emitting short-lived (10.6 hr $T_{1/2}$) test aerosol.

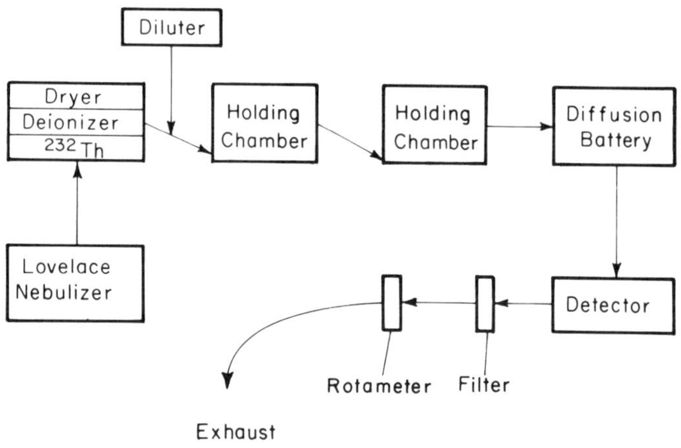

Figure 8. Block diagram illustrating production of short lived alpha-emitting test aerosol based on ^{220}Rn daughters attached to clay particles.

Ultrafine Aerosols

Ultrafine particles of trace metals, soot, polycyclic aromatic hydrocarbons, sulfate, etc., are significant air pollutants associated with combustion, conversion and other uses of fossil fuel. Additionally, high temperature events involving vaporization of nuclear materials may yield condensation aerosols in the ultrafine particle size range (< 100 nm). Deposition, retention and translocation of inhaled ultrafine particles are expected to be unique. Many important questions relating to the inhalation toxicity of ultrafine particles remain to be answered. Among these are: (1) efficiency of deposition in different regions of the respiratory tract, (2) fate, especially long-term disposition of inhaled ultrafine particles, and (3) microdistribution and dosimetry of ultrafine lung burdens of special concern such as $^{239}PuO_2$. To these ends, investigators at the ITRI have been developing generation techniques for ultrafine radiolabeled aerosols for several years.

Kanapilly et al. [34] and Tu and Kanapilly [35,36] generated tagged condensation aerosols in a variety of chemical and physical forms. Basically, the process involves direct vaporization and condensation or vaporization and thermal degradation with subsequent condensation. Ultrafine aerosols of rare earths and other metal oxides were produced by vaporization of a metal chelate of 2,2,6,6, tetramethyl 3,5 heptanedione (THD) [34]. The resultant vapors were oxidized and formed condensation aerosols of the order of 10 nm in size (primary particles). Aerosol growth rates were controlled and yielded ultrafine aggregate aerosols with diffusion diameters or count median diameters of 25-300 nm [37]. Examples of radioactively labeled metal oxide ultrafine aerosols suitable for laboratory animal inhalation studies and other investigations requiring ultrafine tagged aerosol are:

1. $^{239}PuO_2$ [34,37];
2. $^{144}CeO_2$;
3. $^{90}Y_2O_3$ (very high specific activity possible);
4. $^{67}Ga_2O_3$ (potentially useful for human studies);
5. $^{60}Co_3O_4$;
6. $^{57}Co_3O_4$; and
7. $^{140}La_2O_3$.

Many of these aerosols have properties suitable for use in studies as surrogates of environmentally significant aerosols. Examples are $^{67}Ga_2O_3$ chain aggregates for a diesel exhaust surrogate aerosol for use in deposition and short term retention studies in laboratory animals and possibly man. Longer term retention studies in animals could use $^{57}Co_3O_4$ ultrafine particles. Figures 9 and 10 are electron micrographs of aerosols of these types. Table III lists several ultrafine tagged metal oxide aerosols and their characteristics.

Carbon-14 labeled benzo[a]pyrene and pyrene ultrafine aerosols were prepared by Tu and Kanapilly [36] and Mitchell and Tu [38] using the apparatus shown schematically in Figure 11. Similarly, heterogenous condensation aerosols of pyrene coated onto ^{90}Y tagged Y_2O_3 nuclei were produced. The general shape, behavior and size distributions of the pyrene coated $^{90}Y_2O_3$ particles were similar to the pure pyrene aerosol similarly produced, but the tag enabled characterization of the aerosol with instrumental techniques not as readily adaptable to nonradioactive pyrene aerosols.

Ultrafine "sulfate" aerosols (H_2SO_4, $(NH_4)_2SO_4$, and NH_4HSO_4) labeled with ^{35}S have been prepared using controlled condensation techniques [39]. These ^{35}S-tagged aerosols have been used at ITRI for comparison of various

Figure 9. Transmission electron photomicrograph of ultrafine particles of $^{90}Y\text{-}Y_2O_3$ produced by vaporization and heat treatment of ^{90}Y labeled Y-THD.

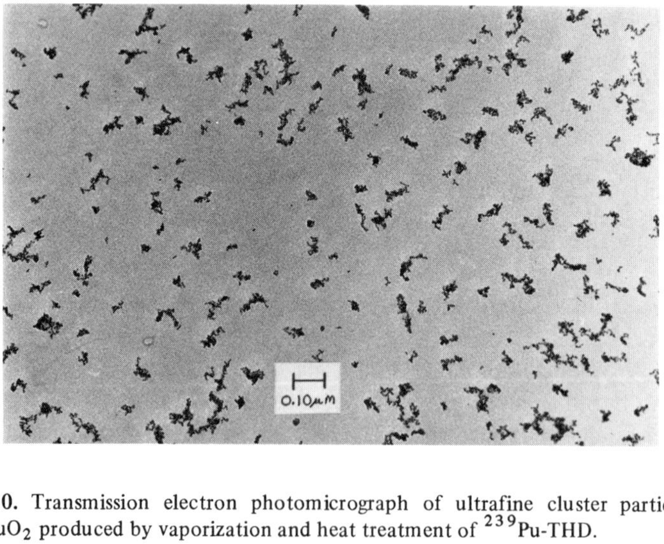

Figure 10. Transmission electron photomicrograph of ultrafine cluster particles of $^{239}PuO_2$ produced by vaporization and heat treatment of ^{239}Pu-THD.

Table III. Radiolabeled Ultrafine Aerosols Generated by Vaporization-Condensation Techniques

Radionuclide and Chemical Form	Size Range (μm)	σ_g	Shape or Type	Comments
^{90}Y-THD	0.1-0.5	1.3	Spherical	Unstable under EM[a] High S.A.[b] obtainable
$^{90}Y_2O_3$	0.05-0.2	1.5-1.7	Spherical or chain aggregates	Stable; high S.A. obtainable
^{90}Y-Ga_2O_3	0.1-0.3	1.3-1.7	Chain aggregates	Larger primaries
^{90}Y-Pyrene	0.1-0.5	1.4	Amorphous	Resembles pyrene except relatively stable
$^{67}Ga_2O_3$	0.01-0.1	1.4-1.5	Chain aggregates	Small primaries
$^{57}Co_3O_4$	0.01-0.3	1.4-1.7	Chain aggregates	Small primaries
$^{144}CeO_2$	0.01-0.3	1.5-1.7	Clusters (10-20)	Larger primaries
$^{239}PuO_2$	0.01-0.1	1.3-1.5	Clusters (10-20)	Larger primaries
^{14}C-B(a)P[c]	0.1-0.5	1.6	Irregular single	Unstable under EM

[a]Electron microscope.

[b]Specific activity.

[c]Benzo[a] pyrene

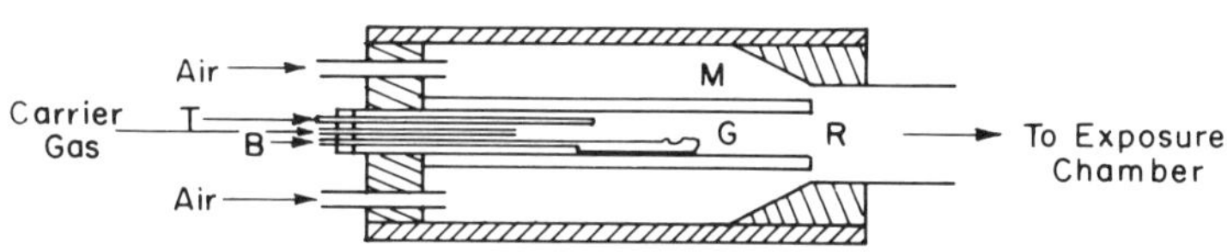

Figure 11. Schematic diagram of aerosol generator used to produce ^{14}C-benzo[a] pyrene and ^{14}C-pyrene ultrafine aerisols, where B is sample boat with provision for carrier gas injection over sample, G is vaporization region, M is dilution air jet region, R is condensation, diluting and mixing region and T is a temperature probe.

sizing instruments and also in laboratory animal inhalation studies to determine regional deposition.

Yttrium-90 tagged monodisperse ultrafine particles of fused aluminosilicate were prepared by size classification with the Electrical Aerosol Classifier (TSI, Inc., Box 43394, St. Paul, MN 55164) and then used to calibrate instruments such as the ten-stage piezoelectric cascade impactor (California Measurements Inc., 150 E. Montecito Avenue, Sierra Madre, CA 91024), and other classification devices based on the aerosol Brownian diffusion mechanism.

Gamma Labels for Plutonium Aerosols

The inhalation toxicity of actinides, principally ^{239}Pu, ^{238}Pu, ^{241}Am, ^{242}Cm and ^{244}Cm derived from nuclear fuel cycles is a fundamental concern in the development of nuclear-powered central electric generating stations. Inhalation toxicity studies of the above listed actinides are expensive and difficult to conduct. A primary problem is quantitating the initial lung burden in laboratory animals exposed by inhalation to insoluble forms of the oxides. Except for ^{241}Am, all of the above listed actinides decay by alpha emission and emit X-rays with insufficient energy for assay of the deposited lung burdens by whole-body counting. Raabe et al. [40] listed criteria for selecting a gamma-emitting radionuclide suitable for labeling $^{238}PuO_2$ monodisperse particles. These criteria include:

1. a specific activity in excess of 460 Ci/g so that the mass of the label is small ($<3\%$) compared to the mass of ^{238}Pu;
2. a half-life sufficiently long for material preparation, use and short-term observation (a few weeks) in animals;
3. the dose to lung or other tissues from the label should be an insignificant fraction of the total dose received;
4. chemical properties that allow uniform labeling of $^{238}PuO_2$ and yield dissolution rates comparable to the dissolution rates of $^{238}PuO_2$ particles;
5. commercial availability at reasonable cost; and
6. the radioactive label should emit an adequate number of gamma photons in the range of 100-200 keV so that in vivo estimates of the deposited lung burdens based on whole-body counting techniques are practical.

Two radionuclides were chosen for study as potential actinide labels: ^{51}Cr and ^{169}Yb. Chromium-51 decays by electron capture with a half-life of 27.8 days and emits 325 keV photons accompanying 9% of the disintegrations. Ytterbium-169 also decays by electron capture, with a half-life of 31.3 days, and among the photons emitted are energies of 310 keV and 200 keV in 11% and 49% of the disintegrations, respectively. Tests on ^{51}Cr labeled $^{238}PuO_2$ particles indicated that tagging was neither uniform nor adequate. Similar studies on uniformity of labeling and dissolution properties of ^{169}Yb-$^{238}PuO_2$ particles indicated that ^{169}Yb was an acceptable short-term tag for $^{238}PuO_2$ particles. About 93% of the original ^{169}Yb radioactivity remained in the $^{238}PuO_2$ particles after 3 weeks of in vitro tests. Subsequent animal inhalation studies indicated a similar relationship between the primary aerosol, $^{238}PuO_2$, and the label.

Commercially available (General Electric Co., Vallecitos Nuclear Center, Pleasanton, CA 94566) ^{169}Yb with specific activity in excess of 1000 Ci/g has been used at ITRI since 1974. Ytterbium-169 is produced by neutron activation of stable ^{168}Yb which has been enriched from the natural abundance of 0.14% to about 22% in Oak Ridge National Laboratory's stable isotope separation program. The high specific activity of ^{169}Yb also make it suited for exchange into montmorillonite clay. Ytterbium-169 labeled monodisperse aluminosilicate particles have been prepared with an activity level of 22 dpm per 1.0-μm diameter particle.

Miscellaneous Methods of Radioactive Labeling Aerosols

Neutron Activation

Induced radiation by neutron activation to create radioactive isotopes of stable nuclides is a potentially useful technique. An 8-hr neutron activation of 1 g of gold (^{197}Au) is illustrated by

$$SA = \sigma f \frac{n(1-e^{-\lambda t_1})\, e^{-\lambda t_2}}{3.7 \times 10^{10}} = 36.94 \text{ Ci/g} \tag{5}$$

where σ = thermal neutron cross section activation area of gold = 9.8×10^{23} cm^2 (98 barns)

f = neutron flux = 8.0×10^{13} $n/cm^2/sec$ (Omega West Reactor at the Los Alamos Scientific Laboratory (LASL))

n = number of target nuclei per gram of gold = 3.06×10^{21}

t_1 = irradiation time = 8 hr

t_2 = time between removal from reactor and use (~ 34 hr)

$$\lambda = \frac{\ln 2}{T_{1/2}} = \frac{0.693}{64.75 \text{ hr}} = \text{gold-198 decay constant}$$

Monidisperse gold particles labeled with ^{198}Au suitable for inhalation studies in laboratory animals and possibly in man have been produced [32]. This procedure involves generation of polydisperse gold aerosols from either "fulminating gold," formed by the reaction

$$4\,AuCl_4^- + 4\,NH_4OH + H_2O \rightarrow \underbrace{[Au_2O_3 \cdot NH_3 + NH(AuNH_2Cl)_2]}_{\text{fulminating gold ppt}} + H_2O \tag{6}$$

or Au_2O_3, collection of monodisperse particles using a spiral duct aerosol centrifuge [40,41] and subsequent neutron activation of gold particles [32]. A recent modification of the procedure used neutron activation of the Au_2O_3 bulk particles and then aerosolization, heat treatment, separation (aerosol centrifugation) and use.

Similar techniques have also been used to label prepared fused aluminosilicate particles containing suitable target nuclei in the exchange sites. If nuclides with large activation cross sections are used, whose activation products have suitable decay characteristics, relatively high specific activities are achievable. An example is activation of natural europium (^{151}Eu and ^{153}Eu) to radioactive ^{152}Eu ($T_{1/2}$ = 12.4 yr) and ^{154}Eu ($T_{1/2}$ = 1.8 yr) respectively. Calculated induced radioactivity in Eu incorporated into fused aluminosilicate particles is 2.32 Ci/g clay for ^{152}Eu and 0.64 Ci/g clay for ^{154}Eu assuming a one-week irradiation of LASL's Omega West reactor.

Neutron activation of coal combustion fly ash has also been used at ITRI to induce a tag for animal inhalation exposure studies [42]. In this example, the irradiated fly ash was stored for 30 days to allow for decay of short-lived products. The actual tag studied was ^{46}Sc. Scandium is a trace constituent in fly ash but behaves similarly to matrix elements. Scandium-46 decays by beta emission to stable ^{46}Ti, has an 84-day half-life and emits gamma photons of 1.12 MeV and 0.89 MeV with every disintegration. Fly ash aerosols used in inhalation studies were in the respirable size range with an AMAD of 2 to 4 μm and σ_g of $\cong$ 2.

A ^{182} Ta tag was induced by neutron activation in dust obtained from machining of tantalum castings. The tantalum dust particles were a potential broncho-imaging material for radiographic purposes. Radioactive tantalum was required to characterize the aerosol and to obtain retention kinetics of inhaled material from animal inhalation studies.

Exploding Wires

Wires exploded by sudden electrical discharge form highly disperse aerosols. Phalen [43] produced aerosols of silver tagged with ^{110}Ag by exploding a wire that had been neutron activated. Explosion was by sudden discharge of a 6000-joule capacitor through the small diameter wire. Others [44-46] have used similar or modified techniques to form a labeled aerosol.

Vaporization

Raabe et al. [47] used laser energy to ignite delta-phase plutonium metal and produced highly dispersed aerosols. An example of one of the aerosols so formed is shown in Figure 12. This technique could be adapted to generate highly dispersed aerosols of suitable pyrophoric material containing a radioactive label.

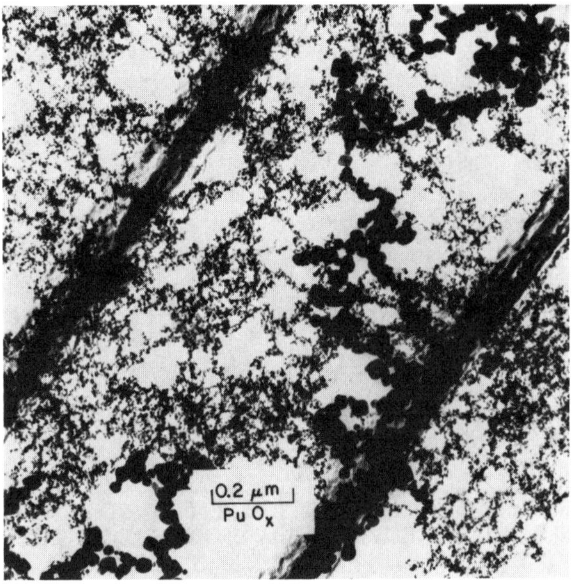

Figure 12. Transmission electron photomicrograph of ultrafine $^{239}PuO_2$ aerosols produced by laser ignition and burning of plutonium metal. Parallel lines are a replica of a diffraction grating used for accurate sizing and are 11,340 lines/cm.

Monodisperse Aerosol Generation

Several methods for generating both soluble and "insoluble" radioactively labeled aerosols have been presented. Various combinations of techniques discussed herein can be used to generate additional tagged aerosols for a variety of applications.

References to monodisperse aerosol generation occur in several methods. Monodisperse aerosols are desirable in many basic and applied studies in aerosol science. Raabe [48] summarized methods of monodisperse particle generation and included:

1. Labeling of preformed monodisperse test particles such as PSL. Techniques include surface attachment of suitable radionuclides.
2. Collection and separation of polydisperse distributions of radioactive spherical uniform density particles using spiral duct aerosol centrifuges and subsequent aerosolization of monodisperse fractions. This technique is limited by requirements to limit generation of multiples of the primary particles. Techniques limited to relatively small number concentrations.

3. Aerosolization of labeled solutions or colloidal suspensions using a vibrating orifice technique [49-52]. This technique is limited to relatively small number concentrations.
4. Spinning disc generators can be used to produce both soluble and insoluble monodisperse particles but have limitations similar to the "vibrating orifice."
5 Electrostatic separation using an electrical mobility analyzer [53], restricted to smaller size ranges (<1.0 μm).
6. Electrostatic dispersion of charged filaments which break up uniformly and must be dispersed and discharged quickly to achieve usable monodisperse droplets.
7. Controlled condensation aerosols of di-2-ethylhexyl sebacate, Carnuba wax, oleic acid, dioctyl phthalate and other materials.

SUMMARY

Inhalation toxicity studies of both fission product radionuclides and actinides derived from nuclear fuel cycles have been facilitated by selection of appropriate radioactive labels to enhance detection of the material under study. In many cases, use of radioactive labels provided the only practical solution to biological radiation dose pattern studies. Many aspects of basic aerosol behavior have also utilized tagged particles.

We believe that radioactive aerosols have not been used in many basic studies for a variety of reasons. Some of these reasons are: lack of familiarity with generation, detection techniques and quantitation, licensing requirements, health physics concerns and long-term disposal. Table IV lists a group of easily obtainable radionuclides, suggested forms and generation techniques and emission characteristics. These aerosols have a potential for enhancing the speed and accuracy of many studies in basic and applied problems.

ACKNOWLEDGMENTS

This study was supported by U.S. Department of Energy Contract Number EY-76-C-04-1013 and reports research conducted in facilities fully accredited by the American Association for Accreditation of Laboratory Animal Care. In addition to the authors, this work represents efforts of numerous professional and technical contributors at ITRI over the past two decades. In particular, the authors recognize contributions from Dr. T. T. Mercer, Dr. R. G. Thomas, Dr. R. O. McClellan, Dr. Mary McKnight, Mr. Jerry Bennick and Mr. R. W. E. Norgon.

Table IV. Radionuclides Useful for Labeling Respirable Sized Aerosols for Animal Inhalation and Basic Aerosol Studies

Radionuclide	T-1/2 (days)	Major Emission Characteristics MeV	γ-Yield (%)	Comments
^{90}Y	2.7	2.24 β	Pure β-emitter	FP[a], as soluble salts, insoluble: oxides, phosphates and in fused clay. Very high S.A. attainable. Separated from ^{90}Sr, inexpensive, reusable source.
^{88}Y	108	1.84 γ	100	$^{88}Sr(p,n)^{88}Y$, can be separated from target, soluble and insoluble forms similar to ^{90}Y.
^{67}Ga	3.25	0.092 γ 0.182 0.296	69 24 22	$^{68}Zn(p,2n)^{67}Ga$, readily available, low dose, human studies. If separated from target nuclei can be exchanged into clay with high S.A.
^{64}Cu	0.54	≅ 0.6 B^+,B^-		$^{63}Cu(n,\gamma)^{64}Cu$, low yield 0.5% γ 1.34 MeV, suitable for low level exchange into clay. Easily handled β emitter. Readily available.
^{57}Co	270	0.122 γ 1.292 γ	87	$^{56}Fe(p,\gamma)^{57}Co$, low energy β emitter, as soluble salts, insoluble clay.
^{59}Fe	45.6	1.095 γ	56	$^{58}Fe(n,\gamma)^{59}Fe$, energetic gamma emitter for use in tagging iron colloids.
^{169}Yb	31.3	4 γs (0.110-0.300) 0.300)	10-35	$^{168}Yb(n,\gamma)^{169}Yb$, very high S.A. available, tag for plutonium, high S.A. in clay (22 dpm/1.0 μm dia. particle).
^{147}Pm	950	(0.224)	Pure β-emitter	FP[a], high S.A. available, difficult to detect.
^{137}Cs	10,957	0.662 γ	85	FP[a], medium energy β-γ-emitter, easily counted, most forms soluble, insoluble in fused clay.
^{134}Cs	747	Many γs = 0.5-1.2	2-99	$^{133}Cs(n,\gamma)^{134}Cs$, higher S.A. available than ^{137}Cs also more countable γs. Soluble, insoluble in fused clay.
^{103}Ru	39.5	γ 0.497	88	FP[a] easily attached to PSL or PVT particles from $^{103}RuO_4$ (gas)

[a]FP = fission product.

REFERENCES

1. Raabe, O. G., G. M. Kanapilly and G. J. Newton. "New Methods for the Generation of Aerosols of Insoluble Particles for Use in Inhalation Studies," *Inhaled Particles III*, Proceedings of an International Symposium, British Occupational Hygiene Association, London, September 14-23, 1970 (Surrey, England: Unwin Bros., Ltd., 1971), pp. 3-18.
2. Kanapilly, G. M., O. G. Raabe and G. J. Newton. "A New Method for the Generation of Aerosols of Insoluble Particles," *Aerosol Sci.* 1:313-323 (1970).
3. Kotrappa, P., and M. E. Light. "Design and Performance of the Lovelace Aerosol Particle Separator," *Rev. Sci. Instrum.* 43:1106-1112 (1972).
4. Mercer, T. T., M. I. Tillery and G. J. Newton. "A Multi-Stage, Low Flow Rate Cascade Impactor," *Aerosol Sci.* 1:9-15 (1970).
5. Kotrappa, P., and C. J. Wilkinson. "Densities in Relation to Size of Spherical Aerosols Produced by Nebulization and Heat Degradation," *Amer. Ind. Hyg. Assoc. J.* 33:449-453 (1972).
6. Mercer, T. T. "On the Role of Particle Size in the Dissolution of Lung Burdens," *Health Phys.* 13:1211-1221 (1967).
7. Kanapilly, G. M., and C. H. T. Goh. "Some Factors Affecting the *In Vitro* Rates of Dissolution of Respirable Particles in Relatively Low Solubility," *Health Phys.* 25:225-237 (1973).
8. Kanapilly, G. M., O. G. Raabe, C. H. T. Goh and R. A. Chimenti. "Measurement of *In Vitro* Dissolution of Aerosol Particles for Comparison to *In Vivo* Dissolution in the Lower Respiratory Tract After Inhalation," *Health Phys.* 24:497-507 (1973).
9. Lundgren, D. L., F. F. Hahn, C. R. Crain and A. Sanchez. "Effect of Influenza Virus Infection on the Pulmonary Retention of Inhaled ^{144}Ce and Subsequent Survival of Mice," *Health Phys.* 34:557-567 (1968).
10. Barnes, J. E., G. M. Kanapilly and G. J. Newton. "Cobalt-60 Oxide Aerosols: Methods of Production and Short-Term Retention and Distribution Kinetics in the Beagle Dog," *Health Phys.* 30:391-398 (1976).
11. Black, A., and M. Walsh. "The Preparation of Bromine-82 and Iodine-131 Labeled Polystyrene Microspheres with Diameters from 0.1 to 30 Microns," *Ann. Occup. Hyg.* 13:87-100 (1970).
12. Bogen, D. D. "Preparation of Radioactive Labeled Polystyrene Latex Monodisperse Submicron Aerosols," *Am. Ind. Hyg. Assoc. J.* 31:349-352 (1970).
13. Hinrichs, R. J., J. L. Kenoyer, R. F. Phalen and T. T. Crocker. "Labeling of Monodisperse Polystyrene Microspheres with Tightly Bound ^{51}Cr," *Am. Ind. Hyg. Assoc. J.* 39:7, 570-575 (1978).
14. Szende, G., and K. Udvarheli. "Production and Labeling of Monodisperse Polystyrene-Vinyl Toluene Copolymer Latexes," *Int. J. Appl. Radiat. Isot.* 26:53 (1975).

15. Hass, P. J., P. S. Lee and R. V. Lourenco. "Tagging of Iron Oxide Particles with ^{99m}Tc for Use in the Study of Deposition and Clearance of Inhaled Particles," *J. Nucl. Med.* 17:122-125 (1976).
16. Yeates, D. B., A. Warbick and N. Aspin. "Production of ^{99m}Tc Labeled Albumin Microspheres for Lung Clearance Studies and Inhalation Scanning," *Int. J. Appl. Radiat. Isot.* 25:578-580 (1974).
17. Few, J. D., M. D. Short and M. L. Thompson. "Preparation of ^{99m}Tc-Labeled Particles for Aerosol Studies," *Radiochem. Radioanal. Letters* 5(6):275-277 (1970).
18. Camner, P., P. A. Hellstrom, M. Lundborg and K. Philipson. "Lung Clearance of 4-μm Particles Coated with Silver, Carbon or Beryllium," *Arch. Environ. Health* 32:58-62 (1977).
19. Camner, P., K. Philipson and J. Svedberg. "Production of 7-μm Monodisperse Fluorocarbon Resin Particles Tagged with ^{18}F," *Int. J. Appl. Radiat. Isot.* 22:349-353 (1971).
20. Ginnell, W. S., and G. P. Simon. "Preparation of Tagged Spherical Clay Particles," *Nucleonics* 2:49 (1953).
21. Amphlett, C. B. *Inorganic Ion Exchangers,* (Amsterdam: Elsevier, 1964).
22. Gaines, G. L. Jr., and H. C. Thomas. "Adsorption Studies on Clay Minerals. V. Montmorillonite-Cesium-Strontium at Several Temperatures," *J. Chem. Phys.* 23:2322 (1955).
23. Eliason, J. R. "Montmorillonite Exchange Equilibrium with Strontium-Sodium-Cesium," *Am. Mineralogist* 51:324 (1966).
24. Lewis, R. J., and H. C. Thomas. "Adsorption Studies on Clay Minerals. VIII. A Consistency Test of Exchange Sorption in the Systems Sodium-Cesium-Barium Montmorillonite," *J. Phys. Chem.* 67:1781 (1963).
25. Frysinger, G. R., and H. C. Thomas. "Adsorption Studies on Clay Minerals. VII. Yttrium-Cesium and Cerium(III)-Cesium on Montmorillonite," *J. Phys. Chem.* 64:224 (1960).
26. Amphlett, C. G. "Ion Exchange in Clay Minerals," *Endeavor* 17:149 (1958).
27. Van Olphen, H. *Clay* (New York: Interscience Publishers, 1963).
28. McKnight, M. E., and R. W. E. Norgon. "A Study of the Exchange Characteristics of Montmorillonite Clay for Fission Product Cations for Use in the Generation of Insoluble Aerosols," AEC Research and Development Report, LF-37, 1967.
29. Boecker, B. B., R. G. Thomas and R. O. McClellan. "Accumulation and Retention of ^{137}Cs-Labeled Fused Aluminosilicate Particles by Beagle Dogs after Repeated Inhalation Exposures," in *Inhaled Particles IV,* W. H. Walton, Ed. (New York: Pergamon Press, 1977), pp. 625-637.
30. Thomas, R. G. "Retention Kinetics of Inhaled Fused Aluminosilicate Particles," in *Inhaled Particles III*, Proceedings of an International Symposium, British Occupational Hygiene Association, London, September 14-23, 1970, (Surrey, England: Unwin Bros., Ltd., 1971), pp. 193-200.

31. Kanapilly, G. M., and G. J. Newton. "A New Method for the Separation of Multicurie Quantities of ^{90}Y from ^{90}Sr," *Int. J. Appl. Radiat. Isotop.* 22:567-575 (1971).
32. Newton, G. J., O. G. Raabe, R. L. Yarwood and G. M. Kanapilly. "Generation of Monodisperse Aerosols of ^{67}Ga-Labeled Aluminosilicate and ^{198}Au-Labeled Gold Spheres," in *Fine Particles, Aerosol Generation, Measurement Sampling and Analysis*, B. Y. H. Liu, Ed. (New York: Academic Press, 1976), pp. 129-144.
33. Agopsowicz, D. E., and G. J. Newton. "A Real-Time Alpha-in-Air Monitoring System," *Health Phys.* (in press).
34. Kanapilly, G. M., K. W. Tu, T. B. Larsen, G. R. Fogel and R. J. Luna. "Controlled Production of Ultrafine Metallic Aerosols by Vaporization of an Organic Chelate of the Metal," *J. Colloid. Interface Sci.* 65: 533-547 (1978).
35. Tu, K. W., and G. M. Kanapilly. "Generation and Characterization of Condensation of Vanadium Pentoxide and Pyrene," *Am. Inst. Hyg. J.* (in press).
36. Tu, K. W., and G. M. Kanapilly. "A High Capacity Condensation Aerosol Generation System," *Environ. Sci. Technol.* 13(6):698-701 (1979).
37. Kanapilly, G. M., and J. H. Diel. "Ultrafine $^{239}PuO_2$ Aerosol Generation Characterization and Short Term Radiation Dose Pattern Study in the Rat," *Health Phys.* (submitted).
38. Mitchell, C. E., and K. W. Tu. "Distribution and Retention of ^{14}C-Benzo[A]Pyrene in Rats Following Inhalation," *Inhalation Toxicology Research Institute Annual Report, 1977-1978*, LF-60, December 1978.
39. Tu, K. W., and G. M. Kanapilly. "Generation and Characterization of Submicron Ammonium Sulfate Aerosols," *Atmos. Environ.* 12:1579-1611 (1978).
40. Raabe, O. G., H. A. Boyd, G. M. Kanapilly, C. J. Wilkinson and G. J. Newton. "Development and Use of a System for Routine Production of Monodisperse Particles of $^{238}PuO_2$ and Evaluation of Gamma Emitting Labels," *Health Phys.* 28:655-667 (1975).
41. Kotrappa, P., and O. R. Moss. "Production of Relatively Monodisperse Aerosols for Inhalation Experiments by Aerosol Centrifugation," *Health Phys.* 21:531-536 (1971).
42. Griffis, L. C. "Inhalation and Clearance of Neutron-Activated Fly Ash in the Rat," *Inhalation Toxicology Research Institute Annual Report, 1977-1978*, LF-60, Albuquerque, NM, 1978.
43. Phalen, R. F. "Evaluation of an Exploded-Wire Aerosol Generator for Use in Inhalation Studies," *Aerosol Sci.* 3:395-406 (1972).
44. Couchman, J. C. "Metallic Microsphere Generation," (Santa Barbara, CA: E. G. & G., Inc., 1966).
45. Spurny, K., and J. P. Lodge, Jr. "Production of Highly Dispersed Model Aerosols," *Staub-Reinhalt Luft* 33:171-173 (1973).

46. Chace, W. G., and H. K. Moore Eds. *Exploding Wires* (New York: Plenum Press, 1962).
47. Raabe, O. G., S. V. Teague, N. L. Richardson and L. S. Nelson. "Aerodynamic and Dissolution Behavior of Fume Aerosols Produced During the Combustion of Laser-Ignited Plutonium Droplets in Air," *Health Phys.* 35:663-674 (1978).
48. Raabe, O. G. In: *Fine Particles, Aerosol Generation Measurement Sampling and Analysis*, B. Y. H. Liu, Ed. (New York: Academic Press, Inc., 1976), pp. 57-110.
49. Castleman, R. A. "The Mechanism of the Atomization of Liquids," *Bur. Std. J. Res.* 6:369-376 (1931).
50. Raabe, O. G., and G. J. Newton. "Development of Techniques for Generating Monodisperse Aerosols with the Fulwyler Droplet Generator," in *Fission Product Inhalation Program Annual Report*, 1969-1970, LF-43, Lovelace Foundation for Medical Education and Research, Albuquerque, New Mexico, pp. 13-17, 1970.
51. Fulwyler, M. J., J. D. Perrings and L. S. Cram. "Production of Uniform Microspheres," *Rev. Sci. Instrum.* 44:205-206 (1973).
52. Berglund, R. N., and B. Y. H. Liu. "Generation of Monodisperse Aerosol Standards," *Env. Sci. Tech.* 7:147-153 (1973).
53. Liu, B. Y. H., D. Y. H. Piu, A. W. Hogen and T. A. Rich. "Calibration of the Pollak Counter with Monodisperse Aerosols," *J. Applied Meteor.* 14:46-51 (1975).
54. Raabe, O. G. "Aerosol Aerodynamic Size Conventions for Inertial Sampler Calibration," *APCA Journal* 26:856-860 (1976).

CHAPTER 20

PARTICLE SIZE-RANGE SELECTION BY OPPOSING-JET CLASSIFICATION

Klaus Willeke and Robert E. Pavlik*

Aerosol Research Laboratory
Department of Environmental Health
University of Cincinnati
Cincinnati, Ohio 45267

INTRODUCTION

In inhalation experiments where airborne particles are ducted to the breathing zone of animal or human subjects, it is frequently desirable to have particles of a rather narrow size range so that the effect of particle size on biological effect may be determined. The particle size plays an important role in the biological response of the test subject because:

1. The chemical composition of airborne particulate contaminants is frequently size-dependent.
2. The fraction of ambient particles entering the respiratory system is particle size dependent.
3. The deposition site in the respiratory system is particle size dependent.
4. The size and shape of insoluble particles, for instance the needle-like shape of fibers, may determine the type and extent of damage at the site of lung deposition.

Monodisperse particles, i.e., particles of a single size, may be generated by a vibrating orifice [1, 2, 3], a spinning disc [4, 5], dispersion of polystyrene latex spheres [6], or evaporation-condensation of nebulized aerosols with subsequent extraction of a narrow size fraction by means of an electrical

*Present address: Naval Regional Medical Center, San Diego, California 92134.

aerosol size classifier [7]. However, most aerosols occurring in industrial or ambient air environments cannot be generated by these techniques. Also, these artificial aerosols are generally produced only in low concentrations. Such techniques are therefore used primarily for the purpose of instrument development and calibration.

In inhalation experiments, dusts larger than about one micrometer in size are generally redispersed dusts of a wide particle-size distribution. Particle-size classification techniques such as cyclones [8] and inertial impactors [9] remove the larger size fraction and retain the lower one. Cyclones and impactors therefore cannot be used to extract a narrow size range from a polydisperse aerosol cloud. In the virtual or dichotomous impactor [10, 11], also called a centripeter [12], the aerosol-laden airjet is impinged into a slowly pumped air mass. However, the large-particle-fraction flow has small particles mixed into it and has a much smaller flowrate than the smaller-particle-fraction flow.

The technique described below separates particles into two aerodynamic size fractions, above and below a desired cut size, and ducts the two airborne size fractions away at equal flowrates. Subsequent size fractionation at a somewhat smaller or larger cut size results in the extraction of the desired particle-size range.

CLASSIFICATION PRINCIPLE

In the opposing-jet classifier, schematically shown in Figure 1, a jet of particle-laden air is directed against an equally strong axisymmetrically-opposed jet of clean air [13-15]. Particles with sufficient inertia deviate from the air streamlines and cross the fluid interface between the jets, while smaller particles continue on with the original air stream. As seen in Figure 1, a thin solid plate is placed between the opposing jets. The plate has a circular cutout which is aligned with the centerline of the opposing jets and separates the impinging air flow into two streams—the upper one containing the unclassified, low-inertia particles, and the lower one containing the classified, high-inertia particles. Once separated, the two particle fractions can be ducted to any desired location for further size classification or for any desired method of analysis, including direct-reading, continuous instrumentation. Once a narrow particle size range has been selected through two successive size classifications, the aerosols may be ducted to exposure chambers for particle size-dependent inhalation experiments.

Earlier attempts by other investigators [16-18] at using the opposing-jet classification principle did not result in satisfactory particle-size classification. This may have been caused by the presence of edgetone effects which occur when a critical H/W ratio is exceeded, where H is the size of the separation

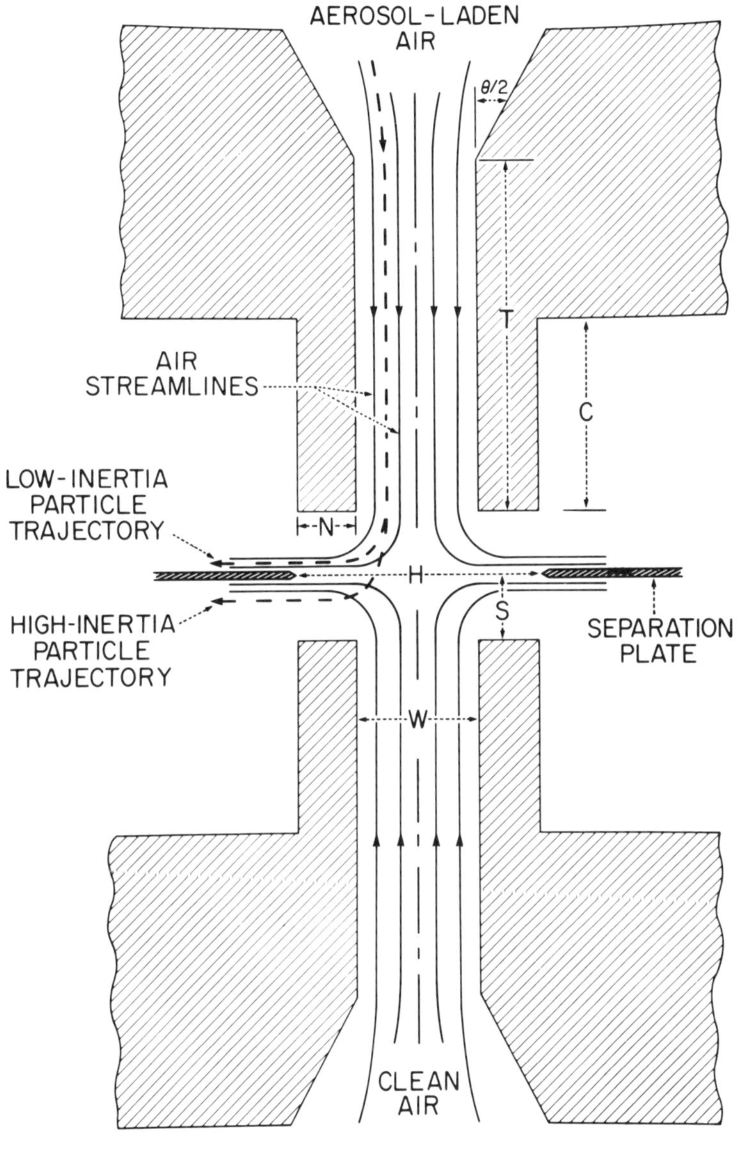

Figure 1. Schematic diagram of the opposing-jet particle-size classifier. Prototype dimensions are: C = 6.7 mm, H = 4 mm, N = 1.6 mm, S = 1.5 mm, T = 13.7 mm, W = 3.2 mm, $\theta = 60^{\circ}$, separation plate thickness = 0.15 mm, inner diameter of collecting chamber = 47.4 mm.

hole and W is the width or diameter of the incident jet. The edgetone effect is a fluid-dynamic instability which causes the interface to oscillate between the upper and lower chambers, thereby alternately transferring particles to either effluent stream irrespective of their size. The authors' investigations of this effect are described in a section below.

EXPERIMENTAL SETUP

The experimental setup used for the development of this technique is represented schematically in Figure 2. The test aerosols were generated by a vibrating-orifice monodisperse-aerosol generator [3], and were discharged to near-Boltzmann equilibrium by a 10-mCi, Kr-85 radioactive source. For the purpose of this technique development, the opposing-jet classifier was operated inside a glove box which was continuously supplied with clean air and maintained at a slightly positive pressure. The particle concentrations in the two flows emanating from the classifier were analyzed by an optical single-particle counter coupled to a multichannel analyzer.

Most experiments were performed with the overall feed device shown on the left in Figure 2. In order to test the hypothesis that some particles may transfer from the upper flow to the lower one and then bounce back into the upper flow, a special radial-scan experiment was designed in which aerosols were isokinetically injected at selected radial positions in the upper airflow. In this experiment, the devices shown in Figure 2 in the middle and on the right in the glove box were used. However, no significant size classification dependence on radial position was found [13].

PERFORMANCE

Figure 3 and Table I display the performance of the particle size classifier at different flowrates. The separation efficiency is obtained by dividing the particle count in the lower outlet by the sum of the counts from the upper and lower outlets. The difference in count between classifier inlet and outlet, divided by the inlet count, gives the particle loss. When a liquid oleic-acid-laden airjet of 7.1 liter/min (0.25 ft^3/min) was directed against a clean air-jet of the same flowrate, the particles were classified with an aerodynamic cut diameter of 2.9 μm, i.e., unit density aerosols of 2.9 μm diameter were equally distributed between both effluent streams when fed into the classifier through the upper inlet. As seen from Figure 3, the particle size classification is very sharp, $\sigma_g = 1.04$, with essentially all particles larger than 3.1 μm transferred to the lower chamber and essentially all particles smaller than 2.7 μm remaining with the upper airstream.

In order to adjust the cut to another size, the flowrate through the classifier

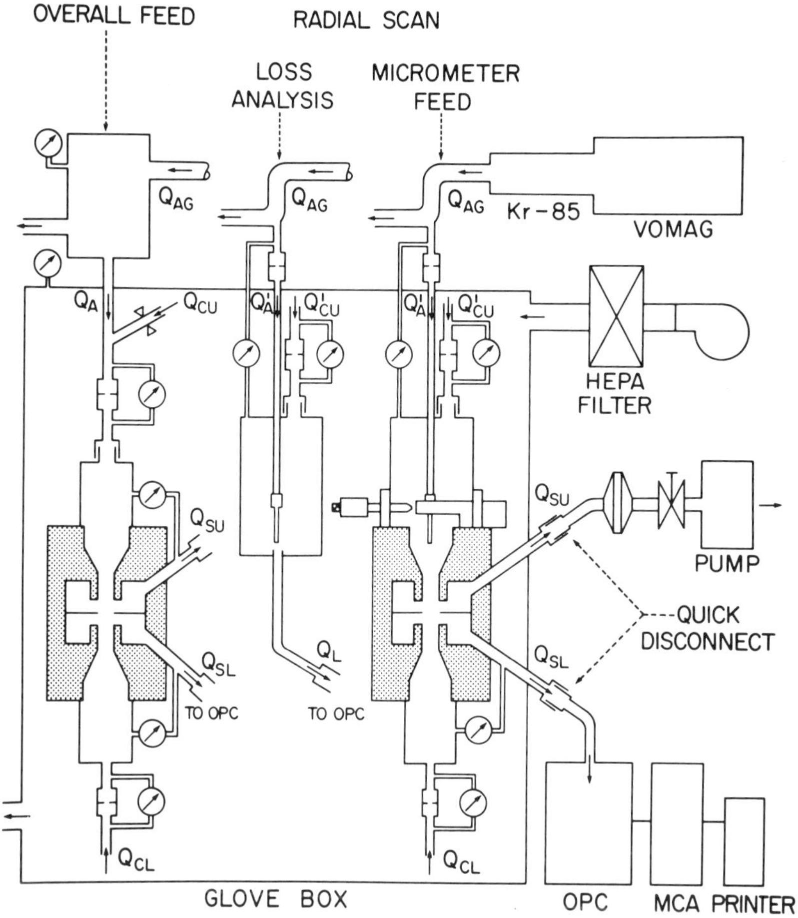

Figure 2. Experimental setup. The loss analysis device, shown in the center, gave the reference count during radial scan experiments. The devices to the left and right differ by their inlet design. During overall feed experiments only the left device was mounted inside the glove box. During radial scan experiments, the other two devices were mounted inside. VOMAG = Vibrating-Orifice Monodisperse Aerosol Generator, OPC = Optical Particle Counter, MCA = Multi-Channel Analyzer.

was varied. Theoretically, through the definition of Stokes number [19], the cut diameter at 50% separation efficiency, D_{50} should be inversely proportional to the square root of flowrate Q:

$$(D_{50})_2/(D_{50})_1 = (Q_1/Q_2)^{1/2}$$

Table I. Opposing-Jet Classification Performance at Different Flowrates[a]

Flowrate[b] (liter/min)	Jet Reynolds Number	50% Cut Size (μm)	Geom. Std. Deviation[c] σ_g	Maximum Particle Loss[d] (%)
3.5	1575	3.6	1.11	48
7.1	3140	2.9	1.04	76
14.2	6300	1.8	1.08	62
28.3	12,600	1.3	1.14	74
53.5	23,800	0.8	1.15	95

[a]W = 3.2 mm, H = 1.25, W = 4.0 mm.
[b]For each inlet or outlet flow.
[c]σ_g = [(separation efficiency)$_{84\%}$/(separation efficiency)$_{16\%}$]$^{1/2}$.
[d]Obtained with liquid oleic acid aerosols.

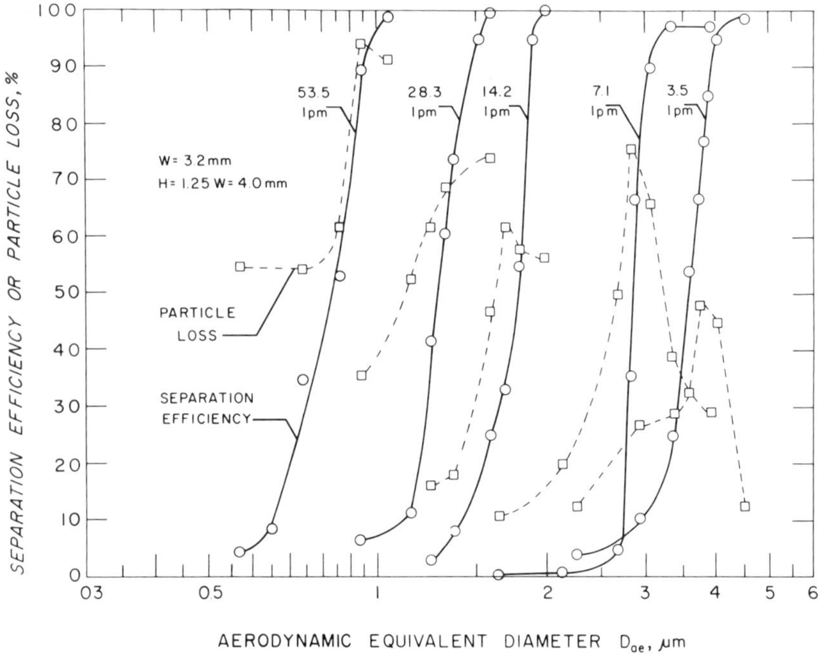

Figure 3. Classifier performance at different flowrates. Indicated flowrate is for each inlet or outlet flow. Tests with liquid oleic acid particles represent worst loss analysis.

For instance, for a given opposing jet geometry, a quadrupling in flowrate should reduce the cut diameter to half of the original value. However, since the Reynolds number varies with flowrate, this is only an approximation.

Experiments [14] showed that the opposing-jet classifier does indeed provide different size cuts simply through adjustments in the flowrates, approximately as predicted by the equation given above. As seen from Figure 3, the present design has a cut diameter of 3.6 μm at Q = 3.5 liter/min, Reynolds number = 1575; and 0.8 μm at Q = 53.5 liter/min, Reynolds number = 23,800. The cut size could not be lowered further for the present design because of high pressure-drops in the inlets and outlets during operation at the high flowrate. The separation efficiency degraded above 4 μm, apparently because the pressure gradient perpendicular to the interface was too low to maintain a stable interface position. It is anticipated that improvements in future classifier designs will extend the useful cut-size range at the upper and lower limits. Figure 3 also shows that the sharpest separation efficiency occurred at a Reynolds number near 3000 as is the case for solid plate impactors [9].

The losses for liquid aerosols are quite high near the 50% cut size. This is the size-range where the aerosols are closest to the fluid interface between the two jets, and may consequently impact and be lost onto the separation plate. Spectrofluorometric measurements of the deposits in the various parts of the classifier show that the area under the sharp spike in the particle loss curve represents the losses onto the separation plate whereas the remaining losses are primarily in the classifier inlet and in its outlets. The latter losses are not truly losses by the classification stage. It is expected that all of these losses can be lowered through improved designs, e.g., by reducing the thickness of the separation plate. When tests were performed with solid aerosols, the particle losses were considerably lower [13].

EDGETONE

Experiments showed that the particle classification performance depends strongly on the size H of the hole in the separation plate. As seen in Figure 4, a hole 25% larger than the jet width W (H = 1.25 W) results in sharp particle size classification, whereas for a larger hole size (H = 1.75 W) almost no size classification occurs. This is attributed to the onset of the edgetone effect [15]. The edgetone is a fluid-dynamic instability which occurs when the laterally deflected thin airjet impinges onto the inner rim of the separation plate, see Figure 5. It produces transverse oscillations which generate a sound field of the same frequency as the fluid oscillations. The effect is a familiar one since it is used to generate the sound in whistles, organ pipes and other musical instruments. In the opposing-jet classifier it degrades all classification

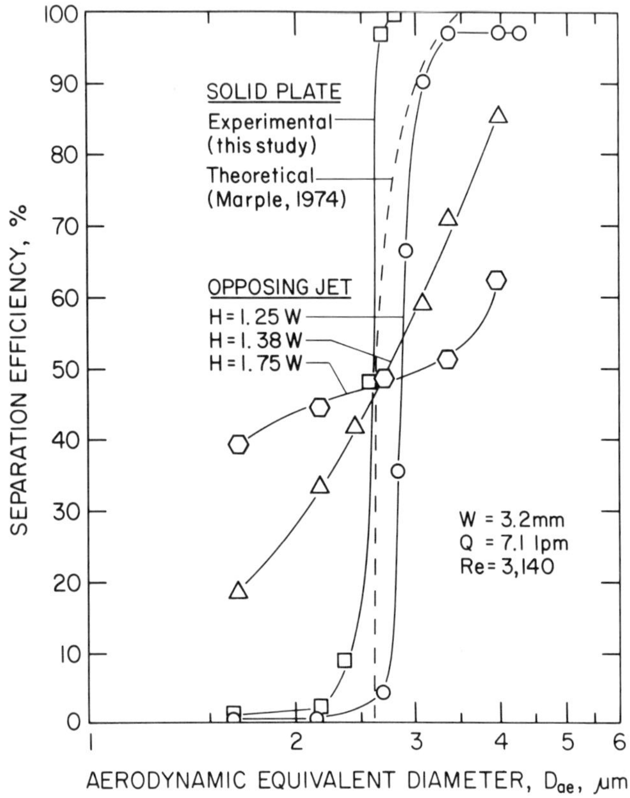

Figure 4. Particle-size classification for different separation-plate hole sizes.

which may have been achieved by the process of inertial impaction since the particles contained in the laterally deflected airstream are alternately deflected along with the airstream to the upper or lower chambers.

In order to study the occurrence and the characteristics of edgetones, sound measurements were taken with the classifier placed in a sound-absorbing room. A 1-inch ceramic microphone, suspended about 25 cm from the classifier, measured the sound pressure levels of edgetones over the frequency range 25 Hz to 25 kHz.

The edgetone intensity was very high for H/W = 1.75. Figure 6 shows how the frequency and intensity of the edgetone vary with flowrate for this hole dimension. Three frequency stages occurred when the flowrate was increased about ninefold from 4.5 liter/min to 40.2 liter/min. Similar frequency shifts have been observed in edgetone studies with rectangular jets [20, 21].

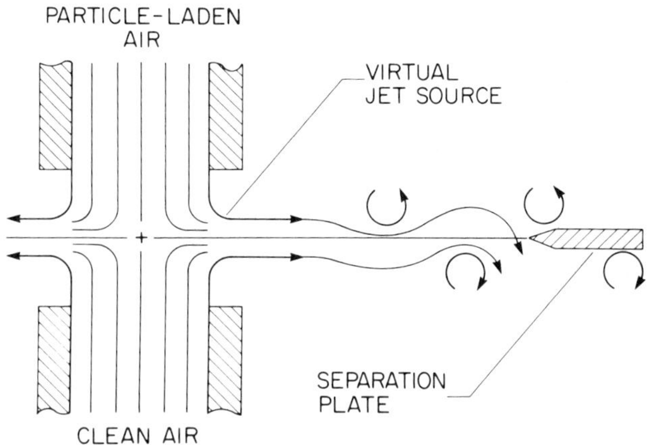

Figure 5. Schematic representation of fluid dynamic oscillations and vortex shedding when opposing-jet classifier is operated in edgetone mode.

The sound pressure levels, displayed in the lower part of Figure 6, exceeded 90 dB relative to 20 μN/m^2 for most of the second frequency stage. The data support previous edgetone studies which have reported that edgetone operation in a closed cavity is characterized by a resonance effect causing a peak in the intensity of the edgetone [22]. Present research focuses on the suppression of the edgetone effect.

PARTICLE LOSSES

Figure 7 illustrates particle deposition patterns observed on the separation plate [15]. The spatial histograms of the deposits distinguish between heavy and moderate particle losses to the separation plate. The locations and intensities of the deposits are approximate and represent averages for more or less ring-structured deposits.

In our studies with the opposing-jet classifier, we found ring or "halo" despoits on the separation plate, similar to those observed in conventional impactors [23]. When we operated the classifier as a conventional impactor by inserting a solid plate between the two airjets, such ring deposits were seen as well. In Figure 7 the ring deposits are shown for different chamber heights B. We have postulated that the deposits are formed as a result of vortex formations along the surfaces of the separation plate. Such vortex formations may be caused by entrainment of air as the high-velocity radial jet enters the expanded volume of the collection chamber, or by boundary layer separation as the radial airflow along the separation plate is decelerated [15]. As seen from Figure 7, the radius of the deposition ring R became

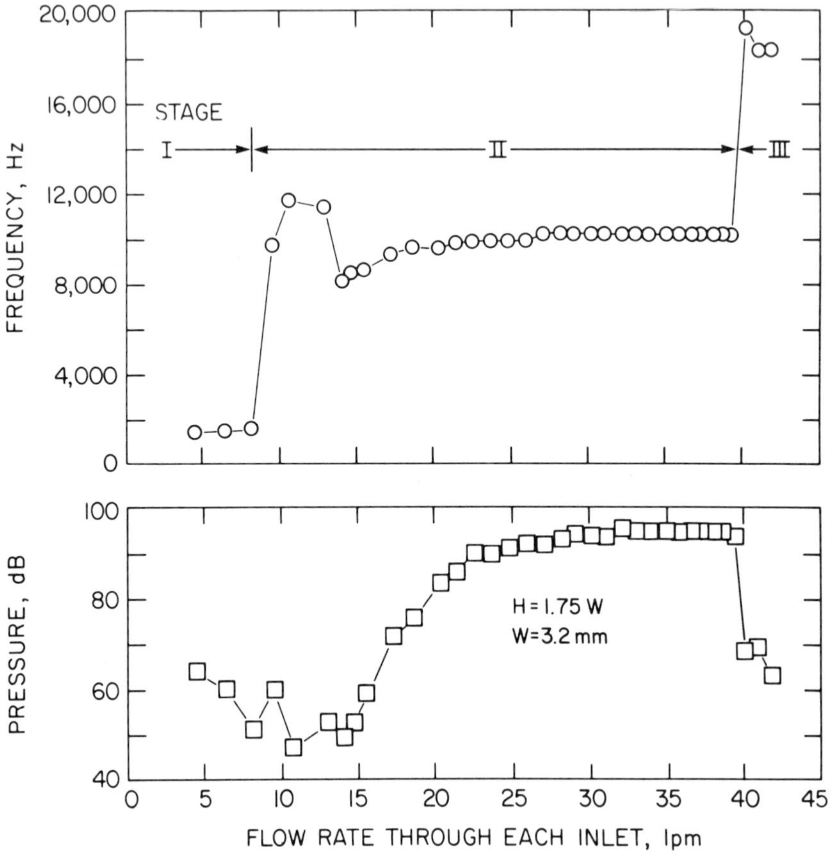

Figure 6. Peak frequencies and sound pressure levels re 20 μN/m^2 for opposing-jet classifier operated in edgetone mode.

smaller as the vertical chamber height B was reduced. For very low chamber height, the second ring was eliminated, probably because of flow attachment to the chamber walls beyond the first vortex. With a smaller chamber height, the secondary flow vortices located between the separation plate and the opposite collection chamber wall are forced to turn within a smaller space and are therefore smaller in size, as one may observe by referring to the R/B scale for the deposits in Figure 7. It should be noted that the ring deposits are not at the same location on each side of the separation plate because vortex formations in decelerated airflows are asymmetrical.

When, in another experiment, the chamber height B was kept constant and the chamber length was reduced, the deposit at about R = B was much

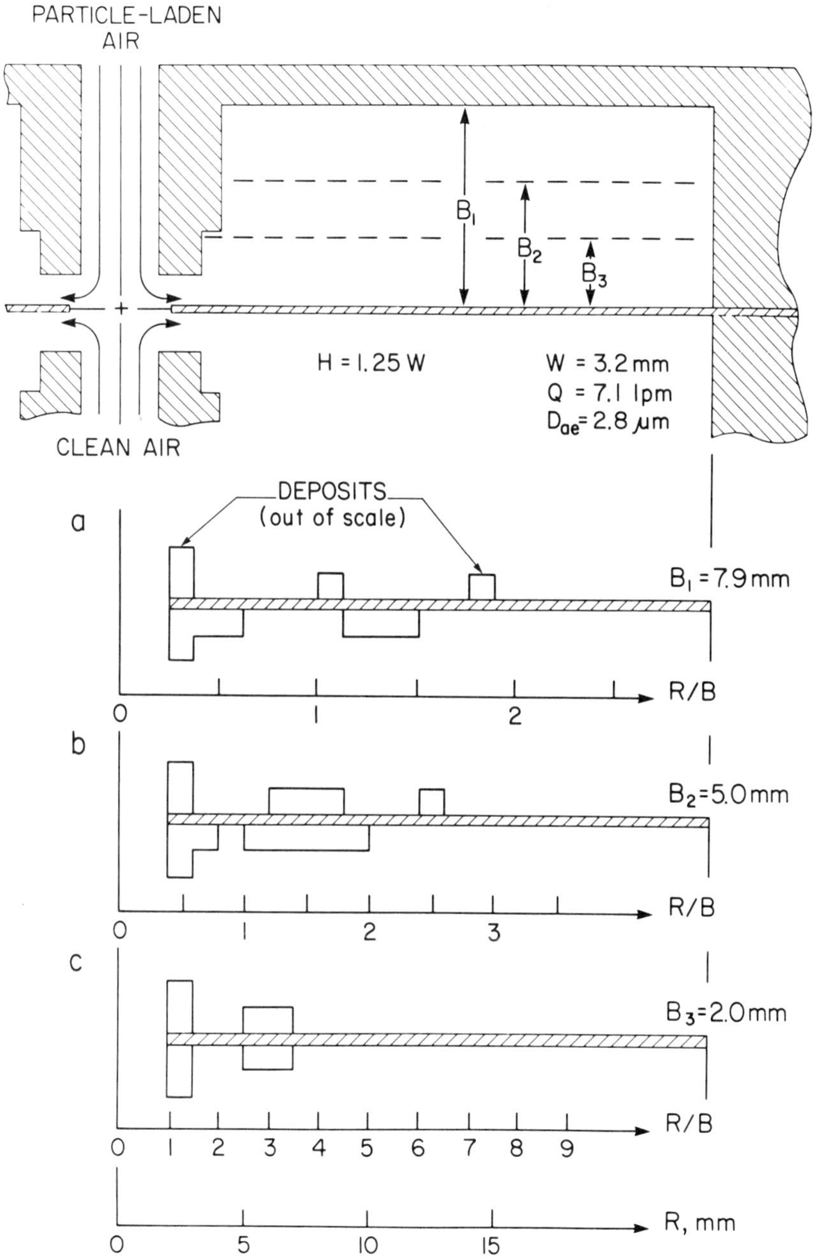

Figure 7. Particle loss deposition patterns for different chamber heights.

heavier than in previous tests and a much more sharply defined circular pattern was observed for the outer ring deposit. This suggests that the reduction in the chamber length made the inner vortex more energetic causing a heavier deposit, while the second, outer vortex was close to the chamber wall, forming a nearly perfect circular deposit.

Further support for the hypothesis that the ring deposits resulted from purely fluid-dynamic secondary flow patterns was provided by tests with particles of different sizes which yielded different separation efficiencies (8%, 36% and 94%) but showed deposition patterns very similar to each other. Research is currently being performed to reduce these losses.

CONCLUSION

The opposing-jet classifier provides sharp particle-size separation above and below a desired aerodynamic cut size with both effluent particle fractions remaining in the airborne state. A second application of this technique permits the extraction of a narrow particle-size range from a polydisperse aerosol cloud. Present research is focused on the suppression of the edgetone effect and the elimination of high particle losses for small separation-plate hole sizes. While the reduction of particle losses is desirable in order to make this technique generally useful, it is not essential for inhalation studies where a constant particle concentration has to be maintained in the exposure chamber and the losses in the classification device can be ignored as long as they are constant and particle deposits do not get reentrained into the effluent airstreams.

ACKNOWLEDGMENTS

This research was supported by grant ENG 77-04667-A01 from the National Science Foundation and is part of a Center Program supported by grant ES-00159 from the National Institute of Environmental Health Sciences. R. E. Pavlik was supported by the Ph.D. Outservice Training Program, Bureau of Medicine and Surgery, U.S. Navy.

REFERENCES

1. Ström, L. "Generation of Monodisperse Aerosols by Means of a Disintegrated Jet of Liquid," *Rev. Sci. Instrum.* 40:778-782 (1969).
2. Raabe, O. G., and G. J. Newton. "Development of Techniques for Generating Monodisperse Aerosols with the Fulwyler Droplet Generator," Fission Product Inhalation Program Annual Report 1969-1970, LF-43 (Albuquerque, NM: Lovelace Foundation, 1970), p. 13.

3. Berglund, R. N., and B. Y. H. Liu. "Generation of Monodisperse Aerosol Standards," *Environ. Sci. Technol.* 7:147-153 (1973).
4. Walton, W. H., and W. C. Prewett. "The Production of Sprays and Mists of Uniform Drop Size by Means of Spinning Disc Type Sprayers," *Proc. Phys. Soc. (London)* 62:341-350 (1949).
5. Lippmann, M., and R. E. Albert. "A Compact Electric-Motor Driven Spinning-Disc Aerosol Generator," *J. Am. Ind. Hyg. Assoc.* 28:501-506 (1967).
6. Raabe, O. G. "Generation of Aerosols of Fine Particles," In: *Fine Particles: Aerosol Generation, Measurement, Sampling and Analysis*, B. Y. H. Liu, Ed. (New York: Academic Press, Inc., 1976), p. 57.
7. Liu, B. Y. H. "Standardization and Calibration of Aerosol Instruments," In: *Fine Particles: Aerosol Generation Measurement, Sampling and Analysis*, B. Y. H. Liu, Ed. (New York: Academic Press, Inc., 1976), p. 39.
8. Chan, T., and M. Lippmann. "Particle Collection Efficiencies of Air Sampling Cyclones: An Emperical Theory," *Environ. Sci. Technol.* 11:377-382 (1977).
9. Marple, V. A., and K. Willeke. "Impactor Design," *Atmos. Environ.* 10:891-896 (1976).
10. Conner, W. D. "An Inertial-Type Particle Separator for Collecting Large Samples," *J. Air Poll. Control Assoc.* 16:35-38 (1966).
11. Dzubay, T. G., and R. K. Stevens. "Ambient Air Analysis with Dichotomous Sampler and X-Ray Fluorescence Spectrometer," *Environ. Sci. Technol.* 9:663-668 (1975).
12. Hounam, R. F., and R. J. Sherwood. "The Cascade Centripeter: A Device for Determining the Concentration and Size Distribution of Aerosols," *Am. Ind. Hyg. Assoc. J.* 26:122-131 (1965).
13. Willeke, K., and R. E. Pavlik. "Size Classification of Fine Particles by Opposing Jets," *Environ. Sci. Technol.* 12:563-566 (1978).
14. Pavlik, R. E., and K. Willeke. "Variable-Cut Particle-Size Classification by Opposing Jets," *Am. Ind. Hyg. Assoc. J.* 39:952-957 (1978).
15. Willeke, K., and R. E. Pavlik. "Secondary Effects in Particle-Size Classification by Opposing Jets," *J. Aerosol Sci.* 10:1-10 (1979).
16. Luna, R. E. "A Study of Impinging Axisymmetric Jets and Their Application to Size Classification of Small Particles," Ph.D. Thesis, Princeton University (1965).
17. Hall, R. E. "Axisymmetric Impinging Jet Flows and Their Application to Particle Size Classification," M.S. Thesis, University of Kentucky (1970).
18. Mears, C. E. "Two-Dimensional Impinging Jet Flows and Their Application to Particle Size Classification," M.S. Thesis, University of Kentucky (1973).
19. Fuchs, N. A. *The Mechanics of Aerosols* (New York: Pergamon Press, 1964), p. 139.

20. Brown, G. G. "The Vortex Motion Causing Edge Tones," *Proc. Phys. Soc. (London)* 49:493-507 (1937).
21. Kirshner, J. M. "Response of a Jet to a Pressure Gradient and its Relation to Edgetones," Proceedings of the Second International JSME Symposium on Fluid Machinery and Fluidics, Tokyo, Sept. 1972.
22. Nyborg, W. L. "Self-Maintained Oscillations of the Jet in a Jet-Edge System," *J. Acoust. Soc. Amer.* 26:174-182 (1954).
23. May, K. R. "Aerosol Impaction Jets," *J. Aerosol Sci.* 6:403-411 (1975).

SECTION III

EXPOSURE FACILITIES

CHAPTER 21

AEROSOL EXPOSURE METHODS

Morton Lippmann
Institute of Environmental Medicine
New York University Medical Center
New York, New York 10016

INTRODUCTION

To determine the nature of the potential hazard arising from the inhalation of aerosols and/or to establish dose-response relationships, it may be necessary to perform controlled inhalation exposure experiments. Studies of transient physiological effects can sometimes be performed on human volunteers, avoiding the uncertainties of interspecies variability. More frequently, the difficulties in obtaining approval for human experimental studies and their generally high costs limit the human studies option to verification of effects seen in experimental animals or make them impractical altogether. Furthermore, human studies are inappropriate whenever there is any forseeable possibility of producing persistent or delayed effects, dysfunction or damage. Thus, most inhalation studies are performed using experimental animals. The species, numbers exposed, concentration, duration and frequency of exposure, aerosol composition and size distribution, and method of exposure all depend upon the purpose of the study and on any special precautions needed for the safe and efficient performance of the study.

This chapter discusses the general features of the various techniques which have been used to expose human volunteers and experimental animals to test aerosols, and the factors influencing the selection of an appropriate exposure technique for a particular study.

EXPOSURE TECHNIQUES

The various exposure techniques can be conveniently subdivided according to the extent to which the parts of the body are exposed. They range from whole body exposure at the one extreme, to exposures limited to the lower respiratory tract at the other.

Whole Body Exposures

In whole body exposures, one or a group of animals (or humans) are contained within an enclosure which is ventilated with the appropriate atmosphere for the appropriate interval. Such enclosures are generally known as exposure or inhalation chambers. In well-designed and operated chambers, the distributions of temperature, humidity, air velocity and concentration of the aerosol which is introduced will be reasonably uniform. The extent of the uniformity will also depend on the density and distribution of the occupants, the aerodynamic properties of the test aerosols, and the placement of obstructions to flow within the chamber, such as the pans for the collection of excreta which are frequently used in animal exposure chambers. Most of the remaining chapters in this section are devoted to more detailed considerations of chamber design and operation; hence these topics will not be discussed further here.

One of the major limitations of most whole body exposures is that there are no measurements of the inhaled doses. The exposed individuals breathe ad lib., with breathing rates and tidal volumes which vary with time and from one individual to another. In many cases, small animals react to the exposure aerosol by altering their normal breathing patterns, and/or by burying their noses into their own fur or that of their cagemates, thereby reducing the inhaled aerosol concentration by filtration. Thus, a uniform chamber atmosphere does not ensure uniformity of the amounts inhaled. Further variability in dose will also arise from variations in regional particle deposition, clearance and translocation among the exposed populations.

An additional complication in some whole body animal exposures arises from the ingestion of particles which deposit upon the pelt or cage. If the ingested particles produce any effects on the gastrointestinal tract or any systemic uptake, it may be difficult to separate them from those attributable to the inhaled particles.

Head Exposures

When deposition on the body results in excessive ingestion, handling or cleanup programs, the exposures can be confined to the head only. This is generally done by confining all or part of the head within an inhalation chamber, while the rest of the body remains outside. The separation can be achieved with a flexible diaphragm containing an elastic orifice which expands to the size of the neck, or by pressing the head into an open-ended conical section which provides an air seal behind the nose and mouth.

Head-only inhalation chambers are generally much smaller than whole body chambers. They can be as simple as a flexible sided hood for human studies or a simple box for animal studies. They can also be relatively complex when designed for uniform exposures of groups of experimental animals, as illustrated in Figures 1 and 2.

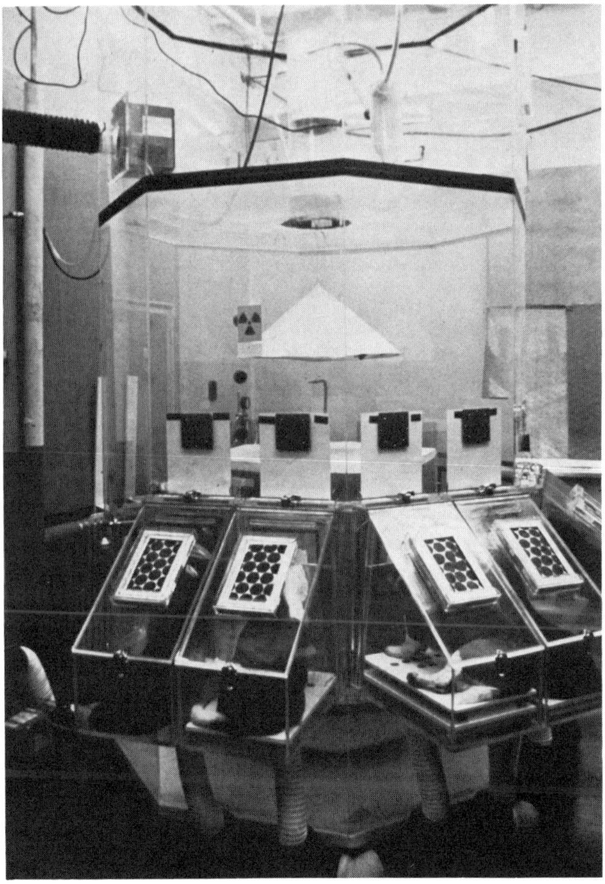

Figure 1. Exposure chamber plus ventilated dog-holding boxes used at Battelle Pacific Northwest Laboratories (courtesy of Dr. Bruce O. Stuart) for lifetime exposure study involving the simultaneous, head-only daily exposures of 10 dogs to radon daughters plus uranium ore dust.

Figure 2. Animal containment and aerosol exposure assembly developed by Laskin at the NYU Institute of Environmental Medicine for the simultaneous exposure of 15 rats, showing rat holders, exposure chamber and aerosol feed. There are four rats on each of three lower levels.

When small animals are exposed in head-only chambers, it is usually necessary to house their bodies within relatively confining support structures such as right cylinders or cut-off soft drink bottles. The confinement may therefore impose thermal and/or other stresses which could affect the response to the aerosol exposure. If sedatives are used to mitigate the stress and self-injury resulting from the confinement, they may also affect the response.

The individual animals in head exposure chambers generally breathe ad lib. and therefore inhale different volumes of aerosol, depending on their patterns and depths of respiration. However, in such a system, it is at least possible to monitor and record the respiration of each animal, and consequently to determine the amount inhaled. This can be done by making the body confinement into a plethysmograph [1]. It could be a closed box with auxiliary cooling, or a capacitance type which could be open to permit the escape of body heat.

Nasal Exposures

Individual human volunteers and animals can be exposed by feeding aerosol directly into their nasal airways. This approach has several advantages:

1. the aerosol goes directly to the individual, and is breathed via the normal physiological route without significant deposition on any other body surface;
2. the flow in and out can be readily monitored and integrated; and
3. rigid confinement is generally not needed.

At the Institute of Environmental Medicine, we have performed nasal inhalation studies in human volunteers using a modified pediatric breathing mask, as illustrated in Figure 3. The soft seal of the mask extends from the upper lip over the bridge of the nose, confining the exposed exterior surface to the outside of the nose and surrounding areas. It does not press on the nose or confine its normal excursions during breathing. The mask was modified by adding a second large entry port which directs the incoming aerosol directly at the nostrils. In the study illustrated in Figure 3, the air was exhaled through the mouth into a mouthpiece [2]. In a sulfuric acid inhalation study currently underway at the Institute, the acid aerosol flows in continuously through one port on the mask and out through the other [3].

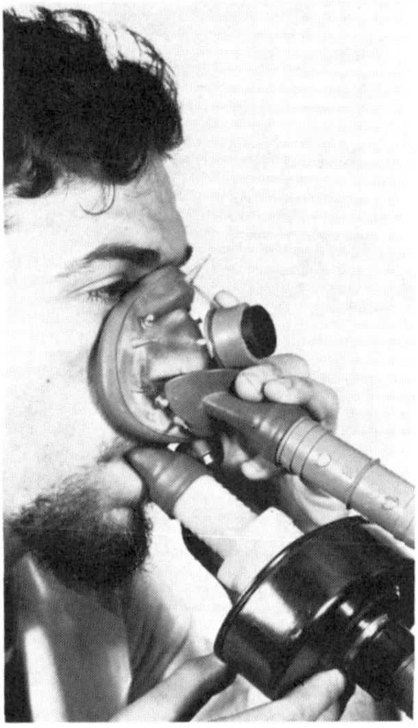

Figure 3. Exposure of a human volunteer to aerosol via a nasal mask, with exhalation via a mouthpiece to a filter canister.

Nasal exposures have been performed in a variety of animal inhalation studies. Exposures of rabbits via the nose can be facilitated by using a form-fitting latex mask which covers and seals the animal's muzzle [1], as illustrated in Figure 4.

Figure 4. Molded latex mask covering mouth and nose of a rabbit for nasal inhalation study.

The nasal airways of most animals function as relatively efficient particle impactors, and it is therefore difficult to deliver particles with aerodynamic diameters larger than ~ 2 μm into their lung airways. An effective way to deliver 2-5-μm particles into the lungs of relatively large animals is to use petroleum jelly-coated nasal catheters to conduct the aerosols past the more constricted portions of the nasal airways. The use of such catheters [4, 5] is illustrated in Figures 5 and 6.

Although nasal exposure techniques are most suitable and most often used for short-term exposures of one hour or less duration, we have exposed donkeys to sulfuric acid mist via nasal catheters for one hour per day, 5 days/week, for 6 months without any adverse effects attributable to the catheters. In other studies of shorter duration, individual exposures via nasal catheters lasted as long as six hours [6].

Oral Exposures

Short-term exposures of human volunteers are frequently performed using a mouthpiece or mouth tube of the type designed for pulmonary function evaluations [7]. In most cases, a nose clip is used to occlude the nasal

Figure 5. Nasal catheters sealed by plastic bag around muzzle of donkey, and connected to a pneumotacograph.

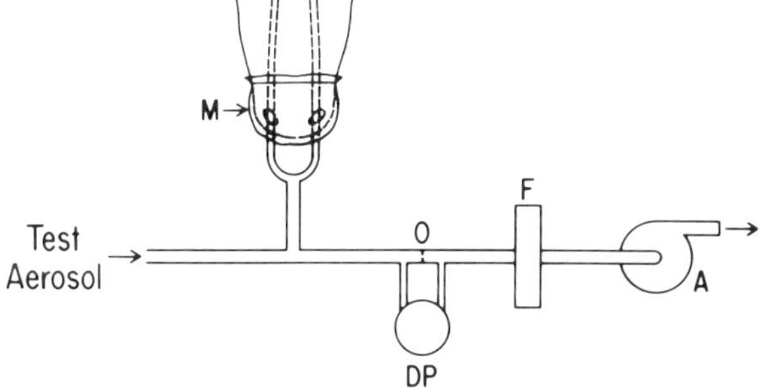

Figure 6. Schematic diagram of apparatus for the exposure of donkeys to test aerosols via nasal catheters. A, exhaust fan; C, tygon catheters; DP, differential pressure transducer to indicate flowrate through orifice (O); F, filter canister to capture excess and exhaled aerosol; M, plastic bag sealing catheters and nostrils.

passages (Figure 7). Oral inhalations can also be performed readily on nonhuman primates and dogs. For such studies, it is convenient to insert a bite plate containing a breathing port into the mouth and to use a tight fitting mask, sleeve or inflatable cuff to keep the mouth closed around the bite plate and to prevent nasal breathing (Figure 8). However, since many experimental animals are obligatory nasal breathers, they cannot readily be exposed to aerosols orally. In order to perform oral exposures on such animals, it is necessary to insert a device which prevents the normal closure of the glottis.

Oral inhalations have some operational advantages. The aerosol penetration into the lungs will be maximal and the flow and respiration can be readily measured. A disadvantage is that sedentary mouth breathing is not the

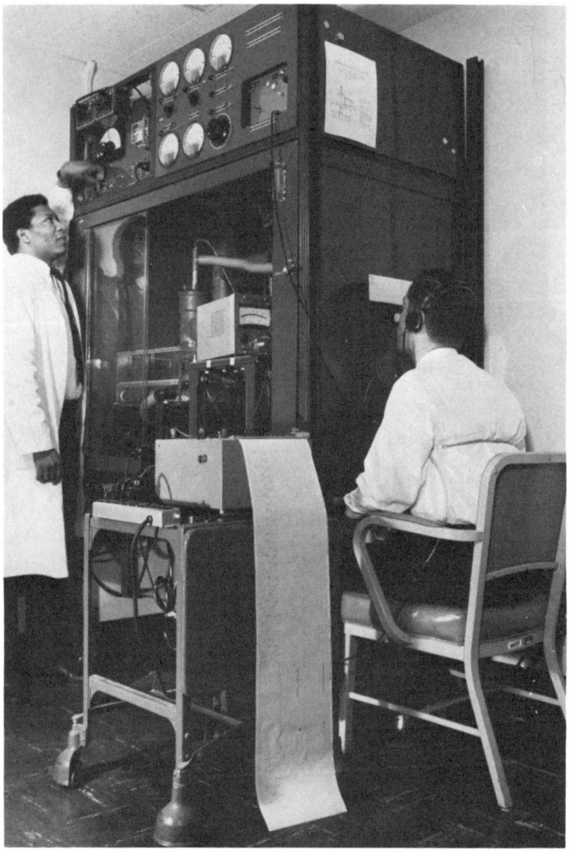

Figure 7. Exposure of human volunteer to γ-tagged monodisperse ferric oxide microspheres via mouthpiece connected to the output of a spinning disc aerosol generator.

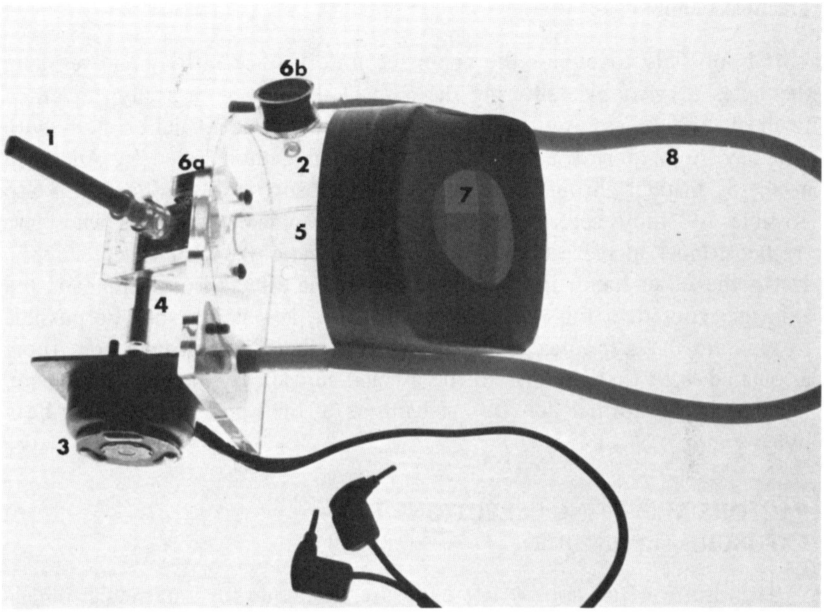

Figure 8. Dog smoking mask used at Battelle Pacific Northwest Laboratory (courtesy of Dr. Bruce O. Stuart):

1. lighted cigarette,
2. tap for pressure transducer (activates at $\sim 0.25''$ H_2O every 10th inspiration),
3. rotary solenoid; activated by signal from pressure transducer via pressure recorder,
4. rotary blade valve. When activated, padded rectangular block rotates downwards from plane of print to open cigarette port and seal fresh air port. Return spring closes cigarette port and opens fresh air port at end of inspiration,
5. flattened tube that fits between dog's canine teeth and over tongue to provide oral inhalation of fresh smoke,
6. one-way valves to release expired air or smoke from mouth (a) and nose (b),
7. rubber gasket to seal mouth,
8. adjustable rubber guard keeps mask positioned on dog's head. Entire assembly is supported in holding position for dog's comfort.

normal condition for either animals or humans. Animals forced to breathe through a mouthpiece will be under abnormal stress or will require sedation in most cases. Although oral inhalations will not usually induce a psychological stress on a human subject, prolonged oral breathing by a habitual nasal breather will often result in a drying and cooling of the larger conductive airways, and may therefore affect the response produced by the inhaled aerosol under evaluation.

Tracheal Cannula

It is possible to bypass the upper respiratory tract entirely and ventilate the lungs in vivo by delivering the aerosol through a surgically implanted tracheal tube. It was even feasible to conduct a chronic inhalation study with 420 days of exposure using this technique, as demonstrated by Auerbach et al. [8] in their chronic cigarette smoke exposure study on dogs. It is also possible to study aerosol deposition and clearance in humans who have tracheostomies in place because of prior traumatic injury or cancer surgery.

On the other hand, the interpretation of the effects seen in humans and animals exposed in this way may be difficult, i.e., it may not be possible to separate the responses to the presence of the tracheostomy from those associated with the response to the inhaled aerosol. It also must be remembered that the normal defense mechanisms of the upper airways have been bypassed.

FACTORS AFFECTING SELECTION OF EXPOSURE TECHNIQUE

The choice of an appropriate exposure technique for a particular inhalation experiment is generally limited by the circumstances and objectives of the study. The discussion which follows outlines some of the basic factors which are usually considered in the selection process.

Purpose of the Study

Depending upon the objective of the study, it may be desirable to approximate normal inhalation conditions or, alternatively, to circumvent the limitations imposed by the normal respiratory system's defense mechanisms. Thus, for conventional dose-response studies, it is generally best to use whole body exposures within inhalation chambers. The humans or animals within the chamber can be sedentary, simulating ambient exposure conditions, or they can be exposed while exercising, simulating vigorous work activities.

When the study is directed toward an evaluation of the deposition of or functional changes produced by therapeutic aerosols or cigarette smoke, which are normally administered via oral inhalation, whole body or nasal exposures of humans or animals would be inappropriate. In this case, oral inhalation represents the most realistic exposure condition.

When the focus of the study is on the mechanism of toxic action or functional alteration at sites of interest, the best method of exposure is the one that most reliably ensures the desired amount and pattern of deposition of inhaled aerosol. To be able to control the deposition sufficiently it may be necessary to use a mouthpiece or nasal catheters, or to control respiration mechanically.

When the purpose of the study is to determine the metabolism of the inhaled particles, it may be necessary to prevent whole body exposure and the ingestion of pelt deposits that results from grooming. In such cases, the options are reduced to head exposures and mask or breathing tube exposures.

A study of the nature of functional responses to a short-term exposure can successfully be performed using a relatively small number of exposed individuals, especially when the response is relatively uniform within the group. On the other hand, when the response is highly varied, or where there is a highly sensitive subgroup among a more resistant general population, it will usually be necessary to provide for a relatively large number of individuals in the study population.

When data are needed to establish population exposure limits or carcinogenic potential, a very large number of experimental animals will usually be needed. The animals must function as surrogates for a large and varied human population. Therefore, such studies will usually require exposures of at least two species, one of which is usually not a rodent, and both males and females. In recent years, many protocols have become standardized, especially when the evaluations are to be used in regulatory decision making. For example, in the National Cancer Institute (NCI) Bioassay Program, chronic toxicity tests for carcinogenicity must be performed in both sexes of two species over the greater part of the animals' life spans, with 50 of each group at each of at least two dose levels. In addition to the 400 animals thereby required for the exposures, there must be additional animals to serve as controls [9].

Large numbers of animals are needed in carcinogenic assays to determine whether the exposures can produce cancers even when they occur at a relatively low incidence rate. Multiple species are needed because no single species is sufficiently similar to humans in the various factors which affect the expression of toxicity or carcinogenicity, i.e., anatomy, physiology, immunology and metabolism. The probability of detecting effects of significance to human health increases with the number of species tested. The actual selection of the particular nonrodent species appropriate to the testing of a particular chemical depends on a careful consideration of the known parameters of the species' responsiveness to related materials.

Types and Numbers of Animals

When studies can be performed on a relatively small number of individuals and are of relatively limited duration, virtually any of the exposure techniques previously described may be suitable, and the selection can be made on the basis of convenience, cost and availability of appropriate facilities. However, when the protocols require exposures of more than a few animals at a time, it usually is necessary to use exposure chambers.

The suitability of the various techniques for individual exposures, i.e., masks, catheters, mouthpieces, etc., depends upon the species under test and the extent to which they can be suitably trained. In general, there are more options for exposing large, relatively docile species such as dogs, sheep and donkeys than for smaller animals such as rabbits, monkeys and rodents. On the other hand, space, cost and facilities limitations usually preclude tests on large numbers of large animals.

Duration of Exposures

For studies involving a single or limited number of inhalation exposures, it may be desirable to use head-only exposures. However, for long-term chronic inhalation studies, it is usually not feasible to provide sufficient animal conditioning and handling time, and such studies are generally limited to whole body exposures.

When the duration of each exposure is 8 hours or less, the animals are generally transferred between housing cages and exposure cages before the exposures, and back again afterwards. While this requires a considerable amount of animal handling, it generally simplifies the procedures needed for cage and chamber cleaning and maintenance. Furthermore, the animals do not have to be fed and watered within the exposure chambers.

Special Considerations

When the animals must remain within the exposure chamber continuously, special provisions must be made for providing food and water, for cage and chamber cleaning and for the removal of excreta. Such special provisions add greatly to the cost and complexity of chamber operations, but are necessary for some specialized studies. These include simulation of continuous air pollution exposures, and exposures to highly toxic or carcinogenic aerosols. Personnel handling or examining animals contaminated by such aerosols would be subjected to inhalation exposures through the resuspension of the particles deposited on the animals' pelts.

Another special consideration is the protection of the animals under test from exposure to airborne pathogens. Epidemic respiratory disease among animal colonies can easily ruin a chronic exposure study and wipe out the extensive investment in it. Transmission of disease between groups of animals can be prevented by isolating them from one another. There must be no exchange of air between the different spaces that they occupy. Personnel entering the animal housing area can also function as disease transmission vectors. They should be provided with clean work clothing and masks to prevent transmission of respiratory diseases to the animals.

There are several approaches to live-in exposure chambers which have proven to be effective. One is the Thomas dome [10], which is illustrated in Figure 9. In order to service the animals, operators wearing protective clothing and respirators enter the chamber through an airlock beneath it.

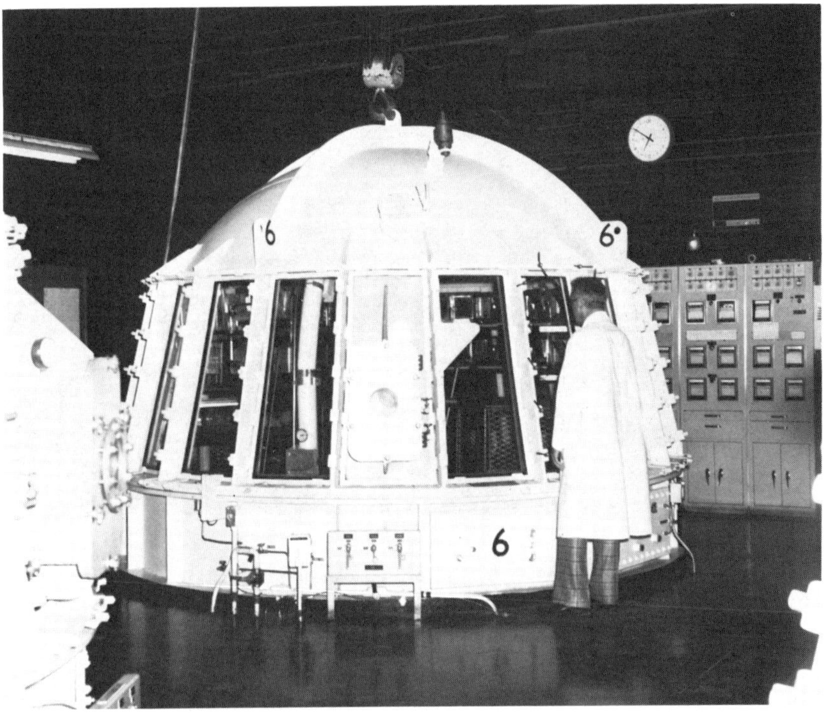

Figure 9. Thomas domes (courtesy of Dr. A. A. Thomas).

Another system, the Laskin line [11, 12], is illustrated in Figure 10. It consists of an interconnected series of glove-boxes. The two on the left are housing units which can also serve for exposure to gases. Animal cages from each housing unit can be transferred into the right hand unit, and placed inside the small aerosol chamber within it, for exposure to carcinogenic aerosols. The glove-box above the aerosol chamber houses the aerosol generation equipment, while the glove-box below contains the wastes. All animal feeding, maintenance, transfers and manipulations within the system are performed using the gloves. The animals remain within the system until they die, and are transferred out for autopsy sealed in two plastic bags.

Figure 10. New York University isolation unit (Laskin line).

When inhalation exposures are to be performed on materials which are either very expensive or very scarce, it becomes necessary to restrict the exposure method options to those that require a minimum of aerosol to be generated. In such cases, it may not be possible to use whole body exposure chambers which have large volumetric airflows. Much less dilution of the aerosol is needed for exposures using head-only chambers or individual masks or breathing tubes.

SUMMARY AND CONCLUSIONS

Inhalation exposure experiments can be performed using a variety of exposure methods. Individuals can be exposed via tracheal tubes, mouthpieces, nasal masks or head enclosures. Groups can be exposed to a common aerosol using chambers which contain either the whole body or which have ports that permit the exposure of part or all of the head. The major advantages of individual exposure methods and head-only exposure chambers are that the amount inhaled can be determined, and that skin exposure and ingestion of surface deposits can be controlled. The major advantages of whole body exposure chambers are that stresses due to physical restraint are largely avoided, and that much larger numbers of individuals can be studied concurrently.

The selection of an exposure technique and equipment for a particular study must be made on the basis of a thorough evaluation of the compatibility of the technique with the objectives and operational efficiency of the study, and with the necessity to insure safe working conditions and to control infectious diseases in the study population.

ACKNOWLEDGMENTS

This work is part of center programs supported by Grant #ES 00260 from the National Institute of Environmental Health Sciences and Grant #CA-13343 from the National Cancer Institute.

REFERENCES

1. Boecker, B. B., F. L. Aguilar and T. T. Mercer. "A Canine Inhalation Exposure Apparatus Utilizing a Whole Body Plethysmograph," *Health Physics* 10:1077-1089 (1964).
2. Lippmann, M. "Deposition and Clearance of Inhaled Particles in the Human Nose," *Ann. Otol. Rhinol. and Laryngol.* 79:519-528 (1970).
3. Leifauf, G., D. B. Yeates, K. Wales and M. Lippmann. "Effect of Inhaled Sulfuric Acid Mist on Tracheobronchial Mucociliary Clearance and Respiratory Mechanics in Healthy Nonsmokers," *Am. Rev. Resp. Dis.* 119:227a (1979).
4. Albert, R. E., J. Berger, K. Sanborn and M. Lippmann. "Effects of Cigarette Smoke Components on Bronchial Clearance in the Donkey," *Arch. Environ. Health* 29:96-101 (1974).
5. Schlesinger, R. B., M. Lippmann and R. E. Albert. "Effects of Short-Term Exposures to Sulfuric Acid and Ammonium Sulfate," *Amer. Ind. Hyg. Assoc. J.* 39:275-286 (1978).
6. Bohning, D. E., R. E. Albert, M. Lippmann and V. R. Cohen. "Effects of Fluorocarbons 11 and 12 on Tracheobronchial Particle Deposition and Clearance in Donkeys," *Amer. Ind. Hyg. Assoc. J.* 36:902-908 (1975).
7. Lippmann, M., and R. E. Albert. "The Effects of Particle Size on the Regional Deposition of Inhaled Aerosols in the Human Respiratory Tract," *Amer. Ind. Hyg. Assoc. J.* 30:257-275 (1969).
8. Auerbach, O., E. C. Hammond, D. Kirman and L. Garfinkel. "Emphysema Produced in Dogs by Cigarette Smoke," in *Inhalation Carcinogenesis*, M. G. Hanna, Jr., P. Nettesheim and J. R. Gilbert, Eds. AEC

Symposium Series 19, CONF-691001 (Springfield, VA: Clearinghouse for Federal Scientific and Technical Information, NBS, 1970), pp. 375-387.

9. Sontag, J. M., N. P. Page and U. Saffiotti. "Guidelines for Carcinogen Bioassay in Small Rodents," NCI-CG-TR-1, DHEW Publication No. (NIH) 76-801 U.S. Government Printing Office, (1976).

10. Thomas, A. "Low Ambient Pressure Environments and Toxicity," *Arch. Environ. Health* 11:316-322 (1965).

11. Laskin, S., M. Kuschner and R. T. Drew. "Studies in Pulmonary Carcinogenesis," in *Inhalation Carcinogenesis* M. G. Hanna, Jr., P. Nettesheim and J. R. Gilbert, Eds. AEC Symposium Series 18, CONF-69-1001 (Springfield, VA: Clearinghouse for Federal Scientific and Technical Information, NBS, 1970), pp. 321-351.

12. Drew, R. T., and S. Laskin. "Environmental Inhalation Chambers," in *Methods of Animal Experimentation, Vol. IV*, W. I. Gray, Ed. (New York: Academic Press, Inc., 1973), pp. 1-41.

13. Stuart, B. O., D. H. Willard and E. B. Howard. "Uranium Mine Air Contaminants in Dogs and Hamsters," in *Inhalation Carcinogenesis*, M. G. Hanna, Jr., P. Nettesheim and J. R. Gilbert, Eds. AEC Symposium Series 18, CONF-691001 (Springfield, VA: Clearinghouse for Federal Scientific and Technical Information, NBS, 1970), pp. 413-427.

CHAPTER 22

AIRFLOW AND AEROSOL DISTRIBUTION IN ANIMAL EXPOSURE FACILITIES

Robert L. Carpenter and Ronald L. Beethe

Inhalation Toxocology Research Institute
Lovelace Biomedical and Environmental
Research Institute
Albuquerque, New Mexico 87115

INTRODUCTION

Facilities used for chronic exposure to aerosols differ from those usually used for nose-only or short-term whole-body exposure. These differences arise because the chronic exposure facility must provide for long-term toxicant generation and animal care in such a manner that uncontrolled, unintentional release of the toxicant under study is prevented and the animal's health is not adversely affected by variables other than the toxicant. Chronic exposure facilities usually contain a number of large chambers for whole-body animal exposures, which must be supported by extensive air supply and exhaust systems. If the chronic exposure facility is intended to be used for the study of highly toxic materials, provisions may be required for isolation of a number of activities that involve handling toxic substances and support the actual animal exposure studies (animal care and toxicant monitoring, for example).

The chronic exposure facility (CEF) designer must translate the scientific requirements of inhalation toxicology studies into architectural and ventilation system specifications understood by an architectural engineer. To achieve this the CEF designer (who often may not be a skilled engineer) must have knowledge of (1) chamber ventilation system design, (2) whole-body exposure chamber operation and (3) additional CEF activities.

The transformation of work activities into a facility floor plan is a task commonly performed by architectural engineers in building design. However, translation of inhalation toxicology study requirements into concentration specifications is not a common task. The CEF designer must assist in developing exposure operation related facility specifications if costs are to be minimized and maximum efficiency is to be obtained in CEF operations.

The CEF is expensive to construct and operate as is indicated by the construction costs of a typical 6000-ft^2 (557.40-m^2) facility. Table I lists the construction costs per square foot for a CEF and contrasts these figures with similar values for typical and laboratory office space. The CEF is approximately twice as expensive per square foot to construct as a result of

Table I. Exposure Facility Construction Costs[a]

	Typical Exposure Complex ($/ft^2)	Typical Laboratory and Office Space ($/ft^2)
Mechanical	42.19	15.63
Electircal	15.41	7.63
Architectural	18.73	17.34
Total	76.33	40.60

[a]Based on 1976 figures and projected to 1979 by assuming 10% annual increases in building costs. Multiply figures quoted by 10.76 for $/m^2.

increased mechanical and electrical construction costs. It should be noted that the air supply to the CEF averages 3.1 ft^3/min (0.08779 m^3/min) per ft^2 of floor space, three times that of typical laboratory and office space. The increased air volume and careful air handling system regulation required for the CEF are directly reflected in the mechanical and electrical costs of constructing such a facility. For these reasons, special care must be taken in the design and specification of air handling systems for whole-body exposure facilities.

This chapter is intended to provide one approach to specifying CEF requirements and provide references to the available literature on exposure chamber operation and chamber ventilation system design. In addition, data on chamber airflow patterns in pyramidal end cone chambers and one method of calculating chamber pressure drop are presented. These data are essential in mating exposure chambers to the CEF chamber ventilation system.

It cannot be overemphasized that the interpretation of chronic exposure results is highly dependent on knowing the inhaled dose responsible for the observed effects. Determination of inhaled dose requires a detailed knowledge of the toxicant aerosol characteristics. Although these requirements are implicit in chronic inhalation toxicology protocols and in the design of a CEF, they are often not adequately addressed during CEF design. The result is inadequate provision for aerosol generation and characterization. Since several other chapters in this volume, as well as several review papers [4, 5], address the subject of aerosol generation and characterization, it is beyond the scope of this chapter to do so. However, the CEF designer must accommodate these activities in his design, i.e., provide adequate space and utilities in order to ensure optimum design and placement of aerosol generation and characterization systems.

CHAMBER VENTILATION SYSTEM DESIGN

Standard industrial ventilation practices may be applied to the design of a chamber ventilation system. This method consists of computing pressure drops caused by each element of the system and specifying a fan capable of moving the required air volume through the system [1]. Care must be taken that each branch of the system can carry the required flow. For this reason, the designer must know where exposure chambers are to be placed within the CEF and what chamber flowrates will be used.

The major components of the overall CEF chamber ventilation system are divided into supply and exhaust subsystems by the exposure chamber. The supply subsystem is above room pressure and the exhaust subsystem is below room pressure. Figure 1 illustrates the components of a typical chamber ventilation system. Fans are often provided on the intake and exhaust sides of the chamber to maintain adequate flows throughout the system and provide the pressure differential at the chamber inlet and exhaust necessary for required chamber airflow.

The air supply system includes a number of air cleaning devices. Although the order of cleanup devices varies, the air is usually passed through roughing filters prior to the intake fans. After leaving the fans, the chamber air is passed through absolute particle filters (HEPA) and charcoal beds to remove fine particulate material and undesirable gases. The air may next be dehumidified by an air conditioning system and reheated to the desired chamber temperature. After this process the air enters the main CEF distribution duct and is conveyed to each room in the exposure facility. At the entrance point to each room, steam humidifiers may be installed in the duct to provide humidity control on a room-by-room basis. This conditioned air is distributed within the room by secondary distribution ductwork. Exposure chambers may be connected to the air supply in groups or individually. Prior to entering the chamber, the air passes through a flow controlling valve and a mixing chamber into which the toxicant under study is introduced. On exiting the

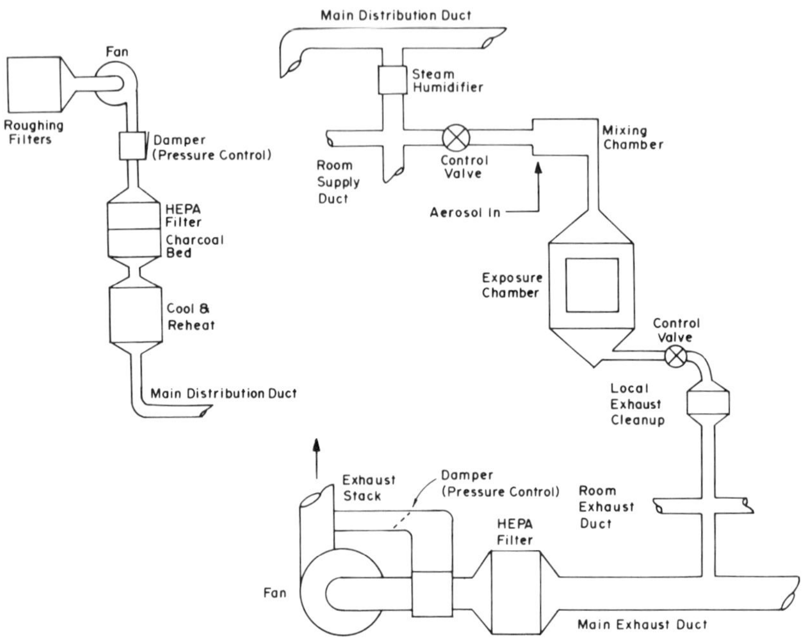

Figure 1. Schematic of a typical chronic exposure facility chamber air supply and exhaust system.

chamber, the air passes through another flow controlling valve and the exhaust system cleanup devices. The settings of flow control valves permit control of chamber flow and pressure. After cleanup, which may be located within each exposure room, the chamber air passes through the exhaust distribution ducting to large HEPA filter banks on the roof. After exiting the HEPA filter, a large fan exhausts the chamber air into the environment.

The exhaust system cleanup devices employed for toxicant removal depend on the nature of the toxicant under study. Highly water-soluble materials may be removed from the chamber exhaust air by high efficiency scrubbers. Particulate matter may be eliminated by passing the exhaust air through absolute filters. Vapors from volatile materials may be adsorbed on charcoal. In extreme cases, the air may be subjected to high temperature incineration to ensure complete toxicant destruction.

In addition to the design and engineering of the components illustrated in Figure 1, chamber ventilation system design must ensure (1) that air flow and pressure drops in the supply and exhaust subsystems are balanced and stable to prevent chamber flow and pressure upsets and (2) that the pressure differential between the supply air and exhaust air systems remains constant for flows ranging from zero to the system maximum. As a result, system ductwork must be leak-free, a specification not usually applied to heating and air conditioning ductwork.

Particular attention must be paid to chamber ventilation system pressure control when a multiroom, multichamber exposure facility is designed. If all chambers share a common inlet and exhaust system, temporary or long-term chamber shutdowns must not upset the pressure and flow balance in those chambers remaining in operation. Pressure controlled dampers or air recirculation loops may be installed in the inlet and outlet system to regulate air pressures, as shown in Figure 1. Ducting may also be installed to bypass the chamber during shutdown.

Within limits, pressure and flow control are available through proper choice of fans and variable speed drives [1, 2]. Figure 2 illustrates operating characteristics of a commonly used centrifugal fan. The shape of these

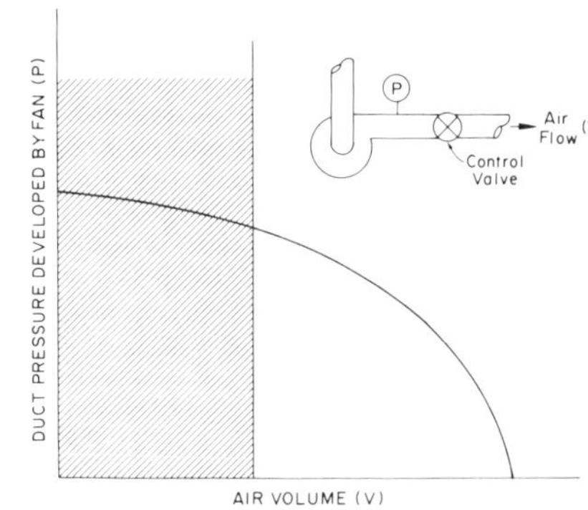

Figure 2. Pressure-volume characteristics of a centrifugal fan. Flow control by increasing pressure is not useful in the shaded region of the fan characteristic curve.

pressure volume curves is a function of fan type and operating speed. As Figure 2 illustrates, there is a range of pressures over which the flow provided by the fan will vary as pressure changes. Within this range of pressure and flow, a restricting orifice or valve may be used as a balancing element in the chamber system. However, as flow is decreased by closing the valve, a region of the fan characteristics is reached where the volume of air delivered by the fan is nearly independent of the pressure drop created by the control valve. In this low delivery volume region, control by throttling is inadequate.

Although flow control is possible by a combination of throttling valves and variable speed drives, excessive fan noise limits the useful low-volume, high-pressure operating region. An alternative to flow control in a single large chamber ventilation system is to subdivide the system into separate supply and exhaust systems for each chamber or group of chambers. A design tradeoff must be made between the high cost of individual chamber air supplies and the limited flexibility of multichamber ventilation systems.

The previous discussion of chamber ventilation system design has been general. It must be recognized that although general design procedures are available [1], the variety of possible ventilation system configurations, exposure chambers, control systems and air cleaning methods require that system design be treated on a case-by-case basis. The result of this design effort should be an optimized chamber air system which meets specifications as to chamber inlet and exhaust connector location, maximum available flow, maximum pressures available, noise level and stability.

WHOLE-BODY EXPOSURE CHAMBER OPERATION

In order to translate inhalation toxicity experimental design into the data required to design the exposure chamber ventilation system, the CEF designer must consider many experimental details. Table II lists a number of factors involving experiment design and exposure chamber operating procedure that affect the chamber ventilation system air requirements. These factors act to determine the number of chambers required for a study; to specify supply

Table II. Experimental and Operational Factors Affecting Chamber Ventilation System Design

Exposure Chamber Factors	Affected Facility Component
Number of Animals Per Chamber	Total number of chambers; chamber ventilation system flowrate
Number of Tiers Per Chamber	Total number of chambers; chamber ventilation system flowrate
Optimum Chamber Air Flowrate	Chamber ventilation system flowrate
Chamber Operating Pressure	Chamber air supply and exhaust system pressure drops
Toxicant Hazard	Exhaust system cleanup devices; exhaust system pressure
Amount and Cost of Toxicant	Maximum chamber air flowrate; maximum number of chambers
Exposure Chamber Number, Size and Layout	Number of chamber ventilation system branches; branch air flowrates
Annual Incoming Air Temperature and Humidity Ranges	Chamber ventilating system air conditioner size and humidifier size
Ventilation System Failsafe Requirements	Number, size and location of duplicate fans; ventilation system control design; facility emergency power requirements

air purity, temperature and humidity; or to specify the chamber exhaust cleanup needed.

Many of the items listed in Table II have been discussed in the literature. Silver described inhalation chamber design and operation at the Edgewood Arsenal [3]. He derived an equation illustrating that the chamber toxicant concentration reached an equilibrium value given by the ratio of toxicant generation rate and chamber flowrate. The time to reach 99% of this value is given as 4.6 times the ratio of chamber volume to chamber flowrate. He observed that toxicant concentration losses were caused by excessive numbers of animals in a chamber and recommended that animals occupy no more than 5% of the total chamber volume. Fraser et al. also discussed chamber loading limits [4]. Their experience, based on observations of chamber air distribution, agreed with Silver's limits based on toxicant concentration reductions. They also related animal loading to chamber air conditioning requirements based on (1) the intake air temperature and relative humidity, (2) animal number and basal metabolic rates, and (3) exposure chamber flowrates. They discussed toxicant dispersal methods, summarizing the characteristics of toxicant vapors, gases and aerosols affecting particle size, concentration and dispersal.

McFarland amplified Silver's observations concerning equilibration times [5]. He demonstrated that a chamber having air flow properties which mix incoming air with the chamber contents will fill and empty with toxicant in a symmetrical fashion. For his recommended chamber flowrates (0.1-1.0 chamber volumes/min), McFarland demonstrated that the filling and emptying times range from 46 to 4.6 minutes.

Drew and Laskin surveyed the principles of inhalation toxicity study design, discussing the choice of species and exposure regime [6]. They described the evolution of pyramidal end cone exposure chambers and discussed chamber ventilation system requirements. More recently, Brookhaven National Laboratory (Upton, NY) sponsored an inhalation chamber workshop at which recent advances in inhalation chamber design and current problems in implementing inhalation toxicology studies were discussed [7]. The results of a survey were presented that indicated that few inhalation toxicology laboratories treated any of the factors listed in Table II in similar fashions [8].

During CEF design, an immediate need is a knowledge of the number, size and location of the exposure chambers proposed for use in the CEF. This information must be available before chamber ventilation system design may begin, but is obtainable only after the numbers of experimental animals per chamber, animals per experiment, toxicant levels and control animals have been determined. The latter four items are usually available from experimental protocols, but deciding what number of animals will be placed in a chamber is more difficult. Practice in institutions doing inhalation toxicology varies widely. Single and multiple tiers of animals are used in both individual and gang housing in each tier. Other investigators use a single tier of animals and individual housing. In determining the number of animals per tier, one must follow the guidelines for animal housing set forth by the Animal Welfare Act of 1970 and those recommended by the Institute for Laboratory Animal Resources [9, 10].

The decision to use single or multiple tiers of animals in an exposure chamber usually depends on what variation in toxicant concentration may be permitted. Since multiple tiers of animals and the catch pans used to collect excreta are barriers to uniform air distribution within the chamber, it becomes increasingly difficult to ensure uniform toxicant administration to all animals as the number of tiers in a chamber increases. Recently Moss described a chamber designed to use the animal tiers as aerosol distribution devices [11]. Figure 3 illustrates the salient features of his design, showing animal tier placement to distribute aerosol within the chamber and to provide an exhaust plenum. Tier positions were determined using dye injected into a scale model of the chamber through which water flowed.

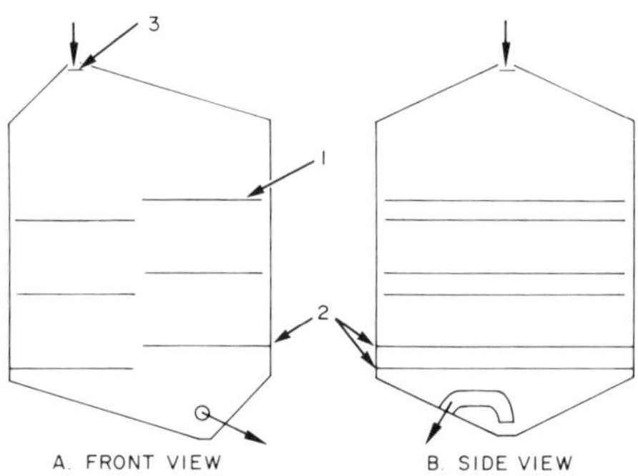

Figure 3. Battelle Northwest Laboratory multilevel whole-body exposure chamber. (1) shelves positioned to distribute incoming aerosol, (2) shelves positioned to provide exhaust plenum, (3) inlet baffle. (Reproduced by permission of the author.)

Moss pointed out that a cost analysis of chronic animal exposure costs indicated that animal care activities are the greatest labor cost associated with chronic exposures. Although few guidelines have been developed, it is becoming apparent that CEF design modifications may reduce operating costs by improving the efficiency of animal care operations. These design modifications must be made on a case-by-case basis. For example, Moss' chamber design includes provisions to permit the chamber and caging to be cleaned as a unit in a large cage washer.

LINKING PYRAMIDAL END CONE CHAMBERS WITH THE CEF CHAMBER VENTILATION SYSTEM

The key step in designing an effective CEF chamber ventilation system while minimizing system cost is to connect properly the exposure chamber and the ventilation system. This requires that chamber air flowrates be established and chamber pressure drops are known. For most chamber designs these data have yet to be determined, leading the CEF designer to overspecify the ventilation system, or to make arbitrary choices for the flows and pressure drops used in system design.

The pressure differential required to produce a given chamber flow must be known in order to:

1. determine if an existing chamber ventilation system will provide adequate chamber flow,
2. determine if an existing system may be throttled to the required flow and still operate on a portion of the fan characteristic curve (Figure 2) that allows adequate system flow control, or
3. design an adequate chamber ventilation system in a new facility.

Although not usually applied to exposure chambers, standard ventilation practices provide a method for estimating chamber pressure/flow relationships if the exposure chamber is assumed to be a part of the chamber air system ductwork [1]. Using this approach to calculate chamber characteristics or to design a chamber, the pressure drop contribution of each chamber element may be added to obtain the pressure/flow data necessary to mate the chamber to the supply and exhaust system.

These pressure drops may be estimated by considering the exposure chamber as an exhaust hood which draws air from a pressurized supply system rather than a room. The data required for these calculations are available if centered chamber air inlets and outlets are used with pyramidal end cones [1].

As an example of pressure drop calculations, Figure 4 illustrates the dimensions and pressure loss regions of a 6-ft, (1.83-m), square cross section, horizontal flow exposure chamber having 4-in. (10.16-cm) pipe [3.5-in. (8.89 cm) i.d.] inlets and outlets in the configuration described above. At the juncture of the inlet plenum and the exposure section, a commercially available screen consisting of 1600 holes/ft^2 of 0.180-in. diameter (17,200 holes/m^2 of 0.45 cm diameter) is installed. Selected holes in the lower third of the chamber are masked off to improve aerosol distribution. From geometrical considerations of air jet spreading at this screen, a uniform velocity face is anticipated approximately 1.5 in. (3.81 cm) from the screen face. This distance is small enough not to reduce the available exposure chamber volume.

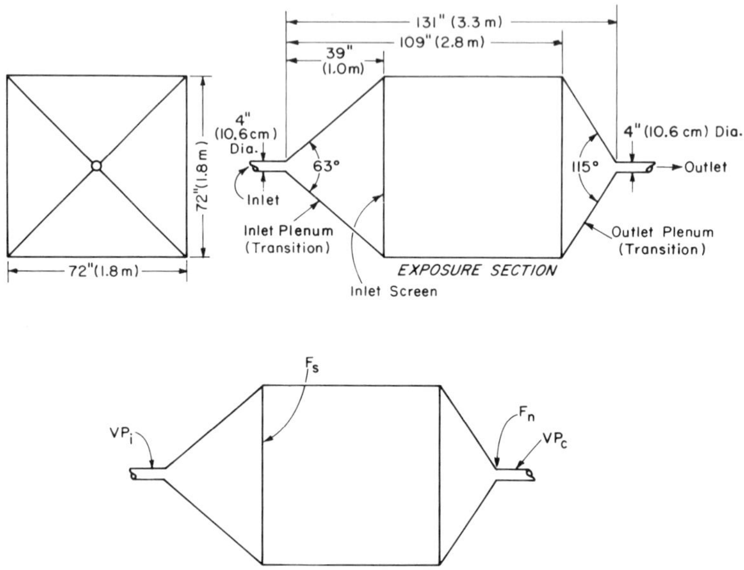

Figure 4. (a) Prototype 6-foot whole-body chamber for pressure drop calculations, (b) schematic of chamber showing inlet and outlet pressures and contributions to pressure drop.

In order to connect the chamber to the CEF ventilation system, the static pressure drop across the chamber as a function of flowrate must be estimated. The inlet, outlet and distribution screen losses may be expressed as loss coefficients that are multiplied by the velocity pressure generated by the air velocity in the ducting and chamber (Figure 4b). Beginning at the chamber outlet, Bernoulli's theorem may be applied in the form $SP_1 + VP_1 = SP_2 + VP_2 + \text{Losses}$, where SP indicates the static pressure and VP indicates velocity pressure of the moving air stream. From this analysis, the outlet static pressure as a function of the chamber static and velocity pressure and the inlet distribution screen losses is given by:

$$SP_e = SP_c + VP_c - VP_e - F_h(VP_e - VP_c) - F_s(VP_s - VP_c) \tag{1}$$

where SP = static pressure
VP = velocity pressure
F_h = pressure loss from turbulence at the outlet
F_s = pressure loss from turbulence at the screen
e, c and s = exhaust, chamber and screen, respectively

The loss coefficients may be evaluated from standard hood design [1]. For this chamber, $F_h = 0.35$ and $F_s = 1.78$.

The outlet static pressure may be related to chamber inlet pressures by a similar analysis. However, loss coefficients for the chamber inlet geometry have not been evaluated. Ventilation data indicate that inlet loss coefficients for passing air into a closed space through one wall and exhausting from the opposite wall equal approximately one. This is probably a worst-case estimate for the pyramidal chamber inlet of Figure 4. Using this estimate for the inlet loss coefficient

$$SP_e = SP_i - VP_e + VP_c(F_h + F_s) - F_h VP_e - F_s VP_s \tag{2}$$

where the subscript i denotes the chamber inlet. Then the static pressure drop across the chamber is

$$\Delta p = SP_e - SP_i = VP_c(F_h + F_s) - (VP_e + F_h VP_e + F_s VP_s) \tag{3}$$

For the chamber in Figure 4, the ratio of exhaust to chamber cross-sectional areas is $1.86 \times 10^{-3}:1$, making VP_c negligible compared to the other terms in Equation 3. The ratio of screen hole area to chamber area indicates that the screen air velocity is 3.5 times the chamber velocity making inlet screen losses negligible. Thus, for this chamber, the pressure drop is generated by exhaust entry cone losses and VP_e,

$$SP_e - SP_i = -(1 + F_h)VP_e \tag{4}$$

For this chamber,

$$SP_e - SP_i = \frac{-1.35 \times 10^{-7} Q^2}{D_e^4} \tag{5}$$

for air at standard conditions with SP in inches of water, D_e (exhaust outlet diameter) in feet and Q (airflow) in cubic feet per minute. If SP is in cm H_2O, D_e in cm and Q in m^3/min, the numerical constant in Equation 5 is -3.69×10^2.

If this chamber were to operate at a flowrate equal to one chamber volume per minute, 216 ft^3/min (6.12 m^3/min), the pressure drop across it would be -0.88 in. H_2O (-2.24 cm H_2O). If the chamber static pressure was to be -1.5 in. H_2O (-3.81 cm H_2O) with respect to ambient, the chamber exhaust system must move 216 ft^3/min (6.12 m^3/min) of air at a static pressure of -2.38 in. H_2O (-6.05 cm H_2O) while the supply system just provides 216 ft^3/min (6.12 m^3/min) of air at 0 in. H_2O (0 cm H_2O) inlet static pressure. At 91 ft^3/min (2.29 m^3/min), the calculated pressure drop is -0.12 in. H_2O (-0.31 cm H_2O) and the measured value is -0.16 in. H_2O (-0.41 cm H_2O).

It should be noted that ventilation practice also may be extrapolated to give some guidelines on shaping chamber elements. For example, exhaust hood design indicates that a centered inlet and exhaust outlet will result in the smallest inlet and exhaust plenums, as well as minimizing pressure drop for a given chamber cross section. Side take-offs or other configurations will require increased plenum size to provide uniform flow distribution. These inlet and outlet configurations will also generate additional pressure drops.

Hood design procedures also provide indications of how baffles may be used to improve aerosol distribution within a chamber. In large hoods, slots or hole patterns are used to distribute air flow uniformly across the hood face. The open area should produce a slot air velocity (V_S) greater than twice the maximum plenum velocity. Under these conditions, the aerosol will exit the inlet plenum face as a jet that will reach 0.1 V_S in approximately 30 hole diameters (or slot widths). This jet will spread to a diameter equal to approximately 13 hole diameters [12]. If the plenum face holes (or slots) are placed 13 times their characteristic dimension apart, a uniform air velocity face will be reached at a distance from the plenum wall equal to approximately 30 times this dimension. The length of this nonuniform velocity area may be varied by increasing or decreasing the hole spacing as chamber space and fabrication costs dictate.

After obtaining a relationship between pressure drop and flow for a given chamber the CEF designer must determine what chamber flowrate should be used. Again, practice seems to vary widely among investigators and work remains to be done to determine optimum flowrates. For 27-in. (68.58 cm) chambers similar to that described by Hinners [13], we have investigated the question of optimum chamber flows [14]. Three criteria were used to define optimum flow.

1. the airflow through the chamber was to be uniform from inlet to outlet, preferably with no rotational component;
2. no recirculation of the aerosol was to be allowed once the material had passed the animal level; and
3. the aerosol distribution was to be uniform with respect to both particle size and concentration.

The chamber exhaust configurations examined are illustrated in Figure 5 and the inlet geometries studied are illustrated in Figure 6. At 8 ft^3/min (0.23 m^3/min), the inlet and outlet configurations of the unmodified chamber (Figure 5) produced the airflow pattern illustrated in Figure 7. As had previously been reported, in this configuration the chamber air spirals down the chamber walls [4]. However, we observed upward air currents along the chamber axis causing a recirculation of chamber air. Modification of the chamber exhaust to provide a distributed exhaust manifold having a 2-in. H_2O (5.08 cm H_2O) pressure drop at 8 ft^3/min (0.23 m^3/min) (Figure 5) eliminates this upward flow. Inlet modification by addition of a screen or elimination of the tangential feed removes the spiral air flow component.

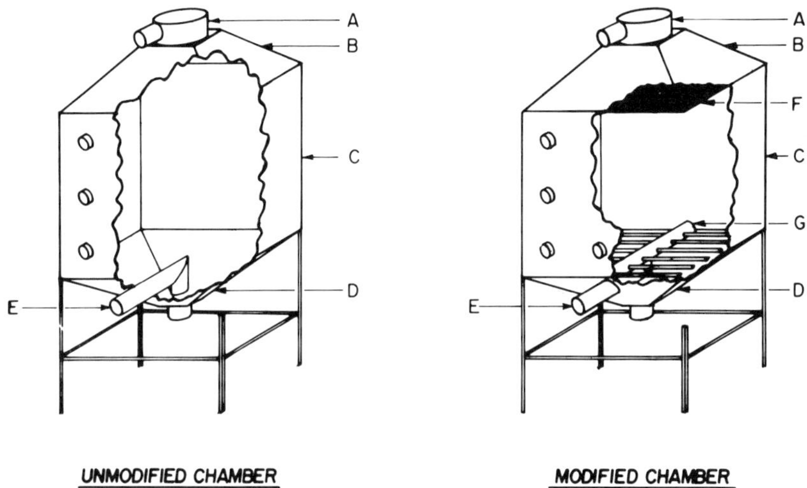

Figure 5. Schematic illustration showing modifications of a 27-in. (68.58 cm), whole-body exposure chamber. (A) inlet cylinder, (B) inlet pyramid, (C) exposure volume, (D) exhaust pyramid, (E) exhaust connection, (F) inlet screen, (G) distributed exhaust manifold.

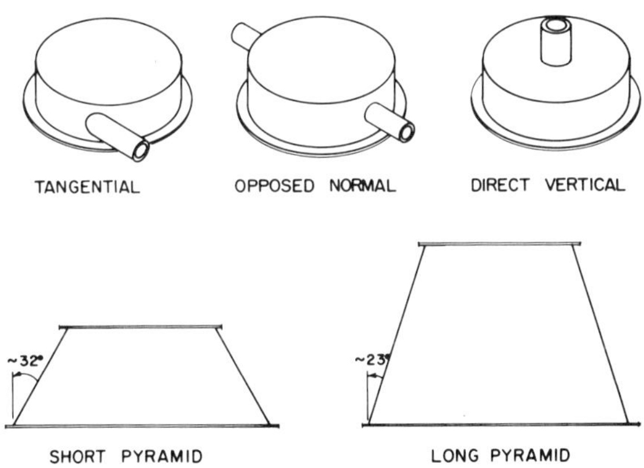

Figure 6. Chamber inlet configurations studied. The short pyramid is ≅ 13 in. (33.02 cm) high and the long pyramid is ≅ 26 in. (66.04 cm) high.

Aerosol distribution within the exposure chamber was seen to be affected by the inlet geometry and the chamber flowrates. All possible combinations of the inlets illustrated in Figure 6 were examined by measuring aerosol concentrations at 14 points in the chamber and expressing concentration variations as a coefficient of variation. As illustrated in Figure 8, the opposed

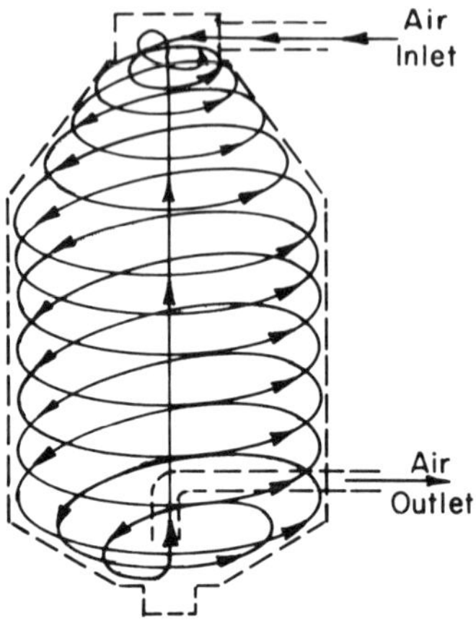

Figure 7. Airflow pattern in unmodified 27-in. (68.58 cm), pyramidal end cone chamber.

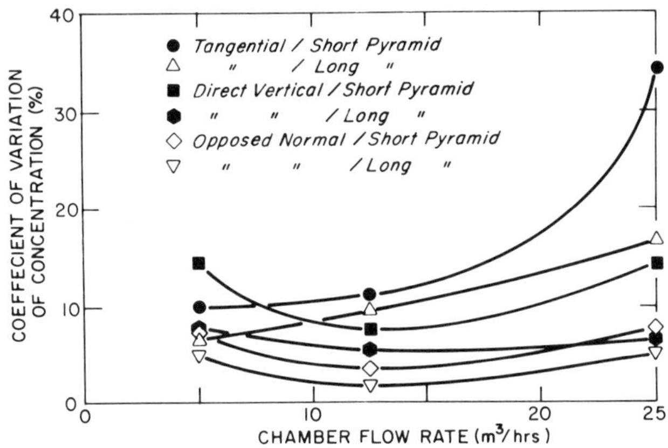

Figure 8. Aerosol concentration variation in a 27-in. (68.58 cm) chamber as a function of chamber flowrate for the inlet geometries described in Figure 6.

normal long pyramid inlet geometry produced the most uniform aerosol distribution. For the direct vertical and opposed normal inlets, minima in the aerosol concentration variation indicate optimum chamber flows corresponding to approximately 0.5 chamber volumes/min.

CONCLUSION

The CEF designer must translate the scientific requirements of an inhalation toxicology study into specifications that may be used in facility design.

To carry out this task, the CEF designer must have knowledge of chamber ventilation system design, exposure chamber operation and any additional CEF activities required to support exposure operations.

A major factor in CEF construction costs is the chamber ventilation system, which contributes greatly to electrical and mechanical construction costs. To optimize this system and thereby minimize costs, the CEF designer must decide ventilation system specifications based on inhalation toxicity experiment design and exposure chamber operational characteristics.

In specifying the chamber ventilation system, the CEF designer must size the exposure chambers and the ventilation system to assure adequate exposure chamber pressure drop and optimum chamber air flowrates. Although these data remain to be determined for many exposure chambers, Although this data remains to be determined for many exposure chambers, standard ventilation practices permit estimation of chamber pressure drops for pyramidal end cone exposure chambers with direct vertical inlets and exhausts.

These calculational procedures permit the CEF designer to design the best exposure chamber ventilation system once chamber flowrates are specified. Using one set of criteria, optimum inlet configurations for pyramidal end cone chambers may be found. For these configurations, chamber flowrates on the order of 0.5 chamber volumes/min provide the most uniform aerosol distribution.

As the need for long-term inhalation toxicity studies grows, new and larger chronic exposure facilities will be constructed. Although the approach to CEF design outlined here is not a panacea for the CEF designer, it may be used as a starting place for design. The wide variety of facility and toxicological requirements placed on a CEF will probably continue to mean that each facility will be designed on a case-by-case basis. Optimum chamber ventilation system design would be encouraged, however, if additional data on chamber pressure-flow characteristics and optimum flowrates were more generally available.

ACKNOWLEDGMENTS

The authors wish to acknowledge Mr. J. Lopez, Drs. B. B. Boecker, J. A. Pickrell, H. C. Yeh, C. H. Hobbs and H. C. Redman for their assistance in preparing this paper.

This study was supported by U.S. Department of Energy Contract Number EY-76-C-04-1013 and reports research conducted in facilities fully accredited by the American Association for Accreditation of Laboratory Animal Care.

REFERENCES

1. ACGIH Committee on Industrial Ventilation. *Industrial Ventilation, 15th Edition* (Lansing, MI: ACGIH, 1978).
2. Jorgensen, R. *Fan Engineering* (Buffalo, NY: Buffalo Forge Company, 1961).
3. Silver, S. D. "Constant Flow Gassing Chambers: Principles Influencing Design and Operation," *J. Lab. Clin. Med.* 31:1153-1161 (1946).
4. Fraser, D. A., R. E. Bales, M. Lippmann and H. E. Stokinger. "Exposure Chambers for Research in Animal Inhalation," U.S. Public Health Service Publication No. 662, U.S. Government Printing Office, (1959).
5. McFarland, H. N. "Respiratory Toxicology," in *Essays in Toxicology, Vol. 7*, W. J. Hays, Jr., Ed. (New York: Academic Press, Inc., 1976), pp. 121-154.
6. Drew, R. T., and S. Laskin. "Environmental Inhalation Chambers," in *Methods of Animal Experimentation IV,* W. I. Gay, Ed. (New York: Academic Press, Inc., 1973), pp. 1-41.
7. Drew, R. T., Ed. *Inhalation Chamber Technology*, Brookhaven National Laboratory Report (in press).
8. Snellings, W. M. "Everything You Always Wanted to Know About Chronic Inhalation Testing But Were Afraid to Ask," in *Inhalation Chamber Technology*, R. T. Drew, Ed., Brookhaven National Laboratory Report (in press).
9. Animal Welfare Act of 1970 (Public Law 91-579; 7 U.S. Code 2121 et seq.); Regulations Found in the Code of Federal Regulations, Title 9: Animals and Animal Products, Chapter 1, Subchapter A (May 1972).
10. National Research Council, Committee on Revision of the Guide for Laboratory Animals Facilities and Care, *Guide for the Care and Use of Laboratory Animals* DHEW (NIH) 73-23, U.S. Government Printing Office, (1972).
11. Moss, O. R. "Can We Design Chambers for Uniform Exposure to Particulates on Several Tiers With Catch Pans?" in *Inhalation Chamber Technology*, R. T. Drew, Ed., Brookhaven National Laboratory Report, (in press).
12. Mutchler, J. E. In: *The Industrial Environment–Its Evaluation and Control*, U.S. Government Printing Office, (1973).
13. Hinners, R. G., J. K. Burkhart and C. L. Punte. "Animal Inhalation Exposure Chambers," *Arch. Environ. Health* 16:194-206 (1968).
14. Beethe, R. L. "Flow Patterns and Aerosol Distribution in Pyramidal End Cone Whole-Body Exposure Chambers," (in preparation).

CHAPTER 23

DESIGN, OPERATION AND DYNAMICS OF AEROSOL EXPOSURE FACILITIES FOR HUMAN SUBJECTS

Karl A. Bell,* Edward L. Avol,
Ronald M. Bailey and Michael T. Kleinman

Environmental Health Laboratory
Rancho Los Amigos Hospital Campus
University of Southern California School of Medicine
Downey, California 90242

Dwight A. Landis† and Steven L. Heisler

Environmental Research and Technology, Inc.
Westlake Village, California 91361

INTRODUCTION

Epidemiological studies have linked sulfur dioxide and total suspended particulates (TSP) in ambient air pollution to increased mortality and increased acute and chronic respiratory morbidity [1]. More recent studies have suggested that particulate sulfate and possibly, particulate nitrate, are critical components of TSP in terms of health effects [2, 3]. They are emitted into the atmosphere directly from fossil fuel combustion sources or produced by atmospheric chemical reactions from primary gaseous pollutants (e.g., sulfur dioxide, nitric oxide, nitrogen dioxide). Controlled animal experiments have shown that submicron aerosols of sulfuric acid and other sulfate compounds may cause increases in the airway resistance during

*Current address: Department of Community and Environmental Medicine, University of California at Irvine, Irvine, California 92717
†Atmospheric Technology, Newbury Park, California 91320

1-hour exposures at concentrations 10-100 times worst-case ambient levels [4]. There is a need for human studies, because we currently do not know how to extrapolate effects observed in animals following controlled pollutant exposures to predict effects in man during ambient exposures.

Such considerations have initiated controlled human exposures which, in turn, have stimulated the development of environmental exposure chambers with high output particle generating and aerosol monitoring systems. For single compounds or simple mixtures, exposure chambers can adequately simulate ambient air pollutant characteristics when appropriate care is taken in the generation and monitoring of the aerosols and in the control of air temperature, relative humidity and pollutant gas concentrations.

The design and operating principles of environmentally controlled chambers which have been used within the United States for exposures of groups of human subjects to clean air and pollutant gases have been recently reviewed [5]. Three of these chambers have now been instrumented for controlled aerosol exposures of small groups of human subjects. The chambers are located at

1. the Environmental Health Laboratory at the Rancho Los Amigos Hospital Campus of the Univeristy of Southern California School of Medicine in Los Angeles;
2. the U.S. Environmental Protection Agency's Clinical Laboratory for the Evaluation and Assessment of Noxious Substances at the University of North Carolina in Chapel Hill; and
3. the University of Maryland Hospital in Baltimore.

Facilities at the University of Rochester, General Motors Research Laboratory and Brookhaven National Laboratory have been or soon will be constructed for human aerosol exposure studies. Some laboratories with plexiglass-type chambers designed for exposing single subjects to gaseous pollutants either by whole body or by mouth are adding aerosol exposure capabilities. These include the University of California, Santa Barbara, the University of Washington, Seattle, and Mount Sinai Hospital, Miami. This chapter is focused primarily on the Rancho Los Amigos Hospital (RLAH) chamber facility, because it has had more operational experience with exposures of groups of human subjects to aerosols and reactive gas mixtures than these other facilities.

ENVIRONMENTAL CONTROL SYSTEM DESIGN

Variables which need to be considered and controlled in experimental aerosol exposures of human subjects include the following:

- the concentration, size, chemical composition and electric charge of airborne particles;
- the concentration of gases which may cause health effects or interact with particles;
- the temperature and humidity of the air; and
- the intensity and frequency of light which may affect subjects directly or indirectly through photochemical reactions of gases and particles.

Noise levels should be minimized to reduce subject stress and discomfort.

Air Purification, Temperature and Humidity Control for the Chamber

Air purification, temperature and relative humidity regulation systems developed for controlled exposures of humans to pollutant gases can be used for aerosol studies. Figure 1 shows a schematic of the air purification and air conditioning systems for the RLAH human exposure chamber and high-capacity aerosol generator. The main chamber air supply is drawn from the outside through a stainless steel duct at a flowrate of 14 m^3/min (500 ft^3/min). Large particles are removed by a prefilter, and a bed of activated carbon removes some pollutant gases. The air is passed through a HEPA filter, designed to remove 99.97% of the particles with diameters of 0.3 μm or greater, and is then heated. The air enters the first Hopcalite* bed (mixture of manganese, copper, cobalt and silver oxides on an activated charcoal support), where carbon monoxide and a large fraction of hydrocarbons are catalytically oxidized at about 250°C to water and carbon dioxide. The air is then cooled to room temperature before passing through one of two beds of Hopcalite, where additional gases are adsorbed. While one Hopcalite bed is adsorbing gases at 40°C, the other bed is being regenerated at 315°C. The partially purified air then passes through beds of activated carbon and beds of Purafil** (potassium permanganate on alumina). At this point most of the remaining ozone, sulfur dioxide, ammonia, moderate to high molecular weight organic vapors and nitrogen oxides are adsorbed and/or oxidized. The regenerative Hopcalite adsorption beds may not be needed in the design of other facilities if a sufficient number of activated carbon and Purafil beds are provided and changed when saturated.

The purified air then passes through an air-conditioning unit consisting of refrigerant coils, heaters and purified steam injectors. A copper-constantan thermocouple and a lithium chloride dew point sensor located inside the chamber continuously monitor the chamber temperature and relative humidity; they are linked to the air-conditioning unit to control automatically the temperature and relative humidity to within 1% and 5% respectively

* Mine Safety Appliances Company, Pittsburgh, PA.

**Purafil, Inc., P.O. Box 80434, Chamblee, GA 30366

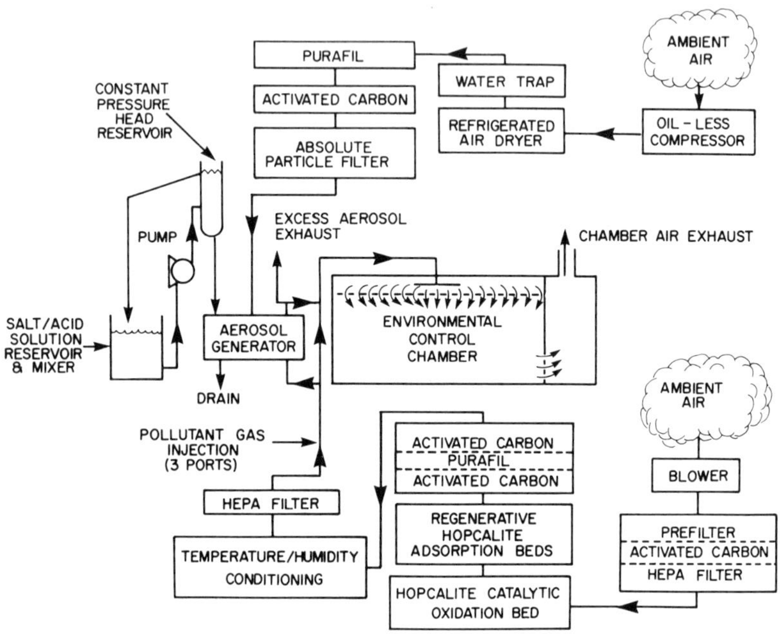

Figure 1. Schematic of the air purification/conditioning system for the aerosol generator and chamber, the nebulizer solution feed and the aerosol flow pattern for the RLAH human environmental exposure chamber.

of the desired values over a 4-6-hour exposure period. The chamber can achieve temperatures in the range of -9°C to 49°C and relative humidities in the range of 10% to 100%. A final absolute filter is used to remove any particulate impurities resulting from steam injection.

The purification system lowers the concentrations of CO and hydrocarbons to less than 2 ppm; O_3, SO_2 and NO_2 to less than 5 ppb; NH_3 and NO to less than 10 ppb; and aerosols to less than 10 $\mu g/m^3$. These concentrations are at least 0.1 of the ambient levels and are near the detection limits of most gas analyzers. When human subjects are present inside the exposure chamber, dust is generated from the chamber floor and their clothing. Most of these dust particles are greater than 1 μm in diameter and their total mass is less than 20 $\mu g/m^3$.

After purification, air is ducted into the environmental control chamber. A series of ports is provided along the ductwork to permit injection of pollutant gases. The system for introducing generated aerosols to the purified air stream will be discussed in the next section.

Air Purification for the Aerosol Generator Compressed Air

Full capacity operation of the aerosol generator constitutes only 4% of the total chamber airflow; thus allowing a less complicated air purification system for the aerosol generator's compressed air supply than that for the main chamber system. An oilless compressor is used to avoid contamination of the air with vapors which are difficult to remove. Removal of CO and light molecular weight hydrocarbons is unnecessary, since the volume of compressed air relative to the purified air is small and would cause an insignificant increase in the CO and hydrocarbon concentration in the purified chamber air. The refrigerated air dryer and water trap removes most of the moisture from the compressed air. The dry compressed air supplies pressure to the nebulizers and is also used to dry rapidly the nebulized droplets when exposures are made with hygroscopic aerosols at low relative humidity levels. The activated charcoal, Purafil and particle filters serve the same function as in the main chamber system.

Chamber and Air Flow Design

The walls of the exposure chamber and all air ducts associated with the exposure facility are constructed of stainless steel to minimize problems of corrosion and reaction of gases and aerosols on the walls. The chamber has two compartments. The main exposure room has a floor area of 25.2 m^2 (4.2 m x 6.0 m) and a volume of 62.3 m^3 (2200 ft^3). The chamber entrance lock and lavatory has a floor area of 4.5 m^2.

As shown in Figure 1, purified air or diluted aerosol is distributed in four directions above a false perforated ceiling (0.317 cm diameter holes spaced 1.27 cm apart) inside the chamber. Air in this plenum is at a higher pressure than the chamber. Consequently, the air and particles are forced through the holes causing good mixing of fine particles and pollutant gases. The air flow is vertically downward and uniform throughout the human subject exposure zone. Near the floor, air is directed toward the outlet grating in the bottom of the wall separating the main chamber from the entrance lock. The air is exhausted to a stack through the ceiling of the entrance lock. Air and particles flow continuously at 14 m^3/min (500 ft^3/min) through the chamber to the exhaust with no recycling during exposure studies. The mean residence time of particles and gas molecules in the chamber is about 4 minutes.

Some particles and molecules may have shorter or longer residence times due to backmixing and recirculation within the chamber. This was demonstrated by separate studies [6] in which SO_2 or O_3 was injected into the intake duct at a steady rate predetermined to give 0.4 ppm in the chamber. About 15-20 minutes, or 4-5 mean residence times, were required for the chamber gas concentration to reach 0.4 ppm. The time period required to raise a submicron aerosol to a stable concentration within the chamber was also about 15-20 minutes, following start-up of the aerosol generator. These data approximate the dynamic flow characteristics of an ideal constant flow stirred tank in which the incoming air mixes completely with the chamber air. Further design details for the chamber, environmental control and gas injection systems are found in previous publications [7, 8].

AEROSOL GENERATION SYSTEM DESIGN

Performance Specifications

The aerosol generators used in earlier animal toxicological studies lacked the capacity and versatility required for the animal screening and human exposure studies planned for the RLAH chamber. The RLAH exposures required maximum sulfate or sulfuric acid concentrations of 1.0 mg/m^3 with mass median aerodynamic diameters (MMAD) of approximately 0.3 μm and geometric standard deviations (σ_g) between 2 and 4. Ammonium nitrate exposures required a maximum concentration of 5 mg/m^3 with a MMAD of about 1 μm and a σ_g between 2 and 3. These size distribution criteria were based on ambient air measurements made with cascade impactors in the Los Angeles Basin [9, 10].

Other design criteria included:

1. production of an aerosol with a stable ($\pm$10%) size distribution and mass concentration over an 8-hour operating period;
2. operation of the particle generator should not affect the regulation of temperature or relative humidity in the chamber;
3. production of particles with low electric charge levels to prevent increased lung deposition resulting from electrostatic forces, and
4. minimization of cost, space and noise.

High Capacity Aerosol Generator Design

To meet these specifications for the RLAH chamber, Environmental Research and Technology, Inc. (Westlake Village, California) developed a high output generator system [8, 11, 12] shown in Figures 1, 2 and 3. It is constructed mainly of stainless steel and consists of 20 nebulizers, a drying chamber, delivery duct and several dampers. An aerosol mist is initially produced by nebulizing a dilute aqueous solution. Clean air (2-3 m^3/min) withdrawn by a blower from the chamber purified air intake duct (Figure 2), dilutes and dries the aerosol mist. The dried aerosol is then injected into the purified air for further dilution before entering the plenum of the exposure chamber (Figure 1).

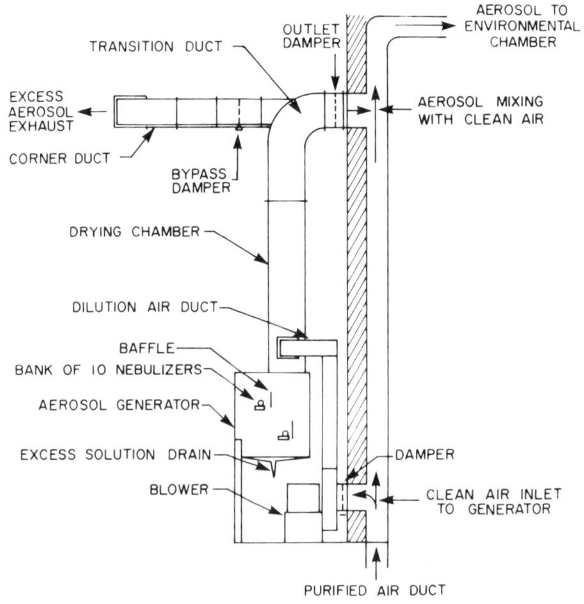

Figure 2. Cross-sectional view of the high capacity aerosol generator system attached to the stainless steel chamber air delivery duct.

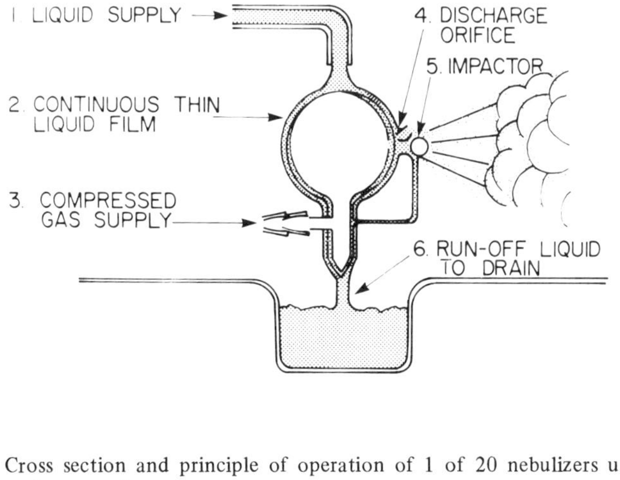

Figure 3. Cross section and principle of operation of 1 of 20 nebulizers used in the high-capacity aerosol generator.

Figure 3 demonstrates the principle of operation [13] of the one HYDROSPHERE and 19 MAXI-COOL* nebulizers. Clean compressed air regulated at 50 psig flows from inside the sphere and exits through a small orifice. A dilute solution of the compound to be nebulized flows over the top of the sphere and uniformly coats the surface. This thin liquid film is ruptured by the compressed air jet, forming a very fine, uniform spray. The larger droplets that are produced are reduced in size by impaction on a small sphere mounted directly in front of the compressed air orifice.

Aerosol Generator Regulation

These nebulizers are arranged in 2 rows of 10 each with adjustable baffles as shown in Figure 2. The baffles are used to remove large droplets and lower the MMAD of the droplet size distribution below 3.5 μm. By using dilute salt or acid solutions and drying the droplets, sulfate and nitrate aerosols with MMAD's between 0.2 and 2 μm may be obtained. By careful regulation of the individual needle valves and accompanying rotameters, the aerosol output from each nebulizer and from the entire generator is fairly constant. Solution flow to the nebulizers is maintained at a constant rate by an elevated reservoir with a constant head pressure produced by the pump and overflow line shown in Figure 1.

The aerosol mass output of the generator is regulated by varying the number of nebulizers used. The MMAD and σ_g of the chamber aerosol are adjusted by varying the nebulizer solution concentration and the baffle position. Additional control of the chamber aerosol mass concentration is accomplished by adjusting the positions of both the output and bypass dampers (Figure 2).

Control of Electrostatic Charge on Exposure Aerosols

Measurements were made of the electrostatic charge on ammonium sulfate aerosols at the position of the outlet damper of the generator and inside the chamber. Charge levels at the outlet (Figure 2) were about 40% of the levels predicted by nebulizer charging theory and levels in the chamber decreased to about one-half of those at the generator outlet [11]. Thus, the initial charge of the generated aerosol is partially neutralized during the 2-sec passage through the stainless steel delivery ducts to the chamber. Based on the results of studies of Melandri et al [14], the electrostatic charge levels on the aerosol in the RLAH chamber would be expected to increase the respiratory tract deposition in human subjects by only a few percent [11]. Consequently, electric charge neutralizers were not added to this aerosol

*McGaw Laboratories, Irvine, California

generator system. If, for some other generation systems the electrostatic charge levels are high enough, compared to an ambient aerosol having a Boltzmann equilibrium charge distribution, to alter human lung deposition significantly, then Kr-85 charge neutralization systems should be included. Continuous radioactivity monitoring and frequent cleaning of the outside of the sealed Kr-85 gas tubes would be required to detect and prevent leaks and inadvertent exposure of human subjects to the Kr-85 gas.

DESIGN OF EXPOSURE CHAMBER ATMOSPHERE MONITORING SYSTEMS

General Principles

For the most reliable results the chamber environment should be monitored continuously with the best available instruments and techniques. Total filter and cascade impactor samples should be taken during the entire exposure period to obtain the time-averaged mass concentration and size distribution of the exposure aerosol. Sample sizes should be sufficient to afford accurate determination of the required analytes at operating concentrations.

Continuous real time analyses are needed to assure the human subjects' safety. An electrical aerosol size analyzer and an optical particle counter can be used to continuously monitor the aerosol mass concentration and size distribution over the 0.01-10 μm range during the exposure period. Alternatively, a light scattering nephelometer can be used to continuously estimate the mass concentration of particles in the 0.1-1 μm range. A condensation nuclei counter may be used to continuously monitor the total number of particles smaller than 1 μm in diameter. Low-pressure, multistage cascade impactors which use piezoelectric crystals as collection stages have recently become available for continuously monitoring the mass concentration and size distribution of aerosols in the 0.05-15 μm range. We have had difficulty in applying this technique, since these impactors may be operated only intermittently when the aerosol is highly concentrated; otherwise the crystals overload frequently.

All of the above instruments require calibration. Factory calibrations for the cutoff diameters and wall losses in cascade impactors need to be verified using monodisperse aerosols of known size, e.g., commercially sized latex spheres [15] or particles produced from vibrating orifice generators [16]. Low-pressure impactors with stage cutoffs below 0.1 μm can be calibrated using monodisperse uranine aerosols and an electrical mobility classifier [17]. Analyzers having optical and electrical components must periodically be calibrated against standard sized aerosols and against cascade impactor and total filter measurements of size distribution and total mass. The accuracy of in-house methods used in chemical analyses of filter samples should periodically be checked against available standard reference materials or by sending samples to outside laboratories for independent analyses by standard methods.

Aerosol and Gas Monitoring System Design for the RLAH Chamber

Aerosol Sampling for Gravimetric and Chemical Analysis

Figure 4 shows a schematic of the aerosol and gas monitoring systems of the RLAH exposure chamber. The 47-mm filter and Sierra 215S cascade impactor are located within the human subject activity zone, a few feet from the chamber air outlet grating with the sampler inlets about the height of the nose and mouth of seated subjects. Sampler position within the subject activity zone is not critical in this chamber. We have demonstrated a uniform mass concentration throughout the human subject exposure zone for heterodispersed sulfate and nitrate aerosols with MMAD's between 0.3 and 1 μm

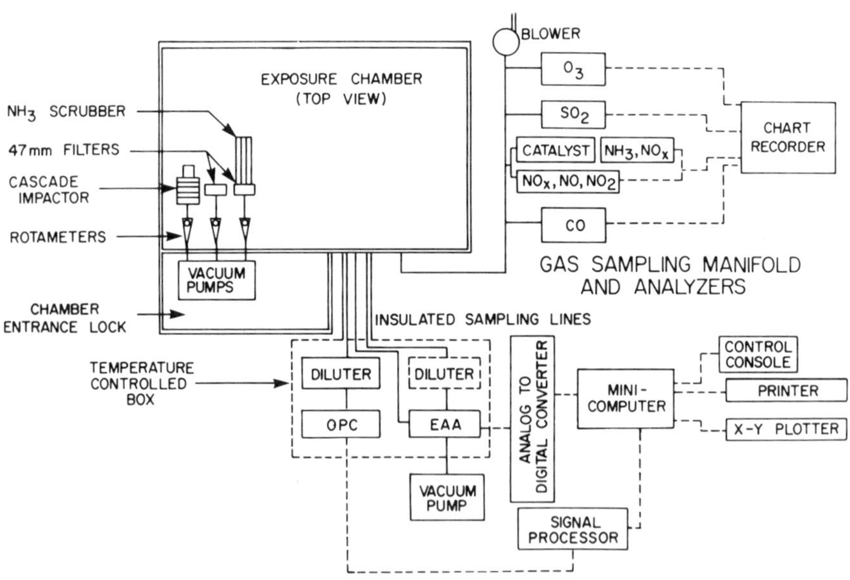

Figure 4. Schematic of the aerosol and gas monitoring systems of the RLAH human environmental exposure chamber.

[8, 18]. The only regions of poor mixing were corners of the chamber, which are inaccessible to the subjects. During exposures with acidic aerosols, an ammonia denuder is used in front of a separate 47-mm filter to prevent neutralization of the collected aerosol by chamber ammonia. The denuder consists of a diffusion battery coated with oxalic acid. The vacuum pumps for these samplers are located in the chamber entrance lock to minimize the sound level within the exposure chamber and possible contamination of the chamber atmosphere by pump exhausts. Filters are analyzed gravimetrically and chemically.

Real Time Monitoring System for Aerosols

The RLAH chamber aerosols are continuously monitored using an optical particle counter (OPC; Climet Instruments Model 208) and electrical aerosol size analyzer (EAA; ThermoSystems Model 3030) with a minicomputer system interface (Digital Equipment Corp., PDP 11/10). The EAA detects particles in the 0.013-0.42 μm size range and the OPC covers the 0.6-3.87 μm range. On command from the computer, the EAA is automatically stepped through a series of preset collector rod voltages and the value of the analyzer current is recorded after it has stabilized. A signal processor has been interfaced with the OPC to group the output pulses by pulse height into 256 channels. The computer then groups the OPC and EAA data into 24 discrete size intervals and calculates the number, concentration, aerosol surface and volume within each size group. The particle volume in the size range (0.42-0.6 μm) between the EAA and OPC is determined through linear interpolation of the volume data. The total mass concentration is determined from the particle density and the sum of the aerosol volume over all the size groups. The aerodynamic median and mean diameters, geometric median and mean diameters, and geometric standard deviation with respect to number, surface area, volume and mass are also calculated.

These data can be displayed at 3-5 minute intervals so the chamber operators can monitor the safety of the atmosphere. If the concentration or size distribution deviates from acceptable limits, the operator may manually adjust the generator. A more advanced system, developed by ERT, Inc. for the EPA-Chapel Hill chamber, can automatically adjust the generator performance to control the chamber aerosol. Concentration and size distribution data may also be averaged and plotted off-line following an exposure with the RLAH system.

A one- or two-stage dilution system [11] is used with the OPC to minimize coincidence errors when an aerosol is sampled. One stage provides a dilution ratio of about 1:30 while two stages provide a dilution ratio of about 1:1000. We have generally used a single stage of dilution for human exposure studies. When toxicological concentrations much higher than ambient worst case conditions are used, an optional 1:30 diluter is required for the EAA.

The OPC, EAA and diluters are mounted inside a metal box maintained at the same temperature as the exposure chamber. The three sampling lines for the OPC and EAA are insulated and enclosed within a 6.4-cm diameter by 2 m long steel tube which pierces the exposure chamber wall at a point 2 m above the floor. This sampling system design minimizes aerosol losses in the lines and assures that the chamber aerosol does not change its size distribution because of relative humidity changes during sampling.

Gas Sampling/Analysis System

Monitors for ozone, sulfur dioxide, carbon monoxide, hydrocarbons, nitric oxide and nitrogen dioxide sample through a common gas sampling manifold. Ammonia concentrations in the exposure atmosphere can also be monitored. These gas analyzers are operated during aerosol exposures as well as during pollutant gas exposures to assure that the chamber air purification system is operating properly.

AEROSOL GENERATION AND MONITORING SYSTEM PERFORMANCE DURING EXPOSURES

The aerosol generation and real time monitoring system has been used for exposure of human volunteer subjects for 2 hr/day on 2 or 3 successive days to nominal concentrations of 100 $\mu g/m^3$ of ammonium sulfate, 85 $\mu g/m^3$ of ammonium bisulfate and 75 $\mu g/m^3$ of sulfuric acid aerosols (MMAD of approximately 0.3 μm and σ_g of 2.5 to 3) at 31°C (88°F) and 40% relative humidity, simulating conditions observed during worst case air pollution episodes in the Los Angeles Basin [18-21]. Total filter and cascade impactor samples were collected during each exposure for gravimetric and chemical analyses to determine the particle mass concentration and size distribution as well as the ammonium, sulfate and hydrogen ion concentrations. The EAA/OPC monitoring system was calibrated before each aerosol study against the data obtained from total filter and cascade impactor samples. The aerosol mass concentration and MMAD varied less than 10% over the 4-6 hour exposure period of a group of 5 to 7 subjects. Typically for sulfuric acid, the MMAD averaged 0.55 μm and repeated measurements taken at 5 minute intervals over a 4 hr period, had a standard deviation of less than 4%. The particle mass concentration averaged 252 $\mu g/m^3$ (at 40% r.h. the aerosol would contain about 50% water by weight) and repeated determinations had a standard deviation of less than 6%.

Particle size distribution data derived from the computer controlled

EAA/OPC system were in good agreement with those from total mass, sulfate ion and ammonium ion analyses of impactor samples. The total aerosol mass concentrations were not always in good agreement between online monitors and gravimetric filter samples, however. The inconsistent manual placement of the instrument diluters appeared responsible for this disparity. This problem is being corrected by permanently locating the diluters in the sampling lines with bypass lines to allow the undiluted sampling option with both the OPC and EAA.

In general the aerosol exposure and monitoring systems have attained most of the originally set criteria. It has not been possible to obtain high ($\geqslant$ 1.0 mg/m^3) aerosol mass concentrations at particle sizes less than 0.5 μm MMAD. We have completed exposure studies of human volunteers to several different sulfate aerosols with an average MMAD of 0.55 μm with a σ_g of about 3.0 at concentrations approximating worst case ambient levels. These studies were carried out generally at low relative humidity (38% r.h.). We also studied the effects of high humidity (85% r.h.) combined with exposure to ammonium sulfate [20,21].

CHEMICAL AND PHYSICAL TRANSFORMATIONS IN EXPOSURE CHAMBERS

In a few studies, unforeseen toxic concentrations of interfering chemical species were formed during controlled exposures of humans and animals to reactive mixtures of gases and/or aerosols. In one study of the effect of a mixture of nitrogen dioxide and ferric oxide aerosol in dogs [22], significant quantities of ammonium and nitrate ions were found in particles filtered from the air of the exposure chamber. Maintaining a continuous flow of clean air through the chamber can help to minimize concentration of ammonia and other interfering bases produced by subjects. Traces of NH_3 penetrating the air purification systems and emitted by human subjects can neutralize acidic aerosols [23]. Consequently, during sulfuric acid exposures at RLAH, subjects were exposed one at a time inside the chamber to reduce the amount of ammonia and to minimize the degree of neutralization. Furthermore, an excess of acid was generated to assure that the unneutralized acid concentration was nominally 75 μg/m^3. The concentration of unneutralized acid to which the subjects were exposed was determined from total filter samples, one of which was fitted with an ammonia denuder (Figure 4). The concentration of neutralized salts of sulfuric acid to which the subjects were exposed ranged from 20-40 μg/m^3, while the unneutralized acid ranged from 93-111 μg/m^3 between exposure days [20,21,24].

Careful design of the pollutant injection, air purification and chamber air delivery systems can also help to minimize adventitious chemical reactions and aerosol formation During exposures of human volunteers to a mixture of 0.37 ppm SO_2 + 0.37 ppm O_3 at McGill University, a sulfur aerosol concentration of about 200 $\mu g/m^3$ was produced, probably by homogeneous nucleation, condensation and growth of the products of chemical reactions of pollutant gases in the air delivery ducts and in the chamber [6]. Trace impurities such as olefinic hydrocarbons and organic gases and ammonia generated by resting and exercising human subjects can participate in the chemical reactions and increase the number of condensation nuclei.

In the RLAH chamber, during similar exposures large numbers of very small particles (0.02-0.05 μm diameters) were formed when O_3 alone or SO_2 + O_3 were added to the purified air, but the mass of sulfate aerosol formed was only about 1 $\mu g/m^3$ (Table I). This occurrence may provide an explanation for the higher incidence of health effects observed at McGill with the SO_2 + O_3 mixture, as compared with no observed effects in the RLAH chamber, other than those attributable to O_3 alone [6].

Table I. Particle Characterization in the RLAH Chamber During Exposures with SO_2 and O_3

Condition	No. of Condensation Nuclei/cm^3		No. of Particles/cm^3 Diameter, (0.5-10 μm)		SO_4^{2-} ($\mu g/m^3$)
Los Angeles Air	30,000		⩾30		5-15
Chamber–no subjects					
Clean air	⩽50		~0.01		–
0.37 ppm SO_2	⩽50		~0.01		–
0.37 ppm O_3	5,000		~0.01		–
0.37 ppm O_3 + 0.37 ppm SO_2	90,000		~0.01		–
Chamber–3 subjects	Rest	Exercise	Rest	Exercise	
Clean air	⩽50		0.3	2	~0
0.37 ppm O_3	8,000	22,000	0.3	2	~0
0.37 ppm O_3 + 0.37 ppm SO_2	100,000	250,000	0.3	2	0.5-1.5

T = 31°C, relative humidity = 35%.

The size of particles and mass of sulfate formed are minimized in the RLAH chamber by separating the injection ports of O_3 and SO_2 by 3 m (so they mix with each other at low concentrations) and by using highly purified air. In the McGill chamber, concentrated streams of SO_2 and O_3

could have reacted rapidly with each other and with ambient impurities such as olefins to form a large number of sulfuric acid nuclei. These nuclei could have grown by homogeneous condensation, coagulation and absorption of ammonia during their 2-min residence time in the chamber [6].

CONCLUSION

Operational experience with the RLAH exposure facility demonstrates that small groups of human subjects can be exposed for periods up to 6 hours to aerosols under well-controlled conditions. Temperature, relative humidity, aerosol mass concentration and size distribution can be controlled to simulate ambient aerosol exposures during air pollution episodes. These aerosol exposure facilities are complex and a great deal of operational experience and training is required before safe, realistic, stable and repeatable exposures of human subjects can be carried out. Carefully controlled and fully characterized exposure facilities are, however, crucial if experimental data concerning possible human health effects of air pollutants are to be of any significant value as scientific criteria for legislative action or legal proceedings.

ACKNOWLEDGMENTS

Development and operation of these aerosol exposure facilities at the Environmental Health Laboratory, Rancho Los Amigos Hospital was supported by the Coordinating Research Council, Inc. (Contracts APRAC-CAPM-20-74 and 27-75), the National Institute of Environmental Health Science (Grant No. RO1-ES01152) and the National Heart and Lung Institute (Specialized Center of Research Grant No. HL-ES-15098). Karl Bell, PhD, designed and coordinated the development of the aerosol generation and aerosol monitoring capability of the laboratory. Dwight Landis and Steven Heisler, PhD, developed most of the high capacity aerosol generation and computer controlled aerosol monitoring system. Jack Hackney, M.D., William Linn, Leonard Wightman, Charles Spier and the other staff of the Environmental Health Laboratory contributed to the successful development and operation of this unique aerosol exposure facility.

REFERENCES

1. Coffin, D. L., and H. E. Stokinger. In: *Air Pollution, Vol. 2, The Effects of Air Pollution*, A. C. Stern, Ed. (New York: Academic Press, Inc., 1974), p. 231.
2. Rall, D. P. "Air Quality and Automobile Emission Control, Vol. 2, Health Effects of Air Pollutants," U.S. Government Printing Office, (1974), p. 413.

3. Knelson, J. "Evidence for the Influence of Sulfur Oxides and Particulates on Morbidity," *Bull. NY Acad. Med.* 54(11):1137-1154 (1978).
4. Amdur, M. O., J. Bayles, V. Ugro and D. W. Underhill. "Comparative Irritant Potency of Sulfate Salts," *Environ. Res.* 16:1-8 (1978).
5. Committee on Medical and Biological Effects of Environmental Pollutants, In: *Ozone and Other Photochemical Oxidants* (Washington, DC: National Academy of Sciences, 1977), p. 388.
6. Bell, K. A., W. S. Linn, M. Hazucha, J. D. Hackney and D. V. Bates. "Respiratory Effects of Exposure to Ozone Plus Sulfur Dioxide in Southern Californians and Eastern Canadians," *Am. Ind. Hyg. Assoc. J.* 38:696-706 (1977).
7. Hackney, J. D., W. S. Linn, R. D. Buckley, E. E. Pedersen, S. K. Karuza, D. C. Law and D. A. Fischer. "Experimental Studies on Human Health Effects of Air Pollutants," *Arch. Environ. Health* 30:373-378 (1975).
8. Bell, K. A., and J. D. Hackney. "Effects of Sulphate Aerosols Upon Human Cardiopulmonary Function," First Year Progress Report, Contract No. CRC-APRAC-CAPM-27-75, U.S. Government Printing Office, No. PB-275 789/6WP (1977).
9. Lundgren, D. A. "Atmospheric Aerosol Composition and Concentration as a Function of Particle Size and Time," *J. Air Poll. Control Assoc.* 20:603-608 (1970).
10. Novakov, T., P. K. Mueller, A. E. Alcocer, et al. "Chemical Composition of Pasadena Aerosol by Particle Size and Time of Day: III. Chemical Status of Nitrogen and Sulfur by Photoelectron Spectroscopy," *J. Colloid Interface Sci.* 39:225-234 (1972).
11. Landis, D. A., S. L. Heisler and K. A. Bell. "The Design and Performance of a High Capacity Aerosol Generation and Monitoring System for the Full Body Exposure of Humans to Sulfate and Nitrate Aerosols," *Am. Indust. Hyg. Assoc. J.* (in press).
12. Landis, D. A., P. K. Mueller, G. M. Hidy and K. A. Bell. "A High Rate Aerosol Generator for Inhalation Studies," presented at the American Industrial Hygiene Conference, Atlanta, Georgia, May 16-21, 1976.
13. Litt, M., and D. Swift. "The Babington Nebulizer: A New Principle for Generation of Therapeutic Aerosols," *Am. Rev. Respir. Dis.* 105: 308-310 (1972).
14. Melandri, C. et al. "On the Deposition of Unipolarly Charged Particles in the Human Respiratory Tract," in *Inhaled Particles IV.* W. H. Walton, Ed. (Oxford: Pergamon Press, 1977), p. 193.
15. Hering, S. V., R. C. Flagan and S. K. Friedlander. "Design and Evaluation of a New Low-Pressure Impactor. I," *Environ. Sci. Technol.* 12: 667-673 (1978).
16. Liu, B. Y. H., Berglund, R. N. and Agarwal, J. K. "Experimental Studies of Optical Particle Counters," *Atmos. Environ.* 8:717 (1974).
17. Hering, S. V., S. K. Friedlander, J. J. Collins and L. W. Richards. "Design and Evaluation of a New Low-Pressure Impactor. 2," *Environ. Sci. Technol.* 13:184-188 (1979).

18. Avol, E. L., R. M. Bailey and K. A. Bell. "Sulfate Aerosol Generation and Characterization for Controlled Human Exposures," *Am. Indust. Hyg. Assoc. J.* 40:619-625 (1979).
19. Hackney, J. D., W. S. Linn and K. A. Bell. "Experimental Studies of the Human Health Effects of Sulfur Oxides," *Bull. NY Acad. Med.* 1177-1185 (1978).
20. Kleinman, M. T., and J. D. Hackney. "Effects of Sulfate Aerosols Upon Human Cardiopulmonary Function," Final Report, Contract No. CRC-APRAC-CAPM-27-75, U.S. Government Printing Office (1978).
21. Avol, E. L., M. P. Jones, R. M. Bailey, N. N. Chang, M. T. Kleinman, W. S. Linn, K. A. Bell and J. D. Hackney. "Controlled Exposures of Human Volunteers to Sulfate Aerosols: Health Effects and Aerosol Characterization," *Am. Rev. Resp. Dis.,* 120:319-327 (1979).
22. Knott, M. J., and M. Malanchuk. "Analysis of Foreign Aerosol Produced in NO_2-Rich Atmospheres of Aerosol Exposure Chambers," *Am. Ind. Hyg. Assoc. J.* 30:147-152 (1969).
23. Larson, T. V., D. S. Covert, R. Frank and R. J. Charlson. "Ammonia in the Human Airways: Neutralization of Inspired Acid Sulfate Aerosols," *Science* 197:161-163 (1977).
24. Bailey, R. M., K. A. Bell, U. D. Karlsen and M. T. Kleinman. "Strong Acid Measurements of Sulfate Exposure Aerosols," presented at the American Industrial Hygiene Conference, Los Angeles, CA, May 1978.

CHAPTER 24

COMPUTER-CONTROLLED HUMAN EXPOSURE FACILITIES UTILIZING WATER-SOLUBLE AEROSOLS

W. L. Crider

U.S. Environmental Protection Agency
Health Effects Research Laboratory
Clinical Studies Division
Research Triangle Park, North Carolina 27711

D. A. Landis*

Environmental Research and Technology
Westlake Village, California 91360

F. H. Weaver

Operating Systems, Inc.
Woodlands Hills, California 91364

INTRODUCTION

The Environmental Protection Agency (EPA) has the responsibility to set standards for air pollution levels that will prevent damage to human health or welfare. The data on which these standards are based come from studies conducted by universities, private industry and government agencies. The EPA sponsors or cosponsors research studies by all three groups. Exposure studies to determine levels of air pollution that can be tolerated without having detrimental effects to human health are conducted through both epidemiological and clinical investigations. A major clinical research effort by EPA was established within its Health Effects Research Laboratory

*Presently with Atmospheric Technology, Newbury Park, California 91320.

(HERL) at Research Triangle Park, NC. This effort is directed towards obtaining data on the subtle health effects of selected air pollutants on volunteer human test subjects. To provide the necessary resources, two computer-automated environmental clinical exposure laboratories were constructed in Chapel Hill, NC, on the campus of the University of North Carolina. This exposure system is identified by the acronym CLEANS, for Clinical Laboratory for Evaluation and Assessment of Noxious Substances.

The two controlled environmental laboratories operate over a wide range of temperatures, humidities and light levels with accurately controlled pollutant levels of carbon monoxide, sulfur dioxide, nitrogen dioxide, nitrogen monoxide, ozone and water-soluble aerosols. Each laboratory operates independently of the other, with the pollutants introduced either singly or in combinations. The period of exposure may last from hours to weeks.

Each laboratory contains the complex physiological instrumentation that is required to acquire, under computer control, the cardiovascular and pulmonary functions of the human test subjects. All measurements are made under the supervision of physicians and medical technicians who use the computer to rapidly acquire, compute and display the physiological measurement data so the control console operator can determine the health status of the subject and evaluate the general reliability of the acquired data.

The design of the CLEANS chamber was based largely on information gathered on visits to five human research facilities at Rancho Los Amigos, Downey, CA; University of Maryland, Baltimore, MD; St. Vincent's Hospital, New York, NY; Hospital for Sick Children, Washington, DC; and New York University, New York, NY. A "Workshop of Controlled Environmental Chambers" held on April 26-27, 1971 provided additional information on the general status of controlled environmental chambers for humans. The CLEANS was completed in May 1976 for gaseous exposure studies. The aerosol pollutant system became functional in March 1978, allowing the full complement of pollutants to be generated and controlled.

The CLEANS system is supported by five service systems: Clinical Environmental Laboratories (CEL); Environmental System; Gaseous Pollutant System; Aerosol Pollutant System; and Physiological Data Acquisition System (PDAS). To understand better the Aerosol Pollutant System, this chapter will briefly describe the Clinical Environmental Laboratories and the environmental subsystem.

CLINICAL ENVIRONMENTAL LABORATORIES

There are two Clinical Environmental Laboratories (CEL). Each CEL consists of an environmentally independent exposure room measuring 4 m (13 ft) long by 6 m (20 ft) wide by 3.2 m (10.5 ft) high. Each CEL was designed to house comfortably three test subjects for extended periods of time. For exposure periods of less than eight hours, the beds are removed as indicated in Figure 1 to make room for a larger number of subjects or special task-related equipment. Other basic equipment located in each CEL

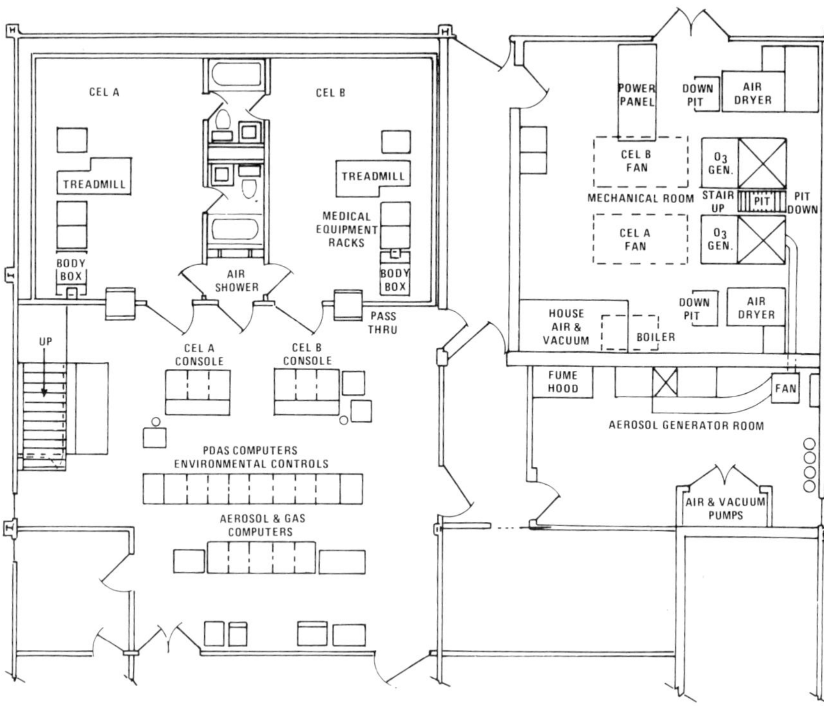

Figure 1. Main floor of CLEANS facility.

consists of the primary task-oriented equipment as shown in Figure 2. Each CEL has a separate air handling and environmental control system. The 0-150 foot-candle lighting in each CEL is controlled in zones to permit separate medical laboratory testing and sleeping areas. CEL A was constructed opposite to CEL B so that one air shower could serve as the means of ingress and egress to both exposure rooms. The air shower, in combination with a shoe sole scrubber, prevents cross-contamination of atmospheres and reduces the entrance of outside dirt. Adjacent to the air shower, emergency doors which open directly into the laboratory control room, are provided for each CEL. The CEL bathrooms are located between the exposure rooms and behind the air shower. Each bathroom is on the same air supply as the connecting

Figure 2. Inside of exposure laboratory.

CEL, but has a separate exhaust. Emergency egress is provided between the back of each CEL through an extra door in the CEL B bathroom opening into CEL A.

Type 304 alloy stainless steel sheet metal covers the CEL interior and air handling system to prevent the outgassing of absorbed pollutants and the static attraction of aerosol particles. The chamber walls are 15 cm (6 in.) thick and are fully insulated. They are constructed with staggered metal studs, foamed insulation, and sheet steel outer panels to reduce the transfer of outside heat and noise. The ceiling is 3.2 m (10.5 ft) high and is constructed of insulated metal decking surfaced on the underside with stainless steel to match the CEL walls. The floor is constructed of removable panels that are perforated to allow passage of the exhaust air from the CEL.

The air with pollutants enters the CEL through six 76-cm (30 in.) by 101.6-cm (40 in.) adjustable vane registers in the ceiling: the air moves downward with a velocity that can be varied between 7.6 m/min (25 ft/min) and 15.3 m/min (50 ft/min). The air is then exhausted through the perforated floor. The pollutants of interest are mixed well with the supply air before it enters the CEL to provide homogenous distribution throughout the CEL. Therefore, the exposure dose is independent of the subject's location.

Each CEL contains a large return air plenum located below the floor. The plenum contains High Efficiency Particle Air (HEPA) and charcoal filters having face areas of 5.9 m^2 (63 ft^2). Presently, only the HEPA filters are in place. The purpose of the filters is to remove aerosols and unwanted contaminants from the recirculating air stream.

CEL A differs from CEL B only in the addition of an extra 5.2 m^2 (56 ft^2) of HEPA filters and an aerosol mixing chamber in the input air duct.

ENVIRONMENTAL SUBSYSTEM

The function of the Environmental Subsystem is to provide and maintain within prescribed limits the air flowrate, temperature, relative humidity and light level in the CEL as specified in the exposure test protocol. Except for the air flowrate, a feedback control system with electronic sensors and pneumatic actuators accomplishes this task. The following range of conditions can be developed in each CEL:

Main Air Flowrate	227-453 m^3/min (8000-16,000 ft^3/min)
Maximum Air Duct Velocity	305 m/min (1000 ft/min)
Make-up Air Flow	34 m^3/min (1200 ft^3/min)
Temperature Range	7-35°C (45-95°F)
Relative Humidity Range	30-70%
Light Intensity Range	0-150 foot-candles
CEL Pressure	2.54 mm (0.1 in.) of water (positive)
Sound Level	NC-40
Clean Room Conditions	4000 particles/m^3

A variable or steady state profile of pollutant gas concentrations within each CEL can be maintained using the Pollutant Gas System PDP-11/40 computer. The profile for each parameter is entered into the computer by the System Operator at the beginning of an exposure experiment, but any profile may be changed during the course of the experiment. The computer adjusts the setting of the Environmental System to maintain the conditions within the CEL to the desired profile. The computer also records actual environmental exposure conditions on a magnetic tape. As a backup, fixed set points can be entered manually into the CEL Environmental Supervisory Controls. If the computer fails, the controls are automatically switched from the computer to the manual settings.

The path of the main recirculated CEL air flow starts at the output of the fan, as shown in Figure 3. It then passes through a sound trap, heating and cooling coils and ceiling registers into the CEL. The air flow continues through the perforated CEL floor, through (or around) the return air plenum filters and finally on through a second sound trap to the input of the fan.

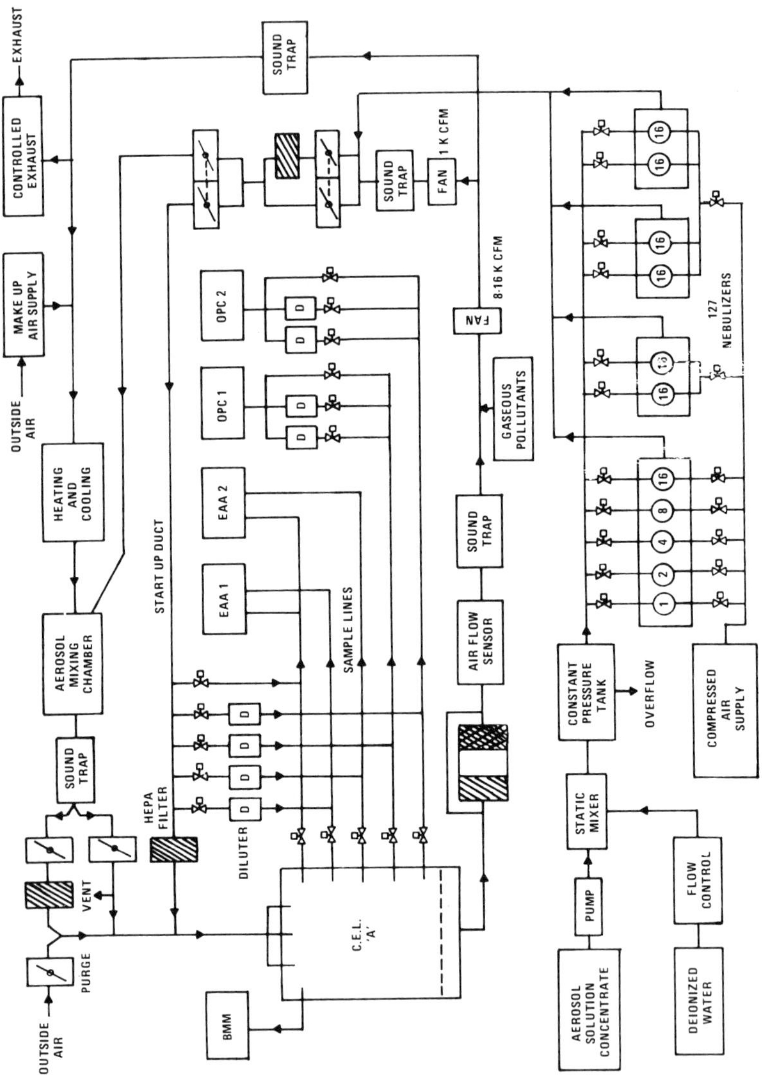

Figure 3. CLEANS and aerosol system diagram.

The air flowrate is varied by remote adjustment of the pitch of the axial vane constant speed fan.

Identical gaseous pollutant atmospheres may be produced in both CEL A and CEL B, but only CEL A has the additional equipment required for the introduction of aerosol pollutants. As a result, CEL A differs from CEL B by the addition of an aerosol pollutant mixing tank, an additional sound trap, and a HEPA prefilter bank with remote controlled motorized filter bypass dampers.

Fresh air is brought into the main air supply of each CEL by adding make-up air at a constant rate of 34 m^3/min. The outside air is prefiltered, charcoal filtered and heated or cooled to 5.3°C dry bulb (db)/4.4°C wet bulb (wb) (41.5°F/23.1°F) and then supplied to an air dryer. The air is cooled and dried to -1°C db/-4.9°C (23.1°F) wb by passing it through a spray of lithium chloride solution in the dryer. The air is then heated to the temperature of the main CEL recirculated air stream. The desired humidity in the recirculated air supply is maintained by heating the make-up air to 37.8°C (100°F) and injecting dry steam at a rate of 78.75 kg/hr (175 lb/hr) on automatic demand for humidification.

The supervisory controllers and indicators for the temperature, humidity and lighting for each CEL are located in the Environmental Control Console equipment racks that are located adjacent to the PDAS computer. Graphic Display Panels, as shown in Figure 4, are located at eye level above the supervisory controls. Each panel provides a schematic diagram of the CEL airflow system and shows the status and alarm conditions for the CEL air flowrate, air pressure, damper position and fan operation.

To prevent outside contaminants from leaking into a CEL, a positive atmospheric pressure of 2.54 mm WG is maintained within each CEL. A pressure modulated exhaust damper maintains this pressure with the make-up air providing the pressurizing force.

As a safety precaution, an emergency purge mode of operation was designed into the CEL airflow system. When a purge mode is initiated, motorized dampers are positioned to allow untreated outside air to be pulled directly into the CEL and exhausted at a rate of 227-453 m^3/min. The CEL pollutant concentration is diluted to 10% of its original concentration after 1-2 minutes of purging.

The gas-fired steam boiler with an output of 10^6 Btu/hr is the central heating system for both CELs. The cooling unit, common to both exposure chambers, uses a chilled glycol system with four compressors to give a total cooling capacity of 55 tons. A larger heating and cooling system was provided so that the temperature and relative humidity for either or both CELs could be changed over the entire operating range within a 3-hr period.

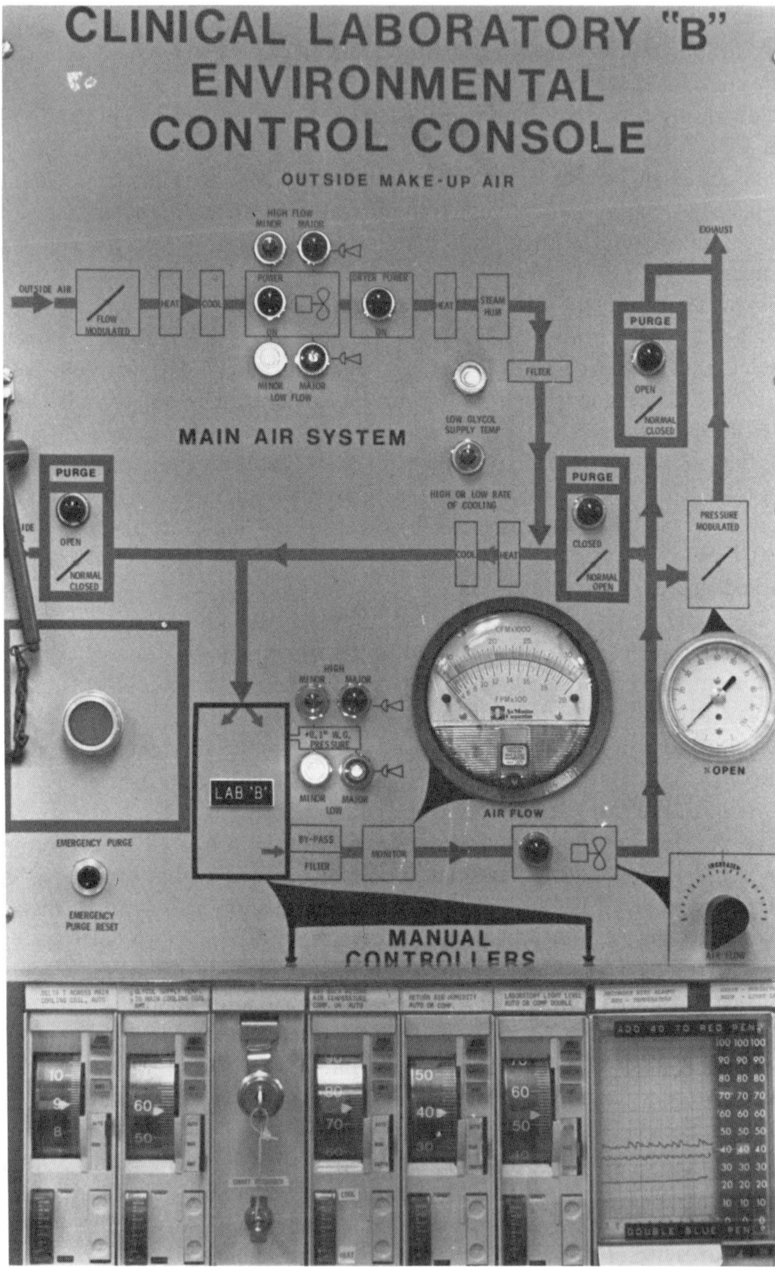

Figure 4. Environmental control console graphic display panel.

AEROSOL POLLUTANT SUBSYSTEM

Aerosols are generated from water soluble materials and introduced into CEL A. Aerosol physical characteristics are controlled within the following limits:

Mass Median Diameter	0.1-1.0 μm
Mass Generated as Usable Aerosol	1-75 mg/min
Measurement Range of Monitors	0.01-10 μm diameter

The Aerosol Subsystem consists of the three complex sections: Aerosol Generation and Delivery, Aerosol Analysis and Data Acquisition and Control. These three sections are controlled using a PDP 11/40 computer. The aerosol mass concentration in the CEL is computed from the output of two aerosol analyzers and the relative humidity in the CEL. The computed mass concentration and MMD are respectively compared to a concentration-time profile and a size-time profile previously entered in the computer. The computer then varies the aerosol generator to minimize the control errors. Once operating conditions are specified and entered in the computer, the Aerosol System is capable of operating during an entire experiment with minimal operator intervention.

Aerosol Generation and Delivery

Aerosols are generated by nebulizing an aqueous solution containing the aerosol material. Among the methods for large output atomization, the Babington technique offers the greatest advantages. Atomizers using this principle have no moving parts and have been proven to be very reliable for the generation of therapeutic aerosols. A commercial Maxi-Cool® is shown in Figure 5. The device pictured consists of a hollow glass sphere with a small hole in one side. Clean compressed air enters the inside of the sphere and is ejected through the small orifice. The solution to be atomized flows over the top of the sphere and forms a thin coating on the surface of the sphere. This coating is ruptured by the high velocity air stream and produces a dense aerosol cloud. A small impaction sphere is mounted in front of the air stream outlet to collect the larger droplets that are produced. Because of the unique design, the generating solution does not pass through a small orifice, and this eliminates the problem of clogging. Figures 6 and 7 show the initial droplet mass distribution from a Babington nebulizer. By dissolving the aerosol material in a volatile solvent, the droplets can be dried to a size much smaller than the primary size produced by the atomizer. This device was selected for this large-scale aerosol generator because it is inexpensive, reliable and very simple in design.

Figure 5. A commercial Babington Maxi-Cool® nebulizer.

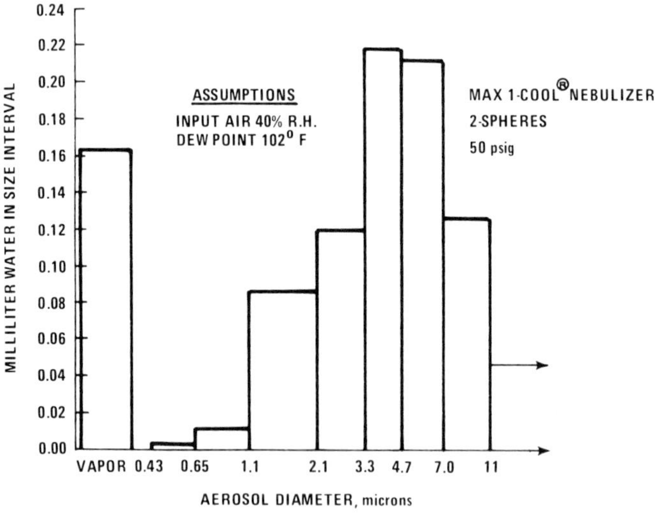

Figure 6. Distribution of one milliliter of water aerosolized by a Maxi-Cool® nebulizer.

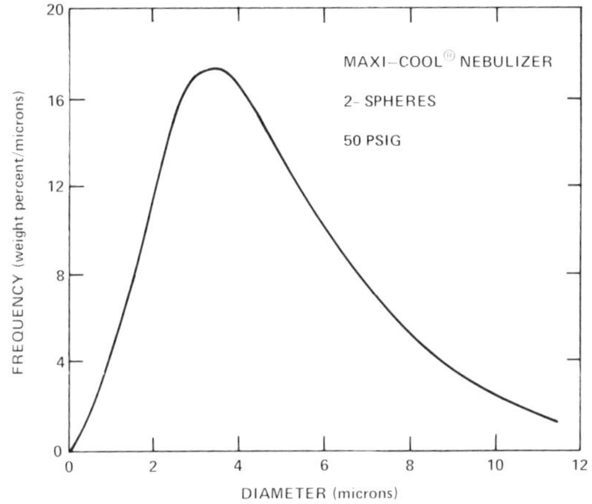

Figure 7. Droplet mass distribution of a Maxi-Cool® nebulizer.

Figure 8 shows a theoretical estimate for the number of Maxi-Cool® nebulizers needed to generate 75 mg/min of aerosol at various aerodynamic mass median diameters (AMMD). This rate of aerosol generation is needed in the CEL system to achieve the exposure concentrations required to simulate most ambient conditions.

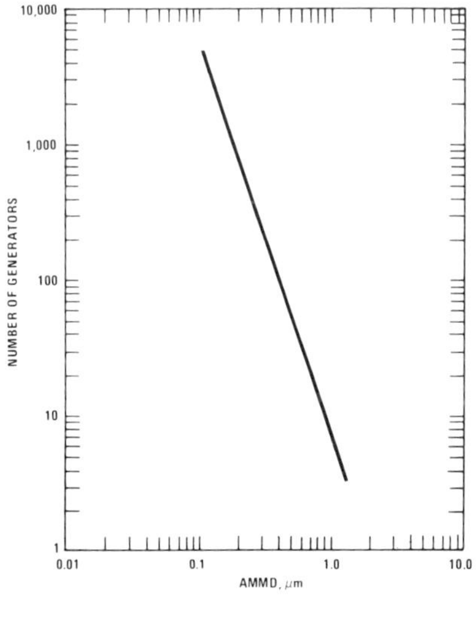

Figure 8. Number of Maxi-Cool® nebulizers required to produce an aerosol generation rate of 75 mg/min as a function of particle diameter.

The amount of water vapor that is added to the CEL during the drying process is illustrated in Figure 9. To maintain the CEL at 40% r.h. using all of the available 1200 ft^3/m of drying air, the rate of water addition must be below 200 g/min. This constant limits the lowest aerosol AMMD generated by this technique to 0.22 μm. The dilution-dryer air flow could be doubled, but this would only decrease the minimum AMMD to 0.18 μm. This addition would be expensive and would only give a small decrease in mean particle size.

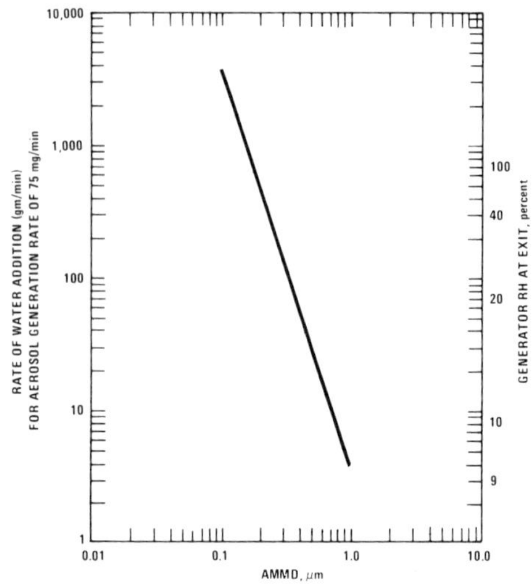

Figure 9. Rate of water addition and generator exit relative humidity as a function of particle size.

If 0.22 μm is chosen as the lowest AMMD that can be generated at 75 mg/min because of relative humidity considerations, then Figure 8 shows that about 550 nebulizers would be required to produce this aerosol. This number is too large for practical consideration. While the Maxi-Cool® nebulizers alone are inexpensive, the support equipment can become expensive. For example:

- Each nebulizer requires about 1 ft^3/min at 50 psig of oilless compressed air.
- Each nebulizer requires 25 cm^3/min solution flowrate (36 liter/day).

Because of this expense, a limit of 127 nebulizers was set. Table I lists the theoretical output characteristics for a generator using this number of Maxi-Cool® units.

Table I. Extrapolated Performance of the Proposed Aerosol Generator Using 127 Maxi-Cool® Nebulizers

Solution Conc. (%)	AMMD (μm)	Output (mg/min)	CEL Conc. at 8000 cfm (μg/m^3)
0.0023	0.10	2.34	10.4
0.01	0.35	97	432
0.5	0.60	505	2256
1	0.76	1008	4496
3	1.08	2946	13145

The 127 nebulizer unit should produce an aerosol with a 0.35 AMMD at a rate of 76 mg/min. The atomization-evaporation technique is not feasible for sizes below this point since the diameter is limited by the relationship $v = \pi d^3/6$, where v is the particle volume and d is the particle diameter.

The atomization-evaporation technique using the Maxi-Cool® nebulizers has been shown to meet the exposure requirements for an AMMD down to 0.35 μm using the 127 nebulizer unit. To generate an aerosol with an AMMD below this point will require other generation techniques.

Advantages of the Babington nebulization technique include:

- There are no moving parts in the nebulizer. The probability of mechanical failure with this approach is minimized. This is an important consideration for long-term exposure studies.
- The generator is readily adapted to computer control. It has been demonstrated that only one variable, the liquid flowrate, must be changed to change the aerosol AMMD. The aerosol output has been shown to be quite stable with time. The particle size can be easily changed to simulate diurnal variations within the exposure chamber. If desired, this change can take place quickly. After the command is given, the aerosol size within the generation section will change within a fraction of a minute. Other lag times in the air dilution and delivery system appear to be about one minute for the aerosol to reach 95% of the desired level. The system lag time, however, is significantly longer.
- The generation technique can be used with any water soluble material. Other generation approaches, while being able to achieve 0.1 μm AMMD at 75 mg/min generation rate (i.e., condensation methods), cannot be used for all water soluble substances.
- The aerosol generation system as illustrated in Figure 10 is a compact design. The actual aerosol generator is relatively small and can be easily installed in a 100-ft^2 area (ca. 10 m^2).
- The aerosol generator can be adapted as a condensation generator. This would enable the generator to be used for condensation, nebulization or as a hybrid generator.

Figure 10. Aerosol generation system.

The droplets that are produced are equilibrated with air at the same relative humidity as maintained in CEL A. The final equilibrated size of the resulting particles is controlled by varying the concentration of the material in the generating solution. The mass generation rate is controlled by varying the number of nebulizers used and by filtering a portion of the aerosol.

A schematic flow diagram of the aerosol subsystem is provided in Figure 3. The aerosol solution concentrate is mixed with deionized water in a static mixing chamber. The solution flows from the static area to a constant pressure tank which is pressurized with compressed air. Excess liquid is drained from this tank through a solenoid valve operated by a float switch. The solution passes from the constant pressure tank through solenoid valves and flow indicators to the nebulizers. The nebulizers are arranged in banks of 16 each. The flow of solution and compressed air to the nebulizers is controlled by solenoid valves which are operated by the computer in the binary format of 1, 2, 4, 8, 16, 32 and 64. Using this arrangement, any number of nebulizers from 1 to 127 can be activated.

The control and adjustments of the mean particle size are achieved by continuously variable dilution in the solution delivery system to the nebulizers. The AMMD of the aerosol that is produced is directly determined by the concentration of solution that is being nebulized. Thus, the particle size is adjusted by changing the relative flowrate of concentrated solution to deionized water.

The droplets produced by the nebulizer banks are equilibrated in a 0.2-m^2 (2-ft^2) drying chamber using a portion of air taken from the recirculated air supply for CEL A. The drying air mixes with the droplets and evaporates them to their equilibrated diameter. Thus, the final size of the aerosol is controlled by the relative humidity in the CEL and the concentration of the solution being nebulized.

From the drying chamber, the aerosol flows to a motorized damper system which permits bypassing a portion of the flow through a HEPA filter. This method is used as a fine control of particle concentration in the CEL. The primary method of controlling the particle concentration is the number of nebulizers receiving aerosol solution flow.

A second motorized damper system permits the choice of allowing the aerosol to flow into the CEL aerosol mixing chamber and into the CEL, or into a start-up duct where the aerosols are filtered out before the air stream enters the CEL. The start-up duct allows the aerosol system to stabilize at the desired concentration and size before actual delivery of the aerosol to the CEL. Up to 28 m^3/min (1000 ft^3/min) of air is diverted from the CEL recirculated air stream through the aerosol delivery system and is returned to the CEL system with or without aerosol particles. Stainless steel was used in the construction of the dampers and ducts to prevent corrosion and static charge buildup that would attract particles to the surface areas.

Aerosol Analysis

Because of the wide range of particle sizes that can be delivered to the CEL, two types of aerosol monitors are used to characterize the size distribution in the CEL. An Electrical Aerosol Analyzer (EAA) with a measurement size range of 0.0032-1.0 micron and an Optical Particle Counter (OPC) with a measurement size range of 0.3-10.0 microns provide the required range span. The interpolation between the EAA and OPC size range is based on a linear interpolation of the distribution of particle mass with respect to the logarithm of the diameter which is performed by the aerosol computer as it receives the output of the analyzers. Particles smaller than 0.01 microns are ignored because of the inherent inaccuracies in the measurement.

Two complete sets of EAA and OPC instruments are used to provide redundancy in case of an instrument failure. The output signals of both the

EAA and OPC are used by the aerosol computer to determine the corrective signals to apply to the aerosol generator to maintain the desired aerosol mass concentration within the CEL. A Beta Mass Monitor (BMM) instrument is used as an independent, noncontrolling measurement of the CEL aerosol concentration. Also, cascade impactor and filter samples are taken to give independent measurements.

The Electrical Aerosol Analyzer (EAA) operates on the principle that the electric mobility of a charged particle is inversely related to the size of the particle. An aerosol sample from the CEL is drawn into the analyzer by a vacuum pump. Once in the analyzer, the particles are charged and then carried in the air stream through a cylindrical condenser. The electric field of the condenser is varied in discrete steps where each charging voltage is selected to precipitate particles smaller than the selected size. The size determines the charged particle mobility in the electric field. The charges on unprecipitated particles having lower mobilities are detected by an electrometer connected to a digital voltmeter and the aerosol computer. Particle size distributions are determined by the differences in the concentrations of unprecipitated particles measured at each successive voltage step. The algorithms and constants used to compute the particle number, surface and volume distributions are stored in the computer. They may also be determined manually from the electrometer current measurements displayed on the digital voltmeter by using a programmable calculator.

The Optical Particle Counter (OPC) measures the intensity of the light scattered by individual particles as they pass through a high intensity light beam. A vacuum pump within the analyzer draws the aerosol sample from the CEL into a detecting chamber. A photomultiplier tube detects the light scattered from each particle and produces an electric pulse. The amplitude and frequency of the pulses correspond to the size and number concentration, respectively, of the particles in the aerosol sample. The aerosol size distribution is determined by sorting the pulses that are produced according to their amplitude.

The Beta Mass Monitor (BMM) measures the density of aerosol deposited on a filter strip. The difference in the beta radiation penetrating a clean filter area and an area where aerosol particles were collected is used to compute the mass of the aerosol sample. The amount of beta radiation attenuation is a function of the mass of collected material over a fixed period of time. A microprocessor in the analyzer also determines a zero reference offset from a clean filter section to compensate for the density of the filter material.

To minimize the effects of temperature and humidity changes on the aerosol samples, the EAA and OPC instruments are housed in closed, temperature-controlled cabinets with insulated stainless steel sample lines to the CEL. Electrical signals from a temperature sensor in the CEL are compared to the cabinet temperature which is maintained at the same temperature as the CEL. A refrigeration unit is attached to each cabinet and compensates for the heat output of the instruments to maintain the required temperature.

Individual sample lines from CEL A were provided to each analyzer because large flow volumes are required, 7-54 liter/min per instrument, and the accuracy of the concentration measurement is affected by the accuracy and stability of the sample flow. During the aerosol start-up period, the sample ports of the analyzers are switched to sample from the start-up duct. Gate type solenoid valves incorporating a straight-through design are used to reduce turbulent aerosol deposition. Passive diluters are automatically switched into each sample line when needed to prevent measurement errors due to sample saturation of the analyzers.

An automatic calibration check is provided for the OPC at predetermined intervals specified by the operator. During the calibration cycle, the sample input port of the OPC is diverted to the output of a monodisperse polystyrene latex aerosol generator. If the response of the OPC to the calibration aerosol is outside the specified tolerance of a computer-stored calibration value, the OPC is automatically removed from online status by the computer and replaced with the back-up OPC.

Due to the measurement ranges and analysis techniques, direct calibrations of the EAA and BMM instruments are performed by the manufacturers. Indirect calibration tests are conducted by comparing the results of measurements made by the instruments on aerosols delivered to the CEL with the chemical and physical analysis of the aerosol material simultaneously collected on filters and cascade impactors.

Data Acquisition and Control

A computerized system provides for automatic data collection, reduction, reporting and feedback control. The system was implemented in a manner which minimizes the requirements for operator interaction. Subject safety and experimental integrity were the primary considerations in the design of the computer application. Provisions have been made to allow for failover to a manual control scheme in the event of a computer outage. The manual controls are normally inoperative in the automatic mode to prevent inadvertent override errors. However, the control mode may be switched from automatic to manual when necessary.

The basic element of an experiment is the System Configuration File (SCF). It consists of data stored on a magnetic disc device. The data include instrument calibration coefficients, aerosol mass and size requirements as a function of time, a physical configuration description and alarm parameters.

In short, it contains all of the information required of the computer to control the complete experiment without operator intervention. Information as to the aerosol size distribution in CEL A or the start-up duct is obtained from the online EAA and OPC and used by the computer to determine the total mass concentration of particles. The relative humidity in the CEL is also used in the determination of the mass concentration because the generated hygroscopic materials accumulate varying amounts of water depending upon the moisture content of the air. The SCF is prepared prior to the start of an experiment. A straightforward prompting language extends the SCF creation/modification capabilities to biomedical personnel and others outside of the computer science disciplines.

Once an SCF is established, an experiment may be conducted under automatic control. A set of 10 simple user communication commands allows the operator to respond to system transients, request reports of experimental data and alter the exposure profiles that were specified in the SCF. The command set was designed to be brief and simple in order to minimize training requirements.

Ten unique reports may be requested. One report, the summary of active alarms, is generated automatically when abnormal conditions of a certain severity are detected. The title block is identical for each block so that the operator always has a consistent, quick visual reference as to the current status of the equipment. A copier is attached to the display device to enable the user to obtain a permanent record of any information displayed. The key vehicle for historical data, however, is a magnetic tape drive. A complete audit trail of each experiment is recorded to allow machine processing of the data after the experiment. This procedure simplifies the preparation of test results as well as providing an auditable base to support conclusions.

Process control is accomplished with the computer through an adaptive feedback control algorithm. The computed aerosol size distribution and mass concentration are compared to the profiles specified in the SCF and new control signals are generated. The variables controlled are water flowrate, concentrated solution flowrate, number of nebulizers active and delivery damper position. Due to the physical characteristics of the generation and delivery system, the computer pauses until the new control signals become effective before beginning new measurements. Thus a cause-pause-effect relationship exists within the computer software rather than an immediate cause and effect relationship.

The computer software was implemented using top-down structured programming techniques. The problem was subdivided into modular areas which were divided into major tasks and so on until a collection of about 150 small routines had been defined. A fill-in form was used to describe the routines in order to guarantee consistency and completeness throughout the documentation. All system constants and variables were parameterized and stored in a single file, avoiding the proliferation of problems that is typically encountered when a number changes. The parameter file was carefully controlled using techniques from the configuration management discipline. Parameterization offered unique advantages during checkout of the system; variables such as delay time between control and measurement could quickly be changed to allow sensitivity analyses and to optimize the overall performance of the system.

Safety Features

Several safety features have been designed into the system to protect laboratory personnel as well as subjects. All components are designed to failsafe in the event of power or equipment outage. The computer continuously monitors the status of key system components and the values of numerous variables. A multilevel and audible alarm system reports abnormal conditions and may even initiate a purge of the CEL. Also, the generation facility is somewhat isolated and designed to handle operational problems.

Safe failure modes typically involve discontinuing delivery of aerosol to the CEL and shutting down the generation process. This sequence will result from power failure, numerous electronic or mechanical component failures, and certain operator errors. The failsafe actions are designed into both the control mechanisms and the Aerosol Computer software. Safety activities initiated by the computer may always be averted if the operator makes a timely response to alarms with the appropriate action. As a part of the safety system, the computer itself is monitored for proper operation. If a malfunction is detected, the computer feedback is automatically eliminated and generation and delivery characteristics are determined by the configuration preset at the control panels.

The computer continuously monitors both digital and analog signals which represent elements of the status of the system. Digital signals include damper and switch positions, sample flowrates in or out of tolerances, and proper or improper operation of the major elements of the measuring instruments. The analog signals provide such parameters as relative humidity, conductivity of the aerosol generator solution, compressed air pressure and measuring instrument data. In addition to safety checking all of the digital and analog parameters, several variables are calculated specifically for verification of safety. For example, the computer calculates the maximum dosage that can be generated with the concentration and number of nebulizers in use. This value is compared with the measured concentration to assure that the proper control is being exercised.

There are two methods for producing alarms to report abnormal conditions. A computer-independent device reports both visually and audibly whenever a computer failure is detected or when the electronics detect a malfunction while not under computer control. The computer implementation includes a multilevel alarm system where each level represents a unique significance. The level of an alarm may advance if the significance of the abnormal situation increases with the length of time that it persists. The severity of the action taken by the computer is determined by the alarm level. For example, a sample flowrate error on an instrument not in use might merit a printed message, whereas total aerosol mass remaining above the tolerance limit for 20 minutes could justify stopping the generation and delivery. System limit specifications entered into the SCF are the triggering points for the alarms. They may be set to updated values as specified by the protocol.

A video display console with a hard-copy option is used by the system operator to interact with the computer. A permanent record of the monitoring data, status information and text entries are computer recorded on a magnetic history tape.

The aerosol generation room is equipped with the standard chemical laboratory safety devices and protective clothing.

SYSTEM PERFORMANCE

The aerosol system described above has been in use for about six months. To date, only fixed set points for aerosol mass concentration and size have been used for exposure periods up to four hours.

Figure 11 illustrates generator performance under manual control but with computer summaries of aerosol mass concentration.

Aerosol mass concentration and size measurements obtained from the OPC-EAA instruments in conjunction with the computer were compared with time-averaged measurements of standard filter samples analyzed first gravimetrically and then chemically. The chemical analysis was done using ion chromatography and was accepted as the standard for mass concentration determinations. The results of eighteen 4-hour exposure experiments are illustrated in Table II. These data show good agreement between the OPC/EAA computed mass concentration and the chemically analyzed filter mass concentrations. The average computed mass was only 1.9% less than the chemical filter, and maximum deviations were +15% and -17%. The average of the gravimetric mass concentration determinations was 5.6% higher than the chemically analyzed filter, and maximum deviations were +34% and -7%. Most of the large gravimetric errors were made during the first few experiments and may be attributed to establishing standard procedures and techniques.

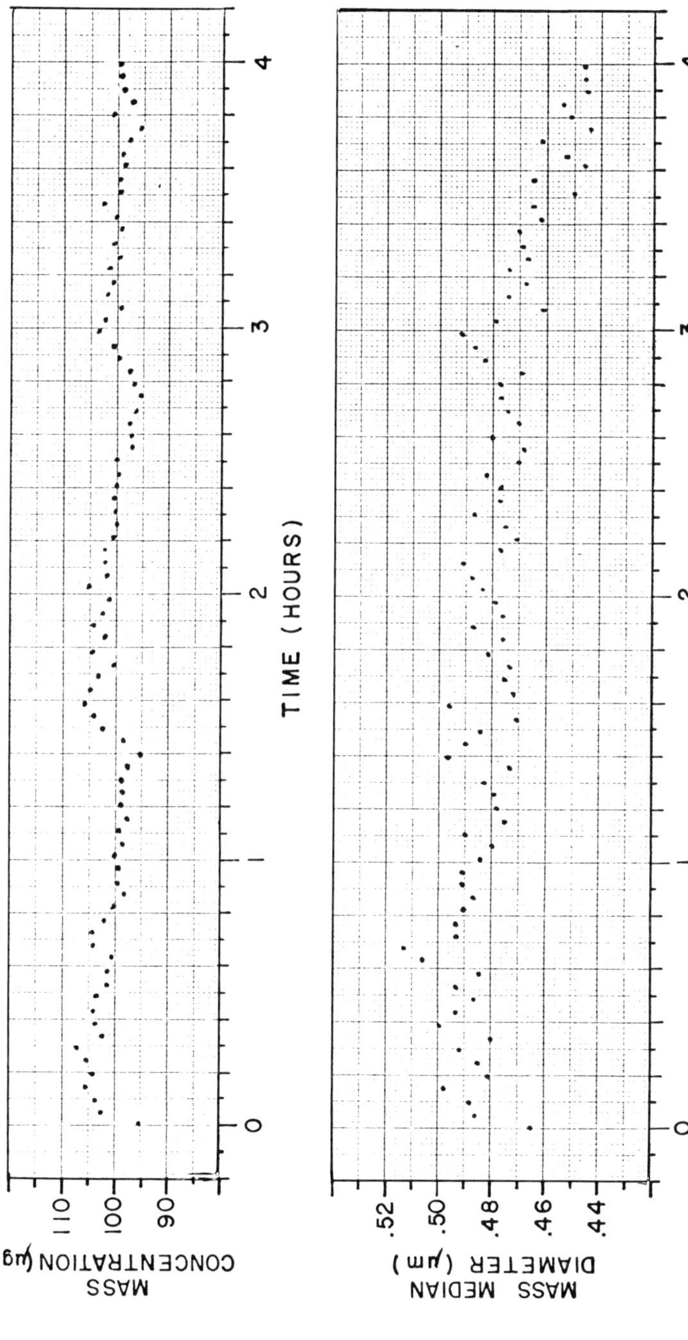

Figure 11. Aerosol generator performance.

Table II. H_2SO_4 Aerosol Size and Mass Concentration Measurements

Filter		Computer			Impactor	
Gravimetric ($\mu g/m^3$)	Chemical ($\mu g/m^3$)	Concentration ($\mu g/m^3$)	Size (dia. μm)	σ_g	Size (dia. μm)	σ_g
112	107	113	0.510	2.66	0.71	2.61
129	96	110	0.496	2.85	0.64	2.97
92	92	106	0.510	2.81	0.48	2.63
117	125	107	0.500	2.65		
120	112	109	0.500	2.62		
100	105	108	0.465	2.56		
109	104	100	0.492	2.49	0.50	2.78
114	123	102	0.486	2.49	0.47	2.50
114	105	102	0.504	2.37	0.68	1.64
131	118	102	0.492	2.55	0.56	3.24
126	106	98	0.490	2.50	0.44	2.72
109	105	102	0.499	2.44	0.52	2.31
115	101	100	0.473	2.39	0.40	2.49
104	101	100	0.492	2.42	0.75	1.95
103	96	102	0.510	2.41	0.61	2.10
106	94	101	0.496	2.36	0.54	2.96
107	105	102	0.490	2.46	0.47	2.70
112	107	105	0.487	2.35	0.43	2.40
Avg. 112	106	104	0.494	2.52	0.55	2.53

A comparison of particle size using a Sierra Model 245 cascade impactor and the OPC/EAA computed values showed that the average of 15 cascade impactor measurements was only 12% larger than the diameters computed from the OPC/EAA in the 0.5 μm diameter range, and maximum deviations were +18% and -35%. Also, the cascade impactor average computed geometric standard deviation (σ_g) of particles sizes was in good agreement with the OPC/EAA computed values 2.53 σ_g vs. 2.52 σ_g.

SUMMARY

A computer-assisted aerosol generation and control system has been used for about six months in human exposure studies to generate aerosols from water-soluble materials at a range up to 78 mg/min to give mass concentrations up to about 300 $\mu g/m^3$ and particle mass median diameters from 0.1 μm to 1.0 μm. During the use of this system, the computed mass concentration has been within ± 16% of time-averaged concentrations determined from chemical analysis of filter samples, and the computed particle size has been within +18% and -35% of measurements made with a cascade impactor.

The convenience of adjusting mass concentration and particle size in combination with short-time measurement data have made the use of this system safe and accurate for human exposure studies.

REFERENCES

1. Strong, A. A. "Description of the CLEANS Human Exposure System," EPA 600/1/78-064, November 1978.

CHAPTER 25

HORIZONTAL AIRFLOW INHALATION EXPOSURE CHAMBER

Juraj Ferin and Leonard J. Leach

Department of Radiation Biology
and Biophysics
School of Medicine and Dentistry
University of Rochester
Rochester, New York 14642

INTRODUCTION

Many investigators have described inhalation exposure chambers of various forms commonly using vertical airflow patterns [1, 2]. In this type of exposure unit, experimental animals are usually placed in vertically stacked cages, allowing those on lower tiers to be in direct contact with excreta expelled from above. There is a good possibility that in addition to the test material, an exposure to animal dander, microorganisms and products of excretion occurs. This fault may become critical in experiments of long duration. Prevention of infection and contamination may also support an integrated exposure and housing facility in which the everyday transfer of animals from the housing to the exposure cages and back becomes unnecessary. For these reasons we have attempted to design and build a new exposure apparatus using horizontal airflow. This allows the use of excreta-collection trays between each level of cages; because the air passes horizontally through a single stack of cages the probability of transmitting airborne disease is lowered.

THE EXPOSURE CHAMBER

Description

The exposure unit schematically shown in Figure 1 consists of four 1.5-ft thick (~ 0.46 m) rectangular modules that are approximately 4 ft (~ 1.22 m) high and 5 ft (~ 1.52 m) wide. The first module supplies particulate-free air to the exposure unit. Module 2 contains the contaminant generating and air

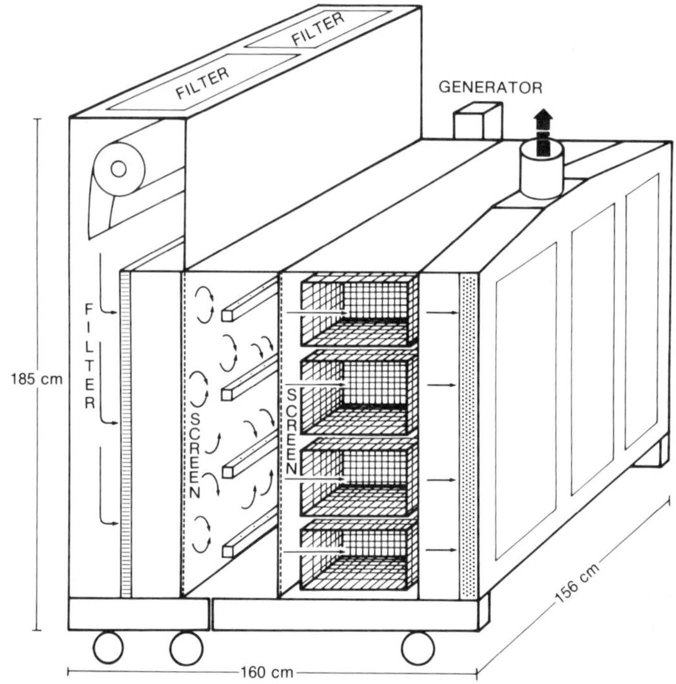

Figure 1. Schematic of horizontal airflow exposure chamber. The entering air is moved through two filters and a slotted diffuser (first screen); the contaminant is injected into the clean air through four horizontal tubes; the contaminant laden air passes through a perforated plate (second screen) and the cage rack; the effluent air is collected, filtered and exhausted.

distribution systems. The third and fourth modules comprise a rat cage rack (this holds 24 cages, each capable of housing 1-3 rats) and an exhaust manifold, respectively. During exposure, the modules are mechanically clamped and sealed together; however, upon completion of the experiment, they can easily be disconnected for rack changing, cleaning or moving.

Operation

With the four modules clamped together, the normal operation procedure is as follows. Metered air from the room is drawn into the air filtration module (module 1) by means of a blower located in the duct attached to the exhaust manifold (module 4). The filtered air passes through a slotted diffuser which induces turbulence in module 2 to enhance mixing. This module contains four horizontal tubes of square cross section and of ~ 1.5 m in length. Each tube has 30 holes of 2 mm diameter evenly spaced along the top and bottom from which test material is dispersed into the turbulent airstream. The four tubes are connected to a vertical manifold. The contaminant generating apparatus is attached to the manifold. Using four valves the airflow through the tubes can be balanced properly. The contaminant laden air then passes through a perforated plate containing 36 holes/cm^2 of 1 mm in diameter. The perforated plate functions as a laminator, enhancing a homogeneous distribution of the airborne contaminant across the entire area of the cage rack (module 3). Finally, the effluent air is collected and filtered in the exhaust manifold (module 4) and subsequently directed to the main exhaust system.

Testing

If not stated otherwise the testing was performed without animals in the chamber.

A Virtis single module generator, operated at 15-20 psi, forced a dioctylphthalate (DOP) aerosol into the system at the top inlet of the manifold. The air velocities in the chamber varied from 22 to 32 m/min. Downstream DOP concentrations were measured with an aerosol photometer through test ports on either side of the exhaust module. Sampler inlets were set up to coincide with centers of the animal cages, when they are in place. The results, which are shown in Figure 2, indicate small variations in aerosol concentrations across the area of the animal cage rack (module 3).

A test aerosol of TiO_2, generated by a modified Wright Dust Feed mechanism [3] was used (gear ratio 3:1, air pressure 10 psi, packing pressure of dust tube ~ 1000 lb). The generator injected the aerosol into the top of the inlet manifold. The four gate valves were completely open. Simultaneous samples were collected at 24 locations in the exposure area (center of each cage) using membrane filters. The TiO_2 deposited in the filters was analyzed chemically [4] or gravimetrically. The results are listed in Table I. Chamber airflow

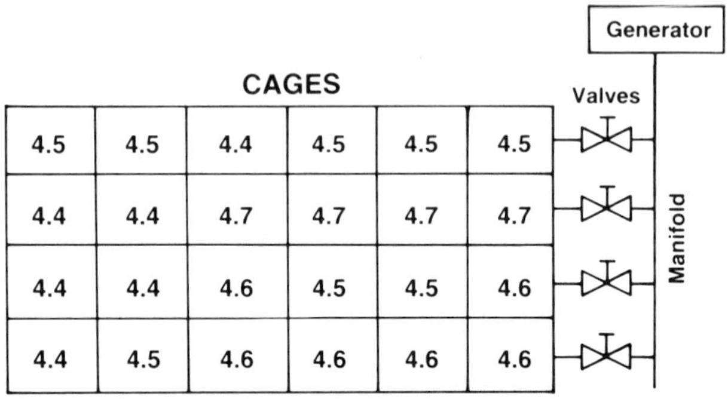

Figure 2. DOP aerosol concentrations at center of each cage (relative log scale).

Table I. Airborne TiO_2 Concentrations at Various Chamber Airflows

NO.	Chamber Airflow (m^3/min)	Sample Time (min)	TiO_2 Mean (mg/m^3)	SD	SD/Mean
1	13.6	36	0.49	0.12	0.24
2	13.6	60	0.65	0.13	0.20
3	6.6	60	1.36	0.32	0.24
4	3.3	30	3.70	0.63	0.17
5	3.1	30	2.91	0.46	0.16
6	1.6	30	9.89	1.18	0.12
7	1.6	15	13.95	1.99	0.14

was varied from 1.6 to 13.6 m^3/min, giving mean aerosol concentrations from ~ 0.5 to 14 mg TiO_2/m^3. With increased airflow the aerosol concentration, because of increased dilution, obviously decreases. The difference in mean concentrations for experiments 6 and 7, despite similar ariflow, is probably caused by variations of the aerosol generator output.

The data sheet reproduced in Figure 3 summarizes typical concentration information obtained from the new exposure chamber utilizing horizontal airflow. In this test, the mean concentration of TiO_2 aerosol was computed to be 6.94 mg/m^3 at a chamber airflow rate of approximately 90 ft^3/min (2.55 m^3/min). When comparing these figures with those obtained in tests with Rochester-type chambers, the aerosol losses in both systems were about 50%. Variations in aerosol concentration at different locations within the animal housing area were also about the same for both systems. For example, the distribution patterns had standard deviations of 6% for the Rochester-type chamber and 8% for the new unit at approximately the same mean TiO_2 aerosol concentration.

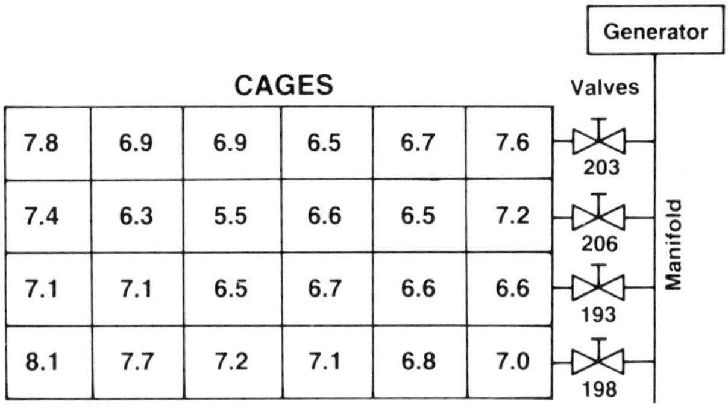

Figure 3. Concentration of airborne TiO_2 (mg/m^3) at the center of each cage and at the valves of the manifold. Sampling time: 1 hr in chamber, 15 min at valves; sampling air flow: ~1 liter/min; generator: modified Wright Dust Feed, gear ratio 3:1, packing 1000 psig, air pressure 10 psig; chamber: airflow ~90 ft^3/min (2.55 m^3/min), duration 2 hr; mean concentration: 6.94 mg TiO_2/m^3, SD = 0.57 (8%).

The respirable fraction of the aerosol was measured at each of the four cage levels using a Harwell Sampler. No appreciable difference was found. The respirable fraction of the TiO_2 aerosol was ~ 36% which is comparable to the values found when using the vertical-flow Rochester chamber. Studies with a Mercer Cascade Impactor revealed the test aerosol to have a MMAD of 1.03 μm with a σ_g of 2.29.

Twenty-four rats (one per cage) were exposed to a TiO_2 aerosol on three consecutive days. The length of the sessions were 6, 6 and 5 hours, respectively. The concentration of the aerosol was ~ 5 mg TiO_2/m^3. After the exposures all animals were sacrificed and the lung content of TiO_2 was determined [4]. The results in Table II show an average lung burden of 260 ± 50 μg TiO_2/lung. In another experiment, after a single exposure an average lung burden of 74 ± 16 mg TiO_2/lung was determined.

A bacterial aerosol of *micrococcus lysodekticus* was generated and injected into the clean airstream at the rear of a cage located near the middle of the

Table II. Lung Burden in Rats after TiO_2 Aerosol Exposure (Three Sessions) (Mean = 260 μg TiO_2/lung, SD = 50 (19%)

	mg TiO_2/lung						
	$<$191	191-225	226-260	261-295	296-330	$>$330	Total
No. of Rats	2	4	6	6	4	2	24
%	8	17	25	25	17	8	100

second level. Petri plates were placed in this and adjacent empty cages to sample the radial distribution of the bacterial aerosol. After exposure and cultivation the number of colonies in the various plates was counted. These experiments showed that the bacteria injected into the airstream in one cage remained in this cage and did not spread to any appreciable degree into the adjacent cages.

DISCUSSION

In the past the common inhalation chamber design featured vertical airflow with a cone at the inlet and one at the outlet end of the chamber. The cones function as transitional ducts and were considered important for good aerosol distribution within the chamber. The cones took up large amounts of space and thus the chambers had to be located in rooms with high ceilings. In our design, a manifold system on both the inlet and the exhaust sides replaced the cones, resulting in appreciable saving of space.

The most important feature of the horizontal airflow is that it makes it possible to position catch pans between levels of animal cages, preventing undesirable contamination. The perforated plate used as a diffuser directs the airflow to an extent that air mixing between adjacent cages is minimal as our testings with bacterial aerosols indicate. Sedimentation of small particles (~ 1 μm MMAD) does not affect appreciably the aerosol concentration in the cages. With the TiO_2 aerosol sedimentation of large amounts of dust in chambers was not observed. However, the manifold and perforated plate require routine cleaning to maintain correct dimensions.

We believe that our aerosol concentration measurements show satisfactory homogeneous particle distributions throughout the chamber. Consistent with that are the results of lung burden analyses. The coefficient of variation of the lung burdens compares favorably with other results [5], with a range factor of about 2, compared to a factor of 4 in Raabe et al. [5].

ACKNOWLEDGMENTS

This paper is based on work performed partially under contract with the U.S. Department of Energy at the University of Rochester Department of Radiation Biology and Biophysics and has been assigned Report No. UR-3490-1471, and partially by the Environmental Health Sciences Center of the National Institute of Environmental Health Sciences (ES01247).

B. Morehouse, E. Renner and R. Gelein provided technical assistance in this study.

REFERENCES

1. Fraser, D. A., A. E. Bales, M. Lippman and H. E. Stockinger. "Exposure Chambers for Research in Animal Inhalation," Public Health Monograph No. 57, U.S. Government Printing Office (1959).
2. Leach, L. J., C. J. Spiegl, R. H. Wilson, G. E. Sylvester and K. E. Lauterbach. "A Multiple Chamber Exposure Unit Designed for Chronic Inhalation Studies," *Am. Ind. Hyg. Assoc. J.* 20(1):13-22 (1959).
3. Wright, B. M. "A New Dust Feed Mechanism," *J. Sci. Inst.* 27:12 (1950).
4. Ferin, J., and M. L. Feldstein. "Pulmonary Clearance and Hilar Lymph Node Content in Rats after Particle Exposure," *Environ. Res.* 16:342-352 (1978).
5. Raabe, O. G., J. E. Bennick, M. E. Light, C. H. Hobbs, R. L. Thomas and M. I. Tillery. "An Improved Apparatus for Acute Inhalation Exposure of Rodents to Radioactive Aerosols," *Toxicol. Appl. Pharmacol.* 26:264-273 (1973).

CHAPTER 26

ANIMAL EXPOSURE FACILITY FOR DIESEL EXHAUST STUDIES*

Robert G. Hinners, Joseph K. Burkart and Myron Malanchuk

Health Effects Research Laboratory
Environmental Research Center
U.S. Environmental Protection Agency
Cincinnati, Ohio 45268

William D. Wagner

Biological Support Branch
Division of Biological and Behavioral Science
National Institute for Occupational Safety and Health
Cincinnati, Ohio 45226

INTRODUCTION

The Environmental Protection Agency (EPA) is charged under the Clean Air Act with the responsibility for regulating emissions from new motor vehicles. It has been estimated that automobiles equipped with diesel engines are likely to capture up to 25% of the United States new car market by 1985 [1]. The concern over fuel economy is the major factor behind the projected increase, since diesel powered vehicles offer approximately a 25% improvement in fuel savings over comparable vehicles powered by spark ignition gasoline engines. With the increase in diesel-powered automobiles,

---

*Any mention of product or company names does not constitute endorsement by the U.S. Environmental Protection Agency.

the anticipated change in the quantities of environmental pollutants emitted to the atmosphere indicates the need for well-instrumented health effects studies to adequately establish their health impact.

Diesel-powered units are used extensively in metal and nonmetal mines in the United States, but their acceptance and use in underground coal mines has been very limited, principally because of concern for potential adverse health effects to miners exposed to the exhaust components. Although there is a substantial body of knowledge relating to coal mine dust and individual constituent pollutants in diesel exhaust, there is very limited information on the biological effects of diesel exhaust in confined spaces such as underground mines. Health hazards may be anticipated due to the known incidence of respiratory disease among coal miners and the potential for interaction between respirable coal mine dust and diesel emissions. Patholigic changes from such combined exposures may differ from those resulting from exposure to either agent alone.

The National Institute for Occupational Safety and Health (NIOSH) is responsible for establishing air standards for coal mines and EPA is responsible for ambient air standards as affected by diesel-powered vehicles among others. Accordingly, the two organizations have signed an interagency agreement to conduct toxicological studies on animals exposed to combined diesel exhaust and coal dust. The EPA facility serves as a center for toxicologic evaluation, in relevant animal models, of potentially hazardous environmental pollutants derived from the exhaust of a light duty diesel automobile engine and is presently being expanded to accommodate the interagency study.

FACILITIES

The System

Briefly, the entire engine exhaust is mixed with filtered and conditioned air in a dilution tube. That diluted exhaust enters a large volume mixing chamber and a portion passes through dynamic flow irradiation chambers (to simulate sunlight) and is then conducted to animal exposure chambers. The system provides nonirradiated exhaust in the same concentration (directly from the mixing chamber) and also filtered, conditioned ambient air for control animal exposure. There are two similar but separate systems for the automotive diesel exhaust and the mine diesel exhaust generation, since they require different engines and operating conditions. The principal difference is the absence of irradiation chambers and the added coal dust generators for certain exposure chambers in the mine diesel system. Figure 1 illustrates the flow and essential equipment in the automotive diesel emission

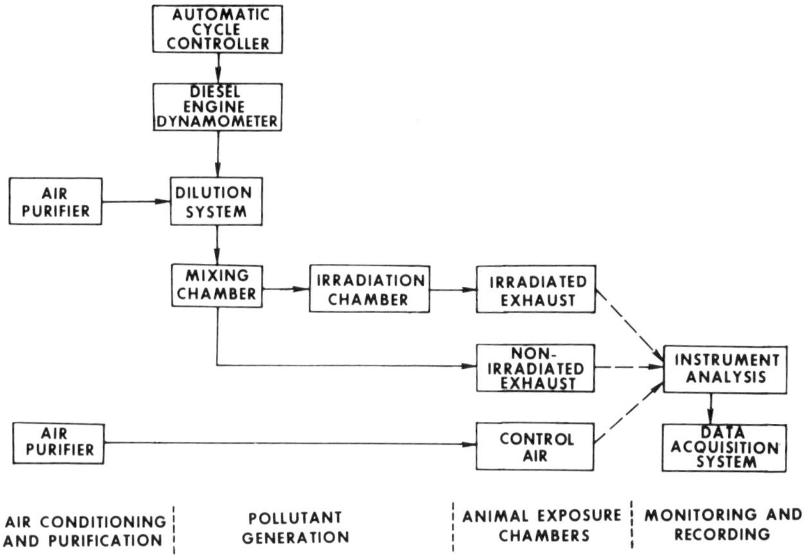

Figure 1. Schematic flow diagram diesel exhaust system.

system. The major physical components of the facility include air purifiers, engine-dynamometers, irradiation chambers, animal exposure chambers, monitoring instrumentation and data acquisition systems. This equipment is further described in the following text.

Air Purifier

An air purifier unit provides, at maximum, 21.2 m^3/min (750 ft^3/min) of CBR (chemical, biological, radiological) filtered and conditioned air to maintain the required temperature and relative humidity in the animal exposure chambers. Two units are required for dilution air for the diesel engine exhaust, and one for the clean air (control) animal exposure chambers. An air purifier system consists of a pressure blower, cooling coils, reheat coils, humidifier and CBR filters combined into a single unit, with a separate refrigeration compressor and outside air-cooled condenser. The sensors for the diluted raw exhaust are located downstream of the dilution tubes in order to correct for temperature and relative humidity changes when the hot and moist engine exhaust enters the system. The filtering system provides a pass through absolute filters to remove particles down to 0.3 micron (99.97% efficient) and pass through activated charcoal of 20.3 cm (8 in.) bed depth. Exposure chambers on control air for both studies are supplied from the same CBR filtered source, with controls set to maintain 24°C (72°F) and 50% relative humidity in the animal chambers. The same air is also ducted to the air filter inlet of the engine being used, since a change in humidity affects the NO_x emissions. There are a total of five air purifiers with two pairs interconnected to provide 1500 ft^3/min of dilution air for each engine study.

Engine

The diesel engine selected for the light duty mobile emissions study is made in Japan by the Nissan Motor Company (makers of Datsun automobiles). For long term animal exposure studies, it is necessary to have automatic cycle control, and this in turn requires an automatic transmission. Mounted on a dynamometer base plate are two Nissan CN6-33 diesel engines coupled to Chrysler Torque-flite automatic transmissions Model A-727. The detailed engine specifications which incorporate the latest low smoke combustion head are as follows:

1. Four cycle, water cooled, naturally aspirated;
2. Six cylinders in-line;
3. Bore and stroke 3.27 in. x 3.94 in. (8.3 mm x 10.01 mm);
4. 198 $in.^3$ displacement (3244 cm^3);
5. 92 maximum brake horsepower at 4000 rpm;
6. Torque, maximum at 2000 rpm–137.5 lb-ft;
7. Compression ratio–22:1; and
8. Approximate weight–662 lb.

Dynamometer

One dynamometer base plate is used for the automotive diesel exhaust studies and another similar installation nearby will be used for the coal mine diesel exhaust study. An Eaton-Dynamatic Model 758-DG dynamometer has been coupled to the transmission of the Nissan engine. The unit is an air gap, eddy current, dual rotational type that employs water cooling and has a power absorption rating of 50 horsepower in a speed range of 1600-10,000 rpm. A 450-lb flywheel is directly coupled to the dynamometer to provide the rotational energy to simulate the inertial mass (load) of a moving 4000-lb vehicle. The dynamometer field current can be set at a constant value or made dependent on the flywheel speed. The dynamometer unit is located between the transmissions of the engines which can be coupled, one at a time, to the double ended dynamometer. This feature provides insurance in case of mechanical breakdown during a chronic study, since another engine can be put on line in about 30 minutes to provide exhaust gas to the animal chambers.

Dilution Tube

The effluent from the engine exhaust system is passed through a muffler and flexible stainless steel (SS) pipe into the dilution tube. The 40.64 cm (16 in.) i.d. dilution tube is made from 16 gauge SS plate and conditioned air enters the tube from a remote supply source. At this point is a mixing baffle plate, with a 20.32 cm (8 in.) diameter hole bored in the center. The incoming dilution air, under pressure, is forced through this hole to mix with the raw exhaust. The tailpipe enters the dilution tube at a right angle to its axis and is elbowed 90° to become concentric with the dilution tube. The exit end of the exhaust elbow is in the same plane as the baffle. Located on the bottom and outside the dilution tube, at the baffle plate, are two quick-disconnect couplings. One allows the end of the flexible exhaust pipe from the muffler to connect with the dilution tube and the other connects to the outside atmosphere. A blank plug is installed in the disconnect to the dilution tube when the exhaust is vented outdoors, which allows interruption of animal exposures and provides opportunity to run the engine for test purposes without contributing exhaust emissions to the animal chambers.

By operating a damper in the air supply line, the dilution ratio can be controlled up to a maximum of 30 to 1. To retain the particulate matter in suspension and prevent condensation, it is necessary to dilute the whole exhaust with no less than 8 parts of air to 1 part of exhaust when cycling. The main portion of the dilution tube consists of a 4.88-m (16.-ft) flanged section before continuing on through a duct of the same diameter and entering the mixing chamber. There are two dilution tubes located above the engine-dynamometer base plates.

Mixing Chamber

The mixing chamber serves to attenuate concentration peaks in the nonirradiated exposure chambers and to avoid the problems associated with stream splitting and proportional sampling.

The diluted diesel exhaust enters the 19.34-m^3 (683-ft^3) mixing chamber through a 40.64 cm (16 in.) diameter SS duct in the side wall. An elbow discharges the exhaust across the front of a tube-axial fan, controlled at a low rpm by a Zero Max unit, to mix the entering exhaust with the chamber atmosphere.

Pipes from the mixing chamber supply the exhaust either direct to the animal exposure chambers or to irradiation chambers enroute to exposure chambers. An adjustable pressure sensor controls a monitorized damper in a vent line to the external atmosphere in order to maintain a positive pressure in the chamber. There are two identical mixing chambers located side by side on the mezzanine over the middle exposure chamber room.

Irradiation Chambers

The photochemical reactions that result from the exposure of diluted raw exhaust to artificial sunlight take place in eight irradiation chambers. Fluorescent lighting panels composed of blue, black and sun lamps outside the chamber pass intense ultraviolet radiation through windows of clear Teflon FEP fluorocarbon film. Normal flow through the irradiation chambers is 0.71 m^3/min (25 ft^3/min) and requires approximately 3.5 hours to reach equilibrium at a constant inlet concentration. Each irradiation chamber has a volume of 19.34 m^3 (683 ft^3) and they are stacked in pairs to conserve floor space. An irradiation chamber has four banks of 66 fluorescent tubes each, mounted in four external cavity reflectors.

Exposure Chambers

There are 36 individual animal inhalation exposure chambers of two different configurations utilized in the diesel exhaust studies [2]. Briefly, they are constructed of 14-gauge type 304 stainless steel and are cubical with formed funnels at the top and bottom of each chamber. Twenty chambers of the 1.37 m (4.5 ft) square size are divided into two groups, with twelve assigned to the mine diesel-coal dust project and eight to the automotive diesel exhaust study. Also, there are sixteen chambers of 1.53 m (5 ft) square size used only on the automotive exhaust study.

Ports are provided with PVC plugs that can be either removed or drilled for different uses that make access to the chamber necessary. Beneath the floor grating is a spray ring with eight wide-angle nozzles to flush the bottom funnel with warm water. A quick-opening level operated gate valve at the bottom of the funnel is connected to a drain line for washing down the chamber.

The exhaust and air mixture is admitted tangentially to the short cylindrical section at the top of the chamber. As a result of a mild swirling motion, the air and test material are mixed and distributed uniformly throughout the chamber. The exhaust air is drawn axially from near the bottom of the chamber through an individual exhaust duct connected to the exhaust system.

A venturi calibrated for 15 air changes per hour is mounted between the flanges of the inlet tube and a butterfly damper. The pressure taps on the venturi are connected to a magnehelic pressure gauge calibrated for flow in ft^3/min. The negative static pressure is measured with a magnehelic gauge connected to one of the port openings and regulated with a globe valve in the exhaust.

Exposure Chamber Room

There are 4 animal exposure chamber rooms with each of 3 rooms containing 8 chambers used in the diesel auto exhaust studies and 1 larger room with 12 chambers for the exhaust and coal dust project. Each chamber room is individually air conditioned by a 5-ton, self-contained air-cooled unit located outdoors. It is necessary to maintain controlled temperature in the exposure rooms in order to prevent heat transfer through the walls of the chambers and to avoid stressing the animals when they are removed from a chamber for examination or chamber cleaning. The floors are sloped to a trench drain at one end of the rooms and have a Dexotex covering to facilitate cleaning with a water hose. Each room is provided with a 24-in. wall exhaust fan for ventilation when chamber doors are opened and room lighting is on a timer for automatic day-night light cycle.

Engine Cycle

The dynamometer driving schedule for the Nissan engine consists of a repetitive series of idle, acceleration, cruise and deceleration modes of fixed time sequences and rates. Table I shows this repetitive series of nine driving modes known as the Federal Short Cycle and used in the present fuel emission studies.

Table I. Federal Short Cycle

Mode	Time in mode (sec)
0-16 mph acceleration	6
16-29 mph acceleration	23
29 mph cruise	10
29-37 mph acceleration	18
37-42 mph acceleration	4.5
42-37 mph deceleration	2.5
37-20 mph deceleration	32
20-0 mph deceleration	7.5
Idle	21.5
Total	125.0

Fuel Selection and Storage

There are numerous federal, military and commercial diesel fuel specifications and classifications used by industry, but those fuels marketed to meet the needs of heavy grade diesel fuel specifications are most commonly used. For automobiles, highway trucks, tractors, marine and off-road equipment, No. 2 diesel is the typical fuel requirement and therefore was selected. The fuel was analyzed to contain 67.4% saturates, 3.4% olefins and 29.2% aromatics with 0.17% sulfur by weight. A fair comparison between the statutory emission levels and those of the engine under test should be made with a fuel that is representative of that in actual commercial use.

Two underground fuel storage tanks of 2250 gal capacity are each equipped with electric fuel pumps to deliver fuel from the main storage supply to 50-gal companion tanks located outside the building near the engine area. These tanks are provided with a special connection to permit the introduction of fuel additives so that 50-gal batches can be prepared without contaminating the main tanks.

Coal Dust Aerosol

Generation and dispersion of the bituminous rank coal dust aerosol at a 2 mg/m^3 respirable dust concentration is accomplished with the Wright-dust-feed mechanism. Freshly micronized coal, reduced to a particle size 50% < 1 μm, 90% < 3 μm, is blown by the feed mechanism directly into the cylindrical section at the top of the exposure chamber, downstream from the diesel exhaust mixing chamber. The diluted exhaust-coal dust-air mixture, or coal dust-air mixture is dispersed uniformly throughout the chamber and drawn down past the animals' breathing zone.

INSTRUMENTATION

Monitoring and Analysis of Gas Concentrations

The atmosphere generated by the combustion of diesel fuel is monitored by measuring:

1. the concentration of six gases in each exposure chamber;
2. the temperature and humidity of the atmosphere in each exposure chamber.

The monitoring is done by two different automatic systems. Both systems are controlled by digital circuitry incorporating computers (microprocessors).

System #1–Online Data Acquisition System for Monitoring Concentration of Gases

As shown in the block diagram (Figure 2), the atmospheric content from each chamber is sampled sequentially into a common manifold via on-off solenoids activated by a digital programmable circuitry and controlled by a Hewlett-Packard 9830 calculator. The manifold is directly connected to

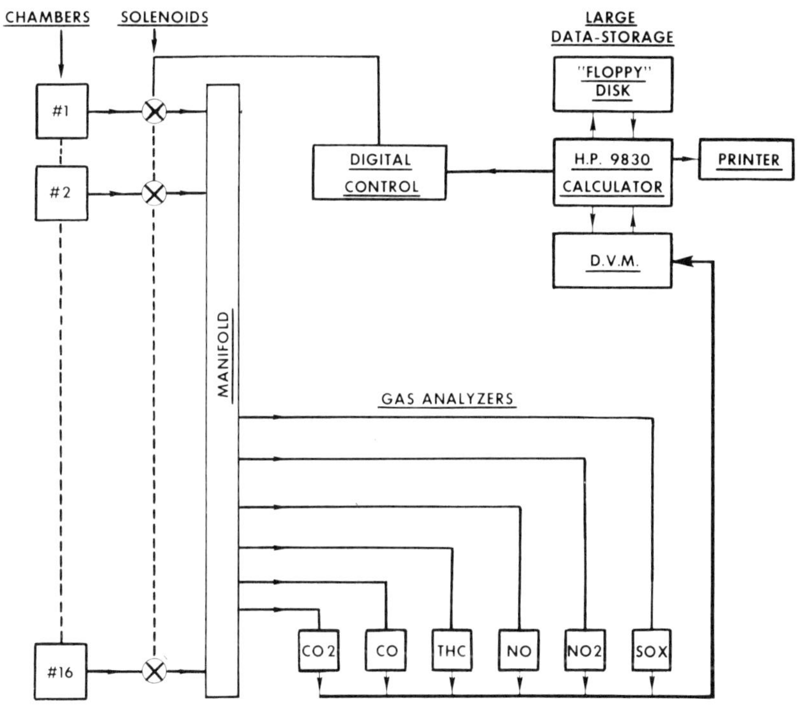

Figure 2. Block diagram data acquisition system.

online analyzers for measuring carbon dioxide (CO_2), carbon monoxide (CO), total hydrocarbons (THC), sulfur oxides (SO_X), nitric oxides (NO) and nitrogen oxides (NO_X). The operation of sampling the analyzers is controlled by the programmed calculator. The solenoid of a chamber is activated, and the atmosphere of that chamber enters the manifold. At this point in time there is a 3-min period allowed for the atmosphere and the analyzers to stabilize. At the end of that time, the outputs of each analyzer are fed sequentially to a digital voltmeter. The data in digital form are transferred to the memory of the calculator where they are stored. When the data from the last analyzer have been stored in the memory, the calculator commands the digital circuitry to switch on the next solenoid which opens the atmosphere of the next chamber to the manifold, and the sequence is repeated.

There are 16 chambers in the automotive engine study to be monitored, and scanning of all the chambers takes 1 hour. Therefore, for an 8-hr day there are 8 scans per day, i.e., 8 measured data points for each gas in each chamber. The statistical analysis of the data can be done at any time for any period of time for any chamber or gas. The analysis program converts the data entered into memory in volts into values in ppm. The printout of analysis contains the following information: date, chamber number, name of the gas, treatment of the atmosphere, number of data points monitored, average value in ppm, maximum value, minimum value, standard deviation, standard error and confidence limits. In general the printout of measured and analyzed data is reported on a weekly basis.

System #2–Monitoring of Temperature and Humidity in the Chambers

The system is based on the psychrometric principle. In each chamber there is a wet and dry bulb using thermistors as sensors. Water and air flow for the wet bulb is controlled by the system. The value of the resistance of wet and dry bulb sensors is monitored via a scanner that samples one chamber at a time (2.5 min/chamber). The values of resistance of each bulb are processed electronically and changed into voltage that is linear with the temperature. The measured voltage is fed into a microprocessor that is programmed to solve a mathematical model that relates dry and wet temperatures to humidity.

EXPOSURE CHAMBER AEROMETRY

Characterization of the pollutant atmospheres helps relate the animal exposure effects to the particular agents present in those atmospheres. On a more general basis, the differences between the irradiated and nonirradiated atmospheres are determined, and involve both the gaseous and particulate phases of each atmosphere. The analyses provide for measurements of ammonia gas and product ammonium compound resulting from chamber animal contributions interacting with the other atmospheric components.

Table II on analytical measurements lists the method or principle and the specific instrument used for each of the components considered. The principal, regulated (i.e., published in the *Federal Register*) components–CO, NO_X and THC–as well as CO_2 and even SO_2 when necessary, are monitored

Table II. Analytical Measurements of Exposure Chamber Atmosphere Components

Component		Principle	Instrument
Carbon monoxide	CO	NDIR	Beckman 865
Carbon dioxide	CO_2	NDIR	Beckman 865
Nitrogen oxides	NO_X	Chemiluminescence	TECo 14T
Nitric oxide	NO		
Nitrogen dioxide	NO_2		
Total hydrocarbons			
Ambient temp.		FID	Beckman 400
High temp. (350°F)		FID	Beckman 402
Hydrocarbons/GC		Gas Chromatography	Beckman 6800
Methane	CH_4		
Non-meth HC	NMHC		
Ethane	C_2H_6		
Ethylene	C_2H_4		
Sulfur dioxide	SO_2	Pulsed UV Fluorescence	TECo 43
Ozone	O_3	Chemiluminescence	Meloy 320
Ammonia	NH_3	Phenol-Hypochlorite	Beckman DB
Aldehydes	CHO's	DNPH/GC	Perkin-Elmer 900
Particulate, mass	M	GF Filter/Gravimetric	
Sulfate	SO_4^{2-}	Ion Chromatography	Dionex 14
Ammonium	NH_4^+	Ion Chromatography	Dionex 14
Nitrate	NO_3^-	Ion Chromatography	Dionex 14
Chloride	Cl^-	Ion Chromatography	Dionex 14
Particulate size	MD	Scanning EM	
Polynuclear aromatics	PAH	Liquid Chromatography	
Coal dust	Total & Respirable Mass	Filter/Gravimetric	

continuously. A system of Teflon tubing lines leading from the chambers to a group of solenoid valves supplies a sample of each chamber atmosphere in turn to the several monitoring instruments. Other atmospheric components are sampled and measured at various intervals.

Methods of Measurement

Gaseous Components

The CO, CO_2, NO and NO_2 measurements are made at regular intervals during the day for each of the exposure chambers with the exhaust emissions atmosphere. The interval between successive samples from one chamber and the next was determined by the rise time in responses collectively considered for the various instruments. Filter holders equipped with glass fiber filters are installed at the probe ends of the Teflon supply lines, i.e., at the exposure chambers, to remove particulates from a gas sample as a means of protecting the supply lines and the instruments against deposition and contamination by such particulate.

In addition to THC values obtained at the room temperature of the instrument room by continuous monitoring, there is provision for daily measurements of heated-probe THC from selected exposure chambers. The flame ionization detector (FID) instrument is fitted with a short, heated (350°F) sampling line in the immediate vicinity of the chamber to minimize line (adsorption) losses.

Individual hydrocarbons (methane, nonmethane hydrocarbons as a group, ethane and ethylene) are monitored at irregular intervals by the gas chromatograph (GC) in the instrument room. Gas sample supply lines, separate from the solenoid-controlled system for the regulated gases mentioned previously, are used to feed the gases continuously from the exposure chamber to the gas chromatograph. These individual organic components may serve as markers for the saturated and the unsaturated organic class groups.

The pulsed UV fluorescence (PF) instrument is convenient for monitoring SO_2. However, because of projected interference from other diesel engine exhaust emissions components, an adsorption device is used to eliminate such interferents and the instrument is used on a discontinuous basis to prolong the life of the adsorptive agent. The West Gaeke [3] colorimetric method applied to bubbler samples serves as a back-up procedure in case of the arrest of the PF instrument.

Ozone (O_3) measurements are taken at the large irradiation chambers, with occasional checks on the exposure chamber atmospheres when applicable.

For determination of ammonia gas (NH_3) concentrations in the more heavily animal-populated exposure chambers, bubbler samples may be collected for analysis by the phenol-hypochlorite method according to Weatherburn [4]. A chemiluminescence (NO_x) instrument provides an alerting system for high (about 1 ppm) peaks of ammonia concentrations.

The dinitrophenyl hydrazine (DNPH) method [5] for individual aldehydes is used in place of the methylbenzothiazolone hydrazone (MBTH) method [6] for aldehydes as a group; the latter has been considered to be inadequate in diesel emissions measurements. Bubbler solutions of the DNPH collecting reagent are immersed in ice baths during the samplings to improve collection efficiency. The bubbler samples are subsequently analyzed by a GC procedure for specific aldehydes, namely, formaldehyde, acetaldehyde, the propionaldehyde-acetone group, crotonaldehyde and hexanaldehyde.

Particulate Components

Engine organic compound particulate mass measurements are made daily from filter samplings obtained from the exposure chambers with the exhaust emissions atmospheres, as well as from the dilution tube.

Aqueous extracts of the particulate filter samples are analyzed by ion chromatography for various cations and anions, namely, ammonium and the sulfate ion concentrations.

Hi-vol samplings are made occasionally to provide large enough samples for analysis of the organic make-up according to functional groups (Figure 3).

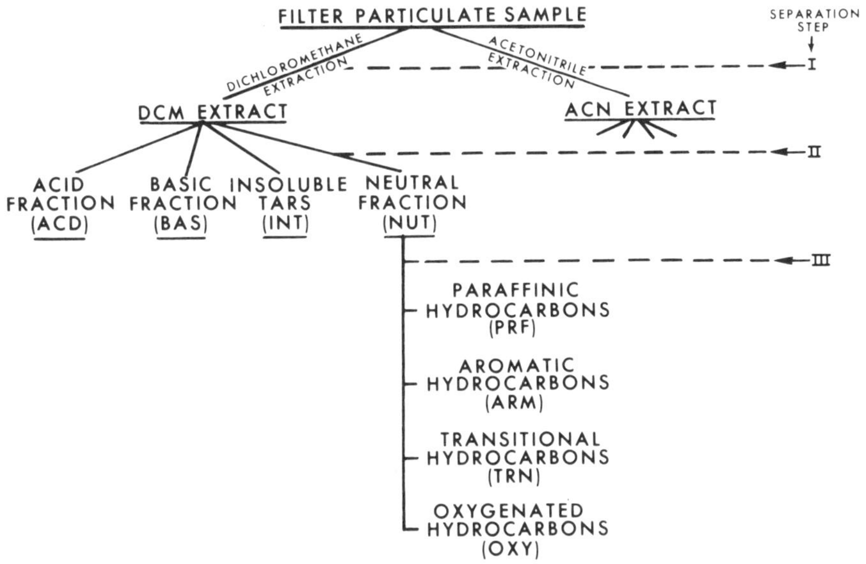

Figure 3. Separation scheme for diesel engine particulate organic compounds.

The filters are extracted by methylene chloride (DCM) and then by acetonitrile (ACM), step I. The extracted organics are treated in a separation scheme, step II, to produce the acidic (ACD), basic (BAS) and neutral (NUT) fractions as well as an insoluble tar (INT) portion. With the major interest in the neutral fraction for mutagenic and carcinogenic potential, that fraction is further treated in gravity flow liquid chromatography, step III, to yield the paraffinic (PRF), aromatic (ARM) and oxygenated (OXY) hydrocarbons groups as well as a color-band visible transitional (TRN) hydrocarbons group. Additional procedures such as high pressure liquid chromatography (HPLC) can be used to resolve individual components in the various fractions as well as class groups in the mother extract (DCM, ACN).

Particle sizing of the particulate is demonstrated in samples collected on Nuclepore filters, which lend themselves to being treated for convenient viewing by a scanning electron microscope.

The Electrical Aerosol Size Analyzer (Thermo-Systems Inc.) is adaptable to particle size distribution measurement, but is used minimally because of the very "dirty" nature of the diesel emissions which can readily foul the instrument flow component parts.

Chemical and physical characterization is made of coal dust before and during the duration of the study. Chemically, inorganic and organic content are determined and include V, As, Pb, Cd, Fe, Ni, Al, Zn, Be, Cr, Cu, Mn, Co, and free and combined SiO_2. Particle-size distributions are determined as well as adsorptive characteristics between coal dust and diesel exhaust.

Particulate respirable and total mass measurements are made daily from periodic samplings obtained from the exposure chambers. Collections are made on Gelman Metricel VM1 filters, 5.0 μm pore size, and analyzed gravimetrically to determine dust concentration.

Analytical Measurement Values

The data from the various analyses is used to supply whole study averages and also periodic averages for the exposure chamber atmospheric pollutant concentrations.

The whole study averages give, for example, a single CO concentration value for all of the CO measurements made on all of the irradiated atmosphere exposure chambers.

The same data bank that supplies the whole study averages is segmented into weekly averages that can readily be plotted as a bar graph, for instance, to show the trend of component concentration levels over the lifetime of the study. Such a trend plot may conveniently show the rate of change as well as the total change in composition as a result of changes, for example, in engine combustion efficiency, in fuel supply (composition), or even in animal loading in the exposure chamber(s). The trend also may confirm, by its consistency in differences between the irradiated and nonirradiated atmosphere concentrations, a real, though subtle difference in the whole study averages for the two types of atmospheres.

HEALTH EFFECTS

The primary health concern regarding the use of diesel engines involves the particulate fraction of diesel exhaust. The nature of the diesel particulate is potentially of more significance than just the total mass of particulate emissions, which are up to 82 times as much as in gasoline exhaust using paired vehicles [7]. The particulate is principally small agglomerated chains of very small carbon spheres which are frequently coated with adsorbed condensed material mostly in the form of high molecular weight organic compounds, sulfur dioxide, sulfates and materials such as polycyclic aromatic hydrocarbons (PAH) [8]. Particulates contribute to health effects by carrying the irritants to the deeper, more sensitive recesses of the lung where they can act to initiate, promote or accelerate emphysematous, fibrotic or carcinogenic processes [9]. There are several important reasons why increased exposure to the particulates in diesel exhaust may constitute a potential health hazard [10]:

1. the carbonaceous particle is basically a high molecular weight PAH;
2. the particles have a high surface area, theoretically enabling them to adsorb large quantities of exhaust gases, most importantly the carcinogenic polycyclic organic matter (POM) such as benzo[a]pyrene;
3. its low settling velocity in air allows time for atmospheric oxidation and degradation to lower-molecular-weight POM which are potentially carcinogenic;
4. the particles are primarily in a size range (0.2-0.3 μm mean diameter) which would allow deep penetration into the lung and may retard the lung's clearance defense mechanisms once deposited.

The use of diesel engines in underground coal mines is being encouraged in order to escape the tragedy of fire and explosions resulting from arcing of presently used electric motors and cables. Organized mine labor is opposed to the use of diesels until biological effects data are available from animal studies using diesel exhaust after it has passed through a scrubber.

To determine if there are potential health hazards arising from the operation of diesel engines in coal mines, this project has four specific objectives. The first is to determine if combined exposure to diesel exhaust and coal dust changes the incidence of neoplastic disease, as compared with controls and to exposure of each alone. The other three objectives are investigation of pulmonary response, immune potential and mutagenic potential of combined exposure to coal dust, diesel exhaust and each agent alone in comparison to control animals.

REFERENCES

1. Barth, D. S., and S. M. Blacker. "The EPA Program to Assess the Public Health Significance of Diesel Emissions," *Air Poll. Control Assoc.* 28(8) 1978.
2. Hinners, R. G., J. K. Burkart and C. L. Punte. "Animal Inhalation Exposure Chambers," *Arch. Environ. Health* 16 (1968).
3. West, P. W., and G. C. Gaeke. "Fixation of Sulfur Dioxide as Disulfitomercurate (II), Subsequent Colorimetric Estimation," *Anal. Chem.* 28: 1816 (1956).

4. Weatherburn, M. W. "Phenol-Hypochlorite Reaction for Determination of Ammonia," *Anal. Chem.* 39:971 (1967).
5. Stump, F. Private communication to N. Clarke, HERL, EPA, Cincinnati from F. Stump, MSERB, EPA, RTP (1977).
6. Cohen, I. R., and A. P. Altshuller. "The MBTH Method for Aldehydes in Air: Collection Efficiencies and Molar Adsorptivities," *Anal. Chem.* 38:1418 (1966).
7. Springer, K. J., and T. M. Baines. "Emissions from Diesel Versions of Production Passenger Cars," *Soc. Auto. Eng.* (1977).
8. Vuk, C. T., M. A. Jones and J. H. Johnson. "The Measurement and Analysis of the Physical Character of Diesel Particulate Emissions," *Soc. Auto. Eng.* (1975).
9. Stokinger, H. E. "Toxicology of Diesel Emissions," Symposium of Diesel Equipment in Underground Mining. No. 8666, U.S. Dept. of Interior (1975).
10. Schreck, R. M. "Health Effects of Diesel Exhaust," Biomedical Sciences Dept., General Motors Research Laboratories, Warren, Mich. (1978).

CHAPTER 27

PARTICULATE AND VAPOR CONTAMINATION IN EXPERIMENTS WITH CARCINOGENS

Eric B. Sansone
Environmental Control and Research Laboratory
Frederick Cancer Research Center
Frederick, Maryland 21701

INTRODUCTION

Biological studies of environmental pollutants are undertaken to assess the risk associated with exposure of a portion of the biosphere to a given pollutant. When the possible combinations of pollutant, biological system, and means for applying the one to the other are considered, the variety and complexity of such studies can well be appreciated. Particulate pollutants to which small laboratory animals are exposed via the respiratory tract will be given attention here.

In biological studies with environmental pollutants, workers and the environment may be exposed to the pollutant under investigation, and the probability of exposure is greatly increased when the animals are exposed to aerosols. Such exposure is particularly undesirable when the pollutant is toxic.

In an analysis of laboratory-acquired infections, Pike [1] reported that in 3218 of 3921 cases (82%), a recognized accident had *not* occurred. This suggests that routine procedures frequently led to unsuspected contamination with infectious material. Of the nonaccidental cases, an aerosol was the probable source of infection in over 500, approximately 16% [1]. Table I lists those human diseases which are believed to be acquired by inhalation.

In addition to microorganisms, both occupational and nonoccupational exposure to nonviable airborne materials can lead to more or less serious

Table I. Human Diseases Probably Acquired by Inhalation[a]

aspergillosis	pneumonic plague
chicken pox	pulmonary anthrax
coccidioidomycosis	Q-fever
common cold	streptococcal sore throat
diphtheria	tuberculosis
influenza	tularemia

[a]Sources: references 2,3

systemic effects. It is probable, in view of Pike's findings, that the majority of adverse effects result from exposures associated with normal (however undesirable they might be) conditions. Table II lists some of the many substances which have been found to cause disease following inhalation as particulates.

Table II. Some Nonviable Substances Which Cause Disease Following Inhalation[a]

aluminum	hematite
4-aminobiphenyl	lead
arsenic compounds	manganese
asbestos	2-naphthylamine
auramine	nickel
benzidine	pesticides
beryllium	radioactive dusts
cadmium oxide and sulphate	silica
chromium	silicon carbide
cotton	talc

[a]Sources: references 4-6

Clearly, investigations of the biological effects of chemicals which may be carcinogenic, aerosolized for administration to experimental animals, can be potentially hazardous to the personnel involved. It is the purpose of this paper to illustrate how experimental facilities may be designed and managed so that contamination is prevented or minimized, and decontamination is facilitated.

DESIGN CONSIDERATIONS

Although the experimental chamber is usually relied upon as the primary barrier to contain the aerosolized material and prevent contamination, certain design features should be considered when working with potentially hazardous materials.

Laboratory Design

The design for the laboratory should be flexible; it should allow for expansion of facilities or change in use. There is not always an opportunity to design the laboratory into which aerosol exposure chambers are to be installed. Nevertheless, to the extent possible, work with potentially hazardous materials should be segregated from other work, personnel and equipment. The laboratory itself should be separated from offices, conference rooms, engineering support areas, etc., by showers and change rooms should the situation require it. To minimize the possible exposure of maintenance personnel to contamination, access to air ducts and utility lines should be via pipe chases in relatively protected areas such as corridors, attics and basements. Gaps in laboratory walls, ceilings and floors around ducts and utility lines should be sealed to prevent contamination of clean areas. Floor plans for microbiological [7-12] and chemical [13,14] laboratories have been reproduced in the literature.

Sealed windows facilitate temperature, humidity and air pressure control. In animal holding areas, diurnal lighting requirements are more easily met if there are no windows. Installation of automatic closing devices on doors will help to maintain air pressure differentials. Doors in laboratory areas should be provided with glass viewing panels and plastic speaking-diaphragms to improve communication while limiting traffic into the area [7,9,15,16].

Experimental Chamber Design

Aside from such obvious considerations as the integrity of the chamber and the influence upon this of piping, ductwork, sampling ports, windows, doors, etc., only two points need be mentioned here: (1) an auxiliary power supply to provide airflow to animals may be desirable in case of power failure; and (2) nose-only or head-only exposure may be advantageous in terms of contamination control when compared with whole-animal exposure.

Experimental chamber designs have been published [17-21]. One of these [19] describes an isolation-exposure system for carcinogen studies. The design, construction and operation of experimental chambers have been reviewed [22,23].

Materials of Construction

The materials and finishes which compose the surfaces of the laboratory and those from which the experimental chamber is made should be selected with care. Factors to consider include strength, expected lifetime, ease of cleaning, corrosion resistance and extent of reaction, absorption and adsorption of test materials to be used.

Glass, plastic, stainless steel, aluminum, neoprene, polytetrafluoroethylene and other materials have been used in the construction of experimental chambers. For use with potential carcinogens which requires corrosion resistance, minimal adsorption capacity and easy complete removal of particulates, stainless steel, glass and polytetrafluoroethylene are probably the preferred materials. Many of the relevant properties of construction and finishing materials for laboratories can be obtained from manufacturer's catalogues.

LABORATORY OPERATION

In this section, those laboratory operations having a major impact on the potential spread of contamination will be discussed. These include airflow systems, animal maintenance, waste management, cleaning and decontamination.

Airflow Systems

To ensure adequate control of airflow, it is recommended that separate air systems be used for clean areas, contaminated areas and exposure chambers. The relative pressures of these systems should be such that the overall direction of airflow is from clean areas through contaminated areas to exposure chambers. These pressure differentials can be maintained by exhausting more air from an area than is supplied. The systems should be interlocked so that if for some reason the chamber exhaust fan stops, the supply to the chamber is cut off.

Supply air to the facility should be filtered to remove at least coarse particulates; specific experimental requirements may call for more complete removal of fine particulates and even, possibly, vapors. Where the potential consequences of contamination are serious, as with carcinogens, no recirculation of air should be permitted. Instruments which measure relative pressures, flow rates, pressure drops, etc., should be provided along with alarms to indicate departure from permissible limits.

To control transfer of contamination, maintain pressure differentials and restrict traffic, airlocks may be desirable at the entrance to the contaminated area. The effect of ventilation per se, of ventilating an airlock, and of the pressure relationships between the airlock and adjacent areas on transfer of contamination have been examined [24,25]. Under the assumption that turbulent air mixing was sufficient to produce uniform concentrations of particles within each room, Lidwell [24] derived expressions for particle concentrations in rooms and corridors as a function of ventilation rates. He was also able to include the effect of airlocks (ventilated, with clean air input and with air extract) between room and corridor. Using this model, Lidwell was able to calculate, for a variety of situations, the ratio of contaminant concentration in the room in which contamination was being generated to that in the corridor or to that in a neighboring room. Thus, he was able to characterize the effectiveness of a particular air system, and to compare different ones. His paper should be consulted for the results of his calculations and the specific conditions under which they were made. In general, if the ratio referred to above were unity for a balanced air supply with no airlocks, it would be (very roughly) ten for directional air flow with no airlocks, one hundred for ventilated or unventilated airlocks, and one thousand for airlocks with recirculation. Reasonable agreement of predicted values with observed data was found.

The pressure differentials established to control airborne transfer of contamination can be lost rapidly if doors are left open. It is also important to note that approximately 0.5 m^3 of air are transferred from one side of a door to the other when a person opens, passes through and closes a door behind him in a period of 6 seconds [26]. The influence of temperature difference, turbulence and ventilation flow on air exchange through doorways has been analyzed [27].

Animal Maintenance

Animals may be housed and maintained in a portion of the experimental apparatus for the duration of the study, or they may be removed after each exposure. The former approach minimizes such problems as cleaning and decontamination, as well as the possibility that the feed will become contaminated with the aerosolized material. Head-only or nose-only exposure, as mentioned above, reduces the likelihood that animals will suffer gastrointestinal tract exposure to test materials from grooming one another. Voegtlin and Hodge [28] estimated, for several uranium compounds, the ratios of oral to inhalation doses for equivalent toxicity and concluded that it was unlikely that ingestion contributed significantly to the results observed. They also found percutaneous absorption insignificant, but warned against inadvertent exposure via the eye. The importance of these alternate routes of exposure, of course, will vary from compound to compound; such factors should be taken into consideration at the experimental design stage.

If animals are housed separately from the exposure chamber, they must be transferred in a manner that avoids contamination of the environment and personnel. Techniques have been described [2,29,30] for animal transfer and transport. Once transferred, they should be housed separately from control (unexposed) animals and from animals that have been exposed to other materials or to the same material but at different doses. Cross-contamination can occur as a result of imporper housing or from poor animal husbandry techniques [31-33]. Brune [34] described the development of nasal and eyelid carcinomas in control animals housed in the same room as animals that had been exposed to *N*-nitrosobutylmethylamine. Particulates can be transferred from exposed to control animals either by contact, for example when a worker tends exposed animals first and does not change gloves, or by the aerosol generated in handling bedding, feed or other materials [35]. In extreme cases, animals can be housed in individually ventilated cages. However, cross-contamination among aerosol-exposed animals housed together is virtually unavoidable. Even air washing is unlikely to decontaminate an animal's fur completely [2].

Because of the potentially hazardous nature of the test material, protocols for handling food, water, animal wastes, bedding and cage changing need to be carefully worked out before experimental work is begun.

Cleaning and Decontamination

The experimental chamber (and often the animal housing area) is usually kept free from fur, excreta, deposited particulates and other debris by frequent washing with an appropriate cleaning solution. This is perfectly satisfactory as long as personnel are not exposed, wastes are properly handled, and absolute cleanliness is not required. However, vapors may be adsorbed on animals' fur [23] and surfaces within the experimental chamber as well as on apparatus and room surfaces [36]. As mentioned above, thorough air washing of animals does not remove all contamination from the fur [2]; it is possible, however, that vacuuming may lead to more complete removal.

Removal of particulates from surfaces, especially fine particulates, can be difficult. Values for the efficiency of removal of particulate contamination from stainless steel, glass and other surfaces have been published [37,38]. The effectiveness of cleaning and decontamination techniques should be validated using appropriate sampling and analysis methods.

Wastes

Exhaust air should be treated to remove potentially harmful contaminants. The approach selected will depend upon the contaminant's properties and its concentration. High-efficiency particulate filters (HEPA) are available from several manufacturers, as are activated carbon filters. Filter housings and the HEPA filters themselves should be leak-tested before going onstream. Techniques for this have been described [39,40]. The service life of a HEPA filter will vary with the conditions of use, but the pressure drop across the filter can be used to indicate when it should be replaced. No such useful technique currently exists for gauging the life of a carbon filter. Filter housings should be equipped with a bag-out system so that when filters must be changed it can be done without contaminating personnel. The particulate filter should be upstream of the carbon filter. If a moderately volatile, highly toxic material is being employed, a second particulate filter might be installed downstream of the carbon filter to prevent escape of any carbon fines onto which the test chemical may be adsorbed. Disposal of used filters should be by incineration or degradation of the test material followed by burial. Other air treatment approaches, such as incineration, absorption or electrostatic precipitation may be useful.

Solid wastes (carcasses, bedding, waste feed, etc.) should be double-bagged, labeled and incinerated. The firebox temperature and retention time should be sufficient to destroy the contaminant completely. Contaminated liquid wastes should be collected and treated to remove or degrade the contaminant before discharge. Applicable degradation or removal techniques may include photodecomposition, oxidation, reduction, filtration using activated carbon or molecular sieves, ozonation, electrical discharge. hydrolysis, application of heat, etc. Slein and Sansone have summarized reports of these and other techniques applied to a wide variety of putative chemical carcinogens [41]. Regardless of the degradation or decontamination technique selected, difficulties will be minimized if the smallest possible volume of waste is generated.

Appropriate gauges and warning signals should be installed on all waste treatment systems so that their performance can be monitored and malfunctions promptly detected and corrected.

SAFETY MANAGEMENT

Work with potential carcinogens can be dangerous, and it is therefore essential that a sound safety program exist at the institution at which work involving carcinogens is contemplated, and that all aspects of the proposed work receive safety approval before the work is undertaken. Many descriptions of safety programs have appeared; these include guidelines for microbiological [8,15,16,42,43] and chemical laboratories [44,45], and for work with chemical carcinogens [46]. It is not necessary here to attempt completely to define the requisite elements of a safety program for an institution in which potentially carcinogenic aerosols are generated for administration to animals. There are a few points, however, that should be emphasized.

The safety program should be advised at the earliest possible stage that inhalation studies involving particulate carcinogens are under consideration. In this way, they may ensure that adequate attention is given to the safety aspects of laboratory design and equipment selection. This is particularly items of equipment that may be needed for waste treatment, for example, may have a long delivery time.

The work practices to be employed will be affected by the physical plan of the laboratory, the experimental design and the requirements for safe operation. These should be jointly developed at an early stage by the investigator and the safety officer, and put in writing. Emergency procedures for such eventualities as fires, spills, equipment malfunctions, etc., should be worked out in advance and included in the standard operating procedures. They may require revision once the experimental work gets under way, but they are useful for planning and training.

Laboratory personnel should be selected with care, not only for their technical ability, but also for their attitude. Employees must be able to maintain superior technical performance and superior safety performance in an environment which does not tolerate departure from standard procedures and practices. To ensure that personnel will perform properly requires that they be thoroughly trained in all aspects of the work, including emergencies, and that they are periodically retrained or refreshed, to prevent any deterioration of performance.

Since the health of the employee is of concern, a preplacement physical examination should be provided for those who may work with carcinogens. This examination should include tests to establish the general health of the prospective employee and any predisposition to ill effects from possible exposure to the agents to be worked with. The examination should ensure that only fit employees are permitted to work with carcinogens and provide a baseline with which future medical evaluations can be compared. Periodic physicals may reveal departures from previously established physical condition, but probably in most cases will verify and extend the employee's baseline. Medical records on employees who work with carcinogens should be retained for an extended period.

An element of safety performance too often overlooked is environmental monitoring. This serves three important purposes: it verifies that the containment systems and work practices are adequate (or alerts one to the fact that they need attention); it demonstrates to laboratory personnel the management's commitment to safety and to the worker's well-being; and it provides a record of performance for the institution and for compliance with regulatory bodies.

ACKNOWLEDGMENTS

Research was sponsored by the National Cancer Institute under Contract No. NO1-CO-75380, with Litton Bionetics, Inc.

REFERENCES

1. Pike, R. M. "Laboratory-associated Infections: Summary and Analysis of 3921 Cases," *Health Lab. Sci.* 13(2):105-114 (1976).
2. Jemski, J. V., and G. B. Phillips. In: *Methods of Animal Experimentation, Vol. I,* W. I. Gay, Ed. (New York: Academic Press, Inc., 1965), pp. 273-341.
3. Williams, R. E. O. In: *Clinical Aspects of Inhaled Particles,* D. C. F. Muir, Ed. (London: Heinemann, 1972), pp. 40-53.
4. Hunter, D. *The Diseases of Occupations* (London: English Universities Press, 1969), pp. 954-1092.
5. Tomatis, L. "The IARC Program on the Evaluation of the Carcinogenic Risk of Chemicals to Man," *Ann. N. Y. Acad. Sci.* 271:396-409 (1976).
6. Bidstrup, P. L. In: *Clinical Aspects of Inhaled Particles,* D. C. F. Muir, Ed. (London: Heinemann, 1972), pp. 156-172.
7. Wedum, A. G., E. Hanel, G. B. Phillips and O. T. Miller. "Laboratory Design for Study of Infectious Disease," *Am. J. Pub. Health* 46:1102-1113 (1956).
8. Phillips, G. B., and R. S. Runkle. "Laboratory Design for Microbiological Safety," *Appl. Microbiol.* 15(2):378-389 (1967).
9. Kuehne, R. W. "Biological Containment Facility for Studying Infectious Disease," *Appl. Microbiol.* 26(3):239-243(1973).
10. Sullivan, J. F. and J. R. Songer. "Role of Differential Air Pressure Zones in the Control of Aerosols in a Large Animal Isolation Facility," *Appl. Microbiol.* 14(4):674-678 (1966).
11. Loudon, R. G., R. A. Frantz, Jr. and C. A. LeMaistre. "Laboratory for the Study of Mycobacterial Aerosols," *Amer. Rev. Resp. Dis.* 100: 160-164 (1969).
12. Barkley, W. E. In: *Biohazards in Biological Research,* A. Hellman, M. N. Oxman and R. Pollack, Eds. (Cold Spring Harbor, NY: Cold Spring Harbor Laboratory, 1973), pp. 327-341.
13. Coleman, H. S., Ed. *Laboratory Design* (New York: Reinhold, 1951).
14. Lewis, H. F., Ed. *Laboratory Planning for Chemistry and Chemical Engineering* (New York: Reinhold, 1962).
15. Chatigny, M. A. In: *Advances in Applied Microbiology, Vol. 3,* W. W. Umbreit, Ed. (New York: Academic Press, Inc., 1961), pp. 131-192.
16. Darlow, H. M. In: *Methods in Microbiology, Vol. 1,* J. R. Norris and D. W. Ribbons, Eds. (New York: Academic Press, Inc., 1969), pp. 169-203.
17. Urban, E. J. C. "Two Chambers for Use in Exposing Laboratory Animals to the Inhalation of Aerosols," *Arch. Ind. Hyg. Occup. Med.* 9:62-68 (1954).
18. Hinners, R. G., J. K. Burkart and C. L. Punte. "Animal Inhalation Exposure Chambers," *Arch. Environ. Health* 16:194-206 (1968).
19. Laskin, S., M. Kuschner and R. T. Drew. In: *Inhalation Carcinogenesis,* M. G. Hanna, Jr., P. Nettesheim and J. R. Gilbert, Eds., USAEC CONF - 691001, (1970), pp. 321-350.

20. Montgomery, M. R., R. E. Anderson and G. A. Mortenson. "A Compact Versatile Inhalation Exposure Chamber for Small Animal Studies," *Lab. Animal Sci.* 26(3):461-464 (1976).
21. Kimmerle, G. In: *Air Pollution and Cancer in Man,* U. Mohr, D. Schmähl and L. Tomatis, Eds. (Lyon, France: IARC, 1977), pp. 49-51.
22. Fraser, D. A., R. E. Bales, M. Lippmann and H. E. Stokinger. "Exposure Chambers for Research in Animal Inhalation," Public Health Monograph No. 57, U.S. Government Printing Office, (1959).
23. Drew, R. T., and S. Laskin. In: *Methods of Animal Experimentation, Vol. IV,* W. I. Gay, Ed. (New York: Academic Press, Inc., 1973), pp. 1-41.
24. Lidwell, O.M. "The Control by Ventilation of Airborne Bacterial Transfer between Hospital Patients, and its Assessment by Means of a Particle Tracer II. Ventilation in Subdivided Isolation Units," *J. Hyg., Camb.* 70:287-297 (1972).
25. Hambraeus, A., and H. F. Sanderson. "The Control by Ventilation of Airborne Bacterial Transfer between Hospital Patients, and its Assessment by Means of a Particle Tracer III. Studies with an Airborne-Particle Tracer in an Isolation Ward for Burned Patients," *J. Hyg., Camb.* 70:299-312 (1972).
26. Whyte, W. and B. H. Shaw. In: *Airborne Transmission and Airborne Infection,* J. F. P. Hers and K. C. Winkler, Eds. (New York: John Wiley and Sons, Inc., 1973), pp. 513-516.
27. Lidwell, O. M. "Air Exchange through Doorways. The Effect of Temperature Difference, Turbulence and Ventilation Flow," *J. Hyg., Camb.* 79:141-154 (1977).
28. Voegtlin, C., and H. C. Hodge, Eds. *Pharmacology and Toxicology of Uranium Compounds* (New York: McGraw-Hill Book Company, 1948), pp. 694-697.
29. Graham, W. R. and E. S. Feenstra. "A Program for the Development of Pathogen-Free Laboratory Animals," *Proc. Anim. Care Panel* 8:54-66 (1958).
30. Jemski, J. V. "Maintenance of Monkeys Experimentally Infected with Organisms Pathogenic for Man," *Proc. Anim. Care Panel* 12(3):89-98 (1962).
31. Phillips, G. B., G. C. Broadwater, M. Reitman and R. L. Alg. "Cross Infections Among Brucella Infected Guinea Pigs," *J. Infect. Dis.* 99:56-59 (1956).
32. Phillips, G. B., J. V. Jemski and H. G. Brant. "Cross Infection Among Animals Challenged with *Bacillus Anthracis,*" *J. Infect. Dis.* 99:222-226 (1956).
33. Kirchheimer, W. F., J. V. Jemski and G. B. Phillips. "Cross-Infection Among Experimental Animals by Organisms Infectious for Man," *Proc. Anim. Care Panel* 11(2):83-92 (1961).
34. Brune, H. "Warning vor Nitrosaminen," *Aktuelle Medizin, Beilage* (Supplement) 44 (November 1966).

35. Sansone, E. B., A. M. Losikoff and R. A. Pendleton. "Potential Hazards from Feeding Test Chemicals in Carcinogen Bioassay Research," *Toxicol. Appl. Pharmacol.* 39:435-450 (1977).
36. Sansone, E. B., and A. M. Losikoff. "Contamination from Feeding Volatile Test Chemicals," *Toxicol. Appl. Pharmacol.* 46:703-708 (1978).
37. Fish, B. R., Ed. *Surface Contamination* (London: Pergamon Press, 1967).
38. Sansone, E. B., and M. W. Slein. "Redispersion of Indoor Surface Contamination: A Review," *J. Haz. Mat.* 2:347-361 (1977/78).
39. American Association for Contamination Control. *Tentative Standard For HEPA Filters: CS-IT* (Boston: AACC, 1968).
40. "Military Specification Filter, Particulate, High-Efficiency, Fire Resistant," MIL-F-51068C, 8 June 1970.
41. Slein, M. W., and E. B. Sansone. *Degradation of Chemical Carcinogens: An Annotated Bibliography* (New York: Van Nostrand Reinhold Company, in press).
42. Wedum, A. G. "Laboratory Safety in Research with Infectious Aerosols," *Pub. Health Rep.* 79(7):619-633 (1964).
43. Runkle, R. S., and G. B. Phillips, Eds. *Microbial Contamination Control Facilities* (New York: Van Nostrand Reinhold Company, 1969).
44. Manufacturing Chemists Association. *Guide for Safety in the Chemical Laboratory* (New York: Van Nostrand Reinhold Company, 1972).
45. Steere, N. V., Ed. *CRC Handbook of Laboratory Safety,* (Cleveland OH: Chemical Rubber Company, 1971).
46. Office of Research Safety. "National Cancer Institute Safety Standards for Research Involving Chemical Carcinogens," DHEW Publication No. (NIH)76-900, U. S. Government Printing Office (1975).

CHAPTER 28

PARTICLE DEPOSITION IN MODEL SYSTEMS OF HUMAN AND EXPERIMENTAL ANIMAL AIRWAYS

Richard B. Schlesinger

New York University Medical Center
Institute of Environmental Medicine
New York, New York 10016

INTRODUCTION

The respiratory tract is both a portal of entry and a target site for a wide variety of airborne pollutants, many of which are in particulate form. The specific pattern of regional deposition of inhaled particles is a main determinant of any local or systemic pathogenic potential, since this pattern not only affects initial dose to the lungs, but also influences subsequent clearance rates and redistribution pathways of retained material. Because of the complex nature of the respiratory tract itself and of the interactions of the determinant factors for the deposition of inhaled particles within it, various model systems have been developed, usually as an approach to aid in the formulation of predictive deposition equations for risk estimates. This chapter provides an overview of the various types of models which have been used for deposition analyses, and also discusses one of the main factors affecting deposition, namely, bronchial tree geometry, in man and experimental animals.

BASIC GEOMETRY OF THE TRACHEOBRONCHIAL TREE

One of the basic parameters of any deposition model is the representation of bronchial branching patterns. Thus, before discussing types of models, it is helpful to distinguish between the different branching patterns upon which these models are based.

The mammalian bronchial tree shows two basic types of branching pattern: dichotomy and trichotomy. In a dichotomous system, one branch (the "parent") divides, giving rise to two branches (the "daughters"). If both of the daughters have the same diameter and length, and branch off the parent at the same angle, the mode of division is known as regular dichotomy. However, in nature, this regular pattern is very rarely found. More often, the two daughters differ from each other in one or more dimensions; in this case, the mode of branching is termed irregular dichotomy. The extreme case of irregular dichotomy is known as monopody. In a monopodial branching system, the larger diameter ("major") daughter may not be easily distinguishable from its parent, since the change in diameter and direction from the parent to the major daughter may be negligible. Thus, the smaller diameter ("minor") branch actually appears to originate laterally from a large stem branch.

Trichotomy is the branching mode characterized by three daughter branches arising from the parent. True trichotomy is quite rare in mammalian lungs, and what seems to be a trichotomous division is usually only apparently so. That is, if one daughter is extremely short, it may appear inconspicuous, so its daughters in turn seem to be arising from the parent via a single division into three bronchi.

In terms of model systems, one important difference between regular and irregular dichotomous branching modes involves the number of airways between the trachea and the terminal bronchioles. In a regular dichotomous branching system, the number of divisions (and, therefore, the path length) between the trachea and the most distal conductive airways is the same along any pathway. This branching mode is, therefore, also termed symmetrical. In an irregular dichotomous, or asymmetrical, system, the number of branch divisions from the trachea to each distal bronchiole is not the same along every pathway. Although the term regular dichotomy is often used interchangeably with symmetrical, and irregular dichotomy with asymmetrical, regular and irregular generally refer to the dimensions of the daughter branches at a specific division, while symmetrical and asymmetrical refer to the number of branch divisions along each pathway.

In many models, the position of a single branch within the bronchial tree is usually indicated by assigning a code number to it. There are two basic ways in which this may be done: divisions up from distal end branches [1, 2] or divisions down from the trachea. The scheme used in this paper is the latter, developed by Weibel [3].

In the Weibel ordering system, each branching division is known as a generation. The trachea is generation 0, and each distal division increases by one generation number. Thus, in terms of this ordering mode, in regular dichotomy, all bronchi of the same diameter would be found in the same generation, and at the same path length from the trachea. On the other hand, in irregular dichotomy, bronchi having the same diameter may be found in various generations, and at different distances from the trachea.

TYPES OF MODEL SYSTEMS

Model systems designed for the study of particle deposition within the tracheobronchial tree may be divided into two broad types: mathematical (or analytical) models, and physical models, the latter used in actual experimental analyses. Although this paper is largely concerned with the physical models, a brief discussion of analytical systems will serve to highlight the reasons for development of experimental models.

Mathematical Models

An analytical model relates the main factors which control deposition, e.g., particle size, airflow, airway anatomy, to various geometric parameters, to quantitatively predict deposition in the respiratory tract.

Since the classic mathematical treatment of regional particle deposition in man performed by Findeisen [4], predictive models have been greatly refined [5-9]. Although these models were designed for spherical particles, others have been developed specifically for prediction of the regional deposition of fibers [10, 11].

Because of their very nature, analytical models involve assumptions and idealizations of almost all aspects of the respiratory tract and of particle dynamics. This results in the loss of some important characteristics of the real system, and often limits the ultimate usefulness and reliability of these models.

One of the major components of a model, and one subject to the greatest oversimplification, is the representation of airway geometry. Most of the early predictive models made use of a very simple lung structure, where each generation consisted of a number of cylindrical tubes of equal diameter connected in series with the preceding and succeeding regions. Recently, however, more realistic anatomic descriptions have been developed, either specifically for use in deposition analyses, or which are easily used in such applications. These descriptive models are usually generalizations derived from experimental data. One of the most widely used structural models is the symmetrically dichotomous Weibel Model "A" [3].

The assumption of regular dichotomy simplifies the treatment of morphometric data. However, the actual bronchial tree is asymmetric, and because the amount of deposition depends upon the path length over which the inhaled aerosol passes [12] precise computations require consideration of asymmetry in development of structural models for use in deposition analysis. Horsfield et al [13] developed two asymmetrical dichotomous bronchial tree models. In order to simplify their models, an average asymmetry was used, a factor which resulted in the loss of the range of variability of the bronchial tree. More recently, Phalen et al. [14, 15] developed a structural model of the human lung which consisted of "typical pathways" based upon geometric mean dimensions, for each lobe within the lung; this accounted for the asymmetry and resultant variable path length within each separate lobe.

Structural models of the bronchial tree have also been developed for experimental animals. Kliment [16-18] devised symmetrical dichotomous models of rabbit, rat and guinea pig lungs, while Raabe [12] applied the "typical pathway" scheme to the rat and hamster.

Most predictive models provide a value for total deposition, and do not allow characterization of local nonuniformities in intrabronchial deposition, such as enhanced deposits experimentally observed at branch points. Recently, however, Shaw et al. [19] developed a mathematical model which does predict peak to average particle deposition flux within bronchi.

In addition to oversimplification of airway geometry, another drawback of analytical models is the oversimplification of airflow pattern, a necessity since there are no exact expressions for flow dynamics in noncylindrical, tapering tubes which undergo repeated branchings and have asymmetric, nonlaminar flow profiles. Analytical models generally assume laminar and/or plug flow, no disturbances produced by bifurcations, and uniform ventilation of airways.

Aside from critical parameters of airway geometry and flow pattern, any predictive model must also contain expressions for deposition probability. However, exact expressions for deposition within a realistic geometry and flow pattern are not available. Thus, semiempirical expressions are often used, based upon analyses using simplified analogues of sections of the bronchial tree. For example, impaction expressions are often obtained from the analytical solution to the equation of motion of a particle in ideal flow in a bent tube, diffusion expressions by the analytical solution for flow in an infinitely long, horizontal tube with ideal flow, and sedimentation expressions by the solution for deposition along a long horizontal tube.

The extreme complexity of bronchial tree structure and the occurrence of complicated flow patterns within the airways make it unlikely that realistic deposition data will be obtained from simplified mathematical models. What are required are experimental systems which attempt to reproduce the complexities inherent within the lungs. Data obtained from these latter systems can then be used to develop more realistic, semiempirical deposition formulations.

Physical Models

The types of physical models which have been used in experimental studies of particle deposition may be divided into two categories: ideal, or conceptual, models, and realistic physical models and casts of portions of the tracheobronchial tree.

Ideal (Conceptual) Models

In a biological system as complex as the lung, it is quite difficult to construct geometrically similar homologous models. Thus, most models are simplified and, therefore, lack some requirements for similarity to the actual lung. Although these ideal models may help to clarify flow dynamics and particle flux at deposition surfaces, and may aid in the understanding of the mechanisms of particle transport and deposition in more complex geometries, they may not be good systems for realistic deposition studies.

Various types of ideal models have been used in experimental deposition studies. These include tubes having a single bend [20], models of single bifurcations [21-24], e.g., Y-models, round nose wedge models, symmetrical Y-shaped models of two or more generations of bifurcations [20, 26], and more extensive plastic or glass models of whole regions of the tracheobronchial tree [25, 27, 28]; these latter often included parts of the upper respiratory tract as well.

Most of the deposition studies using models were performed under steady "inspiratory" flows, generally laminar in nature. A few [23, 29] used simulated cyclic inhalations and some [25, 29] studied expiration as well as inspiration. Furthermore, most of the single bifurcation and multiple generation models had dimensions based upon the Weibel Model "A" and were therefore symmetrical.

Casts of the Tracheobronchial Tree

Solid casts of various parts of the respiratory tract prepared from autopsy specimens have been used for years, primarily for studies of airway anatomy. The earliest of these were produced in the latter part of the 17th century using low melting point alloys introduced into the trachea [30]. Since then, solid airway casts have been produced using a variety of materials, including wax, latex, silicone rubber and various plastic resins.

The use of hollow airway casts allows both the measurement of morphometry and direct experimental investigation of deposition within these same airways. However, until recently, hollow casts were used predominantly for studies of airflow dynamics.

The earliest studies of deposition within hollow casts were concerned with a qualitative description of sites of particle deposition. Proetz [31] studied the deposition by impaction of cigarette smoke and chemical powders introduced via inspiratory currents into casts of the human upper respiratory tract (anterior nares to glottis), and found visible concentrations of deposited particles immediately distal to any bend or constriction in the airway. Ermala and Holsti [32] examined the distribution of tobacco tar in a plaster cast of the human oral cavity and respiratory tract which was exposed to cigarette smoke; tar deposit was found to be localized mainly at airway bifurcations.

Timbrell [33] exposed a hollow silver cast of a pig's lung to asbestos to determine overall penetration characteristics of various fiber types. Schlesinger et al [34] studied the deposition pattern of tantalum powder aerosol in a hollow latex cast of the human larynx and upper tracheobronchial tree.

The most detailed quantitative study of deposition in casts of the human tracheobronchial tree were performed by Schlesinger and Lippmann [35], using a hollow silicone rubber cast extending through the segmental bronchi, and by Schlesinger et al [36], using a cast extending through 2 mm diameter airways. In the latter study, laryngeal casts were used to produce more realistic entry flow conditions.

There are a number of significant differences between ideal models and casts; some of these have effects upon the measured pattern of deposition and, thus, may result in questionable extrapolation of data obtained in ideal casts to the real lung. Some of the more important differences and their effects upon deposition are briefly outlined below:

Single bifurcations, bends and tubes vs. multiple bifurcation models. The presence of repeated bifurcations has an effect upon deposition which is not accounted for in models based upon deposition in isolated airways. Bifurcations disturb the flow, and these disturbances may persist into downstream bronchi. Since the deposition pattern is intimately related to airflow pattern, differences in the latter in different model systems are critical factors in analyzing their usefulness.

Bends and bifurcations also differ from each other in deposition characteristics. Johnston et al [20] found differences in the experimentally derived curves relating deposition efficiency and flowrate for single bifurcation models as opposed to simple, bent tubes; the former was linear, and the latter curvilinear.

Symmetry vs. Asymmetry. Many ideal models have symmetrical branching patterns, usually based upon the Weibel Model "A". However, deposition differs in symmetrical and asymmetrical systems. Ferron [28] studied deposition in glass models which extended from the mouth to the sixth branching generation. One model was copied from a cast, and was, therefore, asymmetrical; the other was based upon the Weibel Model "A". He found there to be differences in deposition in these two systems. For example, impaction was observed to be more prominent in the asymmetric model, where the angles were generally larger than in the symmetric model.

In any branching system, the volumetric flow rate through a specific pathway depends on the path length, the amount of terminal respiratory units served by that pathway, and the flow resistance along the pathway. In a symmetrical system, each path would receive the same proportionate flow. However, in an asymmetrical system, the path lengths to terminal units and the divisions of bronchi within the same generation in each pathway differ, resulting in differences in flow distribution and, therefore, presumably in deposition efficiencies and patterns. Such flow factors also accounted for some of the deposition differences noted by Ferron [28] between his symmetrical and asymmetrical models.

Spatial Orientation of Bronchi. The tracheobronchial tree has a three-dimensional orientation of branches, and their actual spatial orientation may be a critical factor in determining the deposition pattern. In the bronchial tree and, therefore, in hollow casts prepared from it, the planes at which successive branching angles lie (relative to a parent) vary greatly with respect to each other. However, ideal models are usually constructed with all angles in one plane. Cheng [37] and Lee and Wang [26] demonstrated differences in the deposition pattern in symmetrical airway models consisting of two airway generations, depending upon whether the two bifurcations were in the same plane or whether the second bifurcation was normal to the plane of the first. The deposition differences were consistent with variations in flow profiles between these two geometries.

Use of a Larynx. In vivo, air enters the trachea after passing through the larynx, an entry condition which has a great effect upon airflow pattern in the trachea and in a number of succeeding bronchial generations [38, 39]. This effect has, for the most part, been ignored in most deposition studies using model systems.

Laryngeal-induced flow disturbances produce significant effects upon deposition. For example, Chamberlain and Dyson [40] demonstrated that deposition of radon daughters in a hollow plastic cast of the larynx-trachea was higher than that in a smooth straight tube of similar dimensions. The increase in tracheal deposition was attributed to the production of turbulence due to the larynx.

The larynx also affects intrabronchial deposition patterns within the upper airways. Bell [23] studied local deposition within a smooth-walled wedge model of a bifurcation under steady flows, and found intense "hot spots" of deposition near the carina. However, less intense carinal hot spots under one similar condition of particle size and flow were found in our laboratory when a larynx was used. [41]. This result is consistent with the shift of flow streamlines away from the carina due to the influence of the larynx [38]. In one parallel study of deposition in a cast under identical exposure conditions except for the presence of a larynx, lower overall deposition efficiencies were observed in the first few bronchial generations in the test performed using the larynx [41].

The use of any type of simulated larynx appears to be better than using nothing at all. For example, Chang [37] studied deposition in a glass model of the bronchial airways using a Venturi to simulate a larynx. He, too, observed a less intense enhancement of deposition at the bifurcations than did Bell [23].

TECHNIQUES OF DEPOSITION ANALYSIS IN MODEL SYSTEMS

Various techniques have been devised to analyze deposition within model systems. One common procedure involves cutting the model up into sections and measuring the particles deposited within each. This is an especially appropriate method when radioactive test aerosols are being used [25, 27, 42]. Other, more novel techniques include radioautography of an entire model following deposition of a radioactive aerosol [25], X-ray of a cast to visualize the deposition of tantalum aerosols [34], use of a gamma camera to map deposits of radioactive aerosol [35], and use of short-time holography [43]. Another interesting technique was employed by Shaw et al. [24] who studied the deposition of *E. coli* in a round nose wedge model of a bifurcation; the bacteria multiplied at deposition sites, providing an index of relative deposition intensities.

Schlesinger and Lippmann [35], Schlesinger et al. [36] and Chan [44] used scintillation detectors and various collimators to map the deposition pattern of a radioactive aerosol in hollow casts. The detection system in the latter two studies used focusing collimators designed to determine deposition within airways of all sizes and at various branch levels, and to allow selective measurement of deposited activity within bifurcation and length subdivisions of individual bronchi.

Recent work has focused upon the comparison of deposition patterns within casts under constant inspiratory flow with those obtained under simulated cyclic inspiratory flow patterns. Findings of other investigators differ as to whether deposition under these conditions are comparable [23, 29]. In our studies, the mean flow obtained during the cyclic pattern is used as the steady flow condition. The hollow casts used extend through the sixth branching generation and are reproductions produced from a master which was made from a lung removed at autopsy. The experimental procedure involves use of radioactively tagged aerosols and cutting up the cast into small sections for precise measurement of deposition activity at bifurcation and length areas through the fourth generation.

The presence of a larynx is a very critical factor in affecting the deposition pattern in the first few bronchial generations [36]. However, since the glottic opening varies with airflow, use of a cyclic breathing pattern requires a larynx which varies in cross section to maintain relatively realistic entry flow conditions. A variable-orifice larynx model has therefore been developed. Basically, it consists of two brass strips representing the vocal cords, and which move in slits cut into an epoxy cast of an actual larynx (Figure 1). The strips are hinged at one end and attached to a variable speed motor, which regulates the number of inspirations per minute by changing the rate at which the strips open and close. The flow through the entire cast during aerosol exposure is actually regulated by the laryngeal opening.

Figure 2 shows results of four initial test runs, two performed at a steady flow, and the other two at a simulated cyclic inspiratory flow, using this model system. These preliminary tests do not as yet allow a conclusion as to whether cyclic flow produces deposition patterns which are comparable to those obtained at steady flow using the samc mean flow rate. However, tracheal deposition is enhanced during cyclic flow, and is much closer to mean in vivo [45] values than those observed during steady flow tests [36]. In addition, it appears that although total efficiencies may be similar under these two flow regimes, bifurcation efficiencies, at least for these tests, are greater under cyclic flow, an important consideration in terms of initial dose estimates.

In addition to the radiometric analysis, another technique is currently being used in the author's laboratory to map deposition density patterns within select bifurcation regions, since the degree of enhancement of deposition in these regions is an important component of any dosimetric model. Sections of bifurcations which had previously been measured radiometrically are being analyzed using a scanning electron microscope (SEM) and X-ray fluorescence unit. Following suitable preparation for observation, sections are visually scanned. A characteristic iron X-ray is then used to determine sites where deposition has occurred. Figure 3 shows a photograph of one section taken with the SEM, and of the X-ray fluorescence images of deposited iron aerosol in the exact same field. Using the latter, cast surface interferences are eliminated, and maps of surface density may be obtained quite easily.

Figure 1. Photograph of the variable-orifice larynx model coupled to a hollow cast of the upper tracheobronchial tree. The metal strips which represent the vocal cords are seen attached to a motor-driven gear apparatus on one end and running through an epoxy cast of an actual larynx on the other end.

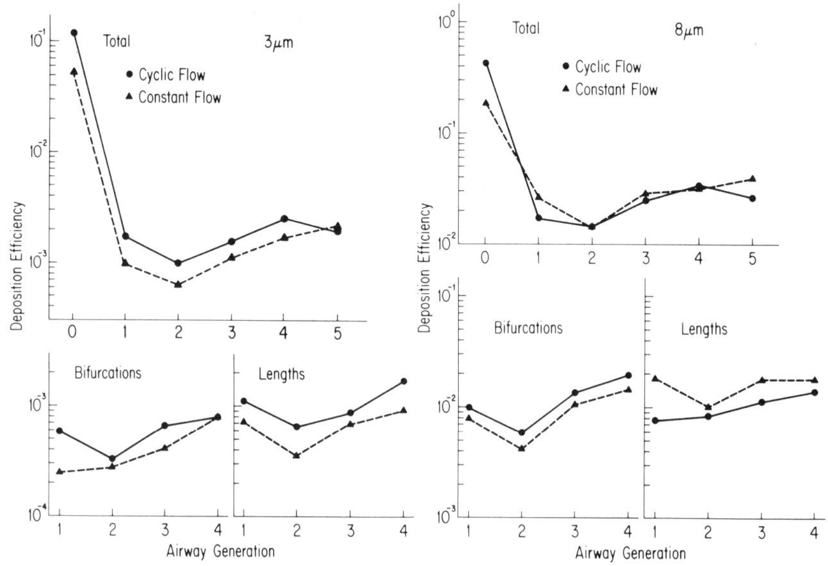

Figure 2. Mean fractional deposition efficiencies obtained in a hollow cast of the upper tracheobronchial tree at two particle sizes. Both the constant flow rate and the mean of the cyclic pattern were 30 liters/min. The top panels represent overall efficiencies of each branching generation, while the bottom panels show efficiencies separately for bifurcation regions and nonbranching (length) sections of each generation.

The discussion so far has concentrated on the types of model systems which have been used in deposition analyses. These were generally based upon human lung dimensions and geometry. However, there has been increasing interest in the use of models of animal lungs in order to analyze deposition obtained in different species, often with the goal of extrapolation of animal data to deposition in the human tracheobronchial tree.

COMPARATIVE BRONCHIAL TREE MORPHOMETRY AND DEPOSITION

Comparative Morphometry

Various experimental animals are used in inhalation toxicology and particle deposition analyses as bioassays of hazards from inhaled particles. Whenever analogues for man are chosen, certain factors must be considered depending upon the specific type of cross-comparison required. In studies of particle deposition, the anatomy of the tracheobronchial tree is a critical

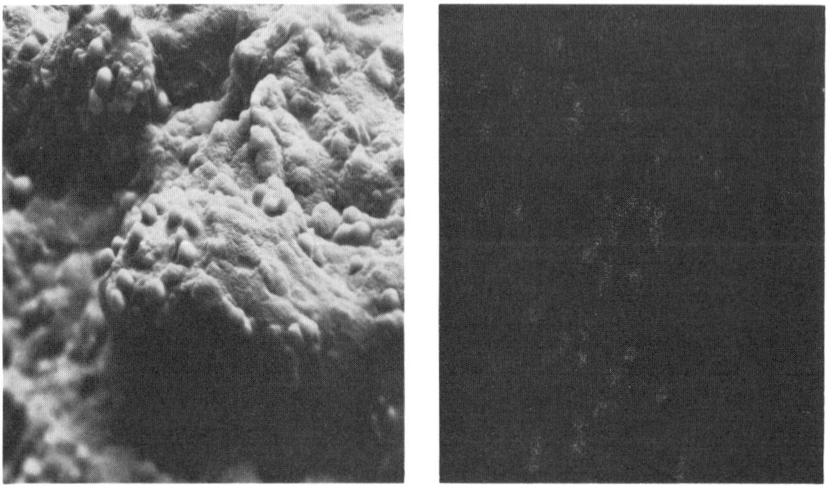

Figure 3. Left, an SEM microcraph of a section of a cast on which ferric oxide microspheres have deposited. Right, X-ray fluorescence images of the deposited ferric oxide particles.

factor, yet it is not usually taken into consideration in the interpretation of animal data. Interspecies differences in susceptibility to various dusts and pulmonary disease are known; these may be related to morphometric and associated physiologic differences, although the significance of the former in terms of pathologic differences between species has not been extensively studied. It is therefore important to determine whether the sites and amount of deposition are similar in experimental animals and man.

Certain aspects of gross and subgross anatomy have been widely studied in numerous species. For example, it is known that there is wide variability in lobation and lobulation, pleural thickness and the relative degree of alveolarization. The discussion in this chapter will, however, be limited to bronchial airway morphometry.

The first comprehensive attempt at a comparative description of the bronchial tree was provided by Aeby [46] in 1880. He prepared metallic casts of over 40 mammalian species, including man. However, because of casting artifacts, some serious errors were made in his anatomic descriptions [30]. It is only recently that detailed, accurate morphometric studies have been performed on man and on animal species commonly used in inhalation studies; the most extensive of these are described by Phalen et al. [47].

Morphometric measurements through the sixth branching generation have been performed in our laboratory on casts of the tracheobronchial tree of man, rabbit and donkey. To characterize bronchial morphometry in these species, plots of the distribution of various branch parameters were developed. One of these, the ratio of branch diameters, is presented in Figure 4. This figure also includes data for the beagle dog, Syrian hamster, hooded rat and human as compiled by Phalen et al [47]. This type of graphic representation allows interspecies characterization of the degree of dichotomy of the bronchial tree. A regular dichotomous branching system would have a major daughter to minor daughter diameter ratio of 1, while perfect monopodiality would appear as a major daughter to parent branch diameter ratio of 1.

The human bronchial tree, at least for the first six generations, exhibits the most symmetrical branching system of all of the species examined; the major/minor ratio is closest to the value of 1. The dog and donkey are almost ideally monopodial, with major/parent ratios generally uniformly near 1. The other species examined exhibit various degrees of irregularity in their airway branching pattern.

Interspecies differences in branching geometry may also be clarified in descriptive as well as quantitative anatomical terms. In all mammals, the trachea divides into two bronchi, which serve the right and left lungs. Each of these main, or stem, bronchi supply a number of lobar bronchi. In man, the main bronchi divide into three right and two left lobar bronchi; these then branch into smaller and smaller bronchi, largely asymmetrically. In other mammals, after the trachea, a long curved airway extends toward the

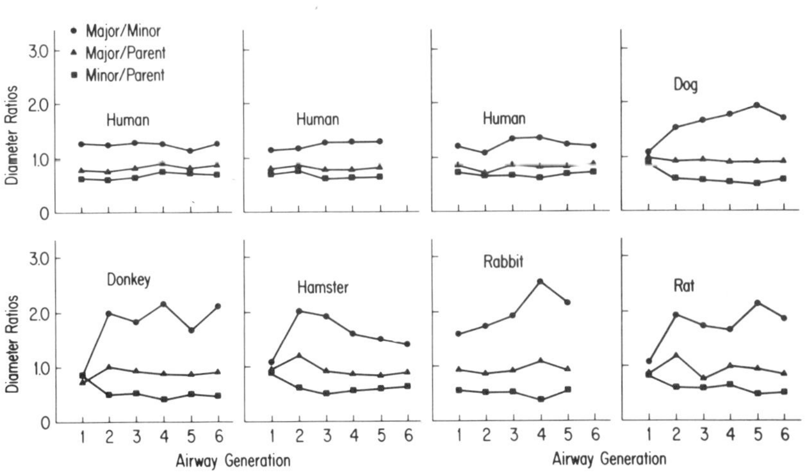

Figure 4. Airway diameter ratios as a function of branching generation in six species. Data of Schlesinger et al. [36], Phalen et al. [47] and this study.

caudal margin of the lowest lobe in both lungs, gradually becoming thinner towards the periphery. From these two bronchi, the lobar bronchi extend laterally and ventrally within the individual lobes, and their subdivisions extend from these, in turn. In the more extreme monopodial systems, such as in the donkey, minor daughters may branch off at nearly 90° angles, while the major daughter angles approach 0°. Figure 5 shows a schematic of the two basic branching patterns of mammalian lungs, while Figure 6 shows photographs of casts of the human, rabbit and donkey bronchial tree prepared in our laboratory.

The concept of the long, curving monopodial stem may be applied, to some extent, to the human bronchial tree. Rather than extending all the way to the lower ends of the lower lobes, the stem bronchus in man may be considered to be that portion of the tree which extends from the tracheal bifurcation to the division of the right and left lower lobar bronchi into the segmental bronchi. Thus, in man, the main bronchi come closest to approaching monopodial branching; however, most larger bronchi exhibit a less irregular branching pattern, while bronchi near the terminal bronchioles may show more regular dichotomy [48]. In the more monopodial systems, a more regular dichotomy is observed in the deeper lung, e.g., bronchioli, and even sometimes in the division of the lobar bronchi from the main bronchi [12, 49]. Thus, neither pure dichotomy nor monopody prevail completely throughout any one lung, with some variations of each occurring. The most striking structural differences in terms of branching pattern occur in the upper airways, since all trees examined approached a more regular symmetrical system in the more distal airways [47].

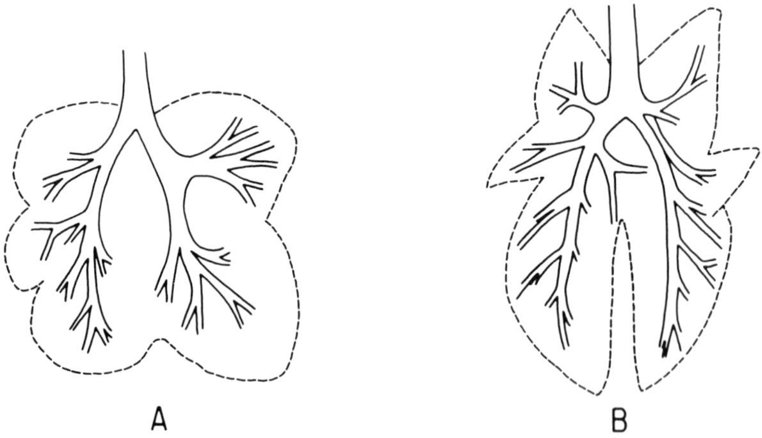

Figure 5. Schematic diagram of bronchial branching patterns; (A) Human lung; (B) Monopodial pattern.

Figure 6. Photographs of casts prepared from the bronchial trees of three species; (A) Human (hollow cast) [36]; (B) Rabbit (solid cast); (C) Donkey (solid cast); (The arrow in B and C points to the long, curving stem bronchus which characterizes monopodial branching).

Bronchial branching schemes have often been compared to botanical trees, with the tracheobronchial tree in man resembling a leafy deciduous tree, while the more monopodial types resemble a conifer [49].

Comparative Deposition of Particles

Detailed studies of the relationship of anatomical differences between species to interspecific deposition patterns have not been performed. This type of analysis is important since the biologic action of an inhaled aerosol may be due to that fraction which preferentially deposits in the tracheobronchial tree; it is therefore essential to know how deposition efficiencies differ in various species of experimental animals as compared with man.

Hollow casts are ideal systems for these studies, since both morphometry and deposition may be measured in the same tracheobronchial tree. However, most studies which have analyzed deposition in experimental animals were performed in vivo, and provide data solely on total respiratory tract deposition. For example, Morrow and Gibb [50] compared the deposition of 0.04-μm (CMD) NaCl aerosols in both dogs and man, and concluded that total lung deposition was similar in these species. But these results may be misleading, in that there may still be differences in fractional tracheobronchial vs. alveolar efficiencies, and this may be critical to any potential hazard. This situation is evident from Table I which compares tracheobronchial deposition for similarly sized particles in man and some other species used in inhalation studies. Median deposition in the rabbit appears to be closest to that in man, at least for larger particles.

Morphometry plays a large role in deposition differences between species. For example, some of the interspecies differences are due to variable tracheal size. Raabe et al. [12] found that in rats the tracheal component accounted for approximately 25% of total tracheobronchial deposition for 0.52-μm particles and approximately 4% for 3.05-μm particles, while in hamsters, the percentages were 11% and 15%.

Interspecific differences in deposition may also be due to variations in branching patterns and/or breathing characteristics, e.g., frequency, tidal volume and regional ventilation. Thus, what is termed "respirable" may differ in various animals. For example, Thomas [51] found that the total deposition of aerosols in mice tended to decrease with increasing particle size in the range of 0.6-2.4 μm (AED). He suggests that this occurs because a size was reached which was large enough to be less respirable by this small animal.

Since animals are of different sizes, it is probable that the bronchial path lengths may differ in different species. This is an important consideration in risk analyses extrapolated from animal data to man. For example, Raabe et al. [12] found in rats and hamsters that the lobes with the smallest number

Table I. Tracheobronchial Deposition in Various Species

Species	Aerosol Particle Size (μm)	% of Total Lung Deposition Occurring in the Tracheobronchial Tree		Reference
		Median	Range	
Human	0.8	19	8-29	44
	3	30	10-57	45
	6	65	28-92	45
Donkey	0.3	25	18-33	55
	3	58	49-68	56
	5	70	68-75	55
Dog	0.5-0.6	∿ 4	---	57
	2.5-3	∿11	---	57
	6.6	∿44	---	57
Rabbit	0.5	33	20-64	58
	3	29	17-55	59
	6	60	40-93	59
	6	77	54-80	58

of divisions and, therefore, shortest average path length between the trachea and terminal bronchioles, had the highest concentration of deposited particles in the alveolar region for particles $\geqslant 1$ μm. The differences were less obvious for submicron particles, which reached lobes more evenly regardless of path length, but appeared to be proportional to relative aeration of lobes. Since the mean number of divisions varies from lobe to lobe, this asymmetry should be taken into consideration in any model. Table II presents the average number of branching generations from the trachea through terminal bronchioles for various species. Although the number of divisions may be similar, actual path lengths are proportional to body size. In some species, shorter bronchial lengths may be compensated by a larger trachea, such that the overall path length from trachea to terminal bronchioles may be similar to man. This is the case, for example, in the beagle dog [52].

Because of the necessity of developing a means to compare deposition data obtained in experimental animals to man in a reliable manner, a number of attempts have been made to develop methods which would allow direct interspecies extrapolation of deposition data. These mainly involve analytical models. For example, Stauffer [53] used the technique of dimensional analysis to develop scaling factors which would allow application of the results of animal experiments to man. The factors were based on particle physics, and the assumption of a geometric similarity between all mammalian

Table II. Typical Airway Path Lengths

Species	Number of Airway Divisions (Generations) (trachea through terminal bronchioles) Mean (of entire lung)	Range of Means for Lobes	Reference
Human	17	---	3
Human	15	14-17	52
Rat (hooded)	15	11-19	12
Hamster (Syrian)	14	10-18	12
Dog (Beagle)	18	15-21	52

lungs. He predicted deposition probabilities to be similar for sedimentation, but a function of body weight for diffusion. McMahon et al. [54] attempted to develop a theory to scale the collection efficiency of the respiratory tract in different species, based on experimental studies of aerosol inhalations in mice, hamsters, rats, rabbits and dogs. He found that certain physiologic parameters changed with body size of the animal, but that competing effects canceled in those controlling dimensionless parameters so the collection efficiency of the lung would be independent of body size.

A second predictive approach for animal-to-man extrapolation makes use of morphometric models in predicting deposition. Kliment [16-18] used latex casts as a basis to develop a dichotomous branching model of the respiratory tract of a few species. He then used predictive deposition equations with this morphometry to calculate deposition in the guinea pig lung, and compared these values with deposition calculated for humans. He concluded that the guinea pig was a good model for deposition in man. He later developed a dichotomous model for the rabbit and rat for use in deposition analyses in these species.

The only direct investigation of intrabronchial particle deposition patterns in hollow casts of bronchial trees which show clearly different geometries was performed by Schlesinger. Figure 7 presents previously unpublished data comparing deposition efficiencies obtained in generations 0 through 5 of the tracheobronchial tree of man and the donkey; particle sizes and flowrates were identical, so differences were due solely to morphometry. Deposition efficiencies are greater in the donkey lung for this comparison test.

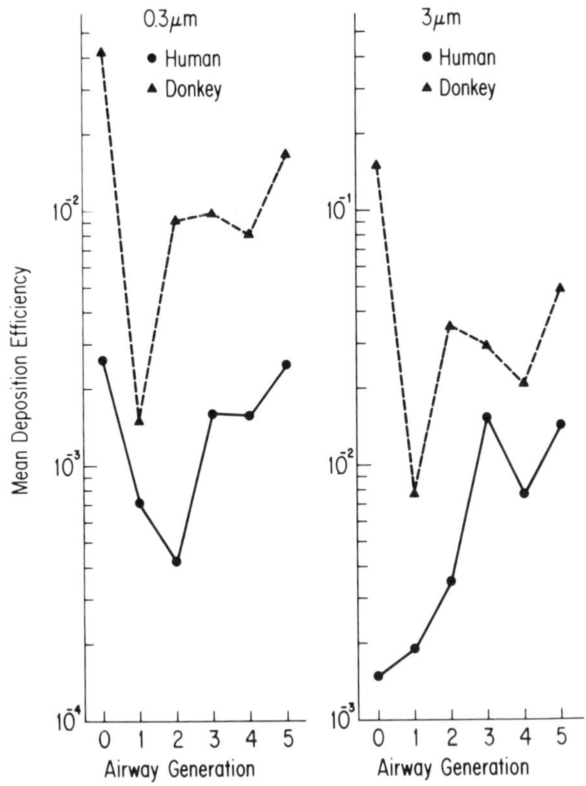

Figure 7. Mean fractional deposition efficiencies obtained in a cast of the human tracheobronchial tree and that of a donkey, for two particle sizes at comparable exposure flow rates.

CONCLUSIONS

Many types of models are used in attempts to describe particle deposition in the tracheobronchial tree. These range from conceptual, ideal models to actual hollow casts of portions of the respiratory tract. Because the complexity of the bronchial tree is a factor in determining deposition, more realistic data may be obtained from the latter type of model system.

Many species of experimental animals are used in deposition studies. But in order to extrapolate deposition data obtained from these to man, it is necessary to understand how differences in lung structure, especially bronchial geometry, affect deposition between species.

ACKNOWLEDGMENTS

The author would like to thank Joshua Gurman for the use of Figures 1, 2 and 3 and Lynn McFadden for performing the tedious morphometric measurements on the rabbit and donkey casts.

This work was supported by Grant #ES 00881 from the U.S. National Institute of Environmental Health Sciences (NIEHS) and is part of center programs supported by Grant #ES 00260 from NIEHS and Grant #CA-13343 from the U.S. National Cancer Institute.

REFERENCES

1. Horsfield, K., and G. Cumming. "Morphology of the Bronchial Tree in Man," *J. Appl. Physiol.* 24:373-383 (1968).
2. Strahler, A. N. "Equilibrium Theory of Erosional Slopes Approached by Frequency Distribution Analysis," *Am. J. Sci.* 248:673-696 (1950).
3. Weibel, E. R. *Morphometry of the Human Lung* (Berlin:Springer-Verlag, 1963).
4. Findeisen, W. "Über das Absetzen Kleiner, in der Luft suspendierten Teilchen in der menschlichen Lunge bei der Atmung," *Arch. Ges. Physiol.* 236:367-379 (1935).
5. Landahl, H. D. "On the Removal of Airborne Droplets by the Human Respiratory Tract. I. The Lung," *Bull. Math. Biophys.* 12:43-56 (1950).
6. Landahl, H. D. "Particle Removal by the Respiratory System: Note on the Removal of Airborne Particulates by the Human Respiratory Tract with Particular Reference to the Role of Diffusion," *Bull. Math. Biophys.* 25:29-39 (1963).
7. Beeckmans, J. M. "The Deposition of Aerosols in the Respiratory Tract. (1) Mathematical Analysis and Comparison with Experimental Data," *Can. J. Physiol. Pharmacol.* 43:157-172 (1965).
8. Task Group on Lung Dynamics. "Deposition and Retention Models for Internal Dosimetry of the Human Respiratory Tract," *Health Phys.* 12:173-207 (1966).
9. Takahashi, K., and H. Ito. "A Computational Model for Regional Deposition of Aerosol Particles in the Human Lung," *Technical Reports of the Institute of Atomic Energy, Kyoto University,* Report No. 170 (Kyoto: Institute of Atomic Energy, 1976).
10. Harris, R. L., and D. A. Fraser. "A Model for Deposition of Fibers in the Human Respiratory System," *Amer. Ind. Hyg. Assoc. J.* 37:73-89 (1976).
11. Harris, R. L., and V. Timbrell. "The Influence of Fiber Shape in Lung Deposition–Mathematical Estimates," in *Inhaled Particles IV,* W. H. Walton, Ed. (Oxford: Pergamon Press, 1977), p. 75.
12. Raabe, O. G., H. -C. Yeh, G. J. Newton, R. F. Phalen and D. J. Velasquez. "Deposition of Inhaled Monodisperse Aerosols in Small Rodents," in *Inhaled Particles IV,* W. H. Walton, Ed. (Oxford: Pergamon Press, 1977), p. 3.

13. Horsfield, K., G. Dart, D. E. Olson, G. Filley and G. Cumming. "Models of the Human Bronchial Tree," *J. Appl. Physiol.* 31:207-217 (1971).
14. Phalen, R. F., H. -C. Yeh and O. G. Raabe. "A Conceptual Model of the Airway Branch for Use in Morphometry of the Mammalian Lung," in *Inhalation Toxicology Research Institute Annual Report 1972-1973,* (Albuquerque, NM: Lovelace Foundation for Medical Education and Research, 1973), p. 281.
15. Phalen, R. F., H. -C. Yeh and O. G. Raabe. "Bronchial Tree Structure in the Human, Beagle, Rat and Hamster," in *Inhalation Toxicology Research Institute Annual Report, 1973-1974,* (Albuquerque, NM: Lovelace Foundation for Medical Education and Research, 1974), p. 289.
16. Kliment, V., J. Libich and V. Kaudersova. "Geometry of Guinea Pig Respiratory Tract and Application of Landahl's Model of Deposition of Aerosol Particles," *J. Hyg. Epid. Microbiol. Immunol.* 16:107-114 (1972).
17. Kliment, V. "Similarity and Dimensional Analysis Evaluation of Aerosol Deposition in the Lungs of Laboratory Animals and Man," *Folia Morph.* 21:59-66 (1973).
18. Kliment, V. "Dichotomical Model of Respiratory Airways of the Rabbit and its Significance for the Construction of Deposition Models," *Folia Morph.* 22:286-290 (1974).
19. Shaw, D. T., N. Rajendran and N. S. Liao. "Theoretical Modeling of Fine-Particle Deposition in 3-Dimensional Bronchial Bifurcations," *Am. Ind. Hyg. Assoc. J.* 39:195-201 (1978).
20. Johnston, J. R., K. D. Isles and D. C. F. Muir. "Inertial Deposition of Particles in Human Branching Airways," in *Inhaled Particles IV,* W. H. Walton, Ed. (Oxford: Pergamon press, 1977), p. 61.
21. Chua, J. H., and C. S. Wang. In: Special Issue No. 9. *J. Res. Assoc. Powder Technology. Japan,* p. 37-45.
22. Yang, Y., W. C. Lee and C. S. Wang. "Deposition of Half Micron Particles from Steady Flows in Models of Human Bronchial Airways," *AICHE Symposium Series No. 147,* 71:33-37 (1975).
23. Bell, K. A. "Aerosol Deposition in Models of a Human Lung Bifurcation" Ph.D. Thesis, California Institute of Technology (1974).
24. Shaw, D. T., N. Rajendran, D. B. Taulbee and K. W. Tu. "The Deposition of Bacterial Aerosol on a Round-Nose Wedge in the Interception Region," *J. Aerosol Sci.* 8:199-204 (1977).
25. Martin, D., and W. Jacobi. "Diffusion Deposition of Small-Sized Particles in the Bronchial Tree," *Health Phys.* 23:23-29 (1972).
26. Lee, W. C., and C. S. Wang. "Particle Deposition in Systems of Repeatedly Bifurcating Tubes," in *Inhaled Particles IV,* W. H. Walton, Ed. (Oxford: Pergamon Press, 1977), p. 49.
27. Nelson, I. C., and B. O. Stuart. "Radon Daughter Deposition in the Respiratory Tract," Battelle Pacific Northwest Laboratory Annual Report for 1970 to the USAEC Division of Biology and Medicine, V. II, Pt. 2, p. 67 (1970).

28. Ferron, G. A. "Deposition of Polydisperse Aerosols in Two Glass Models Representing the Upper Human Airways," *J. Aerosol Sci.* 8:409-427 (1977).
29. Cheng, Y. -S., and C. S. Wang. "Inertial Deposition of Particles in a Bend," *J. Aerosol Sci.* 6:139-146 (1975).
30. Davies, A. "The Evolution of Bronchial Casts," *Med. Hist. (London)* 17:386-391 (1973).
31. Proetz, A. W. "Air Currents in the Upper Respiratory Tract and Their Clinical Importance," *Ann. Otol. Rhinol. Laryngol.* 60:439-451 (1951).
32. Ermala, P., and L. R. Holsti. "Distribution and Absorption of Tobacco Tar in the Organs of the Respiratory Tract," *Cancer* 8:673-682 (1955).
33. Timbrell, V. "Inhalation and Biological Effects of Asbestos," in *Assessment of Airborne Particles,* T. T. Mercer, P. E. Morrow and W. Stober, Eds (Springfield, IL: Charles C. Thomas, 1972), p. 429.
34. Schlesinger, R. B., R. D. Schweizer, T. L. Chan, A. F. Keegan and M. Lippmann. "Controlled Deposition of Tantalum Powder in a Cast of the Human Airways: Applications for Aerosol Bronchography," *Invest. Radiol.* 10:115-123 (1975).
35. Schlesinger, R. B., and M. Lippmann. "Particle Deposition in Casts of the Human Upper Tracheobronchial Tree," *Amer. Ind. Hyg. Assn. J.* 33:237-251 (1972).
36. Schlesinger, R. B., D. E. Bohning, T. L. Chan and M. Lippmann. "Particle Deposition in a Hollow Cast of the Human Tracheobronchial Tree," *J. Aerosol Sci.* 8:429-445 (1977).
37. Cheng, Y. S. "Particle Deposition in Bends and Repeatedly Bifurcating Tubes," Ph.D. Thesis, Syracuse University (1976).
38. Olson, D. E., M. F. Sudlow and K. Horsfield. "Convective Patterns of Flow During Inspiration," *Arch. Intern. Med.* 131:51-59 (1973).
39. Dekker, E. "Transition Between Laminar and Turbulent Flow in Human Trachea," *J. Appl. Physiol.* 16:1060-1064 (1961).
40. Chamberlain, A. C., and E. D. Dyson. "The Dose to the Trachea and Bronchi from the Decay Products of Radon and Thoron," *Brit. J. Radiol.* 29:317-325 (1956).
41. Lippmann, M., and B. Altshuler. "Regional Deposition of Aerosols," in *Air Pollution and the Lung,* E. F. Aharonson, A. Ben-David and M. A. Klingberg, Eds. (New York: John Wiley and Sons, Inc., 1976), p. 25.
42. Gurman, J. L., R. B. Schlesinger and M. Lippmann. "Intrabronchial Deposition Patterns within Replicate Hollow Airway Casts of the Human Lung," presented at American Industrial Hygiene Conference, Chicago, IL, May 1979.
43. Gross, J., and P. Peler. "Holographic Investigation of the Inhalation of Aerosol Particles in a Model of the Human Respiratory Tract," *Staub Reinh. Luft* 34:89-91 (1974).
44. Chan, T. L. "The Characterization of Particle Deposition in the Human Tracheobronchial Tree," Ph.D.Thesis, New York University (1977).

45. Lippmann, M., R. E. Albert and H. T. Peterson, Jr. "The Regional Deposition of Inhaled Aerosols in Man," in *Inhaled Particles III,* W. H. Walton, Ed. (Surrey, England: Unwin, 1971), p. 105.
46. Aeby, C. T. *Der Bronchialbaum der Saugethiere und des Menschen, nebst Bemerkungen über den Bronchialbaum der Vögel and Reptilien* (Leipzig, 1880).
47. Phalen, R. F., H. C. Yeh, G. M. Schum and O. G. Raabe. "Application of an Idealized Model to Morphometry of the Mammalian Tracheobronchial Tree," *Anat. Rec.* 190:167-176.
48. Pump, K. K. "The Morphology of the Finer Branches of the Bronchial Tree of the Human Lung," *Dis. Chest* 46:379-397 (1964).
49. Guzsal, E. "Contributions to the Comparative Anatomy of the Bronchial Tree of Domestic Animals. The Pulmonary Segments," *Acta Vet. Acad. Sci. Hung. (Budapest)* 2:201-223 (1952).
50. Morrow, P. E., and F. R. Gibb. "The Deposition of a Submicronic Aerosol in the Respiratory Tract of Dogs," *Amer. Ind. Hyg. Assoc. J.* 19:196-200 (1958).
51. Thomas, R. L. "Deposition and Initial Translocation of Inhaled Particles in Small Laboratory Animals," *Health Phys.* 16:417-428 (1969).
52. Schum, G. M., M. T. Duggan and H. C. Yeh. "Tracheobronchial Anatomy: Species Differences in Branching Patterns," Inhalation Toxicology Research Institute Annual Report 1975-1976, (Albuquerque, NM: Lovelace BioMedical and Environmental Research Institute, 1976).
53. Stauffer, D. "Scaling Theory for Aerosol Deposition in the Lungs of Different Mammals," *J. Aerosol Sci.* 6:223-225 (1975).
54. McMahon, T. A., J. D. Brain and S. Lemott. "Species Differences in Aerosol Deposition," in *Inhaled Particles IV,* W. H. Walton, Ed. (Oxford: Pergamon Press, 1977), p. 23.
55. Schlesinger, R. B. "The Intrabronchial Pattern of Particle Deposition," Ph.D. Thesis, New York University (1975).
56. Albert, R. E., J. R. Spiegelman, M. Lippmann and R. Bennett. "The Characteristics of Bronchial Clearance in the Miniature Donkey," *Arch. Environ. Health* 17:50 (1968).
57. Cuddihy, R. G., D. G. Brownstein, O. G. Raabe and G. M. Kanapilly. "Respiratory Tract Deposition of Inhaled Polydisperse Aerosols in Beagle Dogs," *J. Aerosol Sci.* 4:35-45 (1973).
58. Leikauf, G. D. "Particle Deposition and Clearance in the Rabbit Respiratory Tract," M. A. Thesis, New York, University (1978).
59. Holma, B. "Short-term Lung Clearance in Rabbits Exposed to a Radioactive Bi-disperse (6 and 3 μ) Polystyrene Aerosol," in *Inhaled Particles and Vapours II,* C. N. Davies, Ed. (Oxford: Pergamon Press, 1967), p. 189.

AUTHOR INDEX

The author's reference number in parentheses follows the page number on which the reference is cited. The number after the parentheses indicates the page on which the reference is listed.

SUBJECT INDEX